# Quantum Theory

Peter Bongaarts

# Quantum Theory

## A Mathematical Approach

 Springer

Peter Bongaarts
Instituut Lorentz
University of Leiden
Leiden
The Netherlands

ISBN 978-3-319-37602-8 ISBN 978-3-319-09561-5 (eBook)
DOI 10.1007/978-3-319-09561-5

Springer Cham Heidelberg New York Dordrecht London

Printed on acid-free paper

Springer is part of Springer Science+Business Media (www.springer.com)

*La filosofia è scritta in questo grandissimo libro che continuamente ci sta aperto innanzi a gli occhi (io dico l'universo), ma non si può intendere se prima non s'impara a intender la lingua, e conoscer i caratteri, ne' quali scritto. Egli è scritto in lingua matematica, e i caratteri son triangoli, cerchi, ed altre figure geometriche, senza i quali mezi è impossibile a intenderne umanamente parola; senza questi è un aggirarsi vanamente per un oscuro laberinto.*

*(Philosophy is written in this vast book, which continuously lies open before our eyes (I mean the universe). But it cannot be understood unless you have first learned to understand the language and recognise the characters in which it is written. It is written in the language of mathematics, and the characters are triangles, circles, and other geometrical figures. Without such means, it is impossible for us humans to understand a word of it, and to be without them is to wander around in vain through a dark labyrinth.)*

Galileo Galilei: Il Saggiatore (The Assayer), 1623

Translation by George MacDonald Ross
http://www.philosophy.leeds.ac.uk/GMR/hmp/texts/modern/galileo/assayer.html

*Voor Piek*

*To the memory of John Cox, my teacher*

Three mathematicians and one mathematical physicist who inspired the author of this book

John von Neumann (1903–1957)

Irving Segal (1918–1998)
*Source* 1964 Summer School, Cargese, France

Alain Connes b.1947
*Source* Workshop "Seminal Interaction between Mathematics and Physics. Rome 2010"

Huzihiro Araki b.1932
*Source* Workshop "Seminal Interaction between Mathematics and Physics. Rome 2010"

*Note* The photographs of Irving Segal, Huzihiro Araki and Alain Connes were taken by the author. Professors Araki and Connes gave permission to use their pictures for this book. The author was unable to trace the descendants of Professor Segal who died in 1998.

# Preface

## Historical Remarks: Mathematics and Physics

Physics, as we know it, began in the sixteenth and seventeenth century, when in the study of natural phenomena empirical observation and mathematical modeling were for the first time systematically and successfully combined. This is exemplified in the person of Galilei, and even more so in Newton, who laid the foundations of our picture of the physical world and who was equally great as a mathematician and as an empirical observer and investigator.

For a long time mathematics and physics formed in an obvious way a single integrated subject. Think of Archimedes, Newton, Lagrange, Gauss, more recently Riemann, Cartan, Poincaré, Hilbert, von Neumann, Weyl, Birkhoff and many others. Lorentz, the great Dutch physicist, was offered a chair in mathematics at the university of Utrecht, simultaneously with a chair in physics at a second Dutch university. He chose the latter and became in 1878 Professor of Theoretical Physics at Leiden university, the first in this subject in Europe.

All this is a thing of the past. From the 1950s onward physics and mathematics parted company, or rather mathematics underwent a drastic change in the way it was formulated, largely due to the Bourbaki movement. It became more abstract, more "formal". In this respect, it should be noted that none of the great Bourbaki mathematicians, Weil, Dieudonné, Grothendieck, for example, had the slightest interest in physics. Another cause was the growing specialization in all of science and, more recently, the increasing publication pressure leading to much narrowly focused short-term research. The language of mathematics is now very different from that of physics.

In the gap between physics and mathematics, mathematical physics as a distinct discipline came into being. Journals were established: Journal of Mathematical Physics (1960), Communications in Mathematical Physics (1965), Letters in Mathematical Physics (1975). Separate conferences were held. The International Association of Mathematical Physics was founded in 1976. All this was and still is very welcome, not because it might lead to a new specialization, which would not be a good thing, but as bridges between mathematics and physics. This book is a modest attempt to contribute to this.

Notwithstanding the present distance between the fields, the interaction between mathematics and physics remains of great importance. Modern theoretical physics cannot exist without advanced mathematics. On the other hand, many ideas in mathematics still have their origin in physics—often in a heuristic form.

The importance of the connection between mathematics and physics is no longer reflected in the curriculum of most universities—certainly not in the Dutch universities. Physics students have to learn a great deal of standard mathematics in their first and second year, mainly calculus and linear algebra, but they are not exposed to the more advanced parts of modern mathematics; its abstract language remains strange in spirit to them, even though they sometimes pick up and learn to use methods based on it. Mathematics, on the other hand, is taught as a self-contained subject, which can be studied for its own sake, without any reference to, or knowledge of physics. Words like "classical mechanics" or "quantum mechanics" have no meaning for most present-day mathematics graduates. Moreover, after a few years of training in rigorously formulated mathematics they will find the loose language of standard physics textbooks very hard to understand.

## Comparable Books

There has always been a strange asymmetry. Numerous books have been written in the past, explaining to physicists advanced topics from mathematics, such as functional analysis, differential geometry, Lie groups and Lie algebras, and algebraic topology. There has been much less in the other direction; books on physics written for mathematicians are relatively rare, even though one would think that there is a need for such books.

However, in the last decade a number of books of this sort have appeared. There is one book that may not have been written with this intent, but it nevertheless clearly stands out in this field. It is Roger Penrose's *The Road to Reality* [1], an amazing book which gives in almost 1,100 pages a wonderful overview of the physical world seen through mathematical eyes. The books that I am about to mention, as well as my own book, can in a certain sense all be seen as providing additional material or more detailed or slightly alternative versions of subjects or aspects discussed in the book of Penrose.

Books that have recently appeared:

– L.A. Takhtajan (2008)
   *Quantum Mechanics for Mathematicians* [2].
– L.D. Faddeev, O.A. Yakubovskii (2009)
   *Lectures on Quantum Mechanics for Mathematics Students* [3].
– K. Hannabuss (1997)
   *An Introduction to Quantum Theory* [4].

- F. Strocchi (2005)
  *An Introduction to the Mathematical Structure of Quantum Mechanics* [5].
- G. Teschl (2009)
  *Mathematical Methods in Quantum Mechanics. With Applications to Schrö-dinger Operators* [6].
- Jonathan Dimock (2011)
  *Quantum Mechanics and Quantum Field Theory. A Mathematical Primer* [7].
- Brian C. Hall (2013)
  *Quantum Theory for Mathematicians* [8].

All these books have some overlap with each other and also with my book. Let me give here the principal points in which my book is different from the books in this list.

- One of the main ideas put forward in my book is that, by using for the description of a physical system the algebra of its observables as basic notion, one can give a uniform formulation of both classical and quantum physics. The algebra of classical observables is commutative, that of quantum observables noncommutative. The states of a system are in both cases the positive normalized linear functionals on the algebra of observables. This idea is also a basis for [3, 5]. Book [3] expresses it in an elegant but very general manner; it is, however, a short book with not much explicit mathematical detail. Book [5] has more such details, but the mathematical framework used for this basic idea is much too narrow. My discussion in Chap. 12 has most of the mathematical details, as far as they are known in the literature at the present time. I, moreover, apply this algebraic framework in Chap. 13 to connecting the various approaches to quantization.
- Most modern applications of quantum theory depend on statistical quantum physics. In [4, 7] there are short and excellent sections on this, but there is nothing in the other books. My book gives in Chap. 11 an extensive discussion of quantum statistical physics, after a review of classical statistical physics in Chap. 10.
- Book [2] describes the mathematics of quantum theory. A superb book, advanced, too difficult even for the average graduate student in mathematics. It explains the mathematics, but not the way it is used in quantum theory. Book [6], another excellent book, gives an account of the mathematics of the Schrödinger equation, an important special topic in quantum theory. My book gives, somewhat different from [2, 6], quantum theory as seen through mathematical eyes, but still as a complete physical theory.
- "Quantization", is the nonunique procedure which, starting from a given classical theory, constructs a corresponding quantum theory. Historically, there is, for example, Born-Jordan quantization and Weyl quantization, in principle different, but for which at present no experimental situations exist that distinguish between the two. Nevertheless, the theoretical problem of the nonuniqueness of quantization remains.

A specific quantum theory may be seen as a deformation of a classical theory, with Planck's $\hbar$ as deformation parameter. Note that $\hbar$ is a constant of nature, but it has a dimension and takes different numerical values for different unit systems. For a macroscopic system of units the value of $\hbar$ is vanishingly small: the classical limit. See Sect. 1.4.4 for a few remarks on dimensions in physics. Deformation quantization has been studied along various lines. There is for instance strict deformation quantization (Marc Rieffel) and formal deformation quantization (Moshe Flato et al.). Strangely enough there is almost no communication between these different schools. None of the books mentioned pays attention to the general problem of quantization, its nonuniqueness and the relation between the different approaches. My Chapter 13 is devoted to all this.

- Almost immediately after the beginning of quantum mechanics in the 1920s, work began on a relativistic version of the theory, culminating in Dirac's great 1928 paper. See Chap. 15, Ref. [6]. It has since then developed into relativistic quantum field theory, today the main theoretical tool for elementary particle physics, very successful, although many interesting and difficult mathematical problems remain. None of the books mentioned above, except [7], has anything on relativistic quantum theory. In my book, Chap. 15 is devoted to relativistic quantum mechanics and Chap. 16 gives a brief introduction to relativistic quantum field theory and elementary particle physics.

- The approach to quantum theory in my book is "axiomatic", which means that I first give the underlying mathematical structure and then the interpretation of the mathematical notions in physical terms, supported by and enlivened with explicit examples. This is the opposite of the historical direction followed by the more traditional textbooks. Most modern books, such as mentioned in my list of references, do the contrary, and present the theory as it is now, which is good, in particular because they do this very well, but they leave out any sort of historical context, refer only to other modern textbooks, never mention original sources. The history of quantum theory is, however, interesting as an intellectual background. Therefore, I believe that some knowledge of it should be a part of a general physics education. For this reason, I supply in my book more historical details than most of the books mentioned.

## Aims of this Book: Potential Readers

The writing of this book was inspired by the general observation that the great theories of modern physics have simple and transparent underlying mathematical structures—usually not emphasized in standard physics textbooks—which makes it easy for mathematicians to understand their basic features. Someone who is familiar with modern differential geometry, in particular Riemannian geometry, will easily grasp the essentials of general relativity. The same is true for familiarity with Hilbert space theory in understanding quantum theory.

This determines, up to a point, the ideal readership for this book. It is a text book on quantum theory, an important topic from modern physics, meant in the first place for advanced undergraduate or graduate students in mathematics, interested in modern physics, and in the second place for physics students with an interest in the mathematical background of physics, unhappy with the level of rigor in standard physics courses. More generally, it should be a useful book for all mathematicians interested in—and sometimes puzzled by—modern physics, and all physicists in search of more mathematical precision in the basic concepts of their field.

Courses that might be given on the basis of this book could be:

– A mathematical approach to quantum theory.
– Quantum theory for mathematicians.

On my book together with one or more other books;

– Modern mathematical physics.
– Modern physics for mathematicians.

New fundamental physical theories are usually not derived from first principles or in a straight manner from earlier theories. Instead they arise as a complex of suggestions, from physical intuition and experiments, mathematical hypotheses, leading finally to a system of precisely formulated postulates, the validity of these guaranteed by the experimental success of the theory that it describes. At the end of this we have the theory in its best possible form, a clear mathematical model, based on a few assumptions, "axioms", together with a physical interpretation, which leads to statements that can be tested experimentally. In this I am much in sympathy with Dirac's point of view that true understanding of a physical theory comes from understanding the beauty of its basic mathematical structure. This is also one of the underlying themes of this book. See for a quote of Dirac on this [9], p. xv, and for a lecture that he gave on the relation of physics and mathematics [10].

## Structure of the Book

The main body of the book consists of two parts:

1. **Part I: The Main Text**. This contains the main story of the book, in 17 chapters, from Chap. 1 "Introduction", to Chap. 17 "Concluding Remarks".
2. **Part II: Supplementary Material**. This consists of a series of chapters with supplementary material. They could have been called appendices, but are in fact called chapters, and run from Supp. Chap. 18 "Topology" to Supp. Chap. 27 "Algebras, States, Representations".

After this comes what in Book Trade jargon is called "back matter", the Subject Index, a List of Authors Cited, and an Index of Persons.

The dual readership that I have in mind—both mathematicians and physicists—determines, for better or worse, the structure and format of this book. Readers with a mathematical background or with a background in physics will reach the core material of the book coming from opposite directions. The series of supplementary

chapters—mainly mathematical—contain much that is already familiar to mathematicians. But modern mathematicians are specialists; a differential geometer, for
example, may not have all necessary details of functional analysis instantly available. So the Supp. Chaps. 18–27 may serve for them as reminders and also will serve
to establish the notation and terminology. For a reader from physics on the other
hand, the mathematical material in these chapters may be not enough. However, the
general ideas there, are more important than the details. In any case, in the supplementary chapters, most of the important mathematical concepts are first introduced in an intuitive manner, and then formulated in an appropriately rigorous and
precise way. It is, moreover, my experience that the physics students interested in the
mathematical aspects of physics are usually bright and quick learners, who can be
expected to find their way using the references that I give. Note, finally, that these
supplementary chapters also make the book reasonably self-contained.

## Summary of the Contents

The book begins in Chap. 1 with an introduction that gives a very brief overview of
present-day physics from a historical point of view.

Quantum physics is more fundamental than classical physics, and therefore does
not rely—in principle at least—on concepts of classical physics. In practice, however, almost all quantum theoretical models have been suggested by corresponding
classical models. This is in particular true for quantum mechanics, which from its
beginning was conceived as obtained by "quantization" of classical mechanics in its
Hamiltonian formulation. This origin is still clearly visible in the structure and the
terminology of modern quantum mechanics. For this reason Chap. 2 is devoted to an
exposition of classical mechanics, with some additional justification for giving this
classical subject so much space in Sect. 2.1. Maybe at least the first half of it should
be read before proceeding to the chapters on quantum theory that follow. The
necessary mathematics can be found in Supp. Chap. 20 "Manifolds".

Chapter 3 is one of the central chapters of this book. It discusses the general
principles of quantum theory. Its emergence against a certain historical background
and its subsequent evolution is sketched in Sects. 3.1–3.3. After some general
remarks in Sect. 3.4, a preview is given in Sect. 3.5 of the modern axiomatic
formulation of quantum theory, based on the mathematical ideas of von Neumann.
This is what will be called the first level of the axiomatization of quantum theory,
introducing the notions of Hilbert space vectors as "pure" quantum states and selfadjoint operators as observables. After this more in detail the notions of states and
observables in Sects. 3.6 and 3.7, time evolution in Sect. 3.8, and finally symmetries in Sect. 3.9. A second and a third level of axiomatization will appear in later
chapters. The mathematical structure of quantum theory is explained in Supp.
Chap. 21 "Functional Analysis: Hilbert Space". Helpful for this are also Supp.
Chap. 18 "Topology", Supp. Chap. 19 "Measure and Integral", and for probabilistic
aspects Supp. Chap. 22 "Probability Theory".

Next, the quantum mechanical description of a single particle moving in a potential is formulated, with Heisenberg's uncertainty relation as important theme in Chap. 4, and the behavior of wave packets describing the motion of a particle in Chap. 5. Chapter 6 treats the particular case of the one dimensional harmonic oscillator, the simplest and at the same time one of the most important nontrivial basic examples of quantum mechanics, the idea of which appears as a first approximation in a wide range of realistic physical situations. The hydrogen atom, and more generally, a particle in a centrally symmetric potential, the discussion of which led historically to quantum mechanics, is discussed in Chap. 7. It provides a good example of the application of symmetry principles in quantum theory. A further typically nonclassical addition to atomic physics is the notion of *spin*. This appears in Chap. 8.

Chapter 9 discusses many-particle systems, in particular systems of identical particles in which typical quantum phenomena appear, such as the Pauli exclusion principle. The second half of this chapter is devoted to the so-called Fock space formalism, or—as it is often called, somewhat unfortunately—"second quantization", a powerful and elegant way—especially in its heuristic form—to describe simultaneously systems of arbitrary many identical particles. Mathematical aspects of the Fock space formalism are explained in Supp. Chap. 23 "Tensor Products".

In the Chaps. 10 and 11 statistical physics is discussed. One of the main themes that is emphasized throughout this book, is that quantum theory is in an essential way a *probabilistic* theory. Therefore 'statistical' in the title of Chap. 11 may seem a bit surprising. The explanation lies in the fact that in the quantum theoretical description of large systems consisting of very many small subsystems there is an additional layer of statistical behavior, very analogous to, but different from what one has in the classical probabilistic description of large systems. It is hard to understand where the basic features of the formalism of quantum statistical physics come from if one does not know something of the classical theory. This will, therefore, be reviewed first in Chap. 10. After that, Chap. 11 will give the general theoretical framework together with some interesting applications of quantum statistical physics. Quantum statistical physics means, that after the first level of quantum theory with Hilbert space vectors as "pure" states, explained in Chap. 3, a second level appears, with a generalized system of axioms, and density operators in an ambient Hilbert space as "mixed" quantum states. For these chapters Supp. Chap. 22 "Probability Theory" is again useful.

Chapter 12 develops a central theme of this book. It presents an algebraic formalism, in which both classical and quantum physical systems fit in a natural manner. First the mathematical framework, then examples of concrete physical systems. This leads to a third level of the axiomatization of quantum theory, algebraic quantum theory, with as primary object an abstract algebra of observables, then states as linear functionals on this algebra, and finally a Hilbert space in which this algebra is represented as operator algebra, a representation which is dependent on the choice of the state functional. In the last section of this chapter, there is a comparison between von Neumann's operator formalism and the lattice formulation of quantum theory by Birkhoff, with the connection of these two by Gleason's theorem. The necessary material on algebras and their representations can be found in Supp. Chap. 27 "Algebras, States, Representations".

Chapter 13 treats quantization in the spirit mentioned above. I try in particular to give an overall picture of various approaches and procedures. In Chap. 14 scattering theory is discussed; the time-dependent and the time-independent formalism, together with the relation between the two. This chapter also contains a section on perturbation theory. Chapters 15 and 16 are devoted to relativistic quantum theory. For both chapters Supp. Chap. 21 "Functional Analysis: Hilbert Space" will be useful. Chapter 15 describes the not wholly successful attempts to find relativistic single particle wave equations, starting from nonrelativistic Schrödinger quantum mechanics, leading finally to relativistic quantum field theory, introduced in Chap. 16. In this chapter, there is also a brief review of elementary particle physics, its history, and its culmination in the so-called Standard Model. Chapter 17 contains concluding remarks, in particular some highly personal remarks on the present and future state of physics—with a very short review of the problems of modern cosmology and of the fascinating subject that can be characterized by the catchword "Einstein-Podolsky-Rosen and All That"—on the difference between physics and science fiction, and finally on sociological aspects of modern physics.

For everything that has to do with symmetry Supp. Chap. 24 "Lie Groups and Lie Algebras" will be useful.

Most of the supplementary chapters explain mathematics, in the first place to physicists. There are two exceptions, where heuristic notions that are very popular in physics textbooks are explained to mathematicians. Supp. Chap. 25 "Generalized Functions" (i.e., Dirac's δ-function and its derivatives), Supp. Chap. 26 Dirac's Bra-Ket notation.

Of course, studying a chapter from the main body of the text should not start with reading the chapter that gives supplementary material, basic as this may be for understanding; it should be used as much as possible alongside and in parallel with the main text.

## The Reference System

There is an extensive system of cross-references, between the chapters and between the chapters and the appendices. Three types of internal references can be distinguished.

1. Almost all chapters have a section with numbered references to books and articles relevant to the chapter, denoted as [1, 2], etc. From within a chapter these are cited as, for instance, See Ref. [2].
2. From now on all the chapters from Part I, together with their sections and subsections, will be denoted with a prefix "Sect." (without parentheses), for instance Sect. 16.9.2 for Subsection 16.9.2 of Chapter 16 . References from one chapter to a second chapter, pointing out, for instance, to a similar situation, will have the form "See for a similar situation Sect. 2.6.1".
3. Chapters, together with their sections and subsections, from Part II will be denoted by the number, preceded by "Supp", for example, as Supp. Sect. 21.9.2.

A reference to a mathematical explanation might read as "For an explanation of this, see Supp. Sect. 21.9.2".

Examples, problems, theorems, etc., will be denoted with parentheses, for instance as (2.3.2,a) **Problem**, (20.5.2,a) **Theorem**.

## Various Remarks

1. Most of the chapters and some of the appendices contain exercises—problems to test the understanding of the reader. They may also provide additional information.

2. *For the readers from physics*: The phrase "if and only if" is often abbreviated as "iff", as is fairly common in mathematical texts. Various standard symbols are used in the mathematical formulas, such as:

   - $\in$  : "is an element of", and $\notin$ : "is not an element of"
   - $\subset$  : "is a subset of", and $\supset$ : "contains"
   - $\prec$  : "precedes"
   - $\leq$  : "less or equal than", and $\geq$  : "greater or equal than"
   - $\neq$  : "is not equal to"
   - $\equiv$  : "is equivalent with", and $\sim$ : "is similar to"
   - $\forall$  : "for all", and $\exists$ : "there exists"
   - $\rightarrow$  : "maps to (sets to sets)"
   - $\mapsto$  : "maps to (elements of sets to elements of sets)"
   - $\Rightarrow$  : "implies", and $\Leftarrow$ : "is implied by"
   
   Etc.

3. Much useful material can be found on the internet. Whenever available I give web addresses for items in the reference list, even though some of these may be ephemeral. For almost any notion from mathematics and physics there exist Wikipedia articles, usually extensive and competent, even though some double-checking with similar sources such that of "mathworld.wolfram" and "nLab" is recommended.

   Also useful although somewhat dated is the *Encyclopaedia of Mathematics*, a multi-volume work, translated from the Russian, originally published by Kluwer and made now freely available in separate web pages, with as starting page http://eom.springer.de/. Authoritative articles on the philosophical background of modern physics are available from the on-line *Stanford Encyclopedia of Philosophy*, with its home page at : http://plato.stanford.edu/.

   There is a website, *HyperPhysics*, consisting of a large number of well-integrated and cross-referenced small pages, with very clear texts and pictures, covering all aspects of standard quantum mechanics. To be found at: http://hyperphysics.phy-astr.gsu.edu/hbase/hframe.html.

   A very useful (printed) general source for information on mathematics, from elementary to advanced, is the *Encyclopedic Dictionary of Mathematics* [11].

Internet sources are ephemere. All the links mentioned in this book were still accessible at the time of writing.

# References

1. Roger Penrose.: The Road to Reality. A Complete Guide to the Physical Universe. BCA (2004) (Penrose has written a series of best-selling books. This book has probably a smaller readership; but it may be considered to be his 'magnum opus'.)

2. Takhtajan, L.A.: Quantum Mechanics for Mathematicians. American Mathematical Society (2008) (A tersely written advanced mathematical textbook. After reading it most mathematicians will still not be able to read physics textbooks on quantum mechanics.)

3. Faddeev, L.D., Yakubovskii, O.A.: Lectures on Quantum Mechanics for Mathematics Students.Translated from the Russian. American Mathematical Society (2009) (Written in a clear and engaging style, with a perfect balance between physics and mathematics, and emphasizing an algebraic point of view that unifies classical and quantum mechanics.)

4. Hannabuss, K.: An Introduction to Quantum Theory. Oxford (1997) (A slightly older book based, as the author says, on a course for mathematics students. It is not written with mathematical concepts as a starting point from which the physical theory can be better understood, something that I feel is particularly important in quantum theory. It seems therefore more suitable for physics students who have a more than average interest in mathematical rigor.) (Nevertheless, a solid, very effective, clear and competent book.)

5. Strocchi, F.: An Introduction to the Mathematical Structure of Quantum Mechanics. World Scientific (2005) (A short series of lectures, containing many interesting details. The mathematical framework that the author employs is unsatisfactory.)

6. Teschl, G.: Mathematical Methods in Quantum Mechanics. With Applications to Schrödinger Operators. American Mathematical Society (2009) (A book that is limited to the discussion of the mathematical basis of what essentially is the nonrelativistic quantum mechanics of a single particle in an external field. It is thorough and written in a clear and precise style.)

7. Dimock, J.: Quantum Mechanics and Quantum Field Theory. A Mathematical Primer. Cambridge University Press (2011) (This recent book is written in a similar style as my book, but without the algebraic point of view. It is less comprehensive in its coverage of important aspects of quantum theory. Nevertheless, a fine book, particular in its discussion of mathematical aspects.)

8. Hall, B.C.: Quantum Theory for Mathematicians. Springer (2013) (This even more recent book has the same purpose as my book: explaining quantum theory to a mathematical readership. This implies a certain amount of overlap. There are, however, important differences. The main one is that one of the main themes of my book is to see quantum theory as a whole in an algebraic context, in which it can be compared with a commutative algebraic formulation of classical physics, in the same way as this is briefly but convincingly done in 3. Any sort of discussion of algebraic aspects of quantum theory is absent from Hall's book. But in its format it is an excellent book, a clearly written, straightforward although somewhat traditional account of nonrelativistic quantum mechanics in a mathematical presentation.)

9. Farmelo, G.: It Must be Beautiful: Great Equations of Modern Science. Granta Books, New edition (2003) (During a seminar in Moscow University in 1955, when Dirac was asked to summarize his philosophy of physics, he wrote at the blackboard in capital letters : Physical laws should have mathematical beauty. This piece of blackboard is still on display.)

10. Dirac, P.A.M.: The relation between mathematics and physics. Proceedings of the Royal Society, vol. 59, pp.122–129. Edingburgh (1938–1939) (Text of a lecture delivered on the presentation of the James Scott prize to Dirac, February 6, 1939. Available at: http://www.damtp.cam.ac.uk/events/strings02/dirac/speach.html)

11. Ito, K.: Encyclopedic Dictionary of Mathematics. 2$^{nd}$ edn. Translated from the Japanese. MIT Press 1977, paperback edition (1993) (This two-volume work can be strongly recommended. When one is new to any topic in mathematics this is the place where to begin. Too many references in the first edition (1977) were to papers in Japanese, but the second edition is much better in this respect.)

# Acknowledgments

I am grateful to:

– Jacek Brodzki, for encouraging me to begin writing this book.

*Two persons read the complete manuscript, checked all the formulas. Their precision, erudition and hard work saved me from many small and a few rather grave errors:*

– Jan Willem Dalhuisen, Henk Pijls.

*Then in alphabetic order:*

– Fatima Azmi, for encouragement from far away Riyadh and Dubai,
– Dirk Bouwmeester, for moral support, and for critical comments on general aspects of the manuscript,
– Richard Gill, for critical remarks on an early version of the manuscript,
– Piet Mulders, for commenting on the section on elementary particle physics in Chap. 16,
– Piek, for checking the grammar and spelling of the complete manuscript,
– Henk Pijls, for drawing the LaTeX Minkowski diagrams,
– Aldo Rampioni, my competent and friendly Springer editor, with Kirsten Theunissen, his assistant,
– Ans van der Vlist, Leiden Science librarian, for obtaining literature from elsewhere,
– the World Wide Web, and in general Internet, for its great possibilities in obtaining information.

*Finally*:

- Irving Segal (1918–1998), whose 1964 Cargèse Summer school lectures saved me from despair when I was first confronted with the mathematical problems of quantum field theory,
- John Cox (1923–2007), from whom I learned that science should be studied as part of a wider intellectual context.

*Of course, all remaining errors are mine.*

# Quote

Whilst writing a book, a book is an adventure. To begin with, it is a toy, then an amusement, then it becomes a mistress, and then it becomes a master, and then it becomes a tyrant, and in the last stage, just as you are to be reconciled to your servitude, you kill the monster and fling him to the public.

Winston Churchill, 2 November 1942

# Contents

# Part I
# The Main Text

# Chapter 1
# Introduction

## 1.1 Introductory Remarks

This chapter gives first in Sects. 1.2 and 1.3 an overview of the historical background
of quantum theory. Then in Sect. 1.4 various general remarks of a methodological
character. Matters touched on are the difference between physics and mathematics (in Sect. 1.4.1), the domain of validity of a physical theory (in Sect. 1.4.2), the
importance of approximative methods in physics (in Sect. 1.4.3), the question of an
axiomatic versus a discursive—historical approach (in Sect. 1.4.4) and the importance of dimensions in a physical theory (in Sect. 1.4.5).

## 1.2 Historical Background

### *1.2.1 Physics Up to the End of the Nineteenth Century*

Ancient and medieval civilizations, those of China, India, Greece and the Arab world,
for instance, were already in the possession of a considerable body of scientific knowledge: insights in certain areas of astronomy and of pure and applied mathematics on
the one hand and empirical knowledge of physical phenomena on the other hand.

On the basis of this, physics as we know it, a successful *combination of mathematical and experimental science*, began in sixteenth and seventeenth century Europe.
Mechanics, describing the action of forces, in particular forces on moving bodies,
was built on the principles laid down first by Galileo and then more systematically
by Newton, and was developed further into a beautiful mathematical theory by—
among others—Lagrange, Laplace, Hamilton and finally Poincaré. Electricity and
magnetism, studied experimentally from the fifteenth century onward, and later more
theoretically, as separate phenomena, were brought together into a single theoretical framework in the second half of the nineteenth century by Maxwell. The basic
notions in his *general theory of electromagnetism* were *electric and magnetic fields*,

© Springer International Publishing Switzerland 2015
P. Bongaarts, *Quantum Theory*, DOI 10.1007/978-3-319-09561-5_1

propagating in space as radiation, with light waves as a special case. In addition to this there was thermodynamics and statistical mechanics, the first a phenomenological framework for describing experimentally observed properties of heat, temperature and energy, the second a way of explaining these 'macroscopic' phenomena by statistical arguments from the 'microscopic' picture of atoms and molecules that gradually became generally accepted.

At the end of the nineteenth century, the result of all this was *classical physics*, a description of the physical world, believed by many to be essentially complete. It consisted of two main components, Newton's classical mechanics, for the description of *matter*, Maxwell's theory of electromagnetism, for *fields* and *radiation*, together with laws governing the interaction between matter and radiation.

The Dutch historian of science E.J. Dijksterhuis characterized the development from Aristotelian to Newtonian physics as the "mechanization of the world picture" [2]. It has now become clear that mechanical models are no longer sufficient to fully understand atoms and molecules, and that we in this situation have to fall back on mathematics. This might be called the "mathematization of the world picture".

### 1.2.2 Problems of Classical Physics

At the beginning of the twentieth century a few small but persistent problems remained, cracks in the walls of the imposing building of classical physics. One of these was the problem posed by the frequency spectra of light emitted by atoms and molecules, measured systematically and with great precision during the last half of the nineteenth century. These spectra were *discrete*; their frequencies followed simple empirical rules, for which no theoretical explanation could be given. There was no way in which the classical picture of atoms and radiation could account for this. A second problem was the *aether*, a special medium that was assumed to fill empty space. The existence of this aether was thought to be necessary for the propagation of light waves through vacuum, but was forced to have very contradictory properties. These problems could not be solved within classical physics; fundamentally new physical ideas were needed, which were found in two new theories which emerged in the first half of the twentieth century.

## 1.3 Physics in the Twentieth Century and Beyond

### 1.3.1 Two Revolutions in Physics

The two new theories that solved the problems of classical physics and broke resolutely with classical notions were *relativity theory* and *quantum mechanics*. They led eventually to a thorough revision of the foundations of physics, with new ideas, in relativity on space and time, and in quantum mechanics on causality and determinism. In this process classical mechanics and classical electromagnetaetherism were relegated to the role of practically useful *approximations* to an underlying more general picture. See [1] for an extensive overview of these developments.

### 1.3.2  The Theory of Relativity

The theory of relativity is the creation of Albert Einstein. In 1905 he wrote a paper of fundamental importance in which he introduced what is now known as the *special theory of relativity*. In this theory space and time are intimately related, forming together a single four dimensional affine space (very roughly: a vector space in which the origin is irrelevant). Which part of this should be considered space and which time depends on the motion of the observer. He also showed that the notion of aether could be dropped altogether. Using differential geometry, then a new part of mathematics, he went on to formulate a *general theory of relativity* in which the four dimensional affine space becomes a four dimensional *Riemannian*, or rather *pseudo-Riemannian manifold*, with the metric and its curvature describing gravity. His main paper on this appeared in 1916. The experimental consequences of the theory of relativity appear only at very high velocities, at the velocity of light; relativity has therefore less consequences for practical applications than quantum mechanics. This does not of course diminish its intellectual importance for our view of the physical world. Relativity theory is discussed in Chap. 15, in particular in Sects. 15.2 and 15.3.

### 1.3.3  Quantum Mechanics

Quantum mechanics was introduced in 1924 by Werner Heisenberg, Erwin Schrödinger, Max Born, and others. It solved the problem of discrete spectra and led to a new and much deeper understanding of physics at the submicroscopic level. It is now seen as a truly fundamental theory, which is assumed to hold for *all* physical phenomena. The technical applications of quantum theory are very important; think of transistors, microprocessors, computers and CD- and DVD-players, mobile phones, all the devices without which modern life seems to be impossible. Without quantum theory the world in which we live would look very different.

### 1.3.4  Outlook for the Twenty First Century

The situation in physics now is reminiscent of that around 1900. Again we have an imposing theoretical framework, giving valid descriptions of physical phenomena, from the small scale of quarks and gluons in elementary particle physics to the large scale of stellar systems in astronomy. Again there are very persistent remaining problems. This time they are not caused by experimental facts that do not fit our experiments, but by the seeming impossibility to integrate conceptually the two pillars of twentieth century physics, quantum theory and relativity, into a single unified theory. So far quantum theory and the theory of relativity are on a very fundamental level *incompatible* with each other. Solving this problem is the main challenge for physics in the twenty first century, a challenge which so far has only been very partially met, despite much interesting and imaginative attempts. We shall make a few further remarks on the present and future state of fundamental physics in Chap. 17, in particular in Sect. 17.2.

### 1.3.5 The Scope of This Book

Quantum theory and the theory of relativity are very different, both in physical content and in the mathematics that they require. They have however in common that they are both based on a coherent and transparent set of mathematical principles, using differential geometry in the case of the theory of relativity, and functional analysis, in particular the theory of operators in Hilbert space, for that of quantum theory—this with the exception of relativistic quantum field theory, a successful more recent branch of quantum physics, the formulation of which is still troubled by grave mathematical problems. Both theories used, when being created, in an essential way mathematics that was new in their time; and both stimulated in turn the development of new branches of mathematics.

The aim of this book is to give an introduction to quantum theory, which makes this background clear and will have therefore a strong emphasis on its underlying mathematical framework, of course without losing sight of its physical meaning.

## 1.4  Methodological Remarks

### 1.4.1 The Difference Between Mathematics and Physics

The absolute requirement for a mathematical theory to be acceptable is that it is correct as a logical theory, as a system of noncontradictory axioms, consistent definitions and rigorously proven theorems. Properties as elegance, usefulness, interest are important but secondary and subjective. For the mathematical framework of a physical theory the same requirement of logical nature holds—in principle, though not always in practice.

There is a second equally important criterion: the theory should describe the physical world in a satisfactory manner. It should give numerical predictions that can be tested by experiments.

According to the philosopher Karl Popper one should not say that a physical theory in this second sense is correct; only that so far it is not yet incorrect.

### 1.4.2 Domain of Validity

Connected with this is the fact that a physical theory has a *domain of applicability*. Classical mechanics, for instance, is valid for situations in which there are no very high velocities involved—near the velocity of light. After that, the theory of relativity takes over. It is also only valid for macroscopic phenomena; in the microworld quantum theory is needed, etc.

### 1.4.3 Approximation

Very few calculations, theoretical procedures, solutions of problems are exact. The famous two dimensional Ising model is a rare exception.

The general situation in physics is *approximation*. Results are calculated by means of expansion in power series in orders of parameters which describe the strength of the interaction, as a deviation from a simple known situation. One usually does not bother too much about the convergence of such a series... Or by an ever finer discretization of a continuum model. Or by computer simulation.

There are physicists who gained a great reputation by concentrating during their whole career on an ever more precise determination, by unbelievably complicated computations, of a single physical constant.

Computers are nowadays extremely important in physics. Nevertheless, interesting new theories are still made by using paper and pen.

### 1.4.4 Dimensions

Most mathematicians do not realize that quantities in a formula in physics are generally not pure numbers, but have a *dimension* and therefore take different numerical values for different systems of units. The general theory which deals with this is called *dimensional analysis*.

In this book we use *the International System of Units (SI)*, or *rationalized MKS system*. This has as basic units: length [L] (meter m), mass [M] (kilogram kg), time [T] (second s). These have obvious multiples or fractions like kilometer km, centimeter cm, gram g, etc. Electric charge is sometimes used as a basic quantitity with unit Coulomb C; by using Coulomb's Law we connect it with the other three and assign it the dimension $[L^{3/2} M^{1/2} T^3]$. The dimension of temperature can also be expressed in the three basic dimensions, although in this case there are several reasonable possibilities. All other quantities have *derived dimensions*, like velocity with the dimension $[LT^{-1}]$, linear momentum with $[MLT^{-1}]$, energy with $[ML^2T^{-2}]$. An older unit system is the so-called *cgs system*, not used in this book.

By fixing the numerical values of certain fundamental physical constants one may reduce the number of basic units. For example, in particle physics one usually takes the velocity of light $c = 1$ and Planck's constant $\hbar = 1$, with the result that in this field energy can be taken as the sole basic unit: all quantities have the dimension of a positive or negative power of the energy.

A physicist will immediately see that certain formulas are incorrect, for instant formulas in which the argument of an exponential or logarithm is a physical quantity which is not dimensionless, as it should be.

### 1.4.5 Axiomatic Versus Historical Approach

In my point of view, following that of Dirac, the best way of understanding fundamental theories of physics is by first understanding their basic mathematical framework, as based on a few precisely formulated assumptions ('axioms'), together with their physical interpretation. This attitude determines the presentation of physics in this book. (For a quote of Dirac on his attitude towards physics, see the remark following [9] in the list of references of the Preface).

Such a formulation does not directly reflect the historical evolution of physics; it is the result of a distillation over many years of this evolution. Nevertheless, the history of this evolution is interesting and important. An awareness of it should be part of the general intellectual habitus of a scientist. Therefore I give in addition to a more or less 'axiomatic' presentation a considerable amount of information on the history of physics.

## References

1. Pais, A.: Inward Bound: Of Matter and Forces in the Physical World. Oxford University Press, Oxford (1988) (A wide ranging history of modern physics, by a man who played an important role in it.)
2. Dijksterhuis, E.J.: The Mechanization of the World Picture: Pythagoras to Newton. Translated from the original Dutch edition of 1950. Princeton University Press, Princeton (1986)

# Chapter 2
# Classical Mechanics

## 2.1 Introduction

In this chapter we review classical mechanics, the theory that describes the behaviour of systems of classical non relativistic particles, as a necessary background for the discussion of quantum mechanics in the next chapters.

Quantum theory as physical theory is at present assumed to be universally valid. This means that it can, in principle, stand on its own. Nevertheless, classical physics, from which it evolved in the twenties, is still very much present in its formulation, formulas and terminology. Moreover classical physics has not lost its value. Classical mechanics remains the appropriate vehicle for the study of most macroscopic situations. Studying, for example, the motion of billiard balls by quantum mechanics, is, of course, possible, but does not make sense. Approximating such a situation by classical mechanics is much simpler and the experimental results are in practice indistinguishable from those obtained by a quantum description. In Chap. 12, we compare classical physics and quantum physics, in particular classical mechanics and quantum mechanics, as 'algebraic dynamical systems', the first one with a commutative, the second with a noncommutative algebra of observables. Following this, we discuss in Chap. 13 quantum physics as a deformation of classical physics with Planck's constant $\hbar$ as a deformation parameter. Finally, even though the overall exposition of quantum theory in this book is 'axiomatic', we think that learning the theory should also include becoming familiar with the main lines of its historical development—as a matter of general education. For all these reasons the fairly extensive treatment of classical mechanics in this chapter precedes the chapters on quantum theory itself.

We start with a short historical overview of the subject in Sect. 2.2. For this Sect. 2.3 treats classical mechanics with Newton's equations as the basis of the subject; Sect. 2.4 gives the Lagrangian form of classical mechanics. Next, in Sect. 2.5, we review the Hamiltonian formalism, as it is obtained from Newton's equations via the Lagrange formalism. It is the proper vehicle for the transition to quantum mechanics. In Sect. 2.6 we discuss a more intrinsic, geometrical formulation of classical

© Springer International Publishing Switzerland 2015
P. Bongaarts, *Quantum Theory*, DOI 10.1007/978-3-319-09561-5_2

mechanics, with general dynamical systems as a notion defined on manifolds in Sect. 2.6.2, again Lagrangian and Hamiltonian systems in Sect. 2.6.3, and finally in Sect. 2.7 the algebraic version of the Hamiltonian formulation, which defines classical mechanics as an algebraic dynamical system and which will play a role in Chap. 12.

## 2.2 Historical Remarks

### 2.2.1 Aristotelian Physics

Classical mechanics describes the motion of bodies under the influence of forces. Superficial observation leads to the impression that physical objects are normally in rest and start moving only when forces are acting on them. Their velocities seem moreover to increase when the forces increase. Aristotle, the Greek philosopher who gave the first all-encompassing picture of the physical world based on empirical observation instead of pure speculation, followed this train of thought. One may—somewhat anachronistically—formulate his basic dynamical law of motion as

$$F = mv,$$

i.e. *the velocity of a moving body is proportional to the force acting on it.* One should add that to understand what happened when an object was thrown or was allowed to fall freely, it was necessary to devise special explanations, none of which now seem to us very convincing. The ideas of Aristotle dominated physics in the western world and in the world of Islam from classical antiquity until the end of the middle ages and the beginning of the renaissance.

### 2.2.2 Galileo and Newton

If one realizes the importance of friction in the motion of bodies and observes situations in which friction is negligible—think of a stone moving on a surface of ice, a different picture emerges: an object which moves with constant velocity will persist in this motion when left alone. The consequences of such observations were first clearly understood by Galileo. He was led to the general *principle of inertia*, which we may formulate as:

*A physical body which is free, i.e. on which no forces act, is either in rest or moves in a straight line with constant velocity.*

This principle has become the basis of what we at present call classical mechanics. Note that this is the modern form of the principle; Galileo thought of free motion as a motion in a great circle on the surface of the earth, still remaining in this way in the Aristotelian—Platonic view of circular motion as the ideal form of motion.

Galileo, who made many other important contributions to the new post-Aristotelian physics that arose in the sixteenth and seventeenth century, can be seen as the first representative of the method that led to the great successes of modern natural science: the combination of careful empirical observation with the use of precise mathematical models.

Another new and important insight that Galileo helped to establish was that the laws of physics are the same for events on earth and in the heavens. This made in a certain sense astronomy and in particular the study of planetary motion a part of mechanics, which greatly stimulated its further development. The heliocentric picture of the solar system had been put forward already by Copernicus. (Galileo was a strong defender of it, with very unpleasant consequences for him personally, as is well-known.)

Thinking—more or less—in terms of Copernicus' model and using the precise numerical data on planetary positions, collected by Tycho Brahe in years of observation, Johannes Kepler was able to establish that the planets move in ellipses with the sun in one of the focal points, thus finally breaking away from Plato's circles. This set the stage for the fundamental work of Newton. Starting from the principle of inertia he developed mechanics as a complete mathematical theory for the description of the motion of physical bodies under the influence of forces, with as central dynamical law the formula

$$F = ma,$$

stating that instead of the velocity the *acceleration*, i.e. the second derivative of the position with respect to time, should be proportional to the force, developing in the process differential and integral calculus as the appropriate mathematical tools for this. Introducing a universal gravitational force between two arbitrary massive bodies, proportional to the product of the masses of the two bodies and inversely proportional to the square of their distance, he was able to obtain—as a first application of his general ideas—a precise description of the motion of the moon around the earth, essentially in terms of Kepler's laws of planetary motion. All this he developed in his monumental "Philosophiae Naturalis Principia Mathematica" published in 1687, one the founding books of modern physics. See [1] for an English translation.

Newton's mechanics was further developed mathematically during the eighteenth and nineteenth century, by Laplace, Lagrange, Hamilton and Poincaré, however with no changes in its basic laws. It remains today a lively subject of mathematical research, particularly as celestial mechanics, with many interesting unsolved problems. Its modern formulation is geometrical, in terms of vector fields on differential manifolds, in particular so-called symplectic manifolds. Nevertheless, as a part of physics, 'classical mechanics' is essentially complete, a theory belonging to 19th century physics. The reason classical mechanics is discussed here in some detail is that it is necessary for the understanding of much of twentieth century physics, in particular quantum mechanics, the main topic of this book.

It is important to remark that classical mechanics, like much of the physics from the end of the nineteenth century, still describes many of the physical phenomena around us with very high precision. It fails however in situations where velocities comparable with the velocity of light are involved, or in situations in the submicroscopic world of atoms and molecules. In the first case Newton's theory has to be replaced by Einstein's theory of relativity and in the second case classical mechanics is superseded by quantum mechanics. This illustrates what was said about the domain of validity of physical theories in Sect. 1.4.3.

## 2.3 Newtonian Classical Mechanics

### 2.3.1 Newton's Equations for a System of Point Particles

Classical mechanics as it is taught nowadays to physics students is essentially Newton's mechanics, with some further developments that will be discussed in the next sections. Consider the typical situation of a system of $N$ point particles with masses $m_1, \ldots, m_N$, described by Cartesian coordinates $\mathbf{r}_1, \ldots, \mathbf{r}_N$,

$$\mathbf{r}_j = (x_j, y_j, z_j), \quad j = 1, \ldots, N.$$

We assume that there are forces acting on the particles, derived from a potential, i.e. the force $\mathbf{F}_j$ on the $j$th particle is equal to

$$\mathbf{F}_j(\mathbf{r}_1, \ldots, \mathbf{r}_N) = -\frac{\partial}{\partial \mathbf{r}_j} V(\mathbf{r}_1, \ldots, \mathbf{r}_N),$$

with $\frac{\partial}{\partial \mathbf{r}_j}$ denoting the triple of partial differentiations $(\frac{\partial}{\partial x_j}, \frac{\partial}{\partial y_j}, \frac{\partial}{\partial z_j})$, for $j = 1, \ldots, N$, and $V(\mathbf{r}_1, \ldots, \mathbf{r}_N)$ a given real function on $\mathbb{R}^{3N}$, the potential energy of the system. The time evolution of the system is described by the $N$ vector-valued functions $\mathbf{r}_j(t)$, which are solutions of *Newton's equations*, in this case the system of coupled second order ordinary differential equations

$$m_j \frac{d^2 \mathbf{r}_{j(t)}}{dt^2} = -\frac{\partial}{\partial \mathbf{r}_j} V(\mathbf{r}_1(t), \ldots, \mathbf{r}_N(t)),$$

for $j = 1, \ldots, N$. Such a classical mechanical system is *deterministic*: if we mean by the state of the system at time $t = t_1$ the $2N$ positions and velocity vectors $\mathbf{r}_j(t_1)$ and $\mathbf{v}_j(t_1) = \frac{d}{dt}\mathbf{r}_j(t_1)$, for $j = 1, \ldots, N$, then the state of the system at $t = t_1$ completely determines the state at a later time $t = t_2 > t_1$, because one can, for a sufficiently smooth potential function $V$, prove the existence of a unique solution for each given set of initial conditions $\mathbf{r}_j(t_1)$ and $\mathbf{v}_j(t_1)$). This does, of course, not mean that such a solution can always be found in an explicit form. For a system of two particles with masses $m_1$ and $m_2$, interacting through a potential

$$V(\mathbf{r}_1, \mathbf{r}_2) = -g \frac{m_1 m_2}{|\mathbf{r}_2 - \mathbf{r}_1|},$$

with $g$ a constant, Newton's equations can be solved in closed form. This is of course the problem of the sun and a planet attracting each other by gravitation; the periodic solutions are Kepler's elliptic planetary orbits. For a similar system consisting of three bodies Newton's equations cannot be solved exactly; the solutions can however be approximated to arbitrary precision.

## 2.3.2 Newton's Equations: A System of First Order Equations

By using the velocities $\mathbf{v}_j(t)$ as independent variables, Newton's equations can be written as a system of $2N$ vector-valued or $6N$ real-valued first order equations.

The position variables and the velocities can be written as $x_1, \ldots, x_n$ and $x_{n+1}, \ldots, x_{2n}$ instead of $\mathbf{r}_1, \ldots, \mathbf{r}_N$, and $\mathbf{v}_1, \ldots, \mathbf{v}_n$, with $n = 3N$. With these new variables Newton's equations take the form

$$\frac{d}{dt} x_s(t) = X_s(x_1(t), \ldots, x_{2n}(t)), \quad j = 1, \ldots, 2n,$$

with for $s = 1, \ldots, n$,

$$X_s(x_1(t), \ldots, x_{2n}(t)) = x_{s+n}(t), \quad s = 1, \ldots, n,$$

and for $s = n + 1, \ldots, 2n$,

$$X_j(x_1(t), \ldots, x_{2n}(t)) = -\frac{1}{m_{s-n}} \frac{\partial}{\partial x_{s-n}} V(x_1(t), \ldots x_n(t)),$$

with $m_1 = m_2 = m_3, m_4 = m_5 = m_6$, etc. This is the standard form for a general system of first order ordinary differential equations. Such a system is given by $2n$ functions $X_s$, defined on an open set $U \subset \mathbb{R}^{2n}$. It is called a *dynamical system*. With appropriate smoothness properties of the $X_s$ there is a unique solution $x_s(t)$ on $U$ with a prescribed value in $t = t_0$, on an open interval around $t_0$, for every point in $U$.

## 2.4 The Lagrangian Formulation of Classical Mechanics

### 2.4.1 Lagrangian Variational Problems

In the Lagrangian formulation Newtonian mechanics is treated as a particular example of a class of variational problems, i.e. problems in which a certain function, or a set of functions, is determined by finding the extremum of a given functional.

Let $\mathcal{U}$ be an open set in $\mathbb{R}^n$ and let $L$ be a given real-valued function on $\mathcal{U} \times \mathbb{R}^n$. Consider curves in $\mathcal{U}$, i.e. functions $\gamma$ from $\mathbb{R}$ into $\mathcal{U}$. To keep things simple all functions are supposed to be $C^\infty$. We denote the $n$ coordinates on $\mathcal{U}$ by $q^1, \ldots, q^n$ and those on $\mathcal{U} \times \mathbb{R}^n$ by $q^1, \ldots, q^n, \dot{q}^1, \ldots, \dot{q}^n$. Consider a fixed finite interval $[t_1, t_2]$ on the real line. A curve $\gamma$, a set of functions $q^1(t), \ldots, q^n(t)$, determines an integral

$$I_{t_1,t_2}(\gamma) = \int_{t_1}^{t_2} L(q^1(t), \ldots, q^n(t), \frac{d}{dt}q^1(t), \ldots, \frac{d}{dt}q^n(t))\, dt.$$

This integral is a *functional* on the space of curves $\gamma$ parametrized by $t$ from the interval $[t_1, t_2]$.

The variational problem defined by this set-up is to find the curve or curves for which the integral is *extremal* with respect to variations which are arbitrary except that they leave the end points fixed. A curve $\gamma$ is extremal in this sense if the integral is constant up to first order under each one parameter deformation $\gamma \mapsto \gamma_\varepsilon$ of the form

$$q^j(t) \mapsto q^j(t) + \varepsilon\, \eta^j(t),$$

for arbitrary (smooth) functions $\eta^j(t)$ with $\eta^j(t_1) = \eta^j(t_2) = 0$, for $j = 1, \ldots, n$, or

$$\frac{d}{d\varepsilon} I_{t_1,t_2}(\gamma_\varepsilon) = 0$$

in $\varepsilon = 0$ for such deformations.

(2.4.1,a) **Problem** Show that this condition leads to the following system of second order ordinary differential equations for the extremal functions $q^j(t)$, the *Euler-Lagrange equations*,

$$\frac{\partial L}{\partial q^j} - \frac{d}{dt}\left(\frac{\partial L}{\partial \dot{q}^j}\right) = 0, \quad j = 1, \ldots, n.$$

In physics and in some of the older mathematical literature on the variational calculus one employs a symbolic notation for 'variations'. Written in this notation the statement above says that the requirement $\delta I_{t_1,t_2} = 0$ under variations $\delta q^j$ with $\delta q^j(t_1) = \delta q^j(t_2) = 0$ implies the Euler-Lagrange equations.

### *2.4.2 Newton's Equations as Variational Equations*

Consider a system of $N$ point particles, as before. Let $\mathcal{U} = \mathbb{R}^n$, with $n = 3N$. The coordinates $q^1, \ldots, q^n$ are the position variables $x_1, y_1, z_1, \ldots, x_N, y_N, z_N$. Define the function $L$ as

$$L(q^1, \ldots, q^n, \dot{q}^1, \ldots, \dot{q}^n) = \sum_{j=1}^{n} \frac{1}{2} m_j (\dot{q}^j)^2 - V(q^1, \ldots, q^n),$$

the *kinetic energy* minus the *potential energy* of the system, if we interpret the $\dot{q}^j$ as the components of the velocities of the particles. We have again $m_1 = m_2 = m_3, m_4 = m_5 = m_6$, etc. In this context the function $L$ is called the *Lagrangian function*, or *Lagrangian* of the system and the integral $I_{t_1, t_2}(\gamma)$ is called the *action*.

(2.4.2,a) **Problem** Show that the Euler-Lagrange equations for this function $L$ are just Newton's equations.

This result means that the time evolution of the mechanical system from $t_1$ to $t_2$ is described precisely by those curves $\gamma$ for which the action is extremal, in fact *minimal* in this case.

Writing Newton's equations in Lagrangian form in this manner does of course not add anything to their contents, but has nevertheless great advantages:

a. There is no need to restrict oneself to Cartesian coordinates; the formulas hold for arbitrary curvilinear coordinates.
b. The Lagrangian formulation is very useful in situations where there are constraints on the system, for instance when the particles are restricted in their motion to a lower dimensional surface.
c. Symmetries and their implications such as conserved quantities can be easily read off from the Lagrangian.
d. The Lagrangian formulation of classical mechanics is the starting point for Feynman's path integral scheme, a semi-heuristic but very useful quantization scheme, which is discussed in Sect. 13.8.
e. The Lagrange formalism is convenient in relativistic field theory where space and time coordinates are treated on the same footing. Relativity theory will be discussed in Chap. 15 and relativistic field theory in Chap. 16.

## 2.5 The Hamiltonian Formulation of Classical Mechanics

In this section we finally obtain the formulation of classical mechanics which is the proper background for our presentation of quantum mechanics. Let us start from a Lagrangian variational system formulated in terms of local coordinate expressions, i.e. with coordinates $q^1, \ldots, q^n$, not necessarily Cartesian, with the associated velocities $\dot{q}^1, \ldots, \dot{q}^n$ and with a given Lagrange function

$L(q^1, \ldots, q^n, \dot{q}^1, \ldots, \dot{q}^n)$. Using $L$ we introduce new variables, the *canonically conjugated momenta* $p_1, \ldots, p_n$, as

$$p_j = \frac{\partial}{\partial \dot{q}^j} L,$$

for $j = 1, \ldots, n$. We assume that this transformation, which is called the *Legendre transformation*, can be inverted, i.e. that the velocities $\dot{q}^j$ can be written as functions of $q^1, \ldots, q^n$ and the new momenta $p_1, \ldots, p_n$. If this is possible, the Lagrangian $L$ is called *regular* or *nondegenerate*. If $L$ is *singular* or *degenerate* the simple road to a Hamiltonian formalism breaks down at this point. More complicated procedures can be found to overcome the problem of degenerateness of $L$, but this will not be discussed here.

Define next the *Hamiltonian function*, or, for short, *Hamiltonian*, as

$$H = \sum_{j=1}^{n} p_j \dot{q}_j - L.$$

Note that this $H$ should be seen as a function of $q^1, \ldots, q^n$ and $p_1, \ldots, p_n$. The time evolution, given in the Lagrangian formulation by functions $q^1(t), \ldots, q^n(t)$, $\dot{q}^1(t), \ldots, \dot{q}^n(t)$, satisfying the system of $n$ Euler-Lagrange equations, is now given by functions $q^1(t), \ldots, q^n(t)$ and $p_1(t), \ldots, p_n(t)$, which are solutions of $2n$ first order equations involving $H$, namely

$$\frac{dp_j}{dt} = -\frac{\partial H}{\partial q^j} \qquad \frac{dq^j}{dt} = \frac{\partial H}{\partial p_j},$$

for $j = 1, \ldots, n$.

(2.5,a) **Problem** Show that these equations—called, not surprisingly, *Hamilton's equations*—are equivalent to the Euler-Lagrange equations.

Hamilton's equations can be written in a more uniform manner as

$$\frac{dp_j}{dt} = \{H, p_j\} \qquad \frac{dq^j}{dt} = \{H, q^j\},$$

for $j = 1, \ldots, n$, with the *Poisson bracket* $\{\cdot, \cdot\}$ defined for an arbitrary pair of functions $f$ and $g$ of the variables $p_1, \ldots, p_n, q^1, \ldots, q^n$ as

$$\{f, g\} = \sum_{j=1}^{n} \left( \frac{\partial f}{\partial p_j} \frac{\partial g}{\partial q^j} - \frac{\partial g}{\partial p_j} \frac{\partial f}{\partial q^j} \right).$$

See Supp. Sect. 20.2.7.2 for the most important properties of the Poisson bracket. This form of Hamilton's equations, in the more mathematically intrinsic form to be discussed further on, will be of great importance for an algebraic formulation of classical mechanics, and more generally for the common algebraic formulation of classical and quantum mechanics that will be formulated in Chap. 12.

The $2n$-dimensional space of the variables $p_1, \ldots, p_n, q^1, \ldots, q^n$ is called the *phase space* of the classical system. The time evolution, a *flow* in this space, has the property that it leaves the integration measure $dp_1 \ldots dp_n \, dq^1 \ldots dq^n$ invariant. This fact is known as *Liouville's theorem*. It is of basic importance in classical statistical mechanics, as will be shown in Chap. 10.

## 2.6  An Intrinsic Formulation

### 2.6.1  Introduction

So far we have restricted the discussion of classical mechanics to the case in which the phase space is just $\mathbb{R}^{2n}$. It is not hard to think of more general situations, for instance that of a particle moving in a circle or on the surface of a sphere. For a general discussion of classical mechanics in its various forms, as a general dynamical system, in Lagrangian and Hamiltonian form, we need as a mathematical background differential geometry, i.e. the general theory of $C^\infty$-manifolds, and more in particular of symplectic manifolds. An extensive review of this is given in Supp. Chap. 20. A reader who is not familiar with the material appearing in this section should consult Supp. Chap. 20 and read it in parallel with this section.

The more intrinsic differential-geometric formulation has several advantages. It has, of course, great esthetic appeal. The formulation in terms of explicit formulas given so far is completely rigorous, but the differential geometric picture gives more structural insight. One important aspect of it is that this formulation, rather surprisingly, leads in a natural way to an algebraic framework, which can be used to describe both classical and quantum physics. This will be explained in Chap. 12.

### 2.6.2  General Dynamical Systems

A *dynamical system* is given by a pair $(\mathcal{M}, X)$, with $\mathcal{M}$ a smooth $m$-dimensional manifold and $X$ a vector field on $X$. This $X$ assigns to each point $p$ of $\mathcal{M}$ in a smooth manner a tangent vector $X_p$ and determines *integral curves*, i.e. curves described by smooth maps $\gamma$ from an open interval in $\mathbb{R}^1$ into $\mathcal{M}$, such that they are tangent to the vector field in each point $p$ of $\mathcal{M}$. The $\gamma$'s form the solutions of a system of first order ordinary differential equations, given in local coordinates $\{x_s\}_s$, with $X_j$ the component functions, as

$$\frac{dx^j(t)}{dt} = X^j(x^1(t), \ldots, x^m(t)), \quad j = 1, \ldots, m.$$

Newton's equations, as given in Sect. 2.3, form a particular example of such a dynamical system.

### 2.6.3 The Lagrange System

Starting from the position space, an $n$-dimensional manifold $\mathcal{Q}$, one constructs the space of positions and velocities, the tangent manifold $T(\mathcal{Q})$ of $\mathcal{Q}$. This is a vector bundle over $\mathcal{Q}$ with as fibres the tangent spaces at the points $p$ of $\mathcal{Q}$. The dynamics of the system is given by the Lagrangian, a function $L$ on $T(\mathcal{Q})$. Curves $\gamma(t)$ define in an obvious manner curves $\widehat{\gamma}(t)$ on the space of velocities. The action is the integral

$$I(\gamma) = \int_{t_1}^{t_2} L(\widehat{\gamma}(t)) \, dt,$$

which is required to be extremal to give the evolution equations. Here we use the Lagrange formalism only as an intermediate step in the transition to the Hamilton formalism; so there is not much need to discuss it further here. It will however be used in our discussion of path-integral quantization in Sect. 13.8.

### 2.6.4 The Hamilton System

The tangent bundle $T(\mathcal{Q})$ has a dual bundle $T^*(\mathcal{Q})$, the *cotangent bundle*, dual in the sense of $C^\infty(\mathcal{Q})$-modules, constructed by smoothly welding together the real-linear duals $T^*_p(\mathcal{Q})$ of the tangent spaces $T_p(\mathcal{Q})$ at all points $p$ of $\mathcal{Q}$. The cotangent bundle will be denoted as $\mathcal{M}$; it is a $2n$-dimensional manifold, the basic manifold in the Hamiltonian description of classical mechanics, the *phase space* of a classical mechanical system. As a cotangent bundle it has a natural closed nondegenerate 2-form $\omega$, so it is a *symplectic manifold*. See Supp. Sect. 20.7.

In Sect. 2.5 we gave, in local coordinates, the road from the Lagrange to the Hamilton formalism by means of the Legendre transformation. In a more intrinsic picture it is a map $s$

$$s : T(\mathcal{Q}) \to T^*(\mathcal{Q}) = \mathcal{M},$$

a smooth welding of maps $s_p$

$$s_p : T_p(\mathcal{Q}) \to T^*_p(\mathcal{Q}) = \mathcal{M}_p, \quad \forall p \in \mathcal{M}.$$

An intrinsic formulation of this map can be given, but we refrain from doing this, as it involves the introduction of mathematical notions for which we have no further use. What is important is the result, a simple and transparent picture, the Hamilton formulation of classical mechanics, the standard point of departure for the transition to quantum mechanics.

The formalism for a Hamiltonian description of classical mechanics consists of the following elements:

- A symplectic manifold $(\mathcal{M}, \omega)$, the phase space.
- A Hamiltonian vector field $X_H$, with $H$ the Hamiltonian function of the system, a function on $\mathcal{M}$ defined in Sect. 2.5. It determines a Hamiltonian dynamical system. The corresponding flow on $\mathcal{M}$ is the time evolution of the system.
- An important structural element is the Poisson bracket, defined in terms of the local coordinates $\{p_j\}_j$ and $\{q^k\}_k$ in Sect. 2.5, and in an intrinsic manner in Supp. Sect. 20.7.2.
- Symmetries of the system are described by canonical transformations of $\mathcal{M}$ and more particularly by Lie groups of such transformations. See Supp. Sect. 20.7.1. Their infinitesimal generators are Hamiltonian vector fields $X_f$, connected with a function $f$ in the manner explained in Supp. Sect. 20.7.1. Such a function is in general an observable of the system, and in this particular situation a *constant of the motion* or a *conserved quantity*.

(2.6.4,a) **Problem** Show that for such a constant of the motion $f$ the Poisson bracket of $f$ with $H$ vanishes.

Note that the Hamiltonian $H$ itself, as a constant of the motion, is usually the energy of the system.

For an overview of the mathematical description of classical mechanics, see [2] and [3]. Two more elementary but valuable books on classical mechanics are [4] and [5].

## 2.7 An Algebraic Formulation

In Supp. Sect. 20.8 we give an algebraic formulation of differential geometry. Applying this to the geometric picture of classical mechanics, sketched in the preceding section, we obtain classical mechanics as an *algebraic dynamical system*.

This means a pair $(C^\infty(\mathcal{M}), X_H)$, consisting of a commutative algebra $C^\infty(\mathcal{M})$, with has an additional Poisson structure given by a Poisson bracket $(f, g) \mapsto \{f, g\}$, a derivation $X_H$ of $C^\infty(\mathcal{M})$, which may be called a Hamiltonian derivation, as it is associated with an element $H$ of $C^\infty(\mathcal{M})$. (Remember the equivalence between derivations of $C^\infty(\mathcal{M})$ and vector fields on $\mathcal{M}$. See Supp. Sect. 20.2.2.) It generates a one parameter group of time evolution automorphisms of $C^\infty(\mathcal{M})$, leaving the Poisson structure invariant. There may be additional groups of such automorphisms, representing symmetries, with as generators Hamiltonian derivations, associated with functions that are constants of the motion, i.e. have zero Poisson bracket with $H$.

This algebraic picture is particularly useful for comparing classical mechanics with quantum mechanics. It will be one of the main ingredients of Chaps. 12 and 13.

## References

1. Newton, I.: Philosophiae Naturalis Principia Mathematica (1685, 1725) An English translation in two volumes by Andrew Motte, revised by Cajori, F. University of California Press 1934, 1962 (The 1729 edition can be obtained at : http://en.wikisource.org/wiki/The_Mathematical_Principles_of_Natural_Philosophy_(1729).)
2. Abraham, R., Marsden, J.E.: Foundations of Mechanics, 2nd edn. American Mathematical Society, Providence (2008)
3. Arnold, V.I.: Mathematical Methods of Classical Mechanics, 2nd edn. Springer, New York (1997) (The two most obvious basic and comprehensive references on classical mechanics, stressing mathematical aspects. The first one is rather heavy going, mathematically speaking, the second less so and is also more tuned to physical application.)
4. Goldstein, H.: Classical Mechanics, 3rd edn. Addison Wesley, Boston (2001)
5. Kibble, T.W.B.: Classical Mechanics, 4th edn. Longman, New York (1997) (Two basic physics textbooks on the subject.)

# Chapter 3
# Quantum Theory: General Principles

## 3.1 Introduction

In this chapter, which is central to this book, the general principles of quantum theory will be discussed, in its simplest form, i.e. excluding at this point statistical quantum theory, to be discussed in Chap. 11. We start in Sects. 3.2 and 3.3 with the historical context, sketching the problems in classical physics at the end of the nineteenth century, the first attempt of a solution by an ingenious intuitive idea of Niels Bohr in the early twentieth century, and finally, the creation of a fully fledged new theory which took care of the problems. A very successful physical theory, however not very well understood mathematically; neither in some of its details nor in its general structure. It was John von Neumann, stimulated by David Hilbert, two of the greatest mathematicians of the twentieth century, who in the 1920s developed a clear and rigorous framework for quantum theory, in terms of new mathematics, functional analysis, and in particular the theory of operators in Hilbert space, which is by now the generally accepted mathematical basis of quantum theory. For this reason this chapter should be read with Supp. Chap. 21 at hand, in particular by a reader who is not familiar with this material.

## 3.2 Historical Background

### 3.2.1 Introduction

Quantum mechanics as we know it, essentially in the form in which it is now taught as a standard subject to physics undergraduates, came into being in the course of a few years, more or less in the period from 1925 to 1928. Its principal creators were Werner Heisenberg, Erwin Schrödinger, Max Born, Pascual Jordan, Wolfgang Pauli and Paul Dirac. It explained the structure of atoms, providing a general theoretical basis for the model that had been suggested earlier by Niels Bohr and that was based on the experiments of Ernest Rutherford. It solved the problem of the discreteness

© Springer International Publishing Switzerland 2015
P. Bongaarts, *Quantum Theory*, DOI 10.1007/978-3-319-09561-5_3

of atomic spectra, predicting in fact the numerical values of such spectra with great accuracy. Because of this it was immediately accepted by the physics community.

*Quantum mechanics was nevertheless for a long time considered to be an extremely difficult and abstruse theory. This was not only due to its new and unusual physical features, but also, to a great extent, to its use—often in an implicit way—of new mathematical concepts.*

### 3.2.2 Atomic Spectra

Maxwell's theory of electromagnetic phenomena predicts *electromagnetic radiation*, i.e. electromagnetic waves propagating through empty space with a constant velocity $c = 2.99 \times 10^8$ m/s. An electromagnetic wave has a frequency $\nu$ and a wave length $\lambda$; one has of course the relation $c = \nu\lambda$. Radiation with frequencies $\nu$ roughly between the values $40 \times 10^{13}$ and $75 \times 10^{13}$ Hertz (number of oscillations per second) has been known for a long time as visible light. The particular mix of frequencies emitted by the sun in this frequency regime is seen by our eyes as 'white' light. Using refraction by a glass prisma, Newton showed that it can be broken up into a band of coloured light, varying continuously from red (the lowest visible frequency) to violet (the highest visible frequency). We know now that beyond these limits there is on one side first the infrared region, with heat radiation and then in the lower frequencies the electromagnetic waves used for radio and television. On the other side the ultra-violet regime with in the very high frequencies X-rays.

In certain circumstances, for instance by external stimulation such as electric sparks passing through a gas, atoms emit electromagnetic radiation. This radiation consists of a great number of sharply defined frequencies, some in the visible, some in the infra-red or ultra-violet region. Each element has its own very characteristic system of frequencies, its *spectrum*. The simplest is that of hydrogen. Its spectrum can be described by a simple general formula

$$\nu = cR \left( \frac{1}{n_a^2} - \frac{1}{n_b^2} \right),$$

with $c$ the velocity of light, $R$ the so-called Rydberg constant equal to $1.097 \times 10^7$ m$^{-1}$ and the $n_a$ and $n_b$ running through the positive integers, with the restriction $n_a < n_b$. For fixed $n_a = 1$ (and variable $n_b$) this gives what is called the Lyman series, lying in the ultra-violet; for $n_a = 2$ one has the Balmer series, in the visible region, and for $n_a = 3$ the Paschen series in the infra-red. Note that this formula was found in the late nineteenth century, *not* on the basis of theoretical arguments, but deduced from the results of spectroscopic measurements, which gave also the precise data of the spectra of many other types of atoms, much more complicated than the hydrogen spectrum.

A description of an atom in terms of the interaction of Newtonian mechanics and Maxwellian electromagnetism will indeed predict atomic radiation, but only radiation with a continuously varying frequency spectrum. The experimentally established systems of *discrete* atomic spectra, different and characteristic for different types of atoms, cannot be understood from classical physics, i.e. from a combination of classical mechanics and classical electromagnetism. Towards the end of nineteenth century physics this became a major fundamental problem in physics, a problem quantum mechanics was finally able to solve.

### 3.2.3 Earlier Ideas

The roots of quantum mechanics can be found in various earlier ideas:

a. In 1900 Max Planck gave a formula for the frequency distribution of black body radiation which correctly described the experimentally observed distribution. (See Sect. 11.1 and in particular Sect. 11.7). For the derivation of this formula he assumed that radiation with frequency $\nu$ could only be emitted or absorbed in discrete quantities, 'energy quanta' $h\nu$, with $h$ a universal constant. He made this discreteness assumption with great reluctance, because it did not fit in with classical notions of electromagnetic radiation. The only justification for it was that it led to a formula which experimentally was correct. The constant $h$, *Planck's constant* as it is now called, is in MKS (meter–kilogram–second) units equal to $6.626 \times 10^{-34}$ km$^2$/s, and in CGS (centimeter–gram–second) units $6.626 \times 10^{-27}$ gcm$^2$/s, so it is extremely small. Because inconvenient factors $2\pi$ appear in many formulas, one often uses $\hbar = \frac{h}{2\pi}$ instead of $h$. This $\hbar$ is in MKS units equal to $1.054 \times 10^{-34}$ kg m$^2$/s. The appearance of $\hbar$ or $h$ is typical for all formulas in quantum mechanics.

b. Certain metals, when irradiated with light, emit a stream of *electrons*, very light electrically charged particles, which had been discovered in 1897. Albert Einstein gave in 1905 a theoretical explanation of this *photo-electric effect*, by further developing Planck's idea, assuming that light of frequency $\nu$ consisted in fact of a particular kind of particles, with energy $E = h\nu$, and moving with the velocity of light. These particles were later called *photons*.

c. The American physicist Arthur Compton found in 1924 that the results of experiments in which X-rays—electromagnetic radiation with much shorter wave length than visible light—were scattered by matter, could be explained by making the same assumption about the corpuscular nature of radiation. This particle character came out even stronger than in the photo-electric effect: the photons had not only energy but also momentum.

d. Ernest Rutherford, a physicist from New-Zealand working in England, concluded in 1911 from his scattering experiments, that an atom should be thought of as a very small planetary system, consisting of a heavy nucleus, surrounded by a cloud of electrons, compensating with their electrical charges the opposite charge of the nucleus. See [1]. For this, a theoretical model with very unusual and nonclassical features was postulated by his assistant Niels Bohr, a young Danish theoretical physicist. Bohr assumed that the electrons could only move in discrete stationary orbits, circles or ellipses, from time to time jumping between two such orbits and emitting or absorbing in the process a quantum of radiation $E = h\nu$, with $E$ the difference in the energies of the two orbits. With this model he was able to obtain good numerical values for many atomic spectra.

### 3.2.4  The Emergence of Quantum Mechanics Proper

The great achievement of quantum mechanics, as it finally appeared in 1925 in the fundamental papers of Heisenberg, Born, Schrödinger and others, was that it gave a *fundamental* and *general* theoretical framework in which all these ideas found their natural place and in which in particular the ad-hoc atomic model of Rutherford and Bohr could be understood and derived from first principles.

### 3.2.5  Quantum Theory in Twentieth Century Physics

Quantum theory is one of the two revolutions that have deeply changed physics in the twentieth century, the other being Einstein's theory of relativity. Combining the two in a consistent way remains an open problem and is the most important challenge facing present-day theoretical physics.

For the importance of group theory in quantum theory, see the book of Weyl [2]. For the history of quantum theory the series of books by Mehra and Rechenberg [3], together with the source book by Van der Waerden [4] are indispensable. Good standard books on quantum theory are [5], [6] and [7].

## 3.3  The Beginning of Quantum Mechanics

### 3.3.1  Two Different Forms

Quantum mechanics began in two seemingly different forms: Heisenberg's *matrix mechanics* and Schrödinger's *wave mechanics*.

In Heisenberg's approach the basic classical physical variables of position $x_j$ and momentum $p_j$ of a particle were replaced by somewhat mysterious algebraic quantities $\hat{x}_j$ and $\hat{p}_j$, which no longer commuted with each other, but satisfied instead the *commutation relation*

$$[\hat{p}_j, \hat{x}_k] = \hat{p}_j\hat{x}_k - \hat{x}_k\hat{p}_j = \frac{\hbar}{i}\delta_{jk},$$

with $j$, $k = 1, 2, 3$ and with $\hbar$ Planck's constant. Other physical quantities, of which the energy $E$ was the most important, were expressed in the basic variables, using as much as possible the classical expressions. It was soon realized that these algebraic *quantum variables* were in fact *infinite matrices*, known from mathematics, and that the basic problem that had to be solved for each physical system in this approach was to find *eigenvalues* and *eigenvectors* of the matrix corresponding with the energy.

Schrödinger had been influenced by the ideas of the French physicist Louis de Broglie, who had suggested a few years earlier that a possible way to understand the discrete orbits in Bohr's model of an atom was to associate with a moving particle of momentum $p$ a wave with wave length $\lambda = h/p$, with $h$ Planck's (original) constant. Orbits were as a consequence restricted by the requirement that they had to consist of a certain number of wave lengths fitted together. Schrödinger made this rather vague but intuitively appealing idea more precise by his proposal to describe the motion of an electron by a complex-valued function $\psi$ depending on the position variables $x_j$ and the time variable $t$. This 'wave function' had to satisfy a partial differential equation, which for an electron of mass $m$ moving in a potential $V(x_1, x_2, x_3)$ was

$$-\frac{\hbar}{i}\frac{\partial \psi}{\partial t} = -\frac{\hbar^2}{2m}\Delta\psi + V\psi,$$

with $\Delta$ the Laplace operator

$$\frac{\partial^2}{\partial x_1{}^2} + \frac{\partial^2}{\partial x_2{}^2} + \frac{\partial^2}{\partial x_3{}^2}$$

and $\hbar = \frac{h}{2\pi}$. Schrödinger originally thought that the wave function described the electron 'smeared out' in space, but it was soon pointed out by others that it had a *statistical* interpretation, with the square of the absolute value of $\psi$ at each point in space describing the probability of finding the electron—still a point particle—there.


### 3.3.2 Unification

Schrödinger discovered that solving his equation amounted to solving the eigenvalue–eigenvector problem of the energy matrix in Heisenberg's approach; some time after this the full mathematical and physical equivalence of both approaches was established. It was realized by mathematicians such as John von Neumann that Schrödinger's functions $\psi$ could be seen as vectors in an infinite dimensional linear space, the same space in which the matrices of Heisenberg acted as linear operators. It was von Neumann who developed finally, around 1930, a unified and rigorous framework for quantum mechanics in which all this was made explicit and became completely clear.

*Von Neumann's framework is now generally accepted as the standard mathematical formulation of quantum mechanics. It will be the basis for the discussion of quantum mechanics in this book.*

## 3.4 Quantum Theory: General Remarks

### 3.4.1 A Non-historical, 'Axiomatic' Presentation

In the preceding two sections the historical background of quantum mechanics was sketched and a few remarks about its beginning were made. This historical path, which was far from staightforward, will not be followed further. In this and subsequent sections, quantum theory will be discussed as it is now. The presentation will be 'axiomatic'; a small number of principles or 'axioms' will be postulated in a precise mathematical manner, together with their physical interpretation, on the basis of which the full theory, with various explicit examples, can be developed. The mathematics needed for this is reviewed in Supp. Chap. 21 (Functional Analysis: Hilbert space), Supp. Chap. 22 (Probability theory), and Supp. Chap. 24 (Lie groups and Lie algebras) for the discussion of symmetries, with Supp. Chap. 18 (Topology) and Supp. Chap. 19 (Measure and integral) as introductions to Supp. Chap. 21. Extensive cross-references will facilitate parallel reading of these supplementary chapters with the main text.

### 3.4.2 Von Neumann's Mathematical Formulation

The mathematical basis of quantum theory, and in particular quantum mechanics, von Neumann's Hilbert space frame work, is well-understood. Hilbert space theory, like functional analysis in general, is linear algebra in infinite dimensional vector spaces, to which nontrivial topological and measure theoretical notions have been added to cope with limits, with infinite sums and integrals, needed to make it interesting and fruitful. The standard physics textbooks ignore this to a large extent; they can be said to follow in principle von Neumann's scheme—often without mentioning his name, or even the term Hilbert space, reducing it to a heuristic level where problems of convergence and divergence in infinite sums are ignored and where it is considered to be sufficient to substitute integrals for discrete sums, when continuous eigenvalues make this necessary. It has to be admitted that this attitude can be justified, up to a point, on the practical ground that it does not cause great harm in most standard day-to-day applications of quantum theory to physical problems. However, the principal aim of this book is *not* to teach quantum theory to future practitioners of the subject, but to present it as one of the great intellectual achievements of the natural sciences in the twentieth century. For this a satisfactory treatment of its mathematical foundations is required. This has the added advantage, in fact, that many notions and arguments become clearer or seem less arbitrary when the underlying mathematical framework of quantum mechanics is kept in sight.

### 3.4.3 Quantum Theory and Quantum Mechanics

A distinction will be made between the terms 'quantum mechanics' and 'quantum theory'.

a. *Quantum mechanics* is the theory invented by, among others, Heisenberg and Schrödinger, for the description of atomic structure, as was sketched in the preceding sections. It is called quantum *mechanics* because most of its observable quantities, in particular position, linear momentum and the expression for the energy, come from the *classical mechanics* of point particles. It is an important part of modern physics and is used for a wide range of submicroscopic phenomena; it is particularly important for its applications in many areas of modern technology. Its mathematical basis—essentially von Neumann's Hilbert space formalism—is completely understood.

   In the course of the past seventy-five years quantum mechanics has been generalized to cover other physical phenomena. We now have *quantum statistical mechanics* or more generally *quantum statistical physics* for collective phenomena such as observed in solids and fluids, *relativistic quantum field theory* for the description of subnuclear particles, various attempts at *quantum gravitation*, etc. Not all of this is at present completely successful or well-understood, certainly not its mathematical aspects.

b. By *quantum theory* I will mean the *general* theory which encompasses all these developments, and which is believed to describe in principle *all* known physical phenomena. It is mathematically formulated in a general manner in terms of *states*, as vectors in Hilbert space, *observables*, as selfadjoint operators, and *time evolution and symmetries*, as groups of unitary operators, even though much of this has not been made precise or explicit. This means that there are challenging fundamental mathematical problems that remain to be solved. This is true in particular for *relativistic quantum field theory* and its derivates.

   In this book the formalism of quantum theory, in terms of Hilbert space vectors and operators, will be presented in its general form, together with its physical interpretation. Most of the explicit examples comes from quantum mechanics, but not those from quantum field theory, briefly discussed in Chap. 16.

## 3.5 The Basic Concepts of Quantum Theory: A Preview

From its early beginnings in the mid-twenties to its more or less definitive form, reached in the late thirties, quantum theory has followed a sometimes tortuous historical path. Not a series of successive logical derivations from results established earlier, but more often leaps to new concepts, brilliantly guessed, not too well understood, but successful.

To solve problems of atomic spectra Heisenberg looked at position and momentum as in some sense noncommuting algebraic variables; Max Born suggested that these were infinite matrices. Schrödinger introduced a function of space and time to describe the state of a quantum mechanical particle, satisfying a partial differential equation later named after him. It was soon realized that such 'wave functions' could be seen as vectors in an infinite dimensional linear space. Von Neumann developed a Hilbert space picture for all this, with wave functions at a given time as state vectors, the matrices of Born and Jordan representing operators. His formalism encapsulated in an elegant manner Born's statistical interpretation of the wave function and the time dependence of the solutions of Schrödinger's equation, as the result of the action of an evolution operator which had to be unitary to express conservation of probability. Finally, symmetry principles and their role in analyzing and understanding complex atomic spectra were clarified by the work in group theory by mathematicians such as Herman Weyl.

It took considerable time before quantum theory reached its final form, and was accepted and found its place in textbooks. From the vantage point of the knowledge and insights that we have now, it appears to us as a coherent physical theory, based on a transparent underlying mathematical formalism. It is with an emphasis on this formalism that quantum theory is presented in this book.

In this book, directed to mathematical readers and to physicists with interest in mathematical aspects of physical theory, the basics of quantum theory will be presented in the form of a number of general principles, 'axioms', i.e. precise mathematical statements with an equally precise physical interpretation. A small but important part of the interpretation of quantum theory is still under discussion, a discussion, partly of a philosophical nature, which has recently acquired new life and has led to some new important physical ideas—with 'Bell's theorem' and 'quantum teleportation' as some of the key words. This will be briefly discussed in the last chapter.

It is useful to distinguish three levels in the 'axiomatization' of quantum theory. The first level covers the elementary quantum mechanics of one or more particles. For lack of a generally accepted name it may be called "pure state Hilbert space quantum theory". Its starting point is the assumption that a quantum state is represented mathematically by a unit vector in a Hilbert space and this is sufficient for many purposes. It will be the basis for our discussions up to Chap. 10. After that we shall need a more general sort of quantum theory with a more general axiom system, which contains the 'pure state quantum theory' as a special case. It may be called "mixed state quantum theory". It covers quantum statistical physics in which a quantum state is represented by a so-called density operator. Statistical concepts play a fundamental role in all of quantum theory, so this may seem a bit surprising. However, this point will become clear from Chap. 10 onwards. We shall later, in Chap. 12, discuss a rather theoretical but aesthetically satisfying further generalization of the preceding second approach. It is called 'algebraic quantum theory' and takes as its departure an 'abstract' algebra of observables. For now we start with 'pure state quantum theory'.

The following system of five 'axioms' is the basis of our presentation and will be discussed in more detail in the next sections:

**Axiom I**: Unit vectors $\psi$ in a fixed Hilbert space $\mathcal{H}$ describe the possible *states* of a physical system.

**Axiom II**: Selfadjoint operators $A$ in $\mathcal{H}$ appear as *observables* of the system.

**Axiom III**: Combination of Axioms I and II gives a physical interpretation in terms of probability theory. A pair $(A, \psi)$ determines a *probability distribution* for measurement of the observable $a$ in the state $\psi$. This probabilistic interpretation is the main point of difference between classical and quantum physics particle. More generally we consider the combination of $\psi$ and a system of commuting operators $A_1, \ldots, A_n$, representing so-called commensurable physical observables $a_1, \ldots, a_n$, leading to a joint probability distribution in $n$ variables.

Note that these first three axioms contain what is essentially new and surprising in quantum theory when seen against the historical background of classical physics. Axioms I and II introduce Hilbert space notions, new and unexpected in physics; their combination in III provides a mathematical idea that lends itself to a probabilistic interpretation with a clear physical meaning. The last two axioms, IV and V, are in comparison rather obvious, given the fact that we are in a Hilbert space context where unitary operators are the natural isomorphisms.

**Axiom IV**: The time evolution of a system is described by a one parameter group $\{U(t)\}_{t \in R^1}$ of unitary operators in $\mathcal{H}$, as $\psi(t_2) = U(t_2 - t_1)\psi(t_1)$. This is for the case in which the interaction does not depend explicitly on time. For explicit time dependence, one has to use a system $\{U(t_2, t_1)\}_{t_1, t_2 \in R^1}$ which acts according to $\psi(t_2) = U(t_2, t_1)\psi(t_1)$. The time evolution of the state vector is completely deterministic; a given initial state vector at time $t_1$ will lead to a well-determined state vector at $t_2$. However, the predictions for measurements at $t_2$ come from the probability distribution derived from the state vector at $t_2$ and is as such probabilistic in character.

**Axiom V**: Symmetries, such as symmetries under spatial translations and rotations, are represented by *unitary representations* of symmetry groups, usually Lie groups.

In Sect. 9.3 a sixth axiom will be added.

This ends our short review of the axioms on which quantum theory is based. In the next sections the Axioms I–V will be presented in a slightly more formal manner, with a more detailed discussion of their meaning.

## 3.6 States and Observables

### 3.6.1 Two Basic Principles: Axiom I and Axiom II

In classical mechanics the *state* of a system of $N$ point particles, can be described completely at each moment in time by $6N$ numbers, the values of $3N$ coordinates and $3N$ momenta. *Observables* are functions of these variables. In classical field theories, such as electromagnetism or general relativity, the state of the system is in general described in an equally complete manner by the values of the various fields and their time derivatives at all points of space. These values are the observables, together with expressions in the fields.

This is typical for all situations in classical physics. In quantum mechanics, and in quantum theory in general, the characterization of the state of a system and the description of observables is radically different; it includes a notion of indeterminacy described in probabilistic terms. We formulate the first two principles as axioms I and II, on which quantum theory is built. For this one needs the mathematics of *Hilbert space*, the generalization of the notion of complex inner product vector space to infinite dimension, and *selfadjoint operator*, a generalization and refinement of symmetric operator. These are extensively discussed in Supp. Chap. 21.

**Axiom I**: *The state of a physical system, at each moment in time, is represented by a unit vector in a Hilbert space, the space of state vectors of the system.*

**Axiom II**: *Observable quantities are represented by selfadjoint operators in the Hilbert space of state vectors.*

Hilbert spaces in quantum theory are always over the complex numbers and are separable, i.e. have a countably infinite orthonormal basis. The selfadjoint operators that represent observables in quantum theory are usually unbounded. See for this Supp. Chap. 21, in particular Supp. Sect. 21.9.2.

We give as basic example the quantum mechanical description of a single point particle moving in a potential.

(3.6.1,a) *Example (Point particle in a potential)* In classical mechanics a point particle is described by the variables of position $\mathbf{x} = (x_1, x_2, x_3)$ and momentum $\mathbf{p} = (p_1, p_2, p_3)$. In quantum mechanics the Hilbert space $\mathcal{H}$ for the state of a single point particle is the complex function space $L^2(\mathbb{R}^3, \mathbf{x})$, the space of square integrable complex-valued functions in the real variables $\mathbf{x} = (x_1, x_2, x_3)$. The functions that represent states have unit norm and are usually called *wave functions*, for reasons to be discussed later. The selfadjoint operators representing the basic observables are the operators $(Q_1, Q_2, Q_3)$ for the position $\mathbf{x} = (x_1, x_2, x_3)$ of the particle and $(P_1, P_2, P_3)$ for the momentum, acting on functions $\psi$ in $\mathcal{H}$, for $j = 1, 2, 3$, as

$$(Q_j\psi)(\mathbf{x}) = x_j\psi(\mathbf{x}),$$

$$(P_j\psi)(\mathbf{x}) = \frac{\hbar}{i}\frac{\partial}{\partial x_j}\psi(\mathbf{x}),$$

**(3.6.1,b) Problem** Show that the operators $Q_j$ are unbounded (In Supp. Sect. 21.9 it is explained what an unbounded operator is).

In classical mechanics the energy for a particle with mass $m$ moving under the influence of a potential $V(\mathbf{x})$ is given by the expression $E = \frac{p^2}{2m} + V$. The energy operator in quantum mechanics is the classical expression in the position and momentum operators $P_j$ and $Q_j$, i.e. it acts on wave functions $\psi$ as

$$(E\psi)(\mathbf{x}) = -\frac{\hbar^2}{2m}\sum_{j=1}^{3}\frac{\partial^2}{\partial x_j{}^2}\psi(\mathbf{x}) + V(\mathbf{x})\psi(\mathbf{x}).$$

In the classical description one has the three components of angular momentum $\mathbf{l} = (l_1, l_2, l_3)$, expressions in position and momentum given by the vector formula

$$\mathbf{l} = \mathbf{x} \times \mathbf{p},$$

which means for $l_1$

$$l_1 = x_2 p_3 - x_3 p_2,$$

with the other two components obtained as similar expressions by cyclic permutation of the indices 1, 2, 3. In quantum mechanics one has the three operators $(L_1, L_2, L_3)$, the first given as

$$L_1 = Q_2 P_3 - Q_3 P_2,$$

acting as a differential operator on the wave functions as

$$(L_1\psi)(\mathbf{x}) = \frac{\hbar}{i}\left(x_2\frac{\partial\psi(\mathbf{x})}{\partial x_3} - x_3\frac{\partial\psi(\mathbf{x})}{\partial x_2}\right),$$

with the other two formulas again obtained by cyclic permutation.

This quantum mechanical picture of a particle is indeed very different from the classical picture. One therefore cannot say that it was in any way logically derived from the classical theory. It also did not in this precise form suddenly spring to live. As so much in quantum theory it is the end point of a succession of small steps, some obvious, some less so. Bohr was able to obtain the discrete spectrum of the hydrogen atom from his atomic model, a brilliant guess without theoretical underpinning. Schrödinger formulated his equation inspired by ideas of Heisenberg and de Broglie, an equation of which he initially gave an erroneous interpretation. It was von Neumann who finally provided the general mathematical framework, as was noted earlier, together with a physical interpretation, of which this picture of the description of a single particle is an example.

The $L^2(\mathbb{R}^3, d\mathbf{x})$ wave function $\psi$ in (3.6.1,a) is the simplest example of the representation of a state vector. Other physical models lead to wave functions in more variables, or to complicated systems of such functions, as in many particle quantum systems, to be discussed in Chap. 9.

## 3.6.2 Axiom III: Physical Interpretation of Axiom I and Axiom II

The physical interpretation of I and II uses as the main mathematical concept the culmination of von Neumann's theory of operators in Hilbert space, the *general spectral theorem for selfadjoint operators*, with their *spectral resolutions*, generalizing the idea of eigenvalue–eigenvector representations of symmetric matrices in finite dimensions, all explained in Supp. Chap. 21, in particular Supp. Sects. 21.9 and 21.10. One further needs basic notions from probability theory, such as *stochastic variable, distribution function, expected value, higher moments*, reviewed in Supp. Chap. 22, and finally the connection between these operator-theoretic and stochastic notions, typical for quantum theory.

## 3.6.3 The Simplest Case: Axiom III for a Single Observable

In the formulation of the physical interpretation of Axioms I and II as Axiom III, we begin with the simplest case, that of a single observable.

**Axiom III$_0$**: *A pair $(\psi, A)$, consisting of a unit vector $\psi$ and a selfadjoint operator A, representing respectively a state of the system and an observable, determine a stochastic variable in the sense of standard probability theory. The (cumulative) distribution function of this variable is given by $F_\psi(\alpha) = (\psi, E_\alpha \psi)$, with the projection operators $E_\alpha$ belonging to the spectral resolution $\{E_\alpha\}_{\alpha \in R^1}$ of the operator A.*

It follows from the formula for $F_\psi$ that the vector $e^{i\beta}\psi$ gives the same physical results as $\psi$, so strictly speaking Axiom I should say that a state is represented by an *equivalence class* of unit vectors, differing only by a phase factor.

Axiom III$_0$ means that the probability for finding a value of the observable in the half-open interval $(\alpha_1, \alpha_2]$ is equal to $F_\psi(\alpha_2) - F_\psi(\alpha_1)$. See for the definition of distribution function in probability theory Supp. Sect. 22.3, and for the way such a function arises in the context of spectral theory Supp. Sect. 21.10.

It can be shown that III$_0$ implies that for an observable $a$, represented by the operator $A$, the resulting probabilities are concentrated on the spectrum of $A$ and that the mean or expected value—or expectation value as it is called in physics—is equal to

$$\overline{a}_\psi = (\psi, A\psi) = \int\limits_{-\infty}^{+\infty} \alpha \, d(\psi, E_\alpha \psi),$$

with the bar here denoting averages. The second moment is obtained by the formula

$$\overline{a^2}_\psi = (\psi, A^2\psi) = \int\limits_{-\infty}^{+\infty} \alpha^2 \, d(\psi, E_\alpha \psi).$$

It is needed together with the expected value to define the standard deviation

$$\Delta a_\psi = \overline{(\alpha - \overline{\alpha}_\psi)^2}^{1/2}.$$

Higher order moments can be calculated in a similar manner but are in general of no great importance.

(3.6.3,a) *Remark* The distribution function $F_\psi(\alpha)$ exists for an arbitrary unit vector $\psi$. However, the expectation, or more generally the $n$th moment, of a probability distribution may not exist or may not be finite. The $n$th order moment $\overline{\alpha^n}$ exists and is finite whenever $\psi$ is in the domain of the operator $A^n$.

### 3.6.4 The Case of Discrete Spectrum

Consider first the technically simple case of an observable represented by a selfadjoint operator with a purely discrete spectrum. In this case we need only *eigenvectors*, with the projections on its associated eigenspaces, not the more sophisticated notion of *spectral resolution*. This implies that for the probabilistic interpretation the general notion of *distribution function* is superfluous; discrete probabilities are sufficient.

(3.6.4,a) *Example* (*The harmonic oscillator*) The one dimensional classical harmonic oscillator is a particle attached to a fixed point, the origin $x = 0$, by an ideal spring, i.e. the particle is moving in a potential $V(x) = \frac{1}{2}kx^2$, and as a result is oscillating with constant frequency $\omega = \sqrt{k/m}$. In rest its energy is zero; when moving it can have arbitrary positive energy. This model will be discussed in greater detail in Chap. 6. Here we need only the fact that the selfadjoint operator that in the quantum case represents the energy $E$, usually denoted as $H$ and called the *Hamiltonian* for reasons that will be explained later, has a spectrum consisting of discrete nondegenerate eigenvalues $\epsilon_0 < \epsilon_1 < \epsilon_2 < \dots$. This means that the energy of a quantum oscillator can only have these discrete values. The spectral theorem reads simply

$$H = \sum_{j=1}^{\infty} \epsilon_j P_j,$$

with $P_j$ the projections on the eigenspaces associated with the eigenvalues $\epsilon_j$. Because the $\epsilon_j$ are nondegenerate, there is an orthonormal basis of eigenvectors $\phi_0, \phi_1, \ldots$, uniquely determined up to phase factors. An arbitrary state vector $\psi$ can be expanded as

$$\psi = \sum_{j=0}^{\infty} c_j \phi_j,$$

with the complex coefficients $c_j = (\phi_j, \psi)$ satisfying the normalization condition

$$\sum_{j=0}^{\infty} |c_j|^2 = (\psi, \psi) = 1.$$

(3.6.4,b) **Problem** Give the form of the spectral resolution $\{E_\epsilon\}_\epsilon$ for the selfadjoint operator $H$. Derive from this the distribution function $F_\psi$ which according to Axiom III$_0$ follows from combining $H$ with a state vector $\psi$. Show finally that this distribution function leads to a set of discrete probabilities $\{\rho_j\}_j$ given by $\rho_j = |c_j|^2$, for $j = 0, 1, 2, \ldots$.

This result means that the real number $\rho_j$ is the probability that a measurement of the system in the state $\psi$ will give $\varepsilon_j$. The expression for the expectation becomes, for $\psi$ in the domain of definition of $H$,

$$\bar{\varepsilon}_\psi = (\psi, H\psi) = \sum_{j,k=0}^{\infty} c_j^* c_k \, \varepsilon_k (\phi_j, \phi_k) = \sum_{j=0}^{\infty} \varepsilon_j |c_j|^2 = \sum_{j=0}^{\infty} \varepsilon_j \rho_j.$$

For $\psi$ in the domain of $H^2$, this gives together with the formula for the second moment, the square of the standard deviation as

$$(\Delta \varepsilon_\psi)^2 = (\psi, (H - \bar{\varepsilon}_\psi)^2 \psi) = \sum_{j=0}^{\infty} (\varepsilon_j - \bar{\varepsilon}_\psi)^2 \rho_j.$$

If $\psi$ happens to be one of the eigenvectors of the operator $H$, then the probability distribution will be concentrated on the corresponding eigenvalue, i.e. a measurement will give with certainty (i.e. "almost surely") this eigenvalue.

From this example it is clear that for the case of discrete spectrum Axiom III$_0$ takes a much simpler form:

*A pair $(\psi, A)$ with $\psi$ a unit vector and $A$ a selfadjoint operator, with a purely discrete spectrum consisting of nondegenerate eigenvalues $\{\alpha_j\}_j$, determines a discrete probability distribution $\{\rho_j\}_j$, given by $\rho_j = |c_j|^2$, with $c_j$ the $j$-th coefficient in the expansion of $\psi$ with respect to the orthonormal basis of eigenvectors of $A$.*

The energy spectrum of an isotropic three dimensional oscillator, with the same frequency $\omega$ along the three spatial axes, is no longer nondegenerate, as will be shown in Sect. 6.8. As a consequence there is a freedom of choice for the orthonormal basis of eigenvectors in each eigenspace. This leads to an obvious generalization of the above example.

### 3.6.5 The Case of Continuous Spectrum

The case in which the operator representing an observable has continuous spectrum is mathematically more sophisticated. It requires for its physical interpretation the general version of the spectral theorem. Eigenvectors no longer exist, at least not as proper normalized vectors in $\mathcal{H}$; instead of a sequence of projection operators on eigenspaces one has to use a system of spectral projections depending on a continuous variable as index, the spectral resolution of the general form of the spectral theorem, involving an operator-valued Stieltjes integral, as is explained in Supp. Sect. 21.10. See for the notion of Stieltjes integral in general Supp. Sect. 19.5.3.

(3.6.5,a) *Example* Consider again the example of the system of a single point particle, as discussed in Example (3.4.1,a), but now in one dimensional space, i.e. with quantum mechanical state space $\mathcal{H} = L^2(\mathbb{R}^1, dx)$ and as state vectors wave functions $\psi(x)$. There is a single position operator $Q$, with a continuous spectrum consisting of all real numbers. It has a system of spectral projection operators $Q_\lambda$ defined as $(Q_\lambda\psi)(x) = \psi(x)$, for $x < \lambda$ and $(Q_\lambda\psi)(x) = 0$, for $x \geq \lambda$. For a particular choice of state vector $\psi$ one has as distribution function $F_\psi$, in this case

$$F_\psi(\lambda) = (\psi, Q_\lambda\psi) = \int_{-\infty}^{\lambda} |\psi(x)|^2\, dx$$

which means that the probability distribution determined by the position operator $Q$ and the normalized state vector $\psi$ is absolutely continuous with respect to the Lebesgue measure on $\mathbb{R}^1$, i.e. can be given by the probablity density $\rho(x) = |\psi(x)|^2$. (For the notion of absolute continuity of a measure with respect to a second measure see Supp. Sect. 19.6.) The expectation value for $Q$ in the state $\psi$ is

$$(\psi, Q\psi) = \int_{-\infty}^{+\infty} x\rho(x)\, dx = \int_{-\infty}^{+\infty} x|\psi(x)|^2\, dx.$$

Note that this is consistent with the fact that a function $\psi(x)$ is in the domain of the operator $Q$ if and only if $\psi(x)$ and $x\psi(x)$ are both square integrable.

In this way the well-known physical interpretation of the 'wave function' in quantum mechanics is derived from our Axiom III$_0$:

*The square of the absolute value of the wave function of a quantum mechanical particle is the probability density for the measurement of the position of that particle.*

This means of course that the probability of finding the particle within the interval $[x_1, x_2]$ is $\int_{x_1}^{x_2} |\psi(x)|^2 dx$. It also means that there exist states in which the position of the particle is very sharply defined, as in the classical situation. However, the difference with the classical situation will manifest itself when one tries to measure position *and* momentum simultaneously. More sharpness in position will necessarily entail less sharpness in momentum. This phenomenon, culminating in the *uncertainty relation of Heisenberg*, will be discussed in Chap. 4, in particular in Sects. 4.3–4.5.

This interpretation in terms of probabilities is a typical quantum phenomenon which has no classical precedent. It is one of the features that did not emerge immediately. Schrödinger, in fact, first thought that his wave function described a particle that was smeared out in space. In the statistical interpretation introduced somewhat later by Max Born the electron is still a point particle; it is the probability density for the measurement of position in a given state that is smeared out.

In general a selfadjoint operator may have a purely discrete, purely continuous, or a mixed spectrum. An example of the last possibility is the operator that represents the energy of a hydrogen atom. It has an infinite system of discrete values, starting at a lowest negative value, the 'ground state energy' $E_0$, which is nondegenerate, and then converging to $E_\infty = 0$, where a continuous spectrum begins which goes all the way to $+\infty$. A hydrogen atom is thought of as a system consisting of an electron moving in the electrostatic field of a nucleus. The discrete spectrum represents the *bound states* of this system, the continuous spectrum corresponds to the situation where the atom is *ionized*, i.e. with the electron separated from the nucleus. All this will be discussed later, in Chaps. 7 and 8.

As is noted in Supp. Sect. 21.10, the continuous spectrum may be absolutely continuous or singularly continuous. The latter case is extremely rare in physics. For this reason we discuss in this book discrete spectra, leading to discrete probabilities, and continuous spectrum, by which we mean absolutely continuous spectrum, leading to probability densities.

## 3.7  Systems of Observables

### 3.7.1  Commensurable Observables

In this section we have so far only discussed single observables. In the case of *systems of observables*, there are new and interesting phenomena, that are typical for quantum theory. We first consider systems of *commensurable observables*, by which we mean systems of observables described by commuting operators. The reason for the use of the word 'commensurable' will become clear in this and the next subsections. Note in this context that the correct definition for commuting of (possibly unbounded) selfadjoint operators is that all their respective spectral projections commute. In general the relation $[A, B] = 0$, for two unbounded operators, if taken literally, involves domain problems, as was remarked earlier.

For a system of $n$ commensurable observables $a_1, \ldots, a_n$, represented by $n$ commuting selfadjoint operators $A_1, \ldots, A_n$, we formulate the general version of Axiom III which of course includes Axiom $\text{III}_0$ as a special case:

**Axiom III**: *A unit vector $\psi$, representing a state of a quantum system, together with $n$ commuting selfadjoint operators $A_1, \ldots, A_n$, representing observable quantities $a_1, \ldots, a_n$, determine a system of stochastic variables in the sense of standard probability theory. The joint or simultaneous (cumulative) distribution function for these variables is given by*

$$F_\psi(\alpha_1, \ldots, \alpha_n) = (\psi, E_{\alpha_1 \ldots \alpha_n} \psi),$$

*with the projection operators $E_{\alpha_1 \ldots \alpha_n}$ belonging to the n-parameter spectral resolution associated with the system $A_1, \ldots, A_n$, according to the general spectral theorem for systems of commuting selfadjoint operators.*

See for the spectral theorem in general Supp. Sect. 21.10, for the particular case of $n$ commuting selfadjoint operators Supp. Sect. 21.5, and for $n$-parameter spectral resolutions the explanation in Supp. Sect. 21.7. For the notion of joint distribution functions see Supp. Sect. 22.3.4.

Axiom III means, for example, for a pair of commensurable observables are represented by commuting operators $A$ and $B$, that the probability of finding the values of these observables in the rectangle given by $\alpha_1 < \alpha \leq \alpha_2$ and $\beta_1 < \beta \leq \beta_2$ is

$$F(\alpha_2, \beta_2) - F(\alpha_1, \beta_2) - F(\alpha_2, \beta_1) + F(\alpha_1, \beta_1),$$

with the possibility of obtaining the probability for any measurable subset of $\mathbb{R}^2$ from this by approximation, and with an obvious generalization to an arbitrary finite system of commensurable observables. (See for the notion of 'measurable set' Supp. Chap. 19).

If two operators $A$ and $B$ have only discrete eigenvalues, say $\alpha_1 < \alpha_2 < \cdots$ and $\beta_1 < \beta_2 < \cdots$, then the generalization of the spectral theorem to systems of commuting operators means that there exist an orthonormal basis of common eigenvectors of $A$ and $B$. For the case of general spectrum of $A$ and $B$ there is a two parameter spectral resolution, $\{E_{\alpha\beta}\}$, as is explained in Supp. Sect. 21.7. An arbitrary unit vector $\psi$ in $\mathcal{H}$ gives then a joint distribution function $F_\psi$, in the sense of probability theory, according to $F_\psi(\alpha, \beta) = (\psi, E_{\alpha\beta}\psi)$. The obvious physical interpretation of this situation is that one finds the values for $A$ and $B$ in the intervals $(-\infty, \alpha]$ and $(-\infty, \beta]$, in a *simultaneous* measurement of $A$ and $B$ in the state represented by the vector $\psi$.

Axiom III implies that the physical interpretation of a system of commensurable observables is part of standard probability theory. An illustration of this is the covariance of two commensurable observables, which is just the usual expression for the statistical correlation between two random variables. Consider $a$ and $b$, with commuting operators $A$ and $B$, in a state $\psi$, then this covariance is

$$\overline{(a - \overline{a})(b - \overline{b})}_\psi = (\psi, (A - \overline{a})(B - \overline{b})\psi)$$

$$= \int\limits_{-\infty}^{+\infty} \int\limits_{-\infty}^{+\infty} (\alpha - \overline{a})(\beta - \overline{b})\, dF_\psi(\alpha, \beta),$$

and can be written, under the usual assumption of absolute continuity of the spectrum, as an integral over a probability density $\rho_\psi$ as

$$\overline{(a - \overline{a})(b - \overline{b})}_\psi = \int\limits_{-\infty}^{+\infty} \int\limits_{-\infty}^{+\infty} (\alpha - \overline{a})(\beta - \overline{b})\, \rho_\psi(\alpha, \beta)\, d\alpha d\beta.$$

(3.7.1, a) *Example* Consider a point particle in three dimensional space. Its quantum state space $\mathcal{H}$ is $L^2(\mathbb{R}^3, d\mathbf{x})$. The three components of the position are represented by the commuting selfadjoint operators $Q_j$, acting as

$$(Q_j\psi)(\mathbf{x}) = x_j\psi(\mathbf{x}),$$

for $j = 1, 2, 3$. The spectral resolution of this system consists of projection operators $E_{\lambda_1\lambda_2\lambda_3}$ given by

$$(E_{\lambda_1\lambda_2\lambda_3}\psi)(x_1, x_2, x_3) = 0,$$

for $x_j \geq \lambda_j$ and $j = 1, 2, 3$, and

$$(E_{\lambda_1\lambda_2\lambda_3}\psi)(x_1, x_2, x_3) = \psi(x_1, x_2, x_3),$$

for $x_j < \lambda_j$. For a state vector $\psi$ with $\|\psi\| = 1$ this gives a joint distribution function

$$F_\psi(\lambda_1, \lambda_2, \lambda_3) = (\psi, E_{\lambda_1\lambda_2\lambda_3}\psi)$$

$$= \int\limits_{-\infty}^{\lambda_1} \int\limits_{-\infty}^{\lambda_2} \int\limits_{-\infty}^{\lambda_3} |\psi(x_1, x_2, x_3)|^2\, dx_1 dx_2 dx_3$$

Because of absolute continuity everything can be written in terms of the joint probability density $\rho(x_1, x_2, x_3) = |\psi(x_1, x_2, x_3)|^2$. All this means that *as far as the position of the particle is concerned*, the quantum mechanical description remains conceptually within the framework of standard probability theory.

### 3.7.2 *Systems of Incommensurable Observables*

The occurrence of systems of *incommensurable observables* i.e. systems of observables represented by noncommuting selfadjoint operators, is a typical quantum phenomenon. It takes us outside the domain of classical probability theory.

Consider a pair $A$ and $B$ of noncommuting selfadjoint operators. After a choice of a state vector $\psi$, $A$ and $B$ give *separate* stochastic variables; together these do *not* form a *system* of stochastic variables in the classical sense: i.e. there does not exist a common probability measure space on which these variables are measurable functions and there is no joint distribution function or probability density. There are nevertheless relations between the results of measurements of the two observables, relations which are typical for quantum theory.

(3.7.2,a) *Example* The operators $P$ and $Q$ for position and momentum of a one dimensional particle do not commute, are incommensurable. The same is true for $Q_j$ and $P_k$, $j = k$, for a three dimensional particle. There are no common probability densities in these cases. There is however an interesting and important relation between the measurement of (the corresponding components of) position and momentum. It is called the *Heisenberg uncertainty relation* and will be discussed extensively in Chap. 4.

### 3.7.3 *Functions of Observables*

In the finite dimensional case the notion of function $f(a)$ of an observable $a$ is straightforward. Suppose that the selfadjoint operator $A$ that represents such an observable has eigenvalues $\{\alpha_j\}_j$ together with an orthonormal basis $\{\phi_j\}_j$. For a function $f$ the observable $f(a)$ is represented by the operator $f(A)$ acting on the $\phi_j$ as $f(A)\phi_j = f(\alpha_j)\phi_j$. The $\phi_j$ are obviously also eigenvectors of $f(A)$, with eigenvalues $f(\alpha_j)$.

In infinite dimension the operator that represents an observable may have continuous spectrum. This means that there are no eigenvalues with corresponding eigenvectors. Fortunately, there is a more sophisticated way of defining functions of observables, valid for the general finite or infinite dimensional situation. It invokes the *spectral theorem* based on a *spectral resolution* with its projections taking the place of the projections on the eigenspaces of finite dimension, which leads to a notion of function of an observable which is analogous to that of the finite dimensional case. We have the following formulas, for the selfadjoint operator $A$, representing the observable $a$

$$A = \int_{-\infty}^{+\infty} \alpha \, dE_\alpha,$$

and, for the operator representing $f(a)$, for any real-valued measurable $f$,

$$f(A) = \int\limits_{-\infty}^{+\infty} f(\alpha)dE_\alpha.$$

For the mathematics behind these formulas see Supp. Chap. 21, and in particular Supp. Sect. 21.10.9.

For the case of $n$ commensurable observables $a_1, \ldots, a_n$ one has

$$f(A_1, \ldots, A_n) = \int\limits_{-\infty}^{+\infty} \ldots \int\limits_{-\infty}^{+\infty} f(\alpha_1, \ldots, \alpha_n)dE_{\alpha_1,\ldots,\alpha_n}.$$

For systems of incommensurable observables, represented by noncommuting operators, there is no 'functional calculus', like the one for commensurable observables, expressed by the formula above. The best that one can do is to take a polynomial $p(a, b)$ in the incommensurable observables $a$ and $b$, and try to write the analoguous operator-valued polynomial $p(A, B)$, but this immediately raises the problem of ambiguity, because, for instance, in classical mechanics one has $pq = qp$, but in quantum mechanics $PQ \neq QP$.

Quantum mechanics is more fundamental than classical mechanics; in atomic physics where it started, it describes the submicroscopic world of atomic nuclei and electrons. In principle the classical description of a physical system like a macroscopic particle, say a bullet, should follow from the underlying quantum description, in what can be called a 'classical limit'.

### 3.7.4 Complete Systems of Commensurable Observables

Observables can be used to *label* states, in particular eigenstates, usually energy eigenstates. In a system like the harmonic oscillator, introduced in Example (3.6.4,a) and to be discussed more fully in Chap. 6, the energy operator $H$ has a purely discrete nondegenerate spectrum—it is multiplicity free in the language of Supp. Sect. 21.10, and its eigenvalues are therefore sufficient to characterize eigenstates. In most other cases the energy operator alone is not sufficient for this purpose. An example of such a situation is the model of the hydrogen atom, mentioned in the second part of Sect. 6.5 of this chapter, and discussed more fully in Chap. 7. The discrete energy eigenstates, describing Bohr's orbits of the electron moving around the central nucleus, are all degenerate, except for the lowest energy state, the 'groundstate'. To label these states unambiguously one adds to the energy two other observables, commensurable with it, a component of angular momentum, traditionally $L_3$, and the square of the 'absolute value' of the angular momentum $\mathbf{L}^2 = L_1^2 + L_2^2 + L_3^2$, both observables with a clear and important physical meaning. The triple $(H, L_3, \mathbf{L}^2)$ is a multiplicity free system, see Supp. Sect. 21.10.10, and is maximal in the sense that every other observable that is commensurable with these three must be a function of $(H, L_3, \mathbf{L}^2)$. Such a system is called a *complete system*. We give a formal definition of this very important notion:

(3.7.4,a) **Definition** A commensurable system of observable quantities $(a_1, \ldots, a_n)$ is called *complete* iff it is represented by a system of commuting selfadjoint operators $(A_1, \ldots, A_n)$ that is multiplicity free in the sense of Supp. Sect. 21.10.10. The notion of multiplicity free is a generalization to infinite dimension of that of non degeneracy of the eigenvalues of the operators in finite dimension. For the mathematical background of this see Supp. Sect. 21.10.10.

It is useful to include in this definition the requirement that if one of the $a_j$ is omitted from the system, then the corresponding system of operators is no longer multiplicity free, or equivalently, that every observable that is commensurable with the $a_j$ is a unique function of the $a_j$.

Not all complete systems of observables contain the energy. Examples are the system $(x_1, x_2, x_3)$ of the components of position and the system $(p_1, p_2, p_3)$ of the components of momentum, for the model of a three dimensional particle, given in Example (3.6.1,a).

### 3.7.5 The Measurement Process

The prediction of results of measurements in quantum theory in terms of probability theory, as described in the preceding sections, is uncontroversial and generally accepted. This is less so for the description of the actual measurement process. According to the orthodox 'Copenhagen' interpretation of quantum mechanics, of which Bohr was the main advocate, a distinction must be made between the (microscopic) system that is the subject of the measuring operation, and the (macroscopic) measurement apparatus which is seen as classical. As a consequence of the measurement the wave function, which describes the system, is supposed to undergo an instantaneous discontinuous change. If, for instance, the measurement of an observable represented by the operator $A$ gives as result the value $a$, necessarily an eigenvalue of $A$, then the wave function is projected onto the corresponding eigenspace. (A more precise formulation is necessary in the case of continuous spectrum). This idea of the 'reduction of the wave packet' has led to much discussion by philosophers of science. The question is asked whether objective and subjective aspects of reality can be truly distinguished, whether an object exists when it is not observed, etc., In recent years the Copenhagen interpretation has been criticized and alternatives have been put forward. These matters will be briefly discussed in Chap. 17.

## 3.8  Time Evolution

### 3.8.1 Time Evolution of Autonomous Quantum Systems

The evolution of a quantum system from an initial time $t_1$ to a later time $t_2$ is described by an *evolution operator* acting on the state vectors in the Hilbert space. It has to be invertible and has to map unit vectors to unit vectors; this together means that it should be a *unitary* operator. We first consider *autonomous* systems, by which we mean systems with an interaction that does not explicitly depend on time, or in other words, systems that are invariant under time translation.

For the dynamics of such an autonomous system this means that the evolution operator depends only on the *difference* of initial and later times. Moreover, time is a continuously changing parameter. We think therefore of a system of unitary operators $\{U(t)\}_{t \in \mathbb{R}^1}$, continuous in $t$, in the sense of strong operator convergence, acting as

$$U(t_2 - t_1)\psi(t_1) = \psi(t_2),$$

and satisfying the following natural condition:

$$U(t_1 + t_2) = U(t_1)U(t_2),$$

for all real $t_1$ and $t_2$. From this property it follows that $U(0) = 1$ and $U(t)^{-1} = U(-t)$, for all $t$ in $\mathbb{R}$ (Problem (21.11,b) in Supp. Chap. 21).

A system $\{U(t)\}_{t \in \mathbb{R}}$ with these properties is called a *one parameter group of unitary operators*. See Supp. Sect. 21.11.1 for an extensive discussion of such groups. According to a theorem of Stone and von Neumann, Theorem (21.11.2,a), it is a general property of a strongly continuous one parameter group that it has a selfadjoint *generator* by which it is uniquely determined. In this case the generator is called the *Hamiltonian* and is denoted as $H$. One can write, inserting Planck's constant $\hbar$ for dimensional reasons,

$$U(t) = e^{-\frac{i}{\hbar}tH}.$$

All this together gives us the general principle describing time evolution for autonomous systems in quantum theory:

**Axiom IV$_1$**: *Time evolution for autonomous systems in quantum theory is described by a continuous one parameter group of unitary operators $\{U(t)\}_{t \in \mathbb{R}^1}$, acting in the Hilbert space of states $\mathcal{H}$. It can be written as $U(t) = e^{-\frac{i}{\hbar}tH}$, with $H$ a selfadjoint operator which is called the Hamiltonian of the system.*

### 3.8.2 The Schrödinger Equation

To describe the dynamics of a concrete physical situation it is the Hamiltonian operator that is explicitly given, not the one parameter group of evolution operators. Note that it is called 'Hamiltonian' because—at least in quantum mechanics—it is obtained by taking the Hamiltonian function which determines the time evolution of the corresponding classical system and substituting operators for the classical variables of position and momentum, following in this the rules of 'canonical quantization'. Note also that usually the Hamiltonian—both in the classical and the quantum case—determines the time evolution *and* represents also the *energy* as observable. Starting from the Hamiltonian one obtains such time dependent state vectors $\psi(t) = e^{-\frac{i}{\hbar}tH}\psi(0)$ as solutions of the first order vector-valued differential equation

$$\frac{d}{dt}\psi(t) = -\frac{i}{\hbar}H\psi(t).$$

This equation, which is valid for all $\psi(t)$ in the domain of $H$, may be called the *'abstract'* or *'vector-valued'* Schrödinger equation. The time evolution in terms of states, i.e. unit vectors in $\mathcal{H}$, is deterministic; given a state at $t = t_1$, the state at a later time $t_2$ is completely determined. This does not affect the overall probabilistic nature of quantum theory, because the outcome of a measurement at $t_2$ cannot be predicted with certainty, as was already noted.

(3.8.2,a) *Example* For a three dimensional point particle with mass $m$, moving in a centrally symmetric potential $V(r)$, the classical Hamiltonian is

$$H_{\text{class}}(\mathbf{p}, \mathbf{x}) = \frac{\mathbf{p}^2}{2m} + V(r).$$

The Hamiltonian of the corresponding quantum system is therefore given by the operator expression

$$H = \frac{P^2}{2m} + V(Q),$$

acting on the state vectors, i.e. 'wave functions' $\psi(\mathbf{x})$, as

$$(H\psi)(\mathbf{x}) = -\hbar^2 \nabla \psi(\mathbf{x}) + V(r)\psi(\mathbf{x}),$$

with notations the same as in (3.4.1,a) and furthermore with

$$r = \sqrt{\sum_{j=1}^{3} x_j^2}, \quad \nabla = (\partial/\partial x_1, \partial/\partial x_2, \partial/\partial x_3).$$

A time-dependent wave function $\psi(\mathbf{x}, t)$ therefore has to satisfy the equation

$$-\frac{\hbar}{i}\frac{\partial}{\partial t}\psi(\mathbf{x}, t) = -\frac{\hbar^2}{2m}\nabla\psi(\mathbf{x}, t) + V(r)\psi(\mathbf{x}, t).$$

This is, for this situation, the 'concrete' *Schrödinger equation*, the central partial differential equation of standard quantum mechanics.

Solving the Schrödinger equation for the time evolution of state vectors is in principle equivalent to solving the generalized eigenvalue–eigenvector problem for the Hamiltonian operator $H$, i.e. finding its spectral resolution.

To understand the idea behind this statement, consider the simple case of an operator $H$ with a purely discrete nondegenerate spectrum, e.g. the one dimensional harmonic oscillator, to be discussed in Chap. 6. Suppose that its eigenvalues $\{\alpha_n\}_n$ are known, together with an orthonormal system of eigenvectors $\{\phi_n\}_n$. An arbitrary vector $\psi$ can be expanded as

$$\psi = \sum_{n=0}^{\infty} c_n \phi_n,$$

with the coefficients $c_n$ determined by the inverse formula $c_n = (\phi_n, \phi_n)$. For a time-dependent $\psi(t)$ this expansion is

$$\psi(t) = \sum_{n=0}^{\infty} c_n(t) \phi_n,$$

with time-dependent coefficients $c_n(t) = (\phi_n, \psi(t))$. Application of the vector-valued Schrödinger equation gives, for each $n$ separately,

$$\frac{d}{dt} c_n(t) = -\frac{i}{\hbar} E_n c_n(t),$$

with solution

$$c_n(t) = e^{-\frac{i}{\hbar} t E_n} c_n(0).$$

This allows us to express $\psi(t)$ in the initial state $\psi(0)$ as

$$\psi(t) = \sum_{n=0}^{\infty} e^{-\frac{i}{\hbar} t E_n} c_n(0)\, \phi_n.$$

For the case of a general Hamiltonian the result is essentially the same, although more sophistication is needed to formulate it properly. It explains in any case the relation between Schrödinger's wave mechanics, in which a wave equation had to be solved, and Heisenberg's matrix quantum mechanics, which involved solving the eigenvalue–eigenvector problem of an infinite dimensional matrix, the two versions of early quantum mechanics, which looked quite different at first sight.

All this should make it clear that the principal task in the quantum mechanics of autonomous systems is solving the 'eigenvalue–eigenfunction problem' for a given Hamiltonian. In realistic physics exact solutions are rare. For quantum mechanics these are known mainly for relatively simple examples like the harmonic oscillator and the hydrogen atom, models which played an important role in the early history of quantum mechanics, and are still of great pedagogical value. As is usual in actual research in most fields of physics, the development of suitable approximation procedures is what counts.

### 3.8.3 The Heisenberg Picture

All observable effects can be expressed in probabilities of measuring certain observables $a$ in certain states $\psi$. Time evolution is no exception; we do not observe the time-dependence of states directly, but only its consequences for the time-dependence of probabilities. One has for instance for the expectation value of the observable $a$, represented by an operator $A$, in a state $\psi$, the formula

$$\bar{a}_\psi(t) = (\psi(t), A\psi(t)) = (e^{-\frac{i}{\hbar}tH}\psi(0), Ae^{-\frac{i}{\hbar}tH}\psi(0)),$$

with the bar in $\bar{a}$ in this context meaning 'statistical mean' or 'average'. This can be written as

$$\bar{a}_\psi(t) = (\psi(0), e^{\frac{i}{\hbar}tH}Ae^{-\frac{i}{\hbar}tH}\psi(0)) = (\psi(0), A(t)\psi(0)).$$

Similar formulas hold for the time-dependence of general moments and the distribution function itself. This means that instead of putting the time-dependence in the state vectors as

$$\psi(t) = e^{-\frac{i}{\hbar}tH}\psi(0),$$

with the observables remaining constant, one may as well let the one parameter group $U(t) = e^{-\frac{i}{\hbar}tH}$ act on the operators as

$$A(t) = e^{\frac{i}{\hbar}tH}A(0)\,e^{-\frac{i}{\hbar}tH},$$

with the state vectors kept constant. The first procedure, with the time evolution in the state vectors, the usual one in most of elementary quantum mechanics, is called the *Schrödinger picture*. The second way of looking at time evolution, with constant states and time-dependent observables, is known as the *Heisenberg picture*; it is used in some of the more advanced parts of quantum theory. Both pictures are physically completely equivalent; the choice between them is purely a matter of convenience. In the Heisenberg picture the basic dynamical equation is a differential equation for operators. By differentiating $e^{\frac{i}{\hbar}tH}A(0)\,e^{-\frac{i}{\hbar}tH}$ with respect to time one obtains

$$\frac{d}{dt}A(t) = \frac{i}{\hbar}[H, A(t)],$$

the *Heisenberg equation*, which in the Heisenberg picture plays the role of the Schrödinger equation in the Schrödinger picture.

### 3.8.4 Stationary States and Constants of the Motion

(3.8.4,a) **Definition** A *stationary state* of an autonomous quantum system is a state vector $\psi$ for which the probabilities of measurements of *all observables* are constant in time.

(3.8.4,b) **Theorem** *A unit vector in $\mathcal{H}$ is a stationary state if and only if it is an eigenvector of the Hamiltonian.*

*Proof* ($\Rightarrow$) Suppose now that $\phi$ is a stationary state. This means that, with $\phi(t) = e^{-\frac{i}{\hbar}tH}\phi$, the expression $(\phi(t), A\phi(t))$ is independent of $t$, for every selfadjoint operator $A$. Take for $A$ the projection $E_\phi$ on $\phi$. The action of $E_\phi$ on $\phi(t)$ can be written as $E_\phi\phi(t) = (\phi, E_\phi\phi(t))\phi = (E_\phi\phi, \phi(t))\phi = (\phi, \phi(t))\phi$. Taking the inner product with $\phi(t)$ gives

$$(\phi(t), E_\phi\phi(t)) = (\phi(t), (\phi, \phi(t))\phi) = (\phi, \phi(t))(\phi(t), \phi) = |(\phi, \phi(t))|^2,$$

so the expression $|(\phi, \phi(t))|$ is independent of $t$. The Schwarz inequality in 21.3 states that $|(\phi, \phi(t))| \leq ||\phi||\,||\phi(t)||$, with in case of equality that $\phi$ is a scalar multiple of $\phi(t)$. We have equality because $|(\phi, \phi(t))| = |(\phi, \phi)| = 1$, $||\phi|| = 1$, by definition, $||\phi(t)|| = ||e^{-\frac{i}{\hbar}tH}\phi|| = ||\phi|| = 1$, and therefore $|(\phi, \phi(t))| = ||\phi||\,||\phi(t)||$, implying $\phi = \rho(t)\phi(t)$, for some complex number $\rho(t)$. One verifies easily that $\rho(t)$ is a phase factor, continuous in $t$, with $\rho(t_1 + t_2) = \rho(t_1)\rho(t_2)$, for all $t_1$ and $t_2$, so has, according to the remark at the beginning of D.11, the form $\rho(t) = e^{i\alpha t}$, for some real $\alpha$. Applying Theorem (21.11.a) (Stone–von Neumann) gives that $\phi$ is in the domain of $H$ with $H\phi = -\hbar\alpha\phi$, i.e. $\phi$ is an eigenvector of $H$ with eigenvalue $-\hbar\alpha$.

($\Leftarrow$) The action of $U(t) = e^{-\frac{i}{\hbar}tH}$ on an eigenvector $\phi_n$ multiplies it by a phase factor $e^{-\frac{i}{\hbar}tE_n}$. Such a phase factor does not show up in any of the formulas used to calculate probabilities. From this it follows easily that $\phi_n$ is a stationary state.

(3.8.4,c) *Remark* Systems with Hamiltonians with purely continuous spectrum have no stationary states. This is the case for a freely moving point particle. The Hamiltonian of a harmonic oscillator, has a purely discrete spectrum, and has therefore an orthonormal basis consisting of stationary states. The Hamiltonian of the hydrogen atom has both discrete and continuous spectrum, corresponding to stationary states, and non stationary states respectively. The physical picture of a hydrogen atom is that it consists of a positively charged nucleus, a proton, together with a negatively charged electron as a satellite. In a stationary state the electron is bound to the nucleus by the Coulomb force; the electron is in a 'Bohr orbit', with one of the well-known discrete energy values. The non stationary states, the 'ionized states', correspond to the situation in which the electron is no longer bound to the nucleus. The harmonic oscillator will be discussed in detail in Chap. 6, the hydrogen atom in Chap. 7.

(3.8.4,d) **Definition** A *constant of the motion* or *conserved quantity* of an autonomous quantum system is an observable for which the probabilities of measuring it in *all states* are constant in time.

(3.8.4,e) **Theorem** *An observable is a constant of motion if and only if the corresponding operator commutes with the Hamiltonian.*

*Proof* Let $A$ be an observable, i.e. a selfadjoint operator, with spectral resolution $\{E_\alpha\}_{\alpha \in \mathbb{R}}$. An arbitrary state vector $\psi$ determines a distribution function $F_\psi^A(\alpha) = (\psi, E_\alpha \psi)$. At a later time $t$ one has $\psi(t) = e^{-\frac{i}{\hbar}tH}\psi$ with distribution function

$$F_{\psi,t}^A(\alpha) = (e^{-\frac{i}{\hbar}tH}\psi, E_\alpha e^{-\frac{i}{\hbar}tH}\psi) = (\psi, e^{\frac{i}{\hbar}tH}E_\alpha e^{-\frac{i}{\hbar}tH}\psi).$$

According to the definition $A$ is a constant of motion if and only if this $F_{\psi,t}^A(\alpha)$ is constant in $t$, for all $\alpha$ and for arbitrary $\psi$, which is equivalent to

$$(\psi, e^{\frac{i}{\hbar}tH}E_\alpha e^{-\frac{i}{\hbar}tH}\psi) = (\psi, E_\alpha \psi).$$

or $e^{\frac{i}{\hbar}tH}E_\alpha e^{-\frac{i}{\hbar}tH} = E_\alpha$, for all $\alpha$ and all $t$. This means that $A$ and $H$ commute, in the sense of commuting of spectral projections. This proves the theorem.

An example of a constant of motion is the momentum of a free particle. Another important example is formed by the three components of the angular momentum of a particle in three dimensional space moving in a rotationally symmetric potential. Constants of motion usually are connected with *symmetries*, as will be discussed in Sect. 3.9.1.

### 3.8.5 Time Evolution for Nonautonomous Quantum Systems

The standard example of a nonautonomous system in quantum mechanics is an atom, or a systems of atoms, or any piece of material, subjected to a variable external electric or magnetic field. The resulting time evolution no longer fits in the scheme given by Axiom $IV_1$. A generalization to nonautonomous systems is required.

The unitary operator which connects the system at an initial time $t_1$ with a later time $t_2$ no longer depends on the difference $t_2 - t_1$ but separately on $t_1$ and $t_2$. It will be denoted as $U(t_2, t_1)$. These operators, for all pairs $(t_1, t_2)$, should satisfy the following conditions:

a. $U(t, t) = 1$, for all real $t$,
b. $U(t_3, t_2)U(t_2, t_1) = U(t_3, t_1)$, for all real $t_1, t_2, t_3$,

which makes the system $\{U_{t_2,t_1}\}_{t_1,t_2}$ into a very simple example of a *groupoid*, a notion that will not be discussed further.

The two conditions imply
$U(t_2, t_1)^{-1} = U(t_1, t_2)$, $U(t, t) = 1$, for all real $t_1, t_2$ and $t$.

One again requires strong continuity of $U(t_1, t_2)$, this time simultaneously in both variables. We shall call such a system of operators $\{U(t_1, t_2)\}_{t_1, t_2 \in \mathbb{R}}$ a *strongly continuous two parameter system of unitary operators*, for lack of a better name. The operator-valued function $U(t_2, t_1)$ cannot be written as an exponential. Such two parameter systems have not been studied as extensively as the one parameter case. However, using a generalization of part of the proof of the Stone—von Neumann theorem one can prove that if certain additional conditions are fulfilled, there is an 'infinitesimal' generator $H(t)$, the time-dependent Hamiltonian. This allows us to use a time-dependent Schrödinger equation, as will be discussed in the next subsection.

All this together can be formulated as a generalization of Axiom IV$_1$:

**Axiom IV$_2$**: *Time evolution for nonautonomous systems in quantum theory is described by a strongly continuous two parameter system of unitary operators $\{U(t_1, t_2)\}$, acting in the Hilbert space of states $\mathcal{H}$. It has – under certain conditions— a time-dependent generator $H(t)$, which is selfadjoint for all $t$, and which is called the (time-dependent) Hamitonian of the system.*

### 3.8.6 The Schrödinger Equation for Nonautonomous Systems

The time-dependent Hamiltonian operator $H(t)$ determines the time evolution of a nonautonomous quantum system. One can again write a Schrödinger equation, the first order vector valued equation

$$\frac{d}{dt}\psi(t) = -\frac{i}{\hbar}H(t)\psi(t).$$

Note that in general the operators $H(t)$ do not commute for different times. Solving the 'eigenvalue–eigenfunction', the main task in the autonomous case, would not make much sense in this case, as it will be a different problem for each $t$. Moreover the solution would not be of great physical interest. In nonautonomous systems, of the kind mentioned before, some piece of matter in a time-dependent external electromagnetic field, which for the explicit example of a three dimensional quantum mechanical particle in a time-dependent external potential $V(\mathbf{x}, t)$ becomes

$$i\hbar\frac{\partial}{\partial t}\psi(\mathbf{x}, t) = -\frac{\hbar^2}{2m}\Delta\psi(\mathbf{x}, t) + V(\mathbf{x}, t)\psi(\mathbf{x}, t),$$

the same equation as in (3.5.2, a), except that the potential $V$ is now time-dependent. In realistic cases the external field acts only for a finite time. It will then be of physical interest to calculate *transition probabilities*, i.e. the probability that after the field is switched off at $t_2$, the system is found to have a certain energy $E_2$, given that it was in an eigenstate with energy $E_1$, at time $t_1$, before the external field was switched on.

### 3.8.7 The Interaction Picture

It is sometimes useful to describe time evolution of autonomous systems in an appar-
ently time-dependent manner, using a two parameter system of unitary operators.
This idea is the basis of *time-dependent scattering theory* discussed in Chap. 14.

**(3.8.7,a) Definition** The *interaction picture evolution* of a quantum system with
Hamiltonian $H = H_0 + H_1$ is given by the operators

$$U_I(t_2, t_1) = e^{\frac{i}{\hbar} H_0 t_2} e^{-\frac{i}{\hbar} H(t_2 - t_1)} e^{-\frac{i}{\hbar} H_0 t_1},$$

or, more generally formulated,

$$U_I(t_2, t_1) = U_0(0, t_2) U(t_2, t_1) U_0(t_1, 0).$$

**(3.8.7,b) Problem** Show that the system $\{U_I(t_2, t_1)\}_{t_1, t_2 \in \mathbb{R}^1}$ is indeed a strongly
continuous two parameter system of unitary operators in the sense of Sect. 3.6.5.

**(3.8.7,c) Problem** Show that $\{U_I(t_2, t_1)\}_{t_1, t_2}$, as a two parameter time evolution
system, satisfies the operator-valued Schrödinger equation

$$\frac{\partial}{\partial t_2} U_I(t_2, t_1) = -\frac{i}{\hbar} H_I(t_2) U_I(t_2, t_1),$$

with $H_I(t) = e^{\frac{i}{\hbar} t H_0} H_1 e^{-\frac{i}{\hbar} t H_0}$ the *interaction picture Hamiltonian*.
  This Schrödinger differential equation for $U_I(t_2, t_1)$ in the variable $t_2$, together
with the initial condition $U_I(t_2, t_1) = 1$ for $t_2 = t_1$, is equivalent to the integral
equation

$$U_I(t_2, t_1) = 1 - \frac{i}{\hbar} \int_{t_1}^{t_2} dt \, H_I(t) U_I(t, t_1),$$

which can be iterated to give an infinite series, the Dyson series.

$$U(t_2, t_1) = 1 - \frac{i}{\hbar} \int_{t_1}^{t_2} dt \, H_I(t) + (-\frac{i}{\hbar})^2 \int_{t_1}^{t_2} dt \int_{t_1}^{t} dt' \, H_I(t) H_I(t') \ldots .$$

See Sect. 14.2.6.
  The interaction picture can also be used to isolate the relevant physical part from
systems that are already nonautonomous, such as for instance the model of a particle
moving in a time-dependent potential, switched on only during a finite time interval,
such as mentioned at the end of the preceding subsection.

## 3.9 Symmetries

### 3.9.1 Groups of Symmetries

Symmetry is a an important notion in physics. It simplifies the solution of concrete problems and is often a guide in finding new models for physical phenomena. In this section symmetries of autonomous quantum systems are discussed. In fact, in this section quantum theory will mean autonomous quantum system.

The symmetries of a mathematical object are its *automorphisms*, i.e. the invertible maps from the object onto itself, which leave its characteristic structure invariant. For a Hilbert space these are the unitary operators. A quantum theory, as a mathematical object a pair $(\mathcal{H}, \{U(t)\}_{t \in \mathbb{R}})$, consisting of a Hilbert space of states $\mathcal{H}$ and a one parameter group $\{U(t)\}_{t \in \mathbb{R}}$ of unitary time-evolution operators, has as its automorphisms the unitary operators in $\mathcal{H}$ that commute with all the unitary operators $U(t)$ of the time-evolution group, or equivalently with the selfadjoint generator $H$ of this group. We state therefore as basic principle:

**Axiom V**: *Symmetries in quantum theory are described by unitary operators which commute with the one parameter group of time evolution, or equivalently, with its generator, the Hamiltonian.*

Examples of symmetries in quantum mechanics are *space reflection* in the case of a one dimensional point particle, when the particle is moving in a potential $V$ with $V(x) = V(-x)$, *space translation* for a free particle, and *spatial rotation* for a three dimensional particle moving in a rotation invariant potential.

Of particular importance are *groups of symmetries*:

**(3.9.1,a) Definition** The group $\mathcal{G}$ is a symmetry group of a quantum system if there is a unitary representation $\pi$ of $\mathcal{G}$ in $\mathcal{H}$ with the property that, for all $g$ in $\mathcal{G}$, the operators $\pi(g)$ commute with the time evolution operators $U(t)$, for all $t$ in $\mathbb{R}$, or equivalently, with the Hamiltonian $H$.

The use of groups and symmetries in quantum theory has been championed from the beginning by the mathematician Hermann Weyl. His book on this subject was very influential. See Ref. [2].

The basic mathematical properties of groups are reviewed in Supp. Sects. 24.1 and 24.2, the elements of representation theory of groups in Supp. Sect. 24.6 Lie groups and Lie algebras, needed below, are discussed in the remainder of Supp. Chap. 24.

### 3.9.2 Infinitesimal Symmetries

Many of the symmetry groups in physics are 'continuous' groups, i.e. Lie groups. A Lie group $\mathcal{G}$ has a Lie algebra $\mathcal{L}(\mathcal{G})$. Elements of $\mathcal{L}(\mathcal{G})$ can be obtained—roughly— by differentiating one parameter groups in $\mathcal{G}$ at the identity element. A representation

$\pi$ of $\mathcal{G}$ gives a (linear) representation $\hat{\pi}$ of $\mathcal{L}(\mathcal{G})$. If $\mathcal{G}$ is simply connected then $\pi$ can—in principle—be recovered from $\hat{\pi}$ by exponentiation

$$\pi(e^{-i\tau h}) = e^{-i\tau\hat{\pi}(h)},$$

for all $h$ in $\mathcal{L}(\mathcal{G})$ and all real $\tau$. If $\mathcal{G}$, together with the representation $\pi$, is a symmetry, all operators $\hat{\pi}(h)$ will commute with time evolution. This suggests the following definition:

(3.9.2,a) **Definition** A Lie algebra $\mathcal{L}$ is an *infinitesimal symmetry* of a quantum system if there is a representation $\hat{\pi}$ of $\mathcal{L}$ in $\mathcal{H}$ such that, for all $h$ in $\mathcal{L}$, the operators $\hat{\pi}(h)$ commute with the time evolution operators $U(t)$, for all $t$ in $\mathbb{R}$, or equivalently, with the Hamiltonian $H$.

Lie groups are *manifolds*. (See Supp. Chap. 19) The great advantage of Lie algebras over Lie groups is that Lie algebras are *linear spaces*. An $n$-dimensional Lie group $\mathcal{G}$ has an $n$-dimensional Lie algebra $\mathcal{L}(\mathcal{G})$, in which a basis can be chosen, say $e_1, \ldots e_n$. This simplifies the condition for infinitesimal symmetry to a finite set of linear relations, namely,

$$[\hat{\pi}(e_j), H] = 0,$$

for $j = 1, \ldots, n$. The $\hat{\pi}(e_j)$ are usually called the *generators* of the symmetry. Working with Lie algebra generators instead of with the full group is very popular in concrete physical applications of symmetry in quantum theory. Note that the generators are usually unbounded operators, and that therefore some of the above statements should be made more precise by taking into account domain questions. We will not worry about this, as it does not do much harm in practice, at least not in elementary quantum mechanics.

It should finally be remarked that symmetry generators are *constants of motion*. This is another reason why symmetries are important. Some of the best-known constants of motion are connected with symmetry in this way: conservation of linear momentum and angular momentum is a consequence of symmetry under spatial translations and spatial rotations, respectively. This will be discussed later, in the context of concrete quantum mechanical models, in particular in Chaps. 5–7.

Books on the mathematical formalism of quantum theory: See the references for Supp. Chap. 21.

## References

### For books on the mathematical formalism of quantum theory, see the references for Supp. Chap. 21

1. Rutherford, E.: The scattering of $\alpha$ and $\beta$ particles by matter and the structure of the atom. Philos. Mag. **6**, 669–688 (1911). Available at: http://www.ffn.ub.es/luisnavarro/nuevo_maletin/Rutherford%20(1911),%20Structure%20atom%20.pdf

2. Weyl, H.: The Theory of Groups and Quantum Mechanics (Translated from the 2d revised German edition 1931 Gruppentheorie und Quantenmechanik) Dover 2003. Available at: https://ia600807.us.archive.org/20/items/ost-chemistry-quantumtheoryofa029235mbp/quantumtheoryofa029235mbp.pdf

## Books on the History of Quantum Theory

3. Mehra, J., Rechenberg, H.: The Historical Development of Quantum Theory. vol. 1–6, in 9 parts Springer (1982–2000) (Reissued in paperback in 2001. Future research will certainly supply additional material to this monumental history of quantum mechanics from 1900 to 1941 and possibly correct some of its biases, but it will remain unsurpassed in its richness of detail on concepts and persons.)
4. van der Waerden, B.L.: Sources of Quantum Mechanics. North-Holland (1967). Reissued as a paperback by Dover, 1968 (The North-Holland edition is available at: https://ia601208.us.archive.org/14/items/SourcesOfQuantumMechanics/VanDerWaerden-SourcesOfQuantumMechanics.pdf. This book contains seventeen of the most important papers on quantum mechanics from the years 1917–26, in which the theory was developed and formulated, translated into English, together with a 59 page historical introduction).

## There is no shortage of excellent textbooks on quantum mechanics. Here follows a personal choice

5. Griffith, D.J.: Introduction to Quantum Mechanics. Prentice-Hall (1995) (A good modern introduction; a physics textbook with a more then acceptable level of mathematical precision and rigour.)
6. Le Bellac, M.: Quantum Physics. Translated from the French. Cambridge University Press (2006) (This is probably now the best comprehensive general textbook on quantum theory. It is a book for physicists, but its mathematics is careful and modern. It also discusses interesting recent developments.)
7. Messiah, A.: Quantum Mechanics. vol. I and II. Translated from the French. North-Holland (1961) (Reissued in one volume in 1999 by Dover. One of the older books that remains valuable because of its thoroughness and clarity of presentation. Well worth having, especially in the moderately priced Dover edition. The North-Holland edition is available at: https://ia601206.us.archive.org/23/items/QuantumMechanicsVolumeI/Messiah-QuantumMechanicsVolumeI.pdf; https://ia601605.us.archive.org/0/items/QuantumMechanicsVolumeIi/Messiah-QuantumMechanicsVolumeIi.pdf.)
8. Merzbacher, E.: Quantum Mechanics. Wiley (1961), 3rd edn, 1997 (Another older but still very useful textbook.)
9. Hannabuss, K.: An Introduction to Quantum Theory. Oxford University Press (1997) (For a mathematics student, having digested the present book and wanting to know more about quantum mechanics as a physical theory, this lucid book, with its emphasis on mathematical rigour, would be natural as a next step.)
10. Dirac, P.A.M.: The Principles of Quantum Mechanics. Original edition 1930. 4th edn. Cambridge University Press (1982) (A great classic. In his mathematical thinking Dirac was intuitive–and therefore very non-rigorous, but at the same time imaginative and innovative. His contributions to the understanding of the underlying mathematical structure of the emerging quantum theory were crucial, even though they had to be put into the rigorous functional analytic framework developed by von Neumann.)

# Chapter 4
# Quantum Mechanics of a Single Particle I

## 4.1 Introduction

This chapter begins the detailed discussion of the behaviour of a single particle in a given potential—the basic example of elementary quantum mechanics—with the description of the properties of wave functions and operators at a fixed time. The next chapter will be devoted to the time evolution of wave functions.

In this first section we review the necessary material for this, some of which was given already in Sect. 3.6.1 of the preceding chapter, where we used this example briefly to illustrate the general notions of states and observables in quantum theory.

The Hilbert space of states $\mathcal{H}$ of a point particle in three dimensional space is $L^2(\mathbb{R}^3, d\mathbf{r})$, the space of square integrable complex-valued functions on $\mathbb{R}^3$. These *wave functions* will be denoted as $\psi(\mathbf{r})$, $\phi(\mathbf{r})$, etc., with $\mathbf{r} = (x_1, x_2, x_3)$. The inner product on $\mathcal{H}$ between two such functions, $\psi_1$ and $\psi_2$, is given by

$$(\psi_1, \psi_2) = \int\limits_{-\infty}^{+\infty} \overline{\psi_1(\mathbf{r})}\psi_2(\mathbf{r}) \, d\mathbf{r}.$$

Note again that the inner product is conjugate-linear in the *first* variable, the standard convention in physics. The basic observable quantities of the theory are *position* and *(linear) momentum*, represented by the selfadjoint operators $Q_j$ and $P_j$ defined in Example (3.6.1,a) of 3, for $j = 1, 2, 3$, as

$$(Q_j\psi)(\mathbf{r}) = x_j\psi(\mathbf{r}),$$

$$(P_j\psi)(\mathbf{r}) = \frac{\hbar}{i}\frac{\partial}{\partial x_j}\psi(\mathbf{r}),$$

satisfying the *Heisenberg* or *canonical commutation relations*

$$[P_j, P_k] = [Q_j, Q_k] = 0,$$
$$[P_j, Q_k] = -i\hbar\delta_{jk}.$$

© Springer International Publishing Switzerland 2015
P. Bongaarts, *Quantum Theory*, DOI 10.1007/978-3-319-09561-5_4

for $j, k = 1, 2, 3$. These relations express important aspects of the problem of the simultaneous measurement of the observables for position and momentum, typical for quantum theory, as will be discussed later. The form of the Heisenberg relations, as given above, is the one found in all textbooks on quantum mechanics, a form which is however mathematically ambiguous, because the operators $P_j$ and $Q_k$ are unbounded. The correct form, using the two one parameter groups of unitary operators generated by the $P_j$ and $Q_k$ is, for $j, k = 1, 2, 3$ and $\alpha, \beta \in \mathbb{R}$,

$$[e^{\frac{i}{\hbar}\alpha P_j}, e^{\frac{i}{\hbar}\beta P_k}] = [e^{\frac{i}{\hbar}\alpha Q_j}, e^{\frac{i}{\hbar}\beta Q_k}] = 0$$

$$e^{\frac{i}{\hbar}\alpha P_j} e^{\frac{i}{\hbar}\beta Q_k} = e^{\frac{i}{\hbar^2}\alpha\beta} e^{\frac{i}{\hbar}\beta Q_k} e^{\frac{i}{\hbar}\alpha P_j}.$$

These 'exponentiated' Heisenberg relations are sometimes called the *Weyl relations*. Stone and von Neumann proved that two different sets of strongly continuous one parameter groups of unitary operators, satisfying these algebraic relations, together with a certain natural irreducibility condition, are unitarily equivalent. The irreducibility condition requires that every invariant closed linear subspace of $\mathcal{H}$, which is invariant under all $e^{\frac{i}{\hbar}\alpha P_j}$ and $e^{\frac{i}{\hbar}\beta Q_k}$, is equal to $\mathcal{H}$. This theorem means that the quantum mechanical description of a particle is uniquely determined by the Heisenberg relations in this form.

From the $P_j$ and $Q_j$ one constructs all other observables, in the first place the *energy*—usually denoted as $H$ and called the *Hamiltonian operator* or just the *Hamiltonian*, because it is in this example as in many others, also the selfadjoint generator of the one parameter group of unitary operators $\{U(t)\}_{t \in \mathbb{R}^1}$—and the three components of *angular momentum* in general by a prescription which is called 'canonical quantization', which assigns to each classical variable as a function $f(\mathbf{p}, \mathbf{r})$ a corresponding operator expression $f(\mathbf{P}, \mathbf{Q})$. This is the traditional procedure that gives the quantum operators as expressions in position and momentum operators, a procedure which is however in general not unambiguous, because noncommuting quantum observables have to be substituted for commuting classical observables. The functions $p_j x_k^3$ and $x_k^3 p_j$, for example, are obviously the same, the corresponding operators $P_j Q_k^3$ and $Q_k^3 P_j$ are not. They are also not hermitian symmetric, so cannot be extended to selfadjoint operators. One way of overcoming this is by symmetrization of the operator expression, i.e. by using the operator expression $\frac{1}{2}(P_j Q_k^3 + Q_k^3 P_j)$. This problem does not occur here in the simple situation of quantum mechanics of a single particle in a potential. The attempt to make this rather heuristic prescription into a general, precise and unambiguous theoretical procedure is called *Geometric Quantization*, an interesting mathematical topic in its own right, only partially successful so far, that will not be discussed in this book.

In any case, the energy operator for a particle with mass $m$, moving under the influence of a force $\mathbf{F} = (F_1, F_2, F_3)$, derived from a potential $V$ by the formula $F_j(x) = -\frac{\partial}{\partial x_j} V(\mathbf{r})$, becomes in this way

$$H = H_0 + V = \left(\frac{\mathbf{P}^2}{2m} + V(\mathbf{Q})\right),$$

or, more explicitly,

$$(H\psi)(\mathbf{r}) = -\frac{\hbar^2}{2m}(\Delta\psi)(\mathbf{r}) + V(\mathbf{r})\psi(\mathbf{r}),$$

with $\Delta$ the Laplace operator $\Delta = \frac{\partial^2}{\partial x_1{}^2} + \frac{\partial^2}{\partial x_2{}^2} + \frac{\partial^2}{\partial x_3{}^2}$. For $H = H_0$, i.e. for $V(\mathbf{x}) \equiv 0$ the particle is called *free*.

Note that we may consider these unbounded operators in first instance as defined on the common dense domain $\mathcal{S}(\mathbb{R}^3)$, the space of Schwartz functions on $\mathbb{R}^3$, the $C^\infty$-functions that, together with all their derivatives, go to zero at infinity faster than arbitrary inverse polynomials. By this choice we avoid some but not all of the obvious domain problems. One can show for instance that the $P_j$ and $Q_k$ are essentially selfadjoint on this domain, i.e. can be extended uniquely to selfadjoint operators (See for the notion of essential selfadjointness of an unbounded symmetric operator Supp. Sect. 21.9.2.).

Note also that $H$ is easily shown to be essentially selfadjoint on $\mathcal{S}(\mathbb{R}^3)$ for a bounded potential $V$. However, in most of the physically interesting cases the potential is an unbounded function, the most obvious being the electrostatic Coulomb potential for the hydrogen atom, $V(r) \sim \frac{e^2}{r}$, with $r = (x_1^2 + x_2^2 + x_3^2)^{1/2}$ and $e$ the electric charge of the electron, which example will be treated in Chap. 7. Proving that $H = H_0 + V$ is at least essentially selfadjoint on the domain of $H_0$ is in general a nontrivial mathematical problem.

In classical mechanics a rotating body has *angular momentum*: an important physical quantity, the conservation of which explains why the axis of a gyroscope persists in a given initial position and also why one is able to ride a bicycle without falling over. A classical particle moving with linear momentum $\mathbf{p}$ has a vector $\mathbf{l}$ of angular momentum with respect to the origin $\mathbf{r} = \mathbf{0}$ of the coordinate system defined as

$$\mathbf{l} = \mathbf{r} \times \mathbf{p},$$

with the components $l_j$ already given in Sect. 3.4.1 of the preceding chapter as

$$l_1 = x_2 p_3 - x_3 p_2$$
$$l_2 = x_3 p_1 - x_1 p_3$$
$$l_3 = x_1 p_2 - x_2 p_1.$$

The 'canonical quantization' prescription gives three *angular momentum operators* $L_1, L_2, L_3$, defined as

$$L_1 = Q_2 P_3 - Q_3 P_2,$$

with $L_2$ and $L_3$ obtained by cyclic permutation of the indices $1, 2, 3$. The $L_k$ are differential operators, essentially selfadjoint on $\mathcal{S}(\mathbb{R}^3)$. One has for instance for $L_1$

$$(L_1\psi)(\mathbf{r}) = \frac{\hbar}{i}\left(x_2\frac{\partial\psi(\mathbf{r})}{\partial x_3} - x_3\frac{\partial\psi(\mathbf{r})}{\partial x_2}\right).$$

The three angular momentum operators do *not* commute, something that has important physical consequences, typical for quantum theory. The commutation relations for the $L_j$ can be worked out by using their explicit forms as differential operators, but can however more easily be calculated algebraically from the known commutation relations for the position and momentum operators. The result is

$$[L_1, L_2] = i\hbar L_3,$$

with the other two formulas again obtained by cyclic permutation. One also has the operator

$$\mathbf{L}^2 = L_1^2 + L_2^2 + L_3^2,$$

representing the square of the absolute value of angular momentum. It commutes with all three components $L_j$, a very important property. Angular momentum in quantum mechanics has non-classical properties: its measurement gives only discrete values; its three components are incommensurable. It is also strongly connected with symmetry aspects: the three operators $L_j$ span a representation of the Lie algebra of $SO(3)$, the three dimensional rotation group. All this will be discussed in Chap. 7, with the group theoretic aspects in Supp. Chap. 24.

## 4.2 'Diagonalizing' the $P_j$: The Fourier Transform

When discussing the properties of position, momentum and energy in quantum mechanics, it is an inessential but notationally very convenient simplification to do this first for a one dimensional particle, a simplification which will of course not make sense for the discussion of angular momentum.

We use the operator and wave function formulas given in the preceding section, but now for a single position variable $x$, with single operators $P$ and $Q$.

A useful device is the *Fourier transform* which maps a function of one real variable $f$ in a similar function $\hat{f}$ by the integral formula

$$\widehat{f}(v) = \frac{1}{\sqrt{2\pi}}\int\limits_{-\infty}^{+\infty} f(u)\, e^{-iuv}\, du.$$

The minus sign in the exponent is a choice which makes the transition to the convential formulation of quantum mechanics more natural. A normalization factor $(2\pi)^{-1/2}$ is put in front of the integral to make sure that the Fourier transform formula and its inverse have the same form, except for a change of sign in the exponent. The inverse formula is

$$f(u) = \frac{1}{\sqrt{2\pi}} \int\limits_{-\infty}^{+\infty} \hat{f}(v) e^{+iuv} dv.$$

With this normalization we have the Plancherel formula

$$\int\limits_{-\infty}^{+\infty} |f(u)|^2 du = \int\limits_{-\infty}^{+\infty} |\hat{f}(v)|^2 dv.$$

The theory of Fourier transforms is a fairly subtle part of mathematical analysis. Here all technicalities can be avoided; the transformation formulas are well-defined as ordinary integrals on the space of Schwartz functions $S(\mathbb{R}^3)$ and can be extended by taking $L^2$-limits to a unitary transformation and its inverse on the full Hilbert space of all square integrable functions.

Hermitian matrices have an orthogonal basis of eigenvectors and can be diagonalized with respect to this basis by applying a suitable unitary matrix. The spectral theorem, as discussed in Supp. Sect. 21.10, provides a generalization of this for general selfadjoint operators in Hilbert space. This makes it possible to 'diagonalize' the momentum operator $P$, i.e. represent it as a multiplication operator in a Hilbert space of functions on the spectrum of $P$. The Fourier transform is the unitary transformation which does this.

Physical quantities have a *dimension*, as was discussed in Sect. 1.4.4. The dimension of position $x$ is, of course, that of a length $[L]$, of momentum $p = mv$, that of mass times velocity $[MLT^{-1}]$. An exponential function should be dimensionless. Because the expression $ipx$ has dimension $[ML^2T^{-1}]$ it has to be multiplied with a constant with dimension $[M^{-1}L^{-2}T]$, the dimension of an *action*. An obvious candidate is Planck's constant $\hbar$, which has indeed the dimension of an action. One still wants in this context the Fourier formula and its inverse to have the same form. All this together leads to the formulas

$$\hat{\psi}(p) = \frac{1}{(2\pi\hbar)^{1/2}} \int\limits_{-\infty}^{+\infty} \psi(x) e^{-\frac{i}{\hbar}px} dx,$$

and its inverse

$$\psi(x) = \frac{1}{(2\pi\hbar)^{1/2}} \int\limits_{-\infty}^{+\infty} \hat{\psi}(p) e^{+\frac{i}{\hbar}px} dp,$$

with the Plancherel formula

$$\int\limits_{-\infty}^{+\infty} |\psi(x)|^2 \, dx = \int\limits_{-\infty}^{+\infty} |\hat{\psi}(p)|^2 \, dp$$

expressing the unitarity of the Fourier transform as Hilbert space transformation. The transform $\hat{\psi}$ of $\psi$ is sometimes called the *wave function in momentum representation*. Its physical interpretation is that $|\hat{\psi}(p)|^2$ is, in accordance with the general principles explained in Sect. 3.4, the probability density for measurement of the momentum $p$ in the state represented by $\hat{\psi}$.

(4.2,a) **Problem**

a. Use the Fourier transformation formulas to show that a translation over $\alpha$ in the variable $x$ followed by a translation over $\beta$ in the variable $p$ changes a wave function $\psi(x)$ into a wave function $\psi'(x)$ according to

$$\psi'(x) = e^{\frac{i}{\hbar}\beta x} \psi(x - \alpha),$$

with a change in momentum space wave function

$$\hat{\psi}'(p) = e^{-\frac{i}{\hbar}\alpha(p-\beta)} \hat{\psi}(p - \beta).$$

b. Show that for the mean values and standard deviations one has

$$\overline{x}' = \overline{x} + \alpha, \quad \overline{p}' = \overline{p} + \beta, \quad \Delta x' = \Delta x \quad \Delta p' = \Delta p.$$

c. Translation in $x$ and $p$ are noncommuting operations. The result of changing the order of these translations gives an inessential phase factor in front of $\psi'(x)$. Calculate this factor.

Obviously, one does not need the spectral theorem to find the eigenfunctions of $P$ as a differential operator. They are, for each eigenvalue $p \in \mathbb{R}^1$, the functions

$$\psi_p(x) = e^{\frac{i}{\hbar}px},$$

the *plane waves* in (one dimensional) space, with wave length $\lambda = 2\pi\hbar/p$. Physicists use these plane waves for the quantum mechanical description of a particle with sharp momentum $p$. This expresses one of the oldest and most influential general ideas of quantum mechanics, the *particle wave duality*: a point particle of mass $m$ and momentum $p = mv$ is described by a plane wave with wave length

$$\lambda = \frac{2\pi\hbar}{p}.$$

This $\lambda$ is called the *de Broglie wave length* of the particle, after the French physicist Louis de Broglie, one of the pioneers of early quantum mechanics, who was the first to suggest this idea of a 'particle wave duality', as was already mentioned in Sect. 3.3.1. Strictly speaking, the interpretation of the function $\psi_p(x) = e^{\frac{i}{\hbar}px}$ as a physical state is nevertheless incorrect, because it is clearly not a function in $L^2(\mathbb{R}^1, dx)$, the Hilbert space of state vectors. The Fourier transformation provides us with a mathematically more precise way of expressing the above physical idea.

The intuitive meaning of the Fourier formula for the function $\psi(x)$—the 'wave function' as it is called, not surprisingly in view of what has just been said—is that it gives $\psi(x)$ as a 'continuous superposition' of plane waves. To approximate the notion of a sharp momentum state we can use a superposition over a narrow $p$-interval $\varepsilon$ around a $p$-value $p_0$, for instance

$$\psi_{p_0,\varepsilon}(x) = \varepsilon^{-\frac{1}{2}} \int_{p_0-\frac{1}{2}\varepsilon}^{p_0+\frac{1}{2}\varepsilon} e^{\frac{i}{\hbar}px} dp.$$

The constant $\varepsilon^{\frac{1}{2}}$ in front is such that $\psi_{p_0,\varepsilon}$ is normalized to one. An elementary calculation gives

$$\psi_{p_0,\varepsilon}(x) = F_\varepsilon(x) e^{\frac{i}{\hbar}p_0x},$$

with

$$F_\varepsilon(x) = \varepsilon^{\frac{1}{2}} \frac{\sin \frac{\varepsilon}{2\hbar}x}{\frac{\varepsilon}{2\hbar}x}.$$

The function $\psi_{p_0,\varepsilon}(x)$ is a plane wave $\psi_{p_0}(x)$, multiplied with the real function $F_\varepsilon(x)$ which makes it square integrable. $F(x)$ has a maximum $\varepsilon^{\frac{1}{2}}$ in $x = 0$ and goes to 0 for $x$ going to $\pm\infty$, oscillating with decreasing amplitude between the absolute value of the function $\frac{2\hbar}{x}\varepsilon^{-1/2}$. For small $\varepsilon$, i.e. when the momentum lies within a narrow interval, the function flattens out in space, which means physically that the position of the particle becomes very unpredictable. For large $\varepsilon$, i.e. for undetermined momentum, $F_\varepsilon(x)$ becomes sharply peaked around $x = 0$, which means that the position of the particle becomes narrowly determined. This example shows us that we can have quantum mechanical states in which the position of the particle can be predicted with arbitrary precision, that we can also have states in which the momentum is as sharply defined as one wishes, but that we can *not* have both at the same time. This can be shown generally by means of an inequality for the standard deviations of the measurements of position and momentum which will be discussed in the next section.

The fact that one cannot simultaneously measure the position and velocity of a particle with arbitrary precision means that the particle can be observed at discrete times, but not over a *continuous* time interval. It cannot be followed in time; its *orbit* is not observable. There is an even stronger interpretation of this: *the classical notion of orbit of a particle is meaningless in quantum mechanics.*

## 4.3  A General Uncertainty Relation

Consider an arbitrary quantum system with Hilbert space $\mathcal{H}$. Take two physical observables $a$ and $b$, not necessarily commensurable, represented by two (not necessarily commuting) selfadjoint operators $A$ and $B$. Let $\psi$ be a state vector in $\mathcal{H}$, which should be such that $A\psi$, $A^2\psi$, $B\psi$, $B^2\psi$, $AB\psi$ and $BA\psi$ are well-defined. Denote in the usual way the mean values $(\psi, A\psi)$ and $(\psi, B\psi)$ as $\bar{a}_\psi$ and $\bar{b}_\psi$, and the standard deviations $(\psi, (A - \bar{a}_\psi)^2\psi)^{1/2}$ and $(\psi, (B - \bar{b}_\psi)^2\psi)^{1/2}$ as $\Delta a_\psi$ and $\Delta b_\psi$.

(4.3,a) **Problem** Prove, for a pair of arbitrary selfadjoint operators $A$ and $B$, and a vector $\psi$ such that $A\psi$, $A^2\psi$, $B\psi$, $B^2\psi$, $AB\psi$ and $BA\psi$ are well defined, the inequality

$$\Delta a_\psi \, \Delta b_\psi \geq \frac{1}{2} |(\psi, [A, B]\psi)|.$$

Hint: Use the identity $(\psi, [A, B]\psi) = (\psi, [A - \bar{a}_\psi, B - \bar{b}_\psi]\psi)$, together with the Schwarz inequality (Supp. Sect. 21.3).

This is the *general uncertainty relation* for the standard deviations of the observables $a$ and $b$ in the state $\psi$. It is a restriction on the possible outcomes of a simultaneous measurement of $a$ and $b$. For commensurable $a$ and $b$, i.e. for commuting $A$ and $B$, there is of course no restriction.

## 4.4  The Heisenberg Uncertainty Relation

Return now to the case of a one dimensional particle and choose as observables $A = P$ and $B = Q$. The general uncertainty relation in the preceding section gives immediately the *Heisenberg uncertainty relation* for position and momentum

$$\Delta p \, \Delta x \geq \frac{\hbar}{2}.$$

Note that we write here $\Delta x$ and $\Delta p$ instead of $\Delta x_\psi$ and $\Delta p_\psi$ because the right-hand side no longer depends on the state vector $\psi$, even though $\psi$ should be such that $\Delta x$ and $\Delta p$ are meaningful. The Heisenberg uncertainty relation is one of the centre pieces of elementary quantum mechanics, illustrating the fundamental incommensurability of position and momentum.

## 4.5  Minimal Uncertainty States

For an arbitrary state $\psi$ the product of the standard deviations in position and momentum is always larger or equal than $\frac{1}{2}\hbar$. We are interested in states that are as close to the classical situation as possible, i.e. state $\psi$ for which this product is *minimal*, meaning that

$$\Delta p \, \Delta x = \frac{1}{2} \hbar.$$

Such *states of minimal uncertainty in position and momentum* will be called *minimal states* or *minimal wave packets* in this section. In the literature the term *squeezed states* is sometimes used.

Finding all minimal states is simplified by the results of Problem (4.2,a): it is sufficient to find the minimal states with $\bar{x} = 0$ and $\bar{p} = 0$—to be called *mean zero minimal states* here; all others can be obtained from these by translation of the position or momentum wave functions.

(4.5,a) **Problem** Show that the wave functions of the mean zero minimal states, after a convenient choice of a phase factor, have the form

$$\psi(x) = \frac{1}{(\Delta x)^{1/2}(2\pi)^{1/4}} \, e^{-\frac{1}{4(\Delta x)^2} x^2},$$

with probability density, the so-called *Gaussian distribution*,

$$\rho(x) = |\psi(x)|^2 = \frac{1}{\Delta x (2\pi)^{1/2}} \, e^{-\frac{1}{2(\Delta x)^2} x^2},$$

with $\Delta x$ the standard deviation for the variable $x$ in the state $\psi$.
Hint: Use the sequence of inequalities

$$\hbar = 2| \mathrm{Im}\, (P\psi, Q\psi)| \le 2|(P\psi, Q\psi)| \le 2\,||P\psi||\,||Q\psi|| = 2\,\Delta p_\psi \, \Delta x_\psi,$$

a special case of the inequality used in the solution of Problem (4.3,a) to derive a differential equation for $\psi(x)$. Note that the Schwartz inequality for two nonzero vectors $\psi_1$ and $\psi_2$ is an equality if and only if there exists a nonzero complex number $\lambda$ such that $\psi_1 = \lambda\psi_2$. (See Supp. Sect. 21.3). Solve the differential equation.

(4.5,b) **Problem** Show that the momentum space wave function $\hat{\psi}(p)$ of a mean zero minimal state has the form

$$\hat{\psi}(p) = \frac{1}{(\Delta p)^{1/2}(2\pi)^{1/4}} \, e^{-\frac{1}{4(\Delta p)^2} p^2},$$

with probability density

$$\hat{\rho}(p) = |\hat{\psi}(p)|^2 = \frac{1}{\Delta p (2\pi)^{1/2}} \, e^{-\frac{1}{2(\Delta p)^2} p^2},$$

in which $\Delta p = \frac{1}{2}\hbar(\Delta)^{-1}$ is the standard deviation for the variable $p$ in the state $\psi$.
Hint: Use the Fourier transformation formulas given in Sect. 4.2.

Both $\rho(x)$ and $\hat{\rho}(p)$ are probability densities of a *normal* or *Gaussian distribution* centered at $x = 0$, respectively at $p = 0$, and with standard deviation $\Delta x_\psi$, respectively $\Delta p_{\hat{\psi}}$.

All the wave functions of minimal states with arbitrary mean values $\bar{x}$ for position and $\bar{p}$ for momentum, a three parameter collection, can according to the results of (4.2,a) be obtained by translation from the mean zero minimal states, up to an inessential phase factor, as

$$\psi_{\bar{x},\bar{p},\Delta x}(x) = \frac{1}{(\Delta x)^{1/2}(2\pi)^{1/4}} \, e^{\frac{i}{\hbar}\bar{p}x} e^{-\frac{1}{4(\Delta x)^2}(x-\bar{x})^2},$$

with probability densities just shifted over $\bar{x}$ and the corresponding probability densities in momentum shifted over $\bar{p}$.

(4.5,c) **Problem** Show that the Fourier transform of $\psi_{\bar{x},\bar{p},\Delta x}$ is equal to

$$\hat{\psi}_{\bar{x},\bar{p},\Delta x}(p) = \frac{1}{(\Delta p)^{1/2}(2\pi)^{1/4}} \, e^{-\frac{i}{\hbar}(p-\bar{p})\bar{x}} e^{-\frac{1}{4(\Delta p)^2}(p-\bar{p})^2},$$

with the standard deviation $\Delta p$ equal to $\frac{1}{2}\hbar(\Delta x)^{-1}$.

## 4.6 The Heisenberg Inequality: Examples

The numerical consequences of the uncertainty relation can be explicitly demonstrated in appropriately chosen physical systems.

(4.6,a) *Example (microscopic system)* The mass of an electron is roughly $9 \times 10^{-31}$ kg. Assume that it moves with a velocity $v = 10^3$ m/s, so the momentum $p$ is $9 \times 10^{-28}$ kg m/s. (We use KMS (kilogram–meter–second) units.) Suppose again an uncertainty in the momentum of 0.1 %, so $\Delta p = 9 \times 10^{-31}$ kg m/s. This gives for the uncertainty in the position a value $\Delta x = 4.5 \times 10^{-5}$ m. This may be a small length in the macroscopic world, but it is very large in the atomic context, for instance in comparison with the size of the hydrogen atom, in a certain way defined by the radius of the inner electron orbit which is roughly $5 \times 10^{-11}$ m. (With this 'orbit' we mean the space region in which the wave function of the electron in its state of lowest energy is appreciably different from 0. An orbit in the classical sense does not exist, as was noted in Sect. 4.2.)

(4.6,b) *Example (macroscopic system)* Think of a bullet with a size of a centimeter and mass of 10 grammes, moving with a velocity of 200 m/s. (The velocity of sound in air at room temperature is 340 m/s). The momentum $p = mv$ is 2 kg m/s. Suppose that there is an uncertainty in this momentum of 0, 1 %, i.e. the standard deviation $\Delta p$ is equal to $2 \times 10^{-3}$ kg m/s. Planck's constant $\hbar$ is $1.054 \times 10^{-34}$ kg m$^2$/s, so $\frac{1}{2}\hbar$ is roughly $5 \times 10^{-35}$ kg m$^2$/s. This gives the minimal uncertainty in the position of the bullet as $\Delta x = \frac{1}{2}\hbar/\Delta p = 2.5 \times 10^{-33}$ m, a value which is extremely small and would be in this context totally unobservable.

Example (4.6,b) illustrates an important general point in the application of quantum theory to physics. Quantum theory is considered to be valid for *all* physical phenomena. It is indispensable for understanding the world of atomic physics. However, in macroscopic physical situations its predictions are often experimentally indistinguishable from those of classical mechanics. In such cases the continued use of classical theories are justified.

## 4.7 The Three Dimensional Case

The basic formulas for the quantum mechanical description of a three dimensional particle were given in Sect. 4.1. In the subsequent sections we discussed the case of a particle in one dimensional space. This simplified our formulas, while nothing essential was lost. In this section we briefly review some of the remaining aspects of the three dimensional case and give—for the sake of completeness—the three dimensional versions of the appropriate formulas.

The position operators $Q_1$, $Q_2$, $Q_3$ commute; they form a triple of *commensurable operators*, which means that they give a system of three *stochastic variables* in the sense of ordinary probability theory: a normalized state vector $\psi$ defines the *joint probability density*

$$\rho(\mathbf{x}) = \rho(x_1, x_2, x_3) = |\psi(x_1, x_2, x_3)|^2$$

for the three position variables $x_1, x_2, x_3$. The same is true for the three components of momentum, with a simultaneous probability density

$$\hat{\rho}(\mathbf{p}) = \hat{\rho}(p_1, p_2, p_3) = |\hat{\psi}(p_1, p_2, p_3)|^2.$$

For this one needs the three dimensional Fourier transformation. The formulas for this transformation and its inverse are

$$\hat{\psi}(\mathbf{p}) = \frac{1}{(2\pi\hbar)^{3/2}} \int\limits_{-\infty}^{+\infty} \psi(\mathbf{x}) \, e^{-\frac{i}{\hbar}\mathbf{p}\cdot\mathbf{r}} d\mathbf{p},$$

$$\psi(\mathbf{x}) = \frac{1}{(2\pi\hbar)^{3/2}} \int\limits_{-\infty}^{+\infty} \hat{\psi}(\mathbf{x}) \, e^{\frac{i}{\hbar}\mathbf{p}\cdot\mathbf{x}} d\mathbf{x},$$

with the Plancherel formula

$$\int\limits_{-\infty}^{+\infty} |\psi(\mathbf{x})|^2 d\mathbf{x} = \int\limits_{-\infty}^{+\infty} |\hat{\psi}(\mathbf{p})|^2 d\mathbf{p}.$$

Each pair $(P_j, Q_j)$, $j = 1, 2, 3$, consists of two *incommensurable* observables. There are no simultaneous probability distributions; instead we have, for $j = 1, 2, 3$, the *Heisenberg uncertainty relations*

$$\Delta p_j \, \Delta x_j \geq \frac{1}{2}\hbar.$$

The states of minimal uncertainty in position and momentum, i.e. the states which satisfy

$$\Delta p_j \, \Delta x_j = \frac{1}{2}\hbar,$$

for all $j = 1, 2, 3$, are the product wave functions

$$\psi_{\mathbf{p},\mathbf{r},\Delta\mathbf{x}}(\mathbf{x}) = \psi_{\overline{p_1},\overline{x_1},\Delta x_1}(x_1) \, \psi_{\overline{p_2},\overline{x_2},\Delta x_2}(x_2) \, \psi_{\overline{p_3},\overline{x_3},\Delta x_3}(x_3),$$

for abitrary expectation values $\overline{p}_j$, $\overline{x}_j$ and arbitrary positive $\Delta x_j$.

# Chapter 5
# Quantum Mechanics of a Single Particle II

## 5.1 Time Evolution of Wave Functions

In the preceding chapter we discussed the quantum mechanical description of a free particle at a fixed time. In this section its time evolution will be studied, along the lines already indicated in Sect. 3.5 (Axiom IV) en Sect. 3.8. For notational simplicity we again consider first the case of a particle in one dimensional space. The time evolution of a state in quantum theory from an initial time $t_1$ to a later time $t_2$ is in general given by a two parameter system of unitary operators $U(t_2, t_1)$, which in the special but more common case of an *autonomous* system, i.e. a system with time-independent interaction, becomes a one parameter group $U(t) = e^{-\frac{i}{\hbar} tH}$ acting on state vectors as $e^{-\frac{i}{\hbar}(t_2 - t_1)H} \psi(t_1) = \psi(t_2)$. State vectors, as functions of time, are solutions of the 'abstract' Schrödinger equation

$$\frac{d}{dt}\psi(t) = -\frac{i}{\hbar}H\psi(t),$$

all as was discussed in Sect. 3.5. In the special case of a one dimensional particle the 'concrete' Schrödinger equation becomes

$$\frac{\partial}{\partial t}\psi(x, t) = -\frac{i}{\hbar}\left(-\frac{\hbar^2}{2m}\frac{\partial^2}{\partial x^2} + V(x)\right)\psi(x, t),$$

with the state vector now the 'time dependent wave function' $\psi(x, t)$. This differential equation is of course only defined for wave functions in the domain of the (unbounded) operator $H$, for instance for functions $\psi$ that are in $S(\mathbb{R}^1)$, in the variable $x$. The potential $V(x)$ should be a suitable function.

© Springer International Publishing Switzerland 2015
P. Bongaarts, *Quantum Theory*, DOI 10.1007/978-3-319-09561-5_5

## 5.2 Pseudo Classical Behaviour of Expectation Values

A state of a particle may have a wave function which is different from 0 only in a small region of space and which has a Fourier transform concentrated in a small momentum interval. It will have small standard deviations $\Delta x$ and $\Delta p$ from the average values $\bar{x}$ and $\bar{p}$ of position and momentum. Such a state can be called 'quasi classical'. It resembles a classical particle with well-defined position $x = \bar{x}$ and momentum $p = \bar{p}$. This is of course only approximately so, because Heisenberg's uncertainty relation tells us that $\Delta x$ and $\Delta p$ cannot be both arbitrary small; the product $\Delta p\, \Delta x$ has $\frac{1}{2}\hbar$ as lower limit. The minimal uncertainty states discussed in Sect. 4.5 are the best one can have in this respect.

The time evolution of an *arbitrary* wave function has classical aspects. This is in particular true for the time dependence of the average values of position and momentum. One has for the expectation value of position

$$\frac{d\bar{x}(t)}{dt} = \frac{d}{dt}(\psi(t), Q\psi(t))$$

$$= (-\frac{i}{\hbar}H\psi(t), Q\psi(t)) + (\psi(t), -\frac{i}{\hbar}QH\psi(t))$$

$$= \frac{i}{\hbar}(\psi(t), [H, Q]\psi(t)).$$

Using the basic commutation relation $[P, Q] = \frac{i}{\hbar}$, together with the fact that $Q$ commutes with $V(Q)$, one finds

$$[H, Q] = \left[\frac{P^2}{2m} + V(Q), Q\right] = \frac{1}{2m}[P^2, Q] = \frac{\hbar}{im}P,$$

which gives finally

$$\frac{d\bar{x}(t)}{dt} = \frac{1}{m}(\psi(t), P\psi(t)) = \frac{1}{m}\bar{p}(t).$$

For $\bar{p}(t)$ one obtains in a similar manner

$$\frac{d\bar{p}(t)}{dt} = \frac{i}{\hbar}(\psi(t), [H, P]\psi(t)) = -(\psi(t), \frac{dV}{dQ}\psi(t)) = -\overline{\left[\frac{dV}{dQ}\right]},$$

in which $\frac{dV}{dQ}$ is the operator which multiplies a wave function with the function $\frac{dV(x)}{dx}$. The pair of first order equations

$$m\frac{d\bar{x}}{dt} = \bar{p}, \qquad \frac{d\bar{p}}{dt} = -\overline{\left[\frac{dV}{dx}\right]}$$

can be combined to a single second order equation

$$m\frac{d^2\bar{x}}{dt^2} = -\left[\frac{dV}{dx}\right],$$

which brings out even more clearly the analogy with the classical equations of motion discussed in Sect. 2.3. The time evolution of a wave function which is quasi classical at an initial time will resemble in first approximation the motion of a classical particle with position $x = \bar{x}(t)$ and momentum $p = \bar{p}(t)$, at least for a certain short initial period of time, after which the wave packet will spread out, loosing its quasi classical character.

## 5.3 The Free Particle

In this simple case the Schrödinger equation can be explicitly solved, i.e. the wave function at time $t_2$ can be expressed in the wave function at an earlier time $t_1$. For this one can use the Fourier transformation. We demonstrate this for the one dimensional case. Let the initial state at $t = t_1$ be given by the function $\psi(x, t_1)$, a function in which we assume to be in $\mathcal{S}(\mathbb{R}^1)$ in the variable $x$. Its Fourier transform is

$$\hat{\psi}(p, t_1) = \frac{1}{(2\pi\hbar)^{1/2}} \int_{-\infty}^{+\infty} \psi(x, t_1)\, e^{-\frac{i}{\hbar}px}\, dx.$$

The Hamiltonian is a simple expression in the momentum operator $P$, so the action of the unitary evolution operator $e^{-\frac{i}{\hbar}tH}$ in terms of the Fourier transform is also simple and gives

$$\hat{\psi}(p, t_2) = e^{-\frac{i}{\hbar}(t_2-t_1)\frac{p^2}{2m}}\, \hat{\psi}(p, t_1).$$

By using the inverse Fourier transformation one finally obtains the solution at $t = t_2$ as

$$\psi(x, t_2) = \frac{1}{(2\pi\hbar)^{1/2}} \int_{-\infty}^{+\infty} \hat{\psi}(p, t_2)\, e^{\frac{i}{\hbar}px}\, dp$$

$$= \frac{1}{(2\pi\hbar)^{1/2}} \int_{-\infty}^{+\infty} \hat{\psi}(p, t_1)\, e^{-\frac{i}{\hbar}(t_2-t_1)\frac{p^2}{2m}}\, e^{\frac{i}{\hbar}px}\, dp$$

$$= \frac{1}{2\pi\hbar} \int_{-\infty}^{+\infty} dp \int_{\infty}^{+\infty} dx'\, \psi(x', t_1) \left[ e^{-\frac{i}{\hbar}p(x-x')}\, e^{-\frac{i}{\hbar}(t_2-t_1)\frac{p^2}{2m}} \right].$$

This is a well-defined repeated integral which gives $\psi(x, t_2)$ as a function in $S(\mathbb{R}^1)$, the space of functions that are infinitely differentiable and go faster to 0 for $x$ going to $\pm\infty$ than any inverse ponynomial, for all $t_2 \geq t_1$. It is sometimes written as

$$\psi(x, t_2) = \int_{-\infty}^{+\infty} G(x, t_2; x', t_1)\psi(x', t_1)dx',$$

with

$$G(x, t_2; x', t_1) = \frac{1}{2\pi\hbar} \int_{-\infty}^{+\infty} e^{-\frac{i}{\hbar}p(x'-x)} e^{-\frac{i}{\hbar}(t_2-t_1)\frac{p^2}{2m}} dp,$$

the *Green's function* or *evolution kernel* of the differential equation. This involves an interchange in the order of integrations which is strictly speaking not allowed and leads to an integral for $G$ which is singular; the formula, which is very useful, can however be made rigorous by interpreting $G$ as a *distribution* or *generalized function* in the sense of Laurent Schwartz, instead of a as function in the ordinary sense. Generalized functions are discussed in Supp. Chap. 25.

An interesting question is what happens to the minimal uncertainty states that we discussed in Sect. 4.5.

(5.3,a) **Problem** Suppose that a free particle is at time $t = 0$ represented by a minimum uncertainty wave function with $\bar{x} = 0$. Show that at any later time the probability density for its position has spread and that it is no longer in a minimal uncetainty state. Hint: Use the formulas at the end of Sect. 4.5.

Finally, the analogous formula for the time evolution of a free particle in three dimensions can be written as

$$\psi(\mathbf{x}, t_2) = \frac{1}{(2\pi\hbar)^3} \int_{-\infty}^{+\infty} d\mathbf{p} \int_{-\infty}^{+\infty} d\mathbf{x}' \, \psi(\mathbf{x}', t_1) \left[ e^{-\frac{i}{\hbar}\mathbf{p}\cdot(\mathbf{x}-\mathbf{x}')} e^{-\frac{i}{\hbar}(t_2-t_1)\frac{\mathbf{p}^2}{2m}} \right],$$

with the heuristic reformulation

$$\psi(\mathbf{x}, t_2) = \int_{-\infty}^{+\infty} G(\mathbf{x}, t_2; \mathbf{x}', t_1)\psi(\mathbf{x}', t_1)d\mathbf{x}',$$

with

$$G(\mathbf{x}, t_2; \mathbf{x}', t_1) = \frac{1}{(2\pi\hbar)^3} \int_{-\infty}^{+\infty} e^{-\frac{i}{\hbar}\mathbf{p}\cdot(\mathbf{x}'-\mathbf{x})} e^{-\frac{i}{\hbar}(t_2-t_1)\frac{\mathbf{p}^2}{2m}} d\mathbf{p}.$$

Note that the equations for the time evolution of average position and momentum are now precisely the classical equation for position and momentum, because $V(\mathbf{x}) = 0$.

## 5.4 A Particle in a Box

Consider the quantum mechanical description of a one dimensional particle moving freely in a box of length $a$. The Hilbert space is $\mathcal{H} = L^2([0, a], dx)$.

(5.4,a) **Problem** Show that the position operator

$$Q : \psi(x) \mapsto x\,\psi(x)$$

in this model is a bounded operator. What is its norm?

The operator $Q$ is clearly symmetric. Because it is bounded, it is selfadjoint. The momentum operator $P$, the differential operator $\frac{\hbar}{i}\frac{d}{dx}$, is well-defined on the linear subspace $\mathcal{D}^{(1)}$ of all differentiable functions $\psi$ in $L^2([0, a], dx)$ for which the derivative $\psi^{(1)}(x) = \frac{d}{dx}\psi(x)$ is again in $L^2([0, a], dx)$, but it is not *symmetric* on $\mathcal{D}^{(1)}$, because of the condition for symmetry on a domain, $(\psi_1, P\psi_2) = (P\psi_1, \psi_2)$, for all pairs of vectors $\psi_1, \psi_2$ from that domain, implies the boundary condition

$$\overline{\psi_1(a)}\psi_2(a) = \overline{\psi_1(0)}\psi_2(0).$$

This leads to a one parameter family of domains of symmetry

$$\mathcal{D}_\alpha^{(1)} := \{\psi \in \mathcal{D}^{(1)} \mid \psi(a) = e^{i\alpha}\psi(0)\},$$

for each $\alpha \in [0, 2\pi)$. It can be shown that the restriction of $P$ to such a $\mathcal{D}_\alpha^{(1)}$ gives an essentially selfadjoint operator $P_\alpha$. All these $P_\alpha$ should be seen as different operators. Note that the intersection

$$\mathcal{D}_\cap^{(1)} = \cap_{\alpha \in [0, 2\pi)}\mathcal{D}^{(1)},$$

consisting of the functions in $\mathcal{D}^{(1)}$ with $\psi(a) = \psi(0) = 0$, would be too small. Restricting $P$ to this domain gives an example of a symmetric operator which is not essentially selfadjoint, but which has an infinite number of selfadjoint extensions. (See Supp. Sect. 21.9.2 for the notion of essential selfadjointness).

(5.4,b) **Problem** Show that $P_\alpha$ has a purely discrete spectrum, with eigenvalues, for $n = 0, \pm 1, \pm 2, \ldots,$

$$p_n^\alpha = \frac{\hbar}{a}(\alpha + 2\pi n),$$

and a corresponding orthonormal basis of eigenfunctions

$$\phi_n^\alpha(x) = \frac{1}{\sqrt{a}} e^{\frac{i}{\hbar} p_n^\alpha x}.$$

For the discussion of the physical interpretation of this situation we have to consider the possibilities for the energy operator, restrictions of the differential operator $H = -\frac{\hbar^2}{2m}\frac{d^2}{dx^2}$ to suitable domains. Starting from the $P_\alpha$ we find selfadjoint operators $H_\alpha = \frac{(P_\alpha)^2}{2m}$ defined on the domains $\mathcal{D}_\alpha^{(2)} \subset \mathcal{D}_\alpha^{(1)}$ consisting of the functions in $\mathcal{D}_\alpha^{(1)}$ that have second derivatives also in $L^2([0, a], dx)$. The $H_\alpha$ have the same orthonormal systems $\{\phi_n^\alpha\}_n$ of eigenvectors as the $P_\alpha$. The energy eigenvalues are, for $n = 0, \pm 1, \pm 2, \ldots$,

$$E_n^\alpha = \frac{(p_n^\alpha)^2}{2m} = \frac{\hbar^2}{2ma^2}(\alpha + 2\pi n)^2.$$

There is an additional selfadjoint energy operator $H^0$, namely the restriction of $H = -\frac{\hbar^2}{2m}\frac{d^2}{dx^2}$ to the space $\mathcal{D}^{(2),0}$, consisting of all $\psi$ in $L^2([0, a], dx)$ with the second derivative $\psi^{(2)}$ in $L^2([0, a], dx)$ and with $\psi(a) = \psi(0)$. $H^0$ has an orthonormal basis of eigenfunctions, for $n = 1, 2, \ldots$,

$$\psi_n^0(x) = \sqrt{\frac{2}{a}} \sin\left(\frac{n\pi x}{a}\right),$$

with eigenvalues

$$E_n^0 = \frac{\hbar^2}{2m}\left(\frac{\pi n}{a}\right)^2.$$

The various $H$ are different operators and describe different physical situations. Taking $H_\alpha$ for $\alpha = 0$ as energy operator means that when the particle reaches one endpoint of the interval $[0, a]$, it immediately reappears at the other endpoint, with the same velocity. The motion is *periodic* and the points $x = 0$ and $x = a$ can be identified; the particle moves in fact on a circle. The case with $\alpha \neq 0$ has no obvious physical interpretation. This should remind us of the general principle that there should be a mathematical model for every realistic physical situation, but that not every model that can be constructed mathematically needs to have a physical realization. The model with energy operator $H^0$ describes the quantum situation of a classical particle moving in a box in the ordinary sense, which means that it is reflected at the walls of the box with an instant change in the sign of its velocity, even though, strangely enough there is no proper selfadjoint momentum operator for the momentum in this situation. In any case, this quantum mechanical model of a particle in a box illustrates in a simple way the physical meaning of different selfadjoint extensions of a symmetric operator.

## 5.5 The Tunnel Effect

Consider a one dimensional particle moving in a potential $V(x)$ which vanishes for all real $x$, except in two intervals, $[-a-d, -a]$ and $[a, a+d]$, on which one has $V(x) = E_0$, with $E_0$, $a$ and $d$ positive constants and $d$ much smaller than $a$, i.e. a potential $V(x)$ which consists of two walls of height $E_0$ and thickness $d$. In the classical description a particle with kinetic energy $\frac{p^2}{2m} < E_0$ cannot pass the potential walls. If it is inside the interval $[-a, a]$ it will remain inside, when it is outside it will remain outside. A particle with kinetic energy larger than $E_0$ can move freely along the real line; it passes over the walls, its velocity only momentarily perturbed by the potential. The quantum situation is different. Suppose that we are certain that the particle is at a given time $t_1$ inside the interval $[-a, a]$. This means that the support of its wave function $\psi(x, t_1)$ is contained in $[-a, a]$. One can show that the time evolution is such that the wave function will spread instantaneously; at each later time $t_2$ it will be nonvanishing at arbitrary points $x$ outside $[-a, a]$, so there is a nonzero probability of finding the particle outside the walls. Conversely, if we know the particle to be outside the walls at an initial time, the chance that we may find it inside at a later time is nonzero. The reason for this is that a strictly localized wave function, i.e. a wave function with compact support, contains all possible energies in its decomposition with respect to the eigenfunctions of the energy operator. A wave function of which the components are cut off below a certain energy, say the energy $E_0$ of the potential walls, is not localized. The corresponding particle can be found inside as well as outside a given interval. This phenomenon that a quantum particle cannot be contained in a region for an indefinite length of time and that it can always pass through potential walls is called the *tunnel effect*.

# Chapter 6
# The Harmonic Oscillator

## 6.1 Introduction

The great success of quantum mechanics between the years 1924 and 1927 was its ability to give a qualitatively *and* quantitatively satisfactory description of atomic spectra. By this it established itself as *the* theoretical model for submicroscopic physics. A simple but typical example in this context was the *hydrogen atom*. An even simpler quantum mechanical model is that of the *harmonic oscillator*, which will be discussed in this chapter. With the model for the hydrogen atom it is one of the few models that can be explicitly solved by elementary means, i.e. the formulas for its energy levels and energy eigenstates can be found in closed form. The differences between the quantum version and the classical version of the harmonic oscillator are very characteristic for the differences between quantum and classical mechanics in general. This makes it an excellent example for pedagogical purposes. It is also of great practical use, because it gives a good first approximation to many more complicated and more realistic physical systems.

## 6.2 The Classical Harmonic Oscillator

A well-known system in elementary classical physics is that of a particle attached to one end of a metal spring. As long as the spring is not stretched too much, the particle will be driven back, in the direction of the point where the string is fixed, with a force proportional with the distance to this point. This situation is idealized in the mechanical model of a *classical harmonic oscillator*: a point particle of mass $m$ moving in $\mathbb{R}^1$ under the influence of a force $F = -kx$, with $k$ the 'spring constant', implying that Newton's equation $F = ma = m\frac{d^2x}{dt^2}$ is

$$\frac{d^2}{dt^2}x(t) + \frac{k}{m}x(t) = 0.$$

© Springer International Publishing Switzerland 2015
P. Bongaarts, *Quantum Theory*, DOI 10.1007/978-3-319-09561-5_6

Note that the force on the particle comes from the potential $V(x) = \frac{1}{2}kx^2$. The general solution, for arbitrary initial position $x(0)$ and velocity $\dot{x}(0) = \left(\frac{dx}{dt}\right)_{t=0}$ is easily found to be

$$x(t) = \dot{x}(0)\omega^{-1} \sin \omega t + x(0) \cos \omega t,$$

with $\omega$ defined as $\sqrt{k/m}$. The motion is *periodic*; the particle oscillates around the origin $x = 0$ with 'circle frequency' $\omega$, a characteristic constant of the system. Kinetic energy is constantly transformed into potential energy and vice versa, but the total energy remains the same. This energy depends on the initial values of the motion or, equivalently, on the maximal amplitude $|x_{max}| = (2E/m\omega^2)^{1/2}$. The possible energy values of the system form a *continuum*; $E$ can have an arbitrary nonnegative value. For $E = 0$ the particle is at rest in the origin $x = 0$, the only *stationary* state.

The system can be put in Hamiltonian form, with canonical coordinates $p$ and $x$, with a Hamiltonian function

$$H(p, x) = \frac{p^2}{2m} + \frac{1}{2}m\omega^2 x^2$$

and Hamiltonian equations

$$\frac{dx}{dt} = \frac{p}{m}, \quad \frac{dp}{dt} = -m\omega^2 x,$$

together equivalent to the second order equation

$$\frac{d^2 x}{dt^2} + \omega^2 x = 0.$$

## 6.3  The Quantum Oscillator

As was already indicated in the introductory section, many important systems in physics can in first approximation be described by a harmonic oscilator or by systems of harmonic oscillators. The basic reason for this is that a general potential $V(x)$ with a minimum in $x = 0$ can be expanded around $x = 0$ as

$$V(x) = V(0) + V'(0)x + \frac{1}{2}V''(0)x^2 + \cdots$$

with $V'$, $V''$ the first, respectively the second derivative of $V$. The first term is a constant and does therefore not affect the dynamics, the second term vanishes because $V$ has a minimum in $x = 0$, so the first nonzero term is a term quadratic in $x$, i.e. a harmonic potential. A similar argument can of course be used when $V$ has a minimum in an arbitrary $x = x_0$. The role of the harmonic oscillator model, as a simple starting point for the description of situations in which particles are bound to each other by strong forces, is even more important in quantum physics than in classical physics. This gives another justification for the detailed discussion of the quantum mechanical one dimensional harmonic oscillator in this.

The state space $\mathcal{H}$ for the quantum mechanical oscillator is the Hilbert space $L^2(\mathbb{R}^1, dx)$, its Hamiltonian operator

$$H = \frac{1}{2m}P^2 + \frac{1}{2}m\omega^2 Q^2.$$

This operator can be shown to be essentially selfadjoint on a suitable domain of smooth functions, e.g. $\mathcal{S}(\mathbb{R}^1)$.

*Remark* The 'prescription' by which the quantum Hamiltonian is obtained from the classical Hamiltonian is here still unambiguous, even though $P$ does not commute with $Q$. There would be a problem if the classical Hamiltonian function would for instance contain a term $p^3 q^3$. The operator $P^3 Q^3$ is not hermitian symmetric, so is not admissible. Classically one has

$$p^3 q^3 = \frac{1}{2}\left((pq)^3 + (qp)^3\right) = \frac{1}{2}\left(p^3 q^3 + q^3 p^3\right).$$

Both the combinations $\frac{1}{2}\left((PQ)^3 + (QP)^3\right)$ and $\frac{1}{2}\left(P^3 Q^3 + Q^3 P^3\right)$ are hermitian symmetric. They are however not the same, so it is not clear which one should be chosen as quantum observable. It is also not immediately obvious whether both operators are essentially selfadjoint. The 'quantization problem' will be discussed more fully in Chap. 13.

Studying a quantum system the central task is to find the eigenvalues and eigenvectors, or mathematically more rigorous, the spectral resolution, of the Hamiltonian— usually the energy operator—of the system. Only in a few cases this problem can be solved exactly. The hydrogen atom is one of these cases; the harmonic oscillator another.

Solving the eigenvalue–eigenvector problem $H\psi = E\psi$ for the harmonic oscillator means solving the differential equation

$$\left(-\frac{\hbar}{2m}\frac{d^2}{dx^2} + \frac{1}{2}m\omega^2 x^2\right)\psi(x) = E\psi(x),$$

for suitable, i.e. square integrable functions $\psi$. This problem was solved first by a straightforward application of a general method for studying solutions of second order differential equations of a certain type; later a more algebraic method was found which uses in an ingenious way the quantum mechanical commutation relations. The first method will be briefly sketched here; the second will be treated in detail in the next section.

Choose a new dimensionless variable

$$u = \sqrt{\frac{m\omega}{\hbar}}\, x,$$

use instead of $E$ the dimensionless quantity

$$\varepsilon = \frac{E}{\hbar\omega} - \frac{1}{2},$$

and define

$$f(u) = e^{\frac{m\omega}{2\hbar}x^2} \psi(x).$$

This leads to a differential equation for $f$ as a function of $u$, which is equivalent to the original equation and which is

$$\frac{d^2f}{du^2} - 2u\frac{df}{du} + 2\varepsilon f = 0.$$

Expand the function $f$ into a power series

$$f(u) = a_0 + a_1 u + a_2 u^2 + \cdots$$

Substitution of this series in the differential equation gives a simple recursion relation for the coefficients,

$$a_{n+2} = \frac{2(n - \varepsilon)}{(n + 1)(n + 2)} a_n,$$

for $n = 0, 1, 2, \ldots$. Given $a_0$, this relation determines the coefficients for even $n$; given $a_1$ one obtains those for odd $n$. The first case gives an even function of the variable $u$, the second case an odd function. Note that substitution of the power series for $\psi$ in the original equation would not give a useful recursion relation; in fact the nontrivial part of this procedure is to find a transformation to a new equation for which the recursion relation is simple and determines the coefficients. One next studies the asymptotic behaviour of the coefficients $a_n$ for $n \to \infty$ and finds that there are two possibilities. The first is that the series of coefficients $\{a_n\}$ is infinite. In that case the function $f(u)$ behaves as $ce^{u^2}$ for $u \to \pm\infty$. This gives for the original function $\psi$ a total exponential factor $e^{\frac{1}{2}u^2}$. Such a solution $\psi$ is not square integrable and is therefore not acceptable as wave function. We are left with the second possibility: the series breaks off and $f(u)$ is a polynomial. This happens precisely for integer values of $\varepsilon$. For $\varepsilon = n_0$ we clearly have that all $a_n$ are nonzero for $n < n_0$ and vanish for $n \geq n_0 + 2$. These results together give the solution of the original eigenvalue–eigenvector problem $H\psi = E\psi$ as

$$E_n = \hbar\omega(n + \frac{1}{2})$$

for the eigenvalues, and for the eigenfunctions

$$\psi_n(x) = N_n H_n(\sqrt{m\omega/\hbar}\; x)\, e^{-\frac{m\omega}{2\hbar}x^2},$$

for $n = 0, 1, 2, \ldots$. The $N_n$ are normalization constants equal to

$$N_n = \frac{1}{2^{n/2}(n!)^{1/2}}\left(\frac{m\omega}{\hbar\pi}\right)^{1/4}.$$

The $H_n(\cdot)$ are the *Hermite polynomials* well-known from other areas of applied mathematics. The first five Hermite polynomials are

$$H_0(u) = 1,$$
$$H_1(u) = 2u,$$
$$H_2(u) = -2 + 4u^2,$$
$$H_3(u) = -12u + 8u^3,$$
$$H_4(u) = 12 - 48u^2 + 16u^4.$$

One finally shows that this is the complete solution: the spectrum of $H$ is purely discrete and the eigenfunctions $\psi_n$ form an orthonormal basis in the Hilbert space $\mathcal{H}$.

These results should be compared with what we know about the classical situation: A classical oscillator has a *continuum* of possible energy values, the interval $[0, +\infty)$ and a *single* stationary state, the situation in which the particle has energy 0 and is at rest in the origin. The possible energy values of a quantum oscillator are *discrete*, $E_n = \hbar\omega(n + 1/2)$, for $n = 0, 1, \ldots$; there is an *infinite sequence* of stationary states, the corresponding eigenstates $\psi_n$, in which the 'groundstate' $\psi_0$, the state with the lowest energy, has a nonzero 'rest energy' $\frac{1}{2}\hbar\omega$.

## 6.4 Lowering and Raising Operators

The starting point of the second method is a transformation of the operators $P$ and $Q$ into a pair of non-selfadjoint operators $\mathbf{a}$ and $\mathbf{a}^*$. Define

$$\mathbf{a} = \alpha P + i\beta Q, \quad \mathbf{a}^* = \alpha P - i\beta Q.$$

By choosing the real parameters $\alpha$ and $\beta$ as

$$\alpha = (2\hbar)^{-\frac{1}{2}}(m\omega)^{-\frac{1}{2}}, \quad \beta = -(2\hbar)^{-\frac{1}{2}}(m\omega)^{\frac{1}{2}}$$

and using the basic commutation relation

$$[P, Q] = -i\hbar \, 1_\mathcal{H}$$

one finds for the new operators the commutation relation

$$[\mathbf{a}, \mathbf{a}^*] = 1_\mathcal{H}$$

and for the Hamiltonian operator $H$ the expression

$$H = \hbar\omega(\mathbf{a}^*\mathbf{a} + 1/2).$$

In all this we use $\mathcal{S}(\mathbb{R}^1)$ as common invariant domain of definition for the operators. Denote $\mathbf{a}^*\mathbf{a}$ as $N$. The operators $N$ and $H$ have the same eigenvectors; eigenvalues $\varepsilon$ of $N$ correspond with eigenvalues $E$ of $H$ through the relation $E = \hbar\omega(\varepsilon + 1/2)$. We therefore study the slightly more convenient eigenvalue–eigenvector problem

$$N\psi = \varepsilon\psi.$$

The solution of this problem is obtained in a procedure consisting of the following steps:
(a) From the basic commutation relation $[\mathbf{a}, \mathbf{a}^*] = 1$ one easily derives the relations

$$[N, \mathbf{a}] = -\mathbf{a}, \quad [N, \mathbf{a}^*] = \mathbf{a}^*.$$

(For example: $[N, \mathbf{a}] = \mathbf{a}^*\mathbf{a}\mathbf{a} - \mathbf{a}\mathbf{a}^*\mathbf{a} = -[\mathbf{a}, \mathbf{a}^*]\mathbf{a} = -\mathbf{a}$.)
(b) These commutation relations are used to prove a lemma, which is the principal ingredient of the procedure:

**Lemma** *If $\psi$ is an eigenvector of $N$ with eigenvalue $\varepsilon$, then either $\mathbf{a}\psi = 0$ or the vector $\mathbf{a}\psi$ is an eigenvector of $N$ with eigenvalue $\varepsilon - 1$. At the same time either $\mathbf{a}^*\psi = 0$ or the vector $\mathbf{a}^*\psi$ is an eigenvector of $N$ with eigenvalue $\varepsilon + 1$.*

*Proof* $N(\mathbf{a}\psi) = [N, \mathbf{a}]\psi + \mathbf{a}N\psi = -\mathbf{a}\psi + \varepsilon\mathbf{a}\psi = (\varepsilon - 1)(\mathbf{a}\psi)$, which states that $\mathbf{a}\psi$ is an eigenvector of $N$, with eigenvalue $\varepsilon - 1$, unless $\mathbf{a}\psi = 0$. Similarly one has $N(\mathbf{a}^*\psi) = [N, \mathbf{a}^*]\psi + \mathbf{a}^*N\psi = \mathbf{a}^*\psi + \varepsilon\mathbf{a}^*\psi = (\varepsilon + 1)(\mathbf{a}^*\psi)$.

*Remark* Because of these properties $\mathbf{a}$ is called a *lowering operator* and $\mathbf{a}^*$ a *raising operator*. Systems of such lowering and raising operators $\mathbf{a}_1, \mathbf{a}_2, \ldots$ and $\mathbf{a}_1^*, \mathbf{a}_2^*, \ldots$, with commutation relations

$$[\mathbf{a}_j, \mathbf{a}_k^*] = \delta_{jk} 1, \quad [\mathbf{a}_j, \mathbf{a}_k] = [\mathbf{a}_j^*, \mathbf{a}_k^*] = 0,$$

are important in other areas of physics. They are discussed in Supp. Sect. 23.5, in the context of symmetric tensor algebras over given Hilbert spaces. It will be shown that there is a second type of systems of lowering and raising operators, in the context of antisymmetric tensor algebras, with anticommutation relations instead of commutation relations. Both types are applied in Chap. 9, for the description of systems of many identical particles, and from that to elementary particle physics, in particular for the formulation of quantum field theory in Chap. 15. In this context one speaks of *annihilation* and *creation operators*.

(c) Eigenvalues of $N$ are necessarily nonnegative. Let $\varepsilon$ be an eigenvalue, with $\psi$ the corresponding eigenvector. One has

$$0 \leq (\mathbf{a}\psi, \mathbf{a}\psi) = (\psi, \mathbf{a}^*\mathbf{a}\psi) = (\psi, N\psi) = \varepsilon(\psi, \psi).$$

Because $(\psi, \psi) > 0$, $\varepsilon$ must be *nonnegative*.

(d) Suppose there is an eigenvector $\psi$ with eigenvalue $\varepsilon$. Repeated action of $a$ on $\psi$ will give a sequence of new eigenvectors $\mathbf{a}\psi, \mathbf{a}^2\psi, \ldots$ with eigenvalues $\varepsilon - 1, \varepsilon - 2, \ldots$. These eigenvalues cannot become negative, so this sequence has to break off. This means that there is a vector $\psi_0$ and a nonnegative real number $\varepsilon_0$, such that $N\psi_0 = \varepsilon_0\psi_0$ and $\mathbf{a}\psi_0 = 0$. This implies $0 = (\mathbf{a}\psi_0, \mathbf{a}\psi_0) = (\psi_0, N\psi_0) = \varepsilon_0(\psi_0, \psi_0)$. Because $(\psi_0, \psi_0) > 0$ this gives $\varepsilon_0 = 0$. The result of this is that the existence of an eigenvector $\psi$ with eigenvalue $\varepsilon$ entails the existence of an eigenvector $\psi_0$ with a *lowest* eigenvalue $\varepsilon_0 = 0$. Starting from this $\psi_0$ we can apply the operator $\mathbf{a}^*$, and obtain a sequence of eigenvectors $\mathbf{a}^*\psi_0, (\mathbf{a}^*)^2\psi_0, \ldots$, with eigenvalues $1, 2, \ldots$. Suppose that for an $n \in \{0, 1, 2, \ldots\}$ the vector $(\mathbf{a}^*)^{n+1}\psi_0 = 0$, while $(\mathbf{a}^*)^n \neq 0$. Then

$$\begin{aligned}
((\mathbf{a}^*)^{n+1}\psi_0, (\mathbf{a}^*)^{n+1}\psi_0) &= ((\mathbf{a}^*)^n\psi_0, \mathbf{a}\mathbf{a}^*(\mathbf{a}^*)^n\psi_0) \\
&= ((\mathbf{a}^*)^n\psi_0, (N + 1)(\mathbf{a}^*)^n\psi_0) \\
&= (n + 1)((\mathbf{a}^*)^n\psi_0, (\mathbf{a}^*)^n\psi_0) \neq 0.
\end{aligned}$$

This holds for any $n$, which means that the upward sequence

$$\psi_0, \mathbf{a}^*\psi_0, (\mathbf{a}^*)^2\psi_0, (\mathbf{a}^*)^3\psi_0, \ldots$$

does *not* break off. Note that the action of $\mathbf{a}$ and $\mathbf{a}^*$ leaves this sequence invariant, up to real factors, because

$$\begin{aligned}
\mathbf{a}((\mathbf{a}^*)^n\psi_0) &= (\mathbf{a}\mathbf{a}^*)((\mathbf{a}^*)^{n-1}\psi_0) \\
&= (1 + N)((\mathbf{a}^*)^{n-1}\psi_0) = (n + 1)(\mathbf{a}^*)^{n-1}\psi_0
\end{aligned}$$

and $\mathbf{a}^*((\mathbf{a}^*)^n\psi_0) = (\mathbf{a}^*)^{n+1}\psi_0$. The vectors $(\mathbf{a}^*)^n\psi$ can be normalized by using the above formula to normalize the vectors $(\mathbf{a}^*)^n\psi_0$. One has

$$\begin{aligned}
\|(\mathbf{a}^*)^{n+1}\psi_0\|^2 &= ((\mathbf{a}^*)^{n+1}\psi_0, (\mathbf{a}^*)^{n+1}\psi_0) \\
&= (n + 1)((\mathbf{a}^*)^n\psi_0, (\mathbf{a}^*)^n\psi_0) \\
&= (n + 1)\|(\mathbf{a}^*)^n\psi_0\|^2 = \ldots = (n + 1)!\,\|\psi_0\|^2.
\end{aligned}$$

We choose $(\psi_0, \psi_0) = 1$ and obtain then an orthonormal sequence of eigenvectors $\psi_0, \psi_1, \ldots,$ with $\psi_n$ defined as

$$\psi_n = \frac{1}{\sqrt{n!}} (a^*)^n \psi_0.$$

(e) Our result sofar is that the eigenvectors of $N$, in the domain $S(R^1)$, occur in infinite orthonormal sequences $\psi_0, \psi_1, \psi_2, \ldots,$ with corresponding sequences of eigenvalues $\varepsilon = 0, 1, 2, \ldots.$ The number of such sequences depends on the number of (linear independent) 'ground states' $\psi_0$. To find the possible vectors $\psi_0$ one must leave the purely algebraic context and return to the explicit Hilbert space picture. A state $\psi_0$ with lowest eigenvalue $\varepsilon_0 = 0$ satisfies $a\psi_0 = 0$, i.e. is a solution of the equation $(\alpha P + i\beta Q)\psi_0 = 0$, or more explicitly,

$$\frac{d}{dx}\psi_0(x) = \frac{\beta}{\alpha\hbar}x\psi_0(x).$$

This is an elementary differential equation, which has the general solution

$$\psi_0(x) = ce^{\frac{\beta}{2\alpha\hbar}x^2},$$

with $c$ a constant. Substitution of the expressions for $\alpha$ and $\beta$ gives

$$\psi_0(x) = ce^{-\frac{m\omega}{2\hbar}x^2}.$$

The standard Gaussian integral

$$\int\limits_{-\infty}^{+\infty} e^{-y^2} dy = \pi^{1/2}$$

is used to normalize $\psi_0$, which gives for the constant $c$ the condition $|c|^2 = (m\omega/\hbar\pi)^{1/2}$. With the obvious choice for $c$ one obtains finally the single 'ground state' eigenfunction

$$\psi_0(x) = \left(\frac{m\omega}{\hbar\pi}\right)^{\frac{1}{4}} e^{-\frac{m\omega}{2\hbar}x^2}.$$

From this the other eigenfunctions $\psi_n$ are obtained as

$$\psi_n(x) = \frac{1}{\sqrt{n!}} ((a^*)^n \psi_0)(x).$$

One can show, using explicit Hilbert space properties, that the system $\{\psi_n\}$ forms an orthonormal basis for $\mathcal{H} = L^2(\mathbb{R}^1, dx)$. This shows that the $\psi_n$, with their eigenvalues $\varepsilon = n$, give a complete (discrete) spectral resolution for the operator $N$ and therefore for the operator $H = \hbar\omega(N+1/2)$, with eigenvalues $E_n = \hbar\omega(n+1/2)$. One can of course verify that the $\psi_n$ here are the same as the ones found in the preceding section.

## 6.5 Time Evolution

The quantum oscillator has an infinite sequence of stationary states, the energy eigenfunctions $\psi_0, \psi_1, \ldots$, for which the predictions of the results of measurements of arbitrary observables are constant in time. (In the standard interpretation of quantum theory an *actual* measurement transforms the stationary state instantly into an in general non stationary state, in the case of a measurement of position with a wave function concentrated around the observed position value. This procedure involves the so-called '*collapse of the wave function*', a notion on the precise meaning of which there is not yet a complete agreement among various schools of thought.) In Chap. 5, Sect. 5.2, we derived equations for the time evolution of the mean values of position and momentum for a general wave packet, which strongly resemble the classical equations. For the harmonic oscillator these *are* the classical equations; because of $\overline{dV/dx} = \overline{d/dx(1/2m\omega^2)x^2} = m\omega^2\overline{x}$ one obtains

$$m\frac{d}{dt}\overline{x}(t) = \overline{p}(t)$$

$$\frac{d}{dt}\overline{p}(t) = -m\omega^2\overline{x}(t).$$

This implies that all the stationary states $\psi_n$ have $\overline{x} = 0$ and $\overline{p} = 0$, just like the single stationary state $x = 0$, $p = 0$ for the classical oscillator, but at the same time with $\Delta x \neq 0$ and $\Delta p \neq 0$. The ground state $\psi_0$ is a *minimal uncertainty state*, as will be shown in the next section. A general state $\psi$ can be expanded as $\psi = \sum_{n=0}^{\infty} c_n\psi_n$; its time evolution is therefore given by the formula

$$\psi(t) = \sum_{n=0}^{\infty} e^{-i\omega(n+1/2)t} c_n(0)\psi_n.$$

The time evolution is particularly interesting for the so called *coherent states*, a class of minimal uncertainty states which emerge here quite naturally by using the operators $\mathbf{a}$ and $\mathbf{a}^*$. This will be discussed in the next section.

## 6.6 Coherent States

The ground state $\psi$ is a solution of the equation $\mathbf{a}\psi = 0$, i.e. it is an eigenvector of the operator $\mathbf{a}$ with eigenvalue 0; and as such it is unique up to scalar factor. Has $\mathbf{a}$ other eigenvectors? The eigenvalue–eigenvector problem for $\mathbf{a}$ is an unusual one, because

$\mathbf{a}$ is not symmetric or selfadjoint. It has nevertheless a fairly simple solution which is moreover of physical interest. To find this solution one needs the commutation relation $[\mathbf{a}, (\mathbf{a}^*)^n] = n(\mathbf{a}^*)^{n-1}$ which is obtained from the basic relation $[\mathbf{a}, \mathbf{a}^*] = 1$ by induction. Expansion of the left-hand side of the equation $\mathbf{a}\phi_\lambda = \lambda\phi_\lambda$ with respect to the orthonormal basis $\psi_0, \psi_1, \ldots$ gives

$$\mathbf{a}\phi_0 = \mathbf{a}\sum_{n=0}^{\infty} c_n\psi_n = \sum_{n=1}^{\infty} c_n\mathbf{a}\psi_n = \sum_{n=1}^{\infty} \frac{c_n}{\sqrt{n!}}\mathbf{a}(\mathbf{a}^*)^n\psi_0$$

$$= \sum_{n=1}^{\infty} \frac{nc_n}{\sqrt{n!}}(\mathbf{a}^*)^{n-1}\psi_0 = \sum_{n=1}^{\infty} \frac{\sqrt{n}\,c_n}{\sqrt{(n-1)!}}(\mathbf{a}^*)^{n-1}\psi_0$$

$$= \sum_{n=1}^{\infty} \sqrt{n}\,c_n\psi_n.$$

This is well-defined if and only if the coefficients $c_n$ are such that

$$\sum_{n=1}^{\infty} n\,|c_n|^2 < \infty.$$

Expansion of the right-hand side gives

$$\lambda\phi_\lambda = \sum_{n=0}^{\infty} \lambda c_n\,\psi_n.$$

By comparing the two expansions one obtains the simple recursion relation $c_{n+1}\sqrt{n+1} = \lambda c_n$, which, after choosing $c_0 = 1$, gives $c_n = \lambda^n/\sqrt{n!}$, for $n = 0, 1, \ldots$. The series of coefficients $\lambda^n/\sqrt{n!}$ is $l_2$-summable, so the vector $\phi_\lambda$ is well-defined. One has

$$||\phi_\lambda||^2 = \sum_{n=0}^{\infty} \frac{|\lambda|^{2n}}{n!} = e^{|\lambda|^2},$$

and similarly, for two different complex numbers $\lambda_1$ and $\lambda_2$,

$$(\phi_{\lambda_1}, \phi_{\lambda_2}) = \sum_{n=0}^{\infty} \frac{(\overline{\lambda_1}\lambda_2)^n}{n!} = e^{\overline{\lambda_1}\lambda_2}.$$

The vector $\phi_\lambda$ can be written as

$$\phi_\lambda = \sum_{n=0}^{\infty} \frac{\lambda^n}{\sqrt{n!}}\psi_n = \sum_{n=0}^{\infty} \frac{\lambda^n}{n!}(\mathbf{a}^*)^n\psi_0 = e^{\lambda\mathbf{a}^*}\psi_0,$$

a very appealing and convenient expression, although one should realize that this does *not* mean that the operator-valued exponential function $e^{\lambda a^*}$ has in itself a good mathematical meaning, but only that the special sequence of vectors

$$\psi_0 + \lambda \mathbf{a}^* \psi_0 + \frac{1}{2!}\lambda^2 (\mathbf{a}^*)^2 \psi_0 + \frac{1}{3!}\lambda^3 (\mathbf{a}^*)^3 \psi_0 + \cdots$$

is convergent.

In this manner we have found a complete solution of the rather unusual eigenvalue–eigenvector problem $a\phi_\lambda = \lambda \phi_\lambda$: for each complex number $\lambda$ there is a vector

$$\phi_\lambda = \sum_{n=0}^{\infty} \frac{\lambda^n}{\sqrt{n!}} \psi_n = e^{\lambda a^*} \psi_0.$$

The vectors $\phi_\lambda$ are normalizable, with norm $\|\phi_n\| = e^{\frac{1}{2}|\lambda|^2}$, and inner product $(\phi_{\lambda_1}, \phi_{\lambda_2}) = e^{\overline{\lambda_1}\lambda_2}$. The eigenvalues $\lambda$ form a continuum, the complex plane, but the eigenvectors are nevertheless normalizable. The uncountable system $\{\phi_\lambda\}_{\lambda \in \mathbb{C}}$ is linearly independent in the sense that each finite subset is linearly independent, even though the Hilbert space is separable. It is a basis of the Hilbert space $\mathcal{H}$, not in the usual sense, but in a generalized 'continuous' way. An arbitrary vector $\psi$ in $\mathcal{H}$ can be written as

$$\psi = \frac{1}{2\pi i} \int_C F_\psi(\lambda) e^{-|\lambda|^2} \phi_\lambda \, d\overline{\lambda} d\lambda,$$

with $\frac{1}{2\pi i} d\overline{\lambda} d\lambda = \frac{1}{\pi} dx dy$, for $\lambda = x + iy$, and with $F_\psi$ a function which is uniquely determined by

$$F_\psi(\lambda) = (\phi_\lambda, \psi).$$

One can show that $F_\psi(\lambda)$ is *anti-holomorphic*, i.e. its complex conjugate $\overline{F_\psi(\lambda)}$ is holomorphic, and that it satisfies the growth condition

$$\int_C |F_\psi(\lambda)|^2 e^{-|\lambda|^2} d\overline{\lambda} d\lambda < \infty.$$

Such functions form a Hilbert space $\widehat{\mathcal{H}}$ with respect to the inner product

$$(F_1, F_2) = \frac{1}{2\pi i} \int_C \overline{F_1(\lambda)} F_2(\lambda) e^{-|\lambda|^2} d\overline{\lambda} d\lambda.$$

The relation between the vectors $\psi$ in $\mathcal{H}$ and the anti-holomorphic functions $F$ in $\widehat{\mathcal{H}}$ given above is a unitary transformation between $\mathcal{H}$ and $\widehat{\mathcal{H}}$. The space $\widehat{\mathcal{H}}$ is called the *Bargmann Hilbert space of anti-holomorphic functions*—or holomorphic functions, if one uses a slightly different convention. It has many interesting mathematical properties.

*Remark* Note the difference between this notion of a 'continuous base' and the one suggested by the system of nonnormalizable eigenfunctions $\phi_p(x) = ce^{\frac{i}{\hbar}px}$ of the momentum operator P.

In this quantum mechanical context the vectors $\phi_\lambda$, or their normalized versions

$$\tilde{\phi}_\lambda = e^{-\frac{1}{2}|\lambda|^2}\phi_\lambda,$$

are called *coherent states*, a term which has its origin in an application in quantum optics, a subject that cannot be discussed here. (Strictly speaking the term 'state' should be reserved for the normalized vectors $\tilde{\phi}_\lambda$, but the $\phi_\lambda$ are often more convenient to use and are therefore usually also called states.)

To understand what the *physical properties* of (normalized) coherent states are, we calculate the expectation values $\bar{p}, \bar{x}$ and standard deviations $\Delta p, \Delta x$. The inverse transformation formulas

$$P = \frac{1}{2\alpha}(\mathbf{a} + \mathbf{a}^*)$$

$$Q = \frac{1}{2i\beta}(\mathbf{a} - \mathbf{a}^*),$$

with $\alpha = (2m\hbar\omega)^{-1/2}$ and $\beta = -(m\omega/2\hbar)^{1/2}$, allow us to express $P$ and $Q$ in $\mathbf{a}$ and $\mathbf{a}^*$. Note that one has

$$(\tilde{\phi}_\lambda, (\mathbf{a}^*)^m\mathbf{a}^n\tilde{\phi}_\lambda) = (\mathbf{a}^m\tilde{\phi}_\lambda, \mathbf{a}^n\tilde{\phi}_\lambda) = \overline{\lambda}^m\lambda^n,$$

for all $m, n = 0, 1, \ldots$. An arbitrary algebraic expression in the $\mathbf{a}$ and $\mathbf{a}^*$ can be written in what is called *normally ordered form* i.e. as a sum of such double monomials, if necessary after a rearrangement using the basic commutation relation $[\mathbf{a}, \mathbf{a}^*] = 1$. With this one obtains

$$\bar{p} = (\tilde{\phi}_\lambda, P\tilde{\phi}_\lambda) = \frac{1}{2\alpha}(\tilde{\phi}_\lambda, (\mathbf{a} + \mathbf{a}^*)\tilde{\phi}_\lambda) = \frac{1}{2\alpha}(\lambda + \overline{\lambda}) = \frac{1}{\alpha}\operatorname{Re}\lambda,$$

and similarly

$$\bar{x} = (\tilde{\phi}_\lambda, Q\tilde{\phi}_\lambda) = \frac{1}{2i\beta}(\lambda - \overline{\lambda}) = \frac{1}{\beta}\operatorname{Im}\lambda.$$

Together this gives

$$\lambda = \alpha \bar{p} + i\beta \bar{x}.$$

For the calculation of $\Delta p$ one needs

$$
\begin{aligned}
\overline{p^2} = (\tilde{\phi}_\lambda, P^2 \tilde{\phi}_\lambda) &= \frac{1}{4\alpha^2}(\tilde{\phi}_\lambda, (\mathbf{a}^2 + (\mathbf{a}^*)^2 + \mathbf{a}^*\mathbf{a} + \mathbf{aa}^*)\tilde{\phi}_\lambda) \\
&= \frac{1}{4\alpha^2}(\tilde{\phi}_\lambda, (\mathbf{a}^2 + (\mathbf{a}^*)^2 + 2\mathbf{a}^*\mathbf{a} + 1)\tilde{\phi}_\lambda) \\
&= \frac{1}{4\alpha^2}(\lambda^2 + \overline{\lambda}^2 + 2|\lambda|^2 + 1).
\end{aligned}
$$

One finds

$$(\Delta p)^2 = \overline{p^2} - (\overline{p})^2$$

$$= \frac{1}{4\alpha^2}(\lambda^2 + \overline{\lambda}^2 + 2|\lambda|^2 + 1) - \frac{1}{4\alpha^2}(\lambda + \overline{\lambda})^2 = \frac{1}{4\alpha^2},$$

and therefore

$$\Delta p = 1/2\alpha = (m\omega)^{1/2}(\hbar/2)^{1/2}.$$

Similarly one obtains

$$(\Delta x)^2 = 1/4\beta^2$$

or

$$\Delta x = -1/2\beta = (m\omega)^{-1/2}(\hbar/2)^{1/2}.$$

Combining the results on $p$ and $x$ one obtains finally

$$\Delta p \, \Delta x = -\frac{1}{4\alpha\beta} = \frac{1}{2}\hbar,$$

which means that each normalized coherent state $\tilde{\phi}_\lambda$ is a *minimal uncertainty state*.

*Conclusion:* For each value $\bar{p}$ and $\bar{x}$ there is a minimal uncertainty state, the normalized coherent state $\tilde{\phi}_\lambda$, an eigenvector of $a$ with eigenvalue $\lambda$, with

$$\lambda = \alpha \bar{p} + i\beta \bar{x} = (2m\hbar\omega)^{1/2}\bar{p} - i(m\omega/2\hbar)^{1/2}\bar{x}.$$

All these minimal states have the *same* standard deviations

$$\Delta p = 1/2\alpha = (m\omega)^{1/2}(\hbar/2)^{1/2}$$

and

$$\Delta x = -1/2\beta = (m\omega)^{-1/2}(\hbar/2)^{1/2}.$$

The minimal states associated with a quantum harmonic oscillator in this way form a two parameter subset of the three parameter set of minimal states in the Bargmann Hilbert space $\widehat{\mathcal{H}}$ which is isomorphic to $\mathcal{H} = L^2(\mathbb{R}, dx)$, found in Sects. 4.5 and 4.7.

## 6.7 Time Evolution of Coherent States

An interesting physical aspect of the coherent states of an harmonic oscillator is their behaviour under time evolution. For an arbitrary state $\psi = \sum_{n=0}^{\infty} c_n \psi_n$ the time evolution is given by the one parameter group of unitary operators $e^{-\frac{i}{\hbar}tH}$, acting as

$$e^{-\frac{i}{\hbar}tH}\psi = \sum_{n=0}^{\infty} e^{-i\omega(n+1/2)t} c_n \psi_n.$$

For a coherent state

$$\phi_\lambda = \sum_{n=0}^{\infty} \frac{\lambda^n}{\sqrt{n!}} \psi_n$$

the time evolution becomes

$$e^{-\frac{i}{\hbar}tH}\phi_\lambda = \sum_{n=0}^{\infty} \frac{e^{-i\omega(n+1/2)t}\lambda^n}{\sqrt{n!}} \psi_n = e^{-\frac{i}{2}\omega t} \sum_{n=0}^{\infty} \frac{\lambda(t)^n}{\sqrt{n!}} \psi_n,$$

with $\lambda(t) = e^{-i\omega t}\lambda$. This means that the coherent state $\phi_\lambda$ is transformed—up to the phase factor $e^{-\frac{i}{2}\omega t}$—into a new coherent state $\phi_{\lambda(t)}$: the set of coherent states is mapped onto itself.

The wave function of a (normalized) coherent state $\tilde{\phi}_\lambda$ moves periodically, up to an irrelevant phase factor. The corresponding probability density $|\tilde{\phi}_\lambda(x)|^2$ is Gaussian. It moves exactly like a 'smeared' classical oscillator with frequency $\omega$. Its width is constant in time and equal to $(m\omega)^{-1/2}(\hbar/2)^{1/2}$. In the classical limit, the limit of a macroscopic situation, this becomes the harmonic motion of a point particle.

## 6.8 The Three Dimensional Harmonic Oscillator

The classical three dimensional harmonic oscillator is a particle of mass $m$ moving in three dimensional space, connected by an elastic wire with a fixed point—for which we take the origin, driven back by a vector force $\mathbf{F}$ with components $F_j = k_j$ with $k_j = m\omega_j^2$.

The wave functions of the three dimensional harmonic oscillator are the square integrable functions $\psi(x_1, x_2, x_3)$; the Hamiltonian operator is

$$H = \frac{1}{2m}(P_1^2 + P_2^2 + P_3^2) + \frac{1}{2}m(\omega_1^2 Q_1^2 + \omega_2^2 Q_2^2 + \omega_3^2 Q_3^2).$$

(6.8,b) **Problem** Show that the eigenfunctions of $H$ are

$$\psi_{n_1 n_2 n_3}(x_1, x_2, x_3) = \psi_{n_1}(x_1)\, \psi_{n_2}(x_2)\, \psi_{n_3}(x_3),$$

with the functions $\psi_{n_j}(x_j)$ the eigenfunctions of the one dimensional oscillator with corresponding eigenvalues

$$\begin{aligned} E_{n_1 n_2 n_3} &= E_{n_1} + E_{n_2} + E_{n_3} \\ &= \hbar(n_1\omega_1 + n_2\omega_2 + n_3\omega_3 + 3/2), \end{aligned}$$

for $n_1, n_2, n_3 = 0, 1, \ldots$.

The eigenvalues of the one dimensional oscillator are nondegenerate. Whether those of the three dimensional oscillator are, depends on the relations between the three frequencies $\omega_j$.

(6.8,b) **Problem** Let $\omega_1 = \omega_2 = \omega_3 = \omega$. The energy eigenvalues can then be written as $E_N$, with $N = n_1 + n_2 + n_3$. For a given $N$ there are $d_N$ triples $(n_1, n_2, n_3)$ that sum up to $N$. This is the degeneracy of $E_N$. Show that $d_N = \frac{1}{2}(n+1)(n+2)$.

# Chapter 7
# The Hydrogen Atom

## 7.1 Introduction: Historical Remarks

In elementary quantum mechanics there are two important models that are exactly solvable, which means that the solution of the eigenvalue problem (from now on short for eigenvalue–eigenfunction problem) can be found in closed form. Both describe realistic physical situations, of a rather different sort, however, and both are a first step towards far-ranging generalizations. The first is the one dimensional harmonic oscillator, discussed in Chap. 6; this chapter is devoted to the second, more complicated example, the hydrogen atom.

Quantum mechanics as a theory for the description of the structure of atoms can be said to have had its beginning in a paper by Bohr [1]. Scattering experiments by Rutherford, the results of which were published in 1911 [2], had suggested the picture of an atom as a tiny planetary system, consisting of a heavy nucleus encircled by very light particles, electrons in fact. This picture was incompatible with classical physics because of the discrete atomic spectra that had been measured with great precision in the second half of the nineteenth century. To explain Rutherford's results, Bohr postulated a model for the hydrogen atom in which the movement of the electron was restricted to discrete orbits. From this the correct emission spectrum could be derived. Bohr's model was a brilliant, daring intuitive idea, but lacked a theoretical basis. This was finally supplied by quantum mechanics proper, as it was developed between 1925 and 1927 by Heisenberg, Schrödinger, Born and others. All this was discussed in Sects. 3.1 and 3.2.

The first satisfactory explanation of the details of Bohr's model was provided by Schrödinger who, inspired by ideas of de Broglie, formulated a wave equation, and used this to derive Bohr's results from a few general quantum mechanical postulates. His 1926 paper is a landmark of 20th century physics [3].

© Springer International Publishing Switzerland 2015
P. Bongaarts, *Quantum Theory*, DOI 10.1007/978-3-319-09561-5_7

## 7.2 The Classical Model

In the simplest classical model of the hydrogen atom one pictures the nucleus as fixed in the origin; the electron is a point particle that moves in an attractive force field, coming from the rotationally symmetric Coulomb potential. As was described in Sect. 2.5, one considers a Hamiltonian function

$$H(\mathbf{p}, \mathbf{r}) = \frac{\mathbf{p}^2}{2m} + V(\mathbf{r}),$$

with $\mathbf{p} = (p_1, p_2, p_3)$, $\mathbf{p}^2 = \sum_{j=1}^{3} (\mathbf{p_j})^2$, $r = (\sum_{j=1}^{3} x^j)^{1/2}$ and $m$ the mass of the particle.

For this particular case the potential is the Coulomb potential

$$V(\mathbf{r}) = -\frac{e^2}{4\pi\epsilon_0 r},$$

with the $e = 1.6 \times 10^{-19}$ C, minus the electric charge of the electron, $\epsilon_0 = 8.85 \times 10^{-12}$ C/kg, the 'permittivity of space', and $m_e = 9.109 \times 10^{-31}$ kg, the mass of the electron.

## 7.3 The Quantum Model

The standard quantization prescription leads to the equation for the eigenvalue problem, which is the *time-independent Schrödinger equation*. The procedure leading to the solution is described in all physics textbooks on quantum mechanics; it is straightforward but far from simple.

The formulation of this equation and its complete solution for the hydrogen atom was probably Schrödinger's greatest contribution to twentieth century physics. Mathematically speaking his work was a tour de force in classical mathematical physics; his approach is still the basis for the discussion of the hydrogen atom in all modern textbooks.

Using the rotational symmetry of the problem one rewrites the problem in spherical coordinates $(r, \theta, \phi)$. The equation can then be split in two independent equations, one in the angular variables $(\theta, \phi)$, the other in the radial variable $r$. The first is valid for a general rotation invariant potential, has a fairly simple transparent solution, and will therefore be discussed first. The second can only be solved in closed form for the Coulomb potential. In both cases one employs classical orthogonal polynomials, Legendre polynomials for the first and, in addition to that, Laguerre polynomials for the second case.

In this chapter we give the main lines of these arguments, more details can be found in various textbooks. Schrödinger used rotation invariance but not in an explicit way the underlying group theory, developed by Weyl somewhat later. This aspect will be emphasized here; it leads to shortcuts in the derivations and makes the general situation more transparent. For this we discuss first briefly rotational symmetry and provide part of the general mathematical background on the rotation group and its representations.

## 7.4  Rotational Symmetry

### 7.4.1  Introduction

Symmetries are very important in physics, in classical physics, and even more in quantum physics, as was discussed in Sect. 3.7. In this chapter its main function is to guide and simplify the solution of the eigenvalue problem for the Hamiltonian of the hydrogen atom. The Coulomb potential is invariant under rotations in three dimensional space. Rotational symmetry in quantum mechanics means that the Hamiltonian operator $H$ commutes with the unitary operators representing the group $SO(3)$ in the Hilbert space $\mathcal{H}$ of the quantum system. The group $SO(3)$, the group of all real $3 \times 3$ orthogonal matrices with determinant one, is a Lie group, so $H$ will also commute with the operators of the corresponding representation of its Lie algebra $so(3)$. This will be used first to discuss the eigenvalue problem for the general case of a particle in a centrally symmetric potential, and after that to give a complete solution of this problem for the particular case of the hydrogen atom.

### 7.4.2  Representations of the Lie Algebra of $SO(3)$

The group $SO(3)$ is a compact Lie group, so all its representations are completely reducible, i.e. can be written as a direct sum of irreducible representations, unitary— or unitarizable, and finite dimensional. See for all these and other group theoretical notions Supp. Chap. 24. The simplest way to obtain these representations is through the representations of the Lie algebra $so(3)$, which can be found by a simple algebraic procedure, of which the details are given in Supp. Sect. 24.8. The Lie algebra $so(3)$ is three dimensional; it has a standard basis consisting of the elements $(L_1, L_2, L_3)$, which satisfy the commutation relations,

$$[L_1, L_2] = i\hbar L_3, \quad [L_2, L_3] = i\hbar L_1, \quad [L_3, L_1] = i\hbar L_2.$$

A Lie algebra $\mathcal{L}$, as an 'abstract' nonassociative algebra, can uniquely be embedded in a large 'abstract' associative algebra, the *universal enveloping algebra* $U(\mathcal{L})$. There is a one–one correspondence between the representations of a Lie algebra and its universal enveloping algebra. This algebra has a so-called Casimir element

$$\mathbf{L}^2 = L_1^2 + L_2^2 + L_3^2.$$

which commutes with each of the $L_j$. A pair $(\mathbf{L}^2, L_j)$, usually $(\mathbf{L}^2, L_3)$, can be used to characterize a set of basis vectors that span the irreducible representations. One has

the following result, derived in Supp. Sect. 24.8: The irreducible representations of
$so(3)$ are characterized by the number $l$ which takes the values $l = 0, \frac{1}{2}, 1, \frac{3}{2}, 2, \ldots$
The representation spaces have dimension $2l + 1$. In each space the operator $\mathbf{L}^2$
has eigenvalue $l(l + 1)\hbar^2$, and the operator $L_3$ the nondegenerate eigenvalues $-l\hbar$,
$(-l + 1)\hbar, \ldots, (l - 1)\hbar, l\hbar$. All this is derived in Supp. Sect. 24.8.

*Important note.* The Lie algebra $so(3)$ of the Lie group $SO(3)$ is isomorphic to the
Lie algebra $su(2)$ of the Lie group $SU(2)$, the universal covering group of $SO(3)$.
For all the values of $l$, integer or half integer, the representations of $su(2) \equiv so(3)$
can be integrated, i.e. exponentiated, to representations of $SU(2)$. Representations of
$SO(3)$ can only be obtained for integer values of $l$. In this chapter we use therefore
only $l = 0.1, 2, \ldots$; in the next chapter, where the notion of *spin* will appear, the full
range will be used.

The representation space for $l = 0$ is one dimensional, i.e. it is spanned by a single
vector, in the case of a particle the *ground state*, which has zero angular momentum.

### 7.4.3  Action of $SO(3)$ in the Hilbert Space $\mathcal{H} = L^2(\mathbb{R}^3, dx)$

Rotation around the $x_3$-axis over an angle $\alpha$ can be described by a one parameter
group $e^{\alpha\, d_3}$ of $3 \times 3$ orthogonal matrices

$$
e^{\alpha\, d_3} = \begin{pmatrix} \cos\alpha & -\sin\alpha & 0 \\ \sin\alpha & \cos\alpha & 0 \\ 0 & 0 & 1 \end{pmatrix}.
$$

In the context of representation theory this is the case $l = 1$; it is usually called the
'basis representation' or 'ground representation' of $SO(3)$.

(7.4.3,a) **Problem** Show that this formula can be developed up to first order in $\alpha$ as
$e^{\alpha d_3} = 1 + \alpha d_3 + \cdots$, with

$$
d_3 = \begin{pmatrix} 0 & -1 & 0 \\ 1 & 0 & 0 \\ 0 & 0 & 0 \end{pmatrix}.
$$

Show next that this induces an action by unitary operators $U_3(\alpha)$ in the Hilbert space
$\mathcal{H}$ of the particle, in first order in $\alpha$ and written in vector notation as

$$
(U_3(\alpha)\psi)(\mathbf{x}) = \psi(e^{-\alpha\, d_3}\mathbf{x}) = \psi(\mathbf{x} - \alpha\, d_3\mathbf{x}) + \cdots,
$$

or

$$
(U_3(\alpha)\psi)(x_1, x_2, x_3) = \psi(x_1 + \alpha x_2, x_2 - \alpha x_1, x_3) + \cdots = \psi(x_1, x_2, x_3)
$$
$$
+ \alpha\left( x_2 \frac{\partial}{\partial x_1}\psi(x_1, x_2, x_3) - x_1\frac{\partial}{\partial x_2}\psi(x_1, x_2, x_3) \right).
$$

By insertion of an imaginary $i$ and Planck's constant $\hbar$, the first order term in $\alpha$ gives

$$\frac{\hbar}{i}\left(x_1\frac{\partial}{\partial x_2} - x_2\frac{\partial}{\partial x_1}\right) = L_3,$$

i.e. the differential operator which represents the angular momentum of a particle in three dimensional quantum mechanics, as it was defined in Sect. 3.6.1. Note in passing that because of the additional imaginary $i$ we have obtained a selfadjoint operator, which is the natural object in quantum theory; in the context of representation theory it means that we look at $so(3, \mathbb{C})$, the complexification of $so(3)$, a trivial extension of the mathematical framework. By cyclic permutation one obtains similar results for rotations around the $x_1$- and $x_2$-axis and the connection with $L_1$ and $L_2$. All this gives an explicit illustration of the relation between 'infinitesimal symmetries' and constants of motion, discussed in Sect. 3.9.

### 7.4.4 The Mathematical Structure of $\mathcal{H}$

The group $SO(3)$ acts in $\mathcal{H}$ by unitary operators. We should remark first that for a general attractive centrally symmetric potential, and certainly for the case of the Coulomb potential, the spectrum of $H$ can be shown to consist of two parts. A discrete part, bounded from below by the energy of the ground state and from above by zero, the ionization energy, and corresponding with *bound states*, and a continuous part, with positive energy values, corresponding with *scattering states*. In what follows we shall say very little about the scattering states, and concentrate on the bound states, represented by the eigenfunctions associated with the discrete spectrum.

Representations of $SO(3)$ can be written as direct sums of irreducible representations. In the situation here $SO(3)$ acts in $\mathcal{H} = L^2(\mathbb{R}^3, d\mathbf{x})$, which is a direct sum of representation spaces $\mathcal{H}^{(l)}$, each characterized by the eigenvalue $2\pi\hbar^2 \, l(l+1)$ of the operator $\mathbf{L}^2$,

$$\mathcal{H} = L^2(\mathbb{R}^3, d\mathbf{x}) = \oplus_{l=0}^\infty \mathcal{H}^{(l)}.$$

In the following it will become clear that all values of $l$ are indeed present. These representations are reducible; in each $\mathcal{H}^{(l)}$ an infinite number of equivalent irreducible representations for a given $l$ are collected; $\mathcal{H}^{(l)}$ is a tensor product

$$\mathcal{H}^{(l)} = \mathcal{H}_{\alpha,l} \otimes \mathcal{H}_l,$$

with $\alpha$ the quantum number or numbers, of the radial part of $\mathcal{H}$, yet to be determined, and $\mathcal{H}_l$ an irreducible $2l + 1$-dimensional representation space of $\mathbf{L}^2$. As a function space $\mathcal{H}^{(l)}$ is spanned by product functions of the form

$$\psi_{\alpha,l}(r, \theta, \phi) = f_{\alpha,l}(r)g_l(\theta, \phi).$$

The norm of such a function is the product of two integrals

$$||\psi_{\alpha,l}|| = \left( \int\limits_{r=0}^{+\infty} |f_{\alpha,l}(r)|^2 r^2 dr \right)^{1/2} \left( \int\limits_{\theta=0}^{\pi} \int\limits_{\phi=0}^{2\pi} |g_l(\theta, \phi)|^2 \sin\theta \, d\theta \, d\phi \right)^{1/2}.$$

The splitting of $\mathcal{H}$ into the subspaces of bound states and scattering states is invariant under the action of the rotation group—and, of course, under the action of the Hamiltonian operator. Except for a brief remark further on we shall only discuss the bound states; therefore we shall from now on mean with $\mathcal{H}$ the subspace spanned by the bound state eigenfunctions.

## 7.5  The Schrödinger Equation

### 7.5.1  Angular Momentum Eigenfunctions

We start with the Hamiltonian operator for a particle in a centrally symmetric potential

$$H = \frac{\mathbf{p}^2}{2m} + V = -\frac{\hbar^2}{2m}\Delta + V.$$

(7.5.1,a) **Problem** Use spherical coordinates $(r, \phi, \theta)$, defined as

$$x = r\sin\theta\,\cos\phi, \qquad y = r\sin\theta\,\sin\phi, \qquad z = r\cos\theta,$$

to show that $H$ can be written as

$$H = \frac{1}{2mr^2}\mathbf{L}^2 - \frac{\hbar^2}{2mr}\frac{\partial^2}{\partial r^2} + V.$$

Show also that the angular momentum operators $\mathbf{L}^2$ and $L_3$ can be written as

$$L_3 = \frac{\hbar}{i}\frac{\partial}{\partial\phi}, \qquad \mathbf{L}^2 = -\hbar^2\left( \frac{1}{\sin\theta}\frac{\partial}{\partial\theta}\left[\sin\theta\frac{\partial}{\partial\theta}\right] + \frac{1}{\sin^2\theta}\frac{\partial^2}{\partial\phi^2}\right).$$

We have already given the properties of the representation theory for $\mathbf{L}^2$ and also $L_3$ in Sect. 7.4.2. Therefore the solution of the eigenvalue problem $H\psi = E\psi$ is much facilitated by discussing first the angular momentum eigenvalue problem, i.e. the equations

$$\mathbf{L}^2\psi = \hbar^2 l(l+1)\psi, \qquad L_3\psi = \hbar m\psi.$$

The space of wave functions $\psi(r, \theta, \phi)$ is spanned by product functions $\psi(r, \theta, \phi) = f(r)g(\theta, \phi)$. The action of the operators $\mathbf{L}^2$ and $L_3$ do not involve differentiation with respect to the radial variable $r$, so we may conveniently discuss the eigenfunction problem of $\mathbf{L}^2$ and $L_3$ in the space of square integrable functions $g(\theta, \phi)$ on the unit sphere of $\mathbb{R}^3$, with the equations

$$\mathbf{L}^2 g_{l,m} = \hbar^2 l(l+1) g_{l,m}, \qquad L_3 g_{l,m} = \hbar m g_{l,m}.$$

We start with the equation for $\mathbf{L}^2$. An irreducible eigenspace of $\mathbf{L}^2$ is spanned by $2l+1$ angular functions $g_{l,m}(\theta, \phi)$, the joint eigenfunctions of the commuting operators $\mathbf{L}^2$ and $L_3$. The solutions of $L_3 g_{l,m} = \hbar m g_{l,m}$ have the form $g_{l,m}(\theta, \phi) = h_{l,m}(\theta) e^{im\phi}$, with $m = -l, \ldots, +l$.

Using this we write the equation

$$\mathbf{L}^2 g_{l,m} = \hbar^2 l(l+1) g_{l,m}$$

explicitly as

$$\frac{1}{\sin\theta} \frac{\partial}{\partial\theta} \left( \sin\theta \frac{\partial g_{l,m}}{\partial\theta} \right) + \frac{1}{\sin^2\theta} \frac{\partial^2 g_{l,m}}{\partial\phi^2} + l(l+1) g_{l,m} = 0.$$

For the case $m = 0$ this reduces to

$$\frac{1}{\sin\theta} \frac{d}{d\theta} \left( \sin\theta \frac{dh_{l,0}}{d\theta} \right) + l(l+1) h_{l,0} = 0.$$

The substitution $\cos\theta = x$ and some rewriting leads to *Lagrange's equation*, well-known in classical mathematical physics,

$$\frac{d}{dx} \left[ (1 - x^2) \frac{dh_l}{dx} \right] + l(l+1) h_l = 0.$$

One assumes that $h_l$ can be expanded in a power series $h_l(x) = a_0 + a_1 x + a_2 x^2 + \cdots$; $h_l$ is a solution iff the coefficients $\{a_j\}_j$ satisfy a recursion relation. The requirement that the solution is normalizable can be satisfied only if the series breaks of. This gives, for each $l = 0, 1, 2, \ldots$, a real-valued polynomial of degree $l$, which can be written as

$$P_l(x) = \frac{1}{2^l l!} \frac{d^l}{dx^l} [(x^2 - 1)^l],$$

and which is called a *Legendre polynomial*, named after the French mathematician Adrien-Marie Legendre. They are normalized according to

$$\int_{\theta=0}^{+\pi} P_l(\cos\theta) P_{l'}(\cos\theta) \sin\theta \, d\theta = \frac{2}{2l+1} \delta_{ll'}.$$

(7.5.1,b) **Problem** Show that the first three Legendre polynomials are equal to

$$P_0(\cos\theta) = 1, \qquad P_1(\cos\theta) = \cos\theta, \qquad P_2(\cos\theta) = \frac{1}{2}(3\cos^2\theta - 1).$$

The eigenfunctions of $\mathbf{L}^2$ for $m \neq 0$ can be found by differentiating with respect to $x$. On may check that the expression

$$P_l^m(x) = (1 - x^2)^{m/2}\frac{d^m P_l(x)}{dx^m},$$

for $m = 0, 1, \ldots l$, called an *associated Legendre function*, is a solution of the equation for $h_{l,m}$ for $0 < m < l$. The associated Legendre functions for $m < 0$ are defined by a formula which is in fact valid for both cases, namely,

$$P_l^m(x) = (1 - x^2)^{|m|/2}\frac{d^{|m|} P_l(x)}{d^{|m|}x}.$$

This leads to a general formula for the joint eigenfunctions for $\mathbf{L}^2$ and $L_3$ in terms of the so-called *spherical harmonics* $Y_l^m$

$$g_{l,m}(\theta, \phi) = Y_l^m(\theta, \phi) = \epsilon\sqrt{\frac{2l+1}{4\pi}\frac{(l-|m|)!}{(l+|m|)!}}e^{im\phi}P_l^m(\cos\theta),$$

for $l = 0, 1, \ldots, m = -l, \ldots, +l, \epsilon = (-1)^m$ for $m \geq 0$ and $\epsilon = 1$ for $m \leq 0$. Note that the system $\{Y_l^m\}$ is an orthonormal basis for all square integrable functions on the unit sphere of $\mathbb{R}^3$.

The space of eigenfunctions of the energy operator $H$, which commutes with $\mathbf{L}^2$ and $L_3$—the essence of rotation symmetry, is spanned by product functions $\psi(r, \theta, \phi) = f(r)g(\theta, \phi)$, so in this subsection we have found the general form of the angular part $g(\theta, \phi)$; in the next that of the radial parts $f(r)$ will be determined.

## 7.5.2 The Radial Part of the Eigenfunctions

For the angular wave function, rotational invariance of the potential $V$ was sufficient. This is still the case for some general properties of radial eigenfunctions; however in order to find explicit solutions we have to specialize to the Coulomb potential of the hydrogen atom, given in Sect. 7.2 as

$$V(r) = -\frac{e^2}{4\pi\epsilon_0 r}.$$

We start again with the Schrödinger equation. Using the expression for $H$ in terms of $L^2$ and $V$ found in Problem (7.5.1,a), we write for a product function $\psi(r, \theta, \phi) = f(r)g(\theta, \phi)$

$$\left[\frac{1}{2mr}\frac{\partial^2}{\partial r^2}(rf) - \left(\frac{l(l+1)}{2mr^2} + \frac{V(r) - E}{\hbar^2}\right)f\right]g = 0.$$

The only differentiation is with respect to the variable $r$, so we may divide by $g$, define the quantities

$$\rho = \frac{1}{\hbar}\sqrt{2m|E|}\,r, \qquad \rho_0 = \frac{V}{E}\rho,$$

and obtain in the end the ordinary differential equation

$$\frac{d^2u}{d\rho^2} - \frac{l(l+1)}{\rho^2}u + \sigma\frac{\rho_0}{\rho}u + \sigma u = 0,$$

with $\sigma = E/|E| = \pm 1$, the value of which will be determined further on. Note that the Coulomb potential becomes

$$V(\rho) = -\frac{e^2\sqrt{2m|E|}}{4\pi\epsilon_0\hbar\rho}.$$

(7.5.2,a) **Problem** Show that both $\rho$ and $\rho_0$ are dimensionless.

## 7.5.3 Asymptotic Behaviour of the Radial Eigenfunctions

Two cases:

1. $\rho \to \infty$: For large $\rho$ the equation becomes asymptotically

$$\frac{d^2u(\rho)}{d\rho^2} = -\sigma u(\rho).$$

It has the general solution, for $E \neq 0$,

$$u(\rho) = Ce^{\pm\sqrt{E/|E|}\,\rho}$$

Two possibilities:

(a)
$$E > 0 \implies \sigma = 1 \implies u(\rho) = Ce^{\pm i\rho}.$$

This represents for both signs a plane wave, which does not lead to a square integrable solution and is therefore physically unacceptable. Note that to check square integrability of the total wave function one has to integrate the variable $r$ over the integration element $r^2 dr$.

(b)

$$E < 0 \quad \Rightarrow \quad \sigma = -1 \quad \Rightarrow \quad u(\rho) = Ce^{\pm \rho}.$$

The positive sign leads to a solution which goes exponentially to infinity, so is unacceptable. For the negative sign it goes exponentially to zero; this gives the correct asymptotical behaviour at infinity.

2. $\rho \to 0$ : For the wave function to be normalizable, as far as its behaviour for small $\rho$ is concerned, one may require that $\int_0^1 |f(r)|^2 r^2 dr$, or equivalently $\int_0^1 |u(\rho)|^2 d\rho$, are convergent integrals. For this it is sufficient that $u$ has a finite value for $\rho = 0$. The behaviour of the differential equation for $u$ near $\rho = 0$ is dominated by the factor $1/\rho^2$; the potential $V$ which has a factor $1/\rho$ may be neglected. The asymptotic form of the equation is then

$$\frac{d^2 u}{d\rho^2} - \frac{l(l+1)}{\rho^2} u = 0.$$

This equation has two independent solutions, $A_1 \rho^{l+1}$ and $A_2 \rho^{-1}$. Clearly only the first one is physically acceptable.

The results (1) and (2) suggest to write $u(\rho)$ as $u(\rho) = \rho^{l+1} v(\rho) e^{-\rho}$.

(7.5.3,a) **Problem** The differential equation for $u(\rho)$ determines an equation for $v(\rho)$, namely

$$\rho \frac{d^2 v}{d\rho^2} + 2(l+1-\rho)\frac{dv}{d\rho} + [\rho_0 - 2(l+1)]v = 0$$

Derive this equation.

It will be clear that the results from this section remain valid for more general rotationally symmetric potentials $V(r)$, which have the same asymptotic behaviour at $\rho = \infty$ and $\rho = 0$ as the Coulomb potential. The next section applies only to the Coulomb case.

### 7.5.4 A Power Series Expansion for $v(\rho)$

The angular part $g_{l,m}(\theta, \phi)$ of the eigenfunctions were obtained in Sect. 7.5.1 by expanding the function $h_{l,m}(\theta)$ in a power series, which then led from the Legendre equation to the Legendre polynomials. Here we use a similar procedure, that will bring us from the above equation for $v(\rho)$ to another system of polynomials from the classical theories of special functions, the *Laguerre polynomials*. This procedure will lead to the energy eigenvalues of the hydrogen atom, the final goal of this section, and will therefore be treated in somewhat more detail.

**(7.5.4,a) Problem** Assume that a solution of the equation is written as a power series

$$v(\rho) = a_0 + a_1\rho + a_2\rho^2 + \cdots .$$

Derive from this the recursion relation for the coefficients $a_k$, for $k = 0, 1, 2, \ldots$,

$$a_{k+1} = \frac{2(k + l + 1) - \rho_0}{(k + 1)(k + 2l + 2)} a_k.$$

Note that if a coefficient $a_k$ vanishes for a certain $k = k_0$ then all higher coefficients vanish too.

Next the asymptotic behaviour of the coefficients for large $k$. One has for the quotient of two successive coefficients

$$\frac{a_{k+1}}{a_k} = \frac{2(k + l + 1) - \rho_0}{(k + 1)(k + 2l + 2)} \rightarrow \frac{2}{k}.$$

This means that the series has an asymptotic series given by the recursion relation $a_{k+1} = \frac{2}{k}a_k$, for $k = 1, 2, \ldots$

**(7.5.4,b) Problem** Show that this leads to asymptotic coefficients $a_k = \frac{2^k}{k!}$ and therefore to an asymptotic function $u(\rho) = e^{2\rho}$.

The function $e^{2\rho}$ overrules the factor $e^{-\rho}$ appearing in the expression $u(\rho) = \rho^{l+1}v(\rho)e^{-\rho}$ put forward in the preceding subsection. This leads again to a nonnormalizable $f(r)$. The only way out is to assume that the series $a_0, a_1, \ldots$ terminates for a finite $k = k_0$. Suppose that $a_{k_0+1} = 0$ and $a_{k_0} \neq 0$. The recursion relation, which takes care that all higher coefficients also vanish, then implies the relation

$$\rho_0 = 2(k_0 + l + 1).$$

This formula means that for each given $l$ there is a series of polynomials indexed by $k_0 = 1, 2, \ldots$

The energy eigenfunctions $\psi$ and (in principle) the eigenvalues of the H-atom depend on the quantum numbers $k_0, l, m$. It is convenient to use instead of the pair $(k_0, l)$ a pair $(n, l)$, with $n$, the *principal quantum number*, defined as

$$n = k_0 + l + 1,$$

with range $n = 1, 2, \ldots$

## 7.5.5 The Energy Eigenvalues and Eigenfunctions

We have $\rho_0 = 2n$, which, by using the data of Sect. 7.5.2 and the fact that $E$ is negative, leads immediately to the energy eigenvalues

$$E_n = \frac{1}{n^2}E_1, \qquad E_1 = -\frac{m}{2}\left(\frac{e^2}{4\pi\epsilon_0\hbar}\right)^2,$$

for $n = 1, 2, \ldots$, with $m_e$ now the mass of the electron, to distinguish it from the magnetic quantum number $m$. The energy eigenfunctions depend on all three quantum numbers $(n, l, m)$, the eigenvalues only on the principal quantum number $n$.

(7.5.5,a) **Problem** Show that the degeneracy of $E_n$ is $n^2$.

*Remark* The groundstate $\psi_{1,0,0}$, or rather its energy, is nondegenerate. This is a fairly general feature of quantum systems, but there are interesting exceptions.

We have shown in the preceding subsection that the solutions of the equation for $u(\rho)$ derived in problem (7.5.3,a) are polynomials. The explicit form of these polynomials can be found by various methods, which we shall not discuss. They were already found in the second half of the nineteenth century by several mathematicians among them Edmond Laguerre in 1879, and are called after him the *Laguerre polynomials*

$$L_q(z) = e^z \frac{d^q}{dz^q}(e^{-z}z^q),$$

together with *associated or generalized Laguerre polynomials*.

$$L_{q-p}^p(z) = (-1)^p \frac{d^p}{dz^p} L_q(z).$$

With these and the data from Sect. 7.5.1 the complete normalized eigenfunction $\psi_{n,l,m}$ becomes

$$\psi_{n,l,m}(r, \theta, \phi) = f_{n,l}(r)g_{l,m}(\theta, \phi)$$

$$= \sqrt{\left(\frac{2}{na_0}\right)^3 \frac{(n-l-1)!}{2n[(n+l)!]^3}}\, e^{-r/na_0} \left(\frac{2r}{na_0}\right)^l$$

$$L_{n-l-1}^{2l+1}\left(\frac{2r}{na_0}\right) Y_l^m(\theta, \phi),$$

for the *principal quantum number* $n = 1, 2, 3, \ldots$, the *azimuthal quantum number* $l = 0, 1, 2, \ldots, n-1$ and the *magnetic quantum number* $m = -l, \ldots, +l$, and with the *Bohr radius*

$$a_0 = \frac{4\pi\epsilon_0\hbar^2}{m_e e^2} = 0.529172 \times 10^{-10}\, m.$$

(7.5.5,b) **Problem** Show that the ground state $\psi_{1,0,0}$ is

$$\psi_{1,0,0}(r, \theta, \phi) = \frac{1}{\sqrt{\pi a_0^3}} e^{-r/a_0},$$

and the corresponding probability density

$$|\psi_{1,0,0}(r, \theta, \phi)|^2 = \frac{1}{\pi a_0^3} r^2 e^{-2r/a_0}.$$

Show that this density has its minimum at $r = 0$ and its maximum at $r = a_0$.

This result means that the Bohr radius $a_0$ can be interpreted as the 'size' of the hydrogen atom.

## 7.6 Transitions in the Hydrogen Atom

In Bohr's model of the hydrogen atom the revolutionary idea was that the electron moves in discrete orbits around the nucleus. However, to this assumption was added the even more important idea that once in a while the electron 'jumps' between two orbits with energies $E'$ and $E''$, resulting in the emission or absorption of packets of radiation, with an energy $\hbar\nu = E' - E''$.

In this situation there are essentially three types of transitions:

1. Absorption,
2. Stimulated emission,
3. Spontaneous emission.

The first two can be described in an approximate but fairly satisfactory manner as due to interaction of the electron with a given classical electromagnetic field, possibly time-dependent. The method used for this is called *time-dependent perturbation theory*. General perturbation theory will be briefly discussed in Sect. 14.4. Spontaneous emission is due to vacuum excitations in the background, which are always present, even in the vacuum situation. It can only be understood in the context of quantum electrodynamics. This falls outside the scope of this book.

## 7.7 The Hydrogen Atom: More Sophisticated Descriptions

So far the hydrogen atom has been described as a one particle system, an electron moving in the Coulomb field of the nucleus, fixed in the origin of the coordinate system. It is easy to describe it as a two particle system.

The nucleus is a positively charged particle, a proton, with mass $m_1$; the electron is much lighter, negatively charged, with mass $m_2$. The Hamiltonian is

$$H = \frac{\mathbf{p}_1^2}{2m_1} + \frac{\mathbf{p}_2^2}{2m_2} + V(\mathbf{r}_1 - \mathbf{r}_2),$$

with $\mathbf{r}_1$ the position of the nucleus, $\mathbf{r}_2$ that of the electron, and with $V(\mathbf{r}_1 - \mathbf{r}_2)$ the Coulomb potential between the two particles. We introduce centre of mass coordinates

$$\mathbf{r} = \mathbf{r_1} - \mathbf{r_2}, \qquad \mathbf{R} = \frac{m_1 \mathbf{r_1} + m_2 \mathbf{r_2}}{m_1 + m_2}.$$

(7.7,a) **Problem** Show that the Hamiltonian in the space of wave functions $\psi(\mathbf{r}, \mathbf{R})$ becomes

$$H = \frac{\mathbf{p}^2}{2m} + \frac{\mathbf{P}^2}{2M} + V(\mathbf{r}),$$

with the new momenta

$$\mathbf{P} = \mathbf{p_1} + \mathbf{p_2}, \qquad \mathbf{p} = \frac{m_2 \mathbf{p_1} - m_1 \mathbf{p_2}}{m_1 + m_2},$$

and new masses

$$M = m_1 + m_2, \qquad \mu = \frac{m_1 m_2}{m_1 + m_2}.$$

The physical interpretation of the transformation to this new description is obvious. The motion is split up into the motion of the centre of mass of the system, which is free, and the more interesting relative movement of the electron with respect to the centre of mass. The two particle description of the hydrogen atom is slightly more realistic than the description with the proton fixed at the origin that we have given in Sect. 7.5; the mass of the electron is $9.10939 \times 10^{-31}$ kg, that of the proton $1.67262 \times 10^{-27}$ kg, so the electron mass is 0.000544 of the proton mass. This means that the results for the hydrogen spectrum in Sect. 5.5, which are, strictly speaking, based on the reduced or relative mass $\mu$, are only very slightly different from those based on the true mass.

In certain regions of the atom the velocity of the electron may be so high that relativistic effects must be taken into account. The Schrödinger equation is nonrelativistic, so it is obvious to use in this situation a relativistic wave equation like the Klein-Gordon or the Dirac equation, to be discussed in Chap. 15. This is not without its problems; nevertheless it is possible by using these equations, or by more pedestrian methods, to calculate relativistic corrections for the energy levels. The eigenvalues no longer depend only on the principal quantum number $n$, but become degenerate. The first order splitting of the levels is in this case called the *fine structure* of the spectrum.

The most important improvement in the quantum mechanical description of the hydrogen atom is obtained by considering its *spin*, an intrinsic angular momentum with a magnitude given by the number $s = \frac{1}{2}$. The presence of the spin in itself does not change the energy levels. However, a magnetic moment is associated with angular momentum, both with the orbital type, treated in this chapter, as with the spin, to be discussed in the next. In the presence of an external magnetic field, this will lead to various possible directions for the particle, in or against the direction of the field, with as result different energies. The resulting field strength depending level splitting is called the *Zeeman effect*. As this effect is a physically realistic phenomenon only when the spin is included, it will not be described here, but in the next chapter.

The hydrogen atom is one of the central examples in the teaching of quantum mechanics. As such the solution of its eigenvalue and eigenvector problem, details of its further properties, etc., are treated in all quantum mechanics textbooks, for example the books mentioned in the reference section of Chap. 3. Some of these books contain much more details than given here.

Three historically important papers:

# References

1. Bohr, N.: On the constitution of atoms and molecules, Philosophical Magazine **26**, 1–25 (1913). http://hermes.ffn.ub.es/luisnavarro/nuevo_maletin/Bohr_1913.pdf
2. Rutherford, E.: The scattering of $\alpha$ and $\beta$ particles by matter and the structure of the atom, Philosophical Magazine **6**, 669–688 (1911). http://www.ffn.ub.es/luisnavarro/nuevo_maletin/Rutherford%20(1911),%20Structure%20atom%20.pdf
3. Schrödinger, E.: [7] Quantisierung als Eigenwertproblem (Erste Mitteilung). Ann. Phys. **79**(4), 361–376 (1926) English Trans.: Quantisation as a Problem of Proper Values (Part I) Collected Papers on Wave Mechanics, pp. 1–12. Chelsea Publishers (1982). http://einstein.drexel.edu/~bob/Quantum_Papers/Schr_1.pdf

# Chapter 8
# Spin: Atomic Structure

## 8.1 Spin

### 8.1.1 Introduction

The simple quantum mechanical model of the hydrogen atom, discussed in the preceding chapter, explains the essence of the Bohr atomic model, and stands as such at the beginning of atomic physics. Its main feature is the electron, an electrically charged point particle moving in the attractive Coulomb potential of a fixed nucleus. In this chapter we turn to further structural properties of the hydrogen atom and of more general atoms.

The most important additional element of the structure of the hydrogen atom and of atomic physics in general is that in quantum mechanics all particles, elementary or composite, have an additional, non-classical degree of freedom, an intrinsic angular momentum, which is called *spin*. An electron, our obvious example of a particle with spin, has no internal structure; it is a point particle, so the intuitive idea suggested by the term "spin", that of a particle spinning around its axis, is wrong. There is no kinetic energy connected with this 'spinning'.

Spin can only be understood as a purely mathematical notion. As such it has the properties of angular momentum; it is described by a vector operator $\mathbf{S}$, with three components $(S_1, S_2, S_3)$, with the standard angular momentum commutation relations

$$[S_1, S_2] = i\hbar S_3, \qquad [S_2, S_3] = i\hbar S_1, \qquad [S_3, S_1] = i\hbar S_2.$$

There is a total spin operator $\mathbf{S}^2 = S_1^2 + S_2^2 + S_3^2$, which commutes with all three components, with possible eigenvalues $s(s+1)\hbar^2$, for $s = 0, \frac{1}{2}, 1, 1\frac{1}{2}, \ldots$. Note that contrary to the case of orbital angular momentum the quantum number of spin angular momentum can take half-integer values. In fact, the electron has $s = \frac{1}{2}$. Composite particles can have any integer or half-integer value for $s$; the $\pi_0$ meson,

© Springer International Publishing Switzerland 2015
P. Bongaarts, *Quantum Theory*, DOI 10.1007/978-3-319-09561-5_8

a particle found in cosmic radiation and also produced in high energy particle accelerators has $s = 0$, the photon, the particle for the quantum theoretical description of electromagnetic radiation, has $s = 1$; the graviton, a particle that is supposed to describe the gravitational field, has $s = 2$. The great majority of the elementary particles, certainly those that appear in atomic physics, have $s = \frac{1}{2}$.

Due to their angular momentum, both orbital and spin, particles behave as little magnets: they have a *magnetic moment*. This gives rise to an important phenomenon, the Zeeman effect, to be discussed in Sect. 8.4.

There is an important distinction between particles with integer spin and particles with half-integer spin. The first are called *bosons*, the second *fermions*. The difference between these two types of particles has far-reaching physical consequences and will be discussed in the next chapter.

### 8.1.2  Historical Remarks

An early experiment in the direction of the discovery of spin was performed by Otto Stern and Walther Gerlach in 1922. This was before the full development of quantum mechanics. It suggested an additional internal quantum number for particles, but without the idea that some sort of angular momentum was involved. In this experiment a beam of silver atoms pass through a narrow horizontal slit in which there is a strong inhomogeneous magnetic field in vertical direction. The particles then reach a photographic plate on which they deposit two distinct dots. The atoms, as small magnetic dipoles, are deflected by the field. In the language of quantum mechanics, this setup measures their magnetic moment, i.e. their intrinsic angular momentum, in the $x_3$-direction. Due to the structure of its system of electrons, the total spin of a silver atom is that of a single electron, namely $s = \frac{1}{2}$, while the total orbital angular momentum is 0.

Next, Wolfgang Pauli, when formulating his *exclusion principle* (see next chapter), thought that there might indeed be an extra internal quantum number involved. However, he did not pursue this point further. Ralph Kronig, a German-American physicist, had heard about this idea from Pauli, and came up with the idea that this might be due to an additional kind of rotation, an angular momentum. His seniors, Bohr and also Pauli himself, thought this was too fanciful and advised against publishing it. Then in 1925, George Uhlenbeck and Samuel Goudsmit, two Leiden theoreticians, independently had the same idea. According to Paul Ehrenfest, their supervisor, young physicists were allowed to make fools of themselves, so he let them publish their idea, which turned out to have a broad response and which was soon universally accepted. Hendrik Casimir, also from Leiden, finally gave the group theoretical background for the general notion of angular momentum. That is why operators such as $\mathbf{L}^2$ are called *Casimir operators*. See the preceding chapter, Sect. 4.2.

### 8.1.3 The Description of a Single Particle with Spin

All particles, or systems of particles, have spin, i.e. in principle, because it may be zero; in atomic physics usually with $s = \frac{1}{2}$. To understand its properties, in particular its interaction with the orbital angular momentum, we may start with the simple example of the hydrogen atom, as a one particle system, described in the preceding chapter.

The Hilbert space of the system is now no longer the same as in the spinless case of the preceding chapter; in a certain rough sense it is 'larger' because of the addition of operators describing the new spin observable. Therefore first a few remarks about this new variable. As already discussed in Sect. 8.1.1 there is a vector operator $\mathbf{S}$ with three components, with the standard $su(2)$ commutation relations. As a complete commuting system of operators one may use the operator $\mathbf{S}^2$ combined with a $S_j$. We shall use for this $S_3$.

The new Hilbert space can be written as a tensor product

$$\mathcal{H} = \mathcal{H}_0 \otimes \mathcal{H}_{\text{spin}},$$

with $\mathcal{H}_0 = L^2(\mathbb{R}^3, d\mathbf{x})$, the original Hilbert space of the particle without spin, and $\mathcal{H}_{\text{spin}}$ the two dimensional complex Hilbert space on which the operators $S_1$, $S_2$ and $S_3$ act as generators of the basic representations of the group $SU(2)$. The elements of $\mathcal{H}_{\text{spin}}$ are often called *spinors*. We use an orthonormal basis $(\chi_+, \chi_-)$, consisting of eigenvectors of $S_3$ for the eigenvalues $(+\frac{\hbar}{2}, -\frac{\hbar}{2})$. A general vector $\psi$ of $\mathcal{H}$ can be written as

$$\psi = \psi_1 \otimes \chi_+ + \psi_2 \otimes \chi_-,$$

with a wave function of the form $\psi(\mathbf{x}, \rho) = \psi_\rho(\mathbf{x})$, with $\rho$ an additional variable, taking the values $\pm 1$. The inner product of two such wave functions is written as

$$(\psi, \phi) = \sum_{\rho = \pm 1} \int_{\mathbf{x} = -\infty}^{+\infty} \overline{\psi_\rho(\mathbf{x}, \rho)}\, \phi_\rho(\mathbf{x}, \rho)\, d\mathbf{x}.$$

The operators that are expressions in position and momentum operators, such as $H$ and $\mathbf{L}$, act on the first factor in the tensor product, or more explicitly, on the variables $\mathbf{x} = (x_1, x_2, x_3)$. The spin operators act on the second factor, explicitly, with respect to the basis $\chi_\pm$, as two dimensional matrices on what are called *Pauli spinors*

$$S_1 = \frac{1}{2}\hbar \begin{pmatrix} 0 & 1 \\ 1 & 0 \end{pmatrix}, \quad S_2 = \frac{1}{2}\hbar \begin{pmatrix} 0 & -i \\ i & 0 \end{pmatrix}, \quad S_3 = \frac{1}{2}\hbar \begin{pmatrix} 1 & 0 \\ 0 & -1 \end{pmatrix}$$

The three $2 \times 2$ matrices in these formulas are the *Pauli $\sigma$-matrices*, appearing in many places in quantum mechanics, and also mentioned in Supp. Sect. 24.8.4.

In this description one may use a complete system of commuting observables consisting of the Hamiltonian $H$ and the orbital angular momentum operators $\mathbf{L}^2$, $L_3$, acting on the first factor of the tensor product, and the spin angular momentum operator $S_3$, acting on the second factor. Note that the spin operator $\mathbf{S}^2$ can be omitted because in this single electron case it is equal to $\frac{3}{4}\hbar^2$ times the unit operator. The energy eigenfunctions are characterized by the quantum numbers $n$, $l$, $m_l$ and $m_s$. There is a second complete system of commuting observables, namely $H$, $\mathbf{J}^2$, $\mathbf{L}^2$, $J_3$, with $\mathbf{J} = \mathbf{L} + \mathbf{S}$ the total angular momentum operator and $J_3$ its third component; the energy eigenfunctions have quantum numbers $n$, $j$, $l$, $m_j$.

(8.1.3,a) **Problem** Show that, in terms of the underlying tensor product structure of the full Hilbert space $\mathcal{H} = \mathcal{H}_0 \otimes \mathcal{H}_{\text{spin}}$, the operator $\mathbf{J}^2$ is equal to,

$$\mathbf{J}^2 = \sum_{j=1}^{3}(L_j^2 \otimes 1 + 2L_j \otimes S_j + 1 \otimes S_j^2).$$

(8.1.3,b) **Problem** Verify the commutation relations for the $J_j$.

(8.1.3,c) **Problem** Show that the operators of the second set indeed commute.

Which set to use depends on further properties of the situation, as will be discussed in the next sections.

In the simplest description and in the absence of external fields the spin does not contribute to the energy, so the Schrödinger equation remains the same. If there is an external electromagnetic field, described by a scalar potential $\phi$ (Coulomb force) and a vector potential $\mathbf{A} = (A_1, A_2, A_3)$ (Lorentz force) one has to use the Schrödinger-Pauli equation, in the two component form of the wave function,

$$i\hbar\frac{\partial}{\partial t}\psi = \left[\frac{1}{2m}(\sigma \cdot (\mathbf{p} - e\mathbf{A}))^2\right]\psi,$$

with the Pauli matrices $\sigma = (\sigma_1, \sigma_2, \sigma_3)$ and the operator for the momentum three vector $\mathbf{p} = (-i\hbar\nabla_1, -i\hbar\nabla_2, -i\hbar\nabla_3)$. Note that the nonrelativistic spin, discussed in this chapter, is sometimes called the *Pauli spin*. The relativistic *Dirac spin*, connected with the Lorentz group, or rather its universal covering group $SL(2, \mathbb{C})$ instead of $SU(2)$, will be briefly discussed in Chap. 15. See for the notion of covering group and universal covering group Supp. Sect. 24.5.

## 8.2 Structure of Atoms in General

### 8.2.1 Introduction

In this section we look at atoms more complicated than the hydrogen atom, i.e. atoms with more than one electron. The model used for such atoms is inspired by what we know about the hydrogen atom; as such it is valid in an approximation which becomes less good with increasing atomic number $Z$ (Remember that $Z$ is the number of protons in the nucleus, and, of course, also the number of electrons surrounding it).

*Remark* At the present day we have a fairly detailed knowledge of the various sorts of atoms and of their interactions with each other. This knowledge is based on complicated calculations by means of approximate schemes, and on the experimental—e.g. spectroscopic—confirmations of these theoretical predictions.

A classic and still very useful account of the theory of atomic structure is given in the book by Herzberg [1].

### 8.2.2 Electron Shells and Subshells

In the absence of external influences, an electron will be in the lowest energy eigenstate that is available, which for the single electron of hydrogen is its ground state. For an atom with more than one electron the situation is different, because of *Pauli's exclusion principle* according to which two electrons in an atom cannot be in the same state, i.e. with the same quantum numbers, say $n$, $l$, $m_l$ and $m_s$. This means that only two electrons can be in their ground state, i.e. with $n = 1, l = 0, m_l = 0$ and $m_s = \pm 1$, the others will be in states with higher energy.

Pauli's exclusion principle is one of the building blocks of atomic physics; its background is a more general quantum mechanical principle concerning the description of systems of identical particles. All this will be extensively discussed in the next chapter.

Following the above principle the electrons in an atom are grouped in *shells*, one for each principal quantum number $n$. They are denoted by capitals as K, L, M, N, O and Q, for $n = 1, 2, 3, 4, 5, 6$ and 7; the electrons in these shells just by the numbers $1, 2, 3, \ldots$. As the electrons prefer the states with the lowest energy, so the innermost shel—with the lowest energy—is filled first; it can hold only two, the next one eight electrons, then 18, and so on. The outside shell need not be completely filled. The electrons in such an incomplete shell are called *valence electrons*; they can pair with similar electrons from neighbouring atoms; as a result molecules consisting of several atoms can be formed. Atoms with an incomplete highest shell will be chemically active; if this shell is closed, the atom will be chemically inert. In this way the chemical properties of atoms depend very much on their electron configuration. There are exceptions to these rules; atoms sometimes may have more than one incompletely filled shell. Finally, electrons can move between shells, by emission or absorption of a photon.

(8.2.2,a) **Problem** Neon has atomic number $Z = 10$. It is a very inert gas. Explain this by its electron shell system.

(8.2.2,b) **Problem** Show that the number of possible states in the $n$th shell is $2n^2$.

An electron shell consists of *subshells*, in which the electrons with a certain orbital angular moment *l* are collected. The electrons in these subshells are denoted by lower-case letters s, p, d, f, and then g, h, i, . . . . The first four letters have a historical origin, dating from before the advent of quantum mechanics, with purely a spectroscopic meaning: s = sharp, p = principal, d = diffuse, f = fundamental. The others are just an alphabetical continuation of the first four. Note that j is omitted.

## 8.3  Interactions Inside the Atom

### 8.3.1 Introduction

The Hamiltonian of a general many electron atom can be written as

$$H = H_0 + H_1 + H_2 + H_3 + H_4 + \cdots ,$$

with $H_0$ the kinetic energy, $H_1$ the Coulomb attraction between the electrons and the nucleus, $H_2$ and $H_3$ the Coulomb repulsion and spin-orbit interactions between the electrons. The term $H_4$ represents further effects, such as relativistic ones and the interaction between the electrons and the nucleus other than the Coulomb force. These effects will not be discussed here.

The first two terms gives the *central field approximation*. This determines in a general and fairly rough manner the structure of the system of energy levels, as specified by $n$, the principal quantum number, and $l$, the quantum number for the orbital angular momentum. For the hydrogen atom the energy depends only on $n$; for all other atoms it depends also on $l$. The 'magnetic quantum numbers' $m_l$ and $m_s$ do not give further splitting of the energy levels unless there is an external magnetic field present, like in the case of the Zeeman effect (Sect. 8.4).

The term $H_3$ describes the interactions by the orbital and spin angular momenta between the electrons. The particulars of this description depend on the size of the atom. For this there are two coupling schemes, the *Russell–Saunders coupling* or *LS coupling* and the *jj coupling scheme*.

### 8.3.2 The Russell–Saunders Coupling

For light atoms, roughly up to $Z = 30$, the electrons move in the neighbourhood of the nucleus, where their orbital angular momenta $\mathbf{L}^{(j)}$ interact with each other. The same is the case for the spin angular momenta $\mathbf{S}^{(j)}$. For calculations all orbital angular momenta are added to total orbital angular momenta $\mathbf{L} = \sum_j \mathbf{L}^{(j)}$ and all the spin angular momenta to a total spin angular momentum $\mathbf{S} = \sum_j \mathbf{S}^{(j)}$. Then $\mathbf{L}$ and $\mathbf{S}$ interact through the term $\mathbf{L} \cdot \mathbf{S}$. The system of commuting operators that are added to the Hamiltonian consists of

$$\mathbf{L}^2, \quad L_3, \quad \mathbf{S}^2, \quad S_3,$$

with quantum numbers $l, m_l, s, m_s$. Note that in the case of $Z = 1$, the hydrogen atom, the operator $\mathbf{S}^2$ acts as multiplication by a fixed scalar $\frac{3}{4}\hbar^2$; so $\mathbf{S}^2$ and $s$ can be omitted. The quantum numbers $m_l$ and $m_s$ are called 'magnetic quantum numbers'.

## 8.3.3 The $jj$ Coupling

In this scheme, which in the main has to be used for heavier atoms, the individual orbital and spin angular momenta are coupled, according to $\mathbf{J}^{(j)} = \mathbf{L}^{(j)} + \mathbf{S}^{(j)}$. These are then summed to a collective angular momentum $\mathbf{J} = \sum_j \mathbf{J}^{(j)}$. The commuting operators that are now combined with $H$ are

$$\mathbf{J}^2, \quad J_3, \quad \mathbf{L}^2, \quad \mathbf{S}^2,$$

with quantum numbers $j, m_j, l, s$. Note that in this case $m_j$ is the only 'magnetic quantum number'. Again $\mathbf{S}^2$ and $s$ can be omitted in the case of the hydrogen atom.

(8.3.3,a) **Problem** Show that $\mathbf{J}^2$ does not commute with either $L_3$ or $S_3$, implying that $j$ is not a good quantum number in the Russell Saunders coupling scheme.

All this is on the calculation of energy levels. However, what is directly observed are the transitions between these levels, by emission or absorption of radiation, with frequencies $\omega$ given by the relation

$$\hbar\omega = |E_{\text{final}} - E_{\text{initial}}|.$$

Not all combinations of levels lead to transitions. There are so called *selection rules*, which forbid certain transitions. In the case of the Russell–Saunders coupling, for example, one has, because of rotation symmetry, for the hydrogen atom

$$\Delta l = \pm 1, \quad \Delta m_l = 0, \pm 1,$$

which is in agreement with the fact that the unit of electromagnetic radiation, the photon, has spin $s = 1$. Selection rules are rarely absolute. Higher order calculations may lead to weak transitions, which are forbidden in lower order.

In conclusion we may say that the inner structure of an atom, even of a simple atom like the hydrogen, is very complicated. To study its various approximation methods have been developed over the years, from the fairly obvious to very sophisticated ones, only feasible with modern computer power. This is even more true for the next step, the study of the interaction of atoms, which means the vast subject of *quantum chemistry*.

## 8.4 The Zeeman Effect

### 8.4.1 Introduction

Electrons behave like small magnetic dipoles. As such they have magnetic moments $\mu$, connected with the orbital and spin angular momenta. In an external magnetic field they direct themselves, either in the direction of, or opposite to this field. Because of this, energy levels will split, with a corresponding splitting of spectral lines, which is proportional with the strength of the field, as long as it is not too strong. For instance, sharp spectral lines such as the $n = 3 \rightarrow 2$ transition in hydrogen split into a system of closely spaced lines. This is the *Zeeman effect*.

### 8.4.2 Main Properties of the Zeeman Effect

The magnetic moment due to the orbital angular momentum is given by the operator relation

$$\mu_{\mathbf{L}} = -g_l \frac{e}{2m} \mathbf{L},$$

with $g_l$ the *orbital gyromagnetic factor* which is equal to one. One may think of the classical picture where such a magnetic moment is the result of a circular electric current. In addition to this there is the magnetic moment due to the spin angular momentum, for which this classical picture does not make sense. It is given by the relation

$$\mathbf{\bar{}}_{\mathbf{S}} = -g_s \frac{e}{2m} \mathbf{S},$$

with $g_s$ the *spin gyromagnetic factor* which is different from $g_l$ and is in very good approximation equal to 2. After their discovery of the spin of the electron Uhlenbeck and Goudsmit found that assigning the value $g_s = 2$ was in agreement with the observed spectral properties. The small deviation from the value 2 is an effect which can be explained by quantum electrodynamics, which is outside the context of this book. Experimentally $g_s$ has been found to be $g = 2.002319304386$, in precise agreement with the calculated values. This is one of the best tests of the validity of quantum electrodynamics.

An external magnetic field **B** will cause a change in the energy according to

$$\Delta E = \frac{e}{m}(\mathbf{L} + 2\mathbf{S}) \cdot \mathbf{B},$$

or, if we take the field **B** in the $x_3$-direction, for a state with total angular momentum $m_j$ in this direction,

$$\Delta E = \mu_B \, m_j \, B,$$

with $\mu_B$ the *Bohr magneton*, the unit of atomic magnetic moment, equal to $\mu_B = \frac{e\hbar}{2m_e}$. As a result, spectral lines, corresponding with transitions between such levels, will be split, proportional to the strength of $B$—as long as $B$ is not too large. The transitions, such as occur in hydrogen, follow the selection rule $\Delta m_l = \pm 1$.

There is a distinction between *normal* and *anomalous* Zeeman effect. The first occurs in transitions between two singlet states. A singlet state is a state with (total) spin $s = 0$. The most common such state is obtained by reducing the four dimensional tensor product space $V_{s=\frac{1}{2}} \otimes V_{s=\frac{1}{2}}$ of the spins of two $s = \frac{1}{2}$ particles as a direct sum of two irreducible representations of $su(2)$, as $V_{s=0} \oplus V_{s=1}$. The first term is a singlet, the second a triplet. Such a singlet is sometimes called an EPR state, because of its role in the Einstein-Rosen-Podolsky paradox, which will be discussed in Chap. 17. In the case of the hydrogen atom in the ground state there is just the two dimensional representation space, the basic representation of $su(2)$, for the electron spin states with $m_s = \pm \frac{1}{2}$. They form a doublet.

All transitions in an atom in a magnetic field in which either the initial state or the final state, or both, are not singlets, are split according to the anomalous Zeeman effect. This splitting is more complicated than that of the normal case. The terms 'normal' and 'anomalous' have lost much of their original meaning, because there is a single underlying mechanism. The 'anomalous' Zeeman effect is just the Zeeman effect with spin included.

### 8.4.3 Historical Remark

In the years 1896–1897 the Dutch physicist Pieter Zeeman did experiments with sodium chloride in a magnetic field and observed the splitting of each of the two yellow D-lines. The simplest form of this phenomenon—the *normal Zeeman effect*—was immediately well understood in the context of the then emerging theory of the electron, developed in particular by H.A. Lorentz, but the explanation of a second more complicated version—the *anomalous Zeeman effect*—had to wait until the discovery of the spin of the electron in 1925.

All books on quantum mechanics discuss spin, and more generally, the structure of atoms, in various slightly different ways and with slightly different emphasis.

### Reference

1. Herzberg, G.: Atomic Spectra and Atomic Structure, 2nd edn. Dover, New York (2010) (A great classic, of historical value, but still very instructive. The 2010 photocopy edition is rather poor, so the earlier 1944 Dover edition is much to be preferred; it is available at https://ia701208.us.archive.org/24/items/AtomicSpectraAtomicStructure/ Herzberg-AtomicSpectraAtomicStructure.pdf.)

# Chapter 9
# Many Particle Systems

## 9.1 Introduction

So far we have discussed single particles, or systems of two particles, such as the two particle model of the hydrogen atom in Chap. 7 and the properties of spin in Chap. 8. This chapter is devoted to general $n$-particle systems. The mathematical background for this is the theory of $n$-fold tensor products of Hilbert spaces. This together with its physical meaning will be discussed in the next section. After this, in Sect. 9.3, attention is focused on the important special case of identical particles. Mathematically this means symmetric and antisymmetric tensor products. A new important quantum phenomenon appears: the existence of bosonic and fermionic particles. The mathematical and physical aspects of this are treated in Sect. 9.3, and the helium atom, with its system of two electrons, appears as an example in Sect. 9.4. The development of the understanding of the physics of identical particles is one of the great stories in the history of quantum mechanics.

The second half of this chapter, from Sect. 9.5 onwards, is devoted to a formalism for the description of systems of $n$ identical particles, simultaneously for all $n = 0, 1, \ldots$ It is called the "Fock space formalism", after its inventor, the Russian theoretical physicist Fock, or 'second quantization', a conceptually confusing name used in many physics textbooks. This topic has two faces: a mathematical one based on tensor algebras as direct sums of symmetric or antisymmetric tensor products, and a very different looking one, appearing in physics textbooks, heuristic, but elegant and transparent, and moreover pointing to the connection with quantum field theory.

Supp. Chap. 23 contains material on tensor products and tensor algebras necessary for this chapter.

© Springer International Publishing Switzerland 2015
P. Bongaarts, *Quantum Theory*, DOI 10.1007/978-3-319-09561-5_9

## 9.2 Combining Quantum Systems—Systems of $N$ Particles

New Hilbert spaces can be constructed from given ones. In quantum theory two such constructions play a role, direct sums and tensor products. These two correspond to two different ways of looking at composite quantum systems.

1. Let $\mathcal{H}_1$ and $\mathcal{H}_2$ be the Hilbert state spaces of two quantum systems, for instance of two particles. The system in which both particles do not appear simultaneously, but successively or alternatively, is described by the direct sum $\mathcal{H} = \mathcal{H}_1 \oplus \mathcal{H}_2$. This procedure can be generalized to an arbitrary finite or countably infinite number of Hilbert spaces $\mathcal{H}_j$. Direct sums are briefly discussed in Supp. Sect. 21.5.

For the Hamiltonian operators $H_1$ and $H_2$ of the two systems, the Hamiltonian for the composite system is the *sum operator* $H = H_1 \oplus H_2$, which can be written in terms of column vectors with elements from $\mathcal{H}_1$ and $\mathcal{H}_2$ as entries, as

$$H \begin{pmatrix} \psi_1 \\ \psi_2 \end{pmatrix} = \begin{pmatrix} H_1\psi_1 \\ H_2\psi_2 \end{pmatrix}.$$

The total Hamiltonian $H$ is then a $2 \times 2$ matrix with operators as matrix elements:

$$H = \begin{pmatrix} H_1 & 0 \\ 0 & H_2 \end{pmatrix}.$$

General operators in this direct sum have the form

$$A = \begin{pmatrix} A_{11} & A_{12} \\ A_{21} & A_{22} \end{pmatrix},$$

with $A_{jk}$ a linear map from $\mathcal{H}_k$ into $\mathcal{H}_j$. If the Hamiltonian $H$ has the above 'diagonal' form there is no dynamic interaction between the two subsystems; there are no transitions in time. In this case one says that the system has a *selection rule*. Often this is only a first approximation; 'off-diagonal' terms of higher order in an appropriate parameter in the total Hamiltonian will break the selection rule. Remember the role of selection rules in atomic physics, discussed in the preceding chapter, in Sects. 8.3 and 8.4.

It may happen that not only the Hamiltonian but *all* operators that have a physical meaning in $\mathcal{H}_1 \oplus \mathcal{H}_2$ have 'diagonal' form. This would mean, not only that there is no dynamical interaction, but that there is no other meaningful physical relation between the two systems. In general a linear combination of two state vectors represents, after normalization, again a physical state. In this particular case the direct sum Hilbert space $\mathcal{H}_1 \oplus \mathcal{H}_2$ is a mathematical concept, with no physical meaning, i.e. linear combinations $\lambda_1\psi_1 + \lambda_2\psi_2$, $\psi_1 \in \mathcal{H}_1$, $\psi_2 \in \mathcal{H}_2$, are not physical states. We say that there is a *superselection rule* in $\mathcal{H}$, with the subspaces $\mathcal{H}_1$ and $\mathcal{H}_2$ called *superselection subsectors* or *coherent subspaces* of the system. In the next section we shall discuss an important example of this phenomenon. The notion of superselection rule was introduced in a paper by Wick and Wightman [1]. For a more recent exposition, which takes into account modern developments, see the lecture notes of Fredenhagen [2].

Note that all this goes through for an $n$-tuple of Hilbert spaces $(\mathcal{H}_1, \ldots, \mathcal{H}_n)$, and, with a few obvious mathematical refinements, for an infinite sequence $(\mathcal{H}_1, \mathcal{H}_2, \ldots)$.

(9.2,a) *Example* The nucleus of an atom consists of protons and neutrons. These particles have very similar properties, so in nuclear physics one sometimes describes them, in first approximation, as two different states of a single particle, the nucleon. The Hilbert space of the nucleon is the direct sum of the Hilbert space of the proton and the Hilbert space of the neutron. In this direct sum Hilbert space the neutron and the proton are distinguished by 'isobaric spin', a quantum number important in elementary particle physics, in which the word 'spin' has nothing to do with rotations in space like for ordinary spin treated in Chap. 8.

2. *Tensor products.* Let $\mathcal{H}_1$ and $\mathcal{H}_2$ be the Hilbert spaces of two quantum systems, for instance of two particles *that are assumed to be different.* The Hilbert space of the system in which the particles appear simultaneously is the tensor product space $\mathcal{H}_1 \otimes \mathcal{H}_2$. For a quantum system described by an $n$-tuple of Hilbert spaces $(\mathcal{H}_1, \ldots, \mathcal{H}_n)$ one constructs the tensor product $\mathcal{H}_1 \otimes \cdots \otimes \mathcal{H}_n$. The mathematical properties of tensor products are extensively discussed in Supp. Chap. 23.

If system 1 has a time evolution operator $U_1(t) = e^{-\frac{i}{\hbar}tH_1}$ with Hamiltonian operator $H_1$, and system 2 has $U_2(t) = e^{-\frac{i}{\hbar}tH_2}$ with $H_2$, *and if there is no interaction between the two systems*, then the time evolution of the total system is described by the operator $U(t) = e^{-\frac{i}{\hbar}tH}$, with $U(t)$ the tensor product operator $U_1(t) \otimes U_2(t)$ and total Hamiltonian $H = H_1 \otimes 1_{\mathcal{H}_2} + 1_{\mathcal{H}_1} \otimes H_2$. More interesting is the situation in which the two systems interact. In that case an interaction Hamiltonian $H_{12}$ is added to $H = H_1 + H_2$ and the resulting evolution operator has no longer the form of a 'pure tensor' operator.

(9.2,b) *Example* The simplest model for the hydrogen atom (in Chap. 7) is that of a particle, an electron, moving in a fixed central potential $V(|\mathbf{r}|)$. The Hilbert space for this system is $\mathcal{H} = L^2(\mathbb{R}^3, d\mathbf{r})$. In a slightly more realistic description the hydrogen atom is a two particle system consisting of the nucleus, a proton with a large mass $M$ and an electron with a small mass $m$, interacting with each other through the potential $V(|\mathbf{r}_1 - \mathbf{r}_2|)$. The Hilbert space of this two particle system is the (completed) tensor product $\mathcal{H}_1 \otimes \mathcal{H}_2 = L^2(\mathbb{R}^6, d\mathbf{r}_1 d\mathbf{r}_2)$. Without the interaction the Hamiltonian of this composite system would be the sum of the terms $H_1 \otimes 1_2$ and $1_2 \otimes H_2$, written as $H_1$ and $H_2$, with

$$H_1 = -\frac{\hbar^2}{2m}\Delta_1, \quad H_2 = -\frac{\hbar^2}{2M}\Delta_2,$$

in which $\Delta_j$ is the Laplace operator in the variable $\mathbf{r}_j$, for $j = 1, 2$. The interaction gives an additional term $H_{12}$ which multiplies the two particle wave function $\psi(\mathbf{r}_1, \mathbf{r}_2)$ with $V(|\mathbf{r}_1 - \mathbf{r}_2|)$ and which, as operator, has indeed no longer the form of a pure tensor. Of course, in this particular example the translation invariance of the potential makes it possible to reduce this picture to a one particle description, for a single particle with mass $M = \frac{mM}{m+M}$ as discussed in Sect. 7.7.

(9.2,c) *Example* The Pauli spin model of the hydrogen atom also discussed in Chap. 8. The state space is the tensor product $L^2(\mathbb{R}^3, d\mathbf{r}) \otimes \mathbb{C}^2$.

Direct sums of Hilbert spaces are used in this section and the next one for the notion of 'superselection rule'; tensor products are important in all the sections that will follow.

This is enough on the description of arbitrary systems of two and in general $n$ particles. The remainder of this chapter will be devoted to the important special case of *identical particles*.


## 9.3  Systems of Identical Particles

What has been discussed so far in this chapter is not very surprising, given the general principles of quantum theory as they have been presented in the preceding chapters. A new and interesting phenomenon occurs for systems consisting of *identical* particles.

We have argued in Sect. 4.2 that particles in quantum mechanics do not have *orbits* because their position and momentum cannot be measured simultaneously with arbitrary precision. Of course, the probabilities for finding the particles can be concentrated in small disjunct areas of space; however, sooner or later these areas will overlap. All this means that in a system of $n$ identical particles the individual particles cannot be identified. The fact that we cannot observe the difference between the state consisting of a particle (1) concentrated near a point $\mathbf{r}_1$ and a particle (2) near $\mathbf{r}_2$, and a state with particle (1) near $\mathbf{r}_2$ and (2) near $\mathbf{r}_1$, does, strictly speaking, not imply that these two states are the same. However, we take as a fundamental assumption that they are indeed the same. This means, in general, that identical particles in a many particle system have no individual identity. A system of $n$ electrons is just that; speaking of a system consisting of electron 1, electron 2, and so on, does not make sense. We add therefore a sixth general axiom to the five formulated in Chap. 3. We start by splitting it up in three separate but connected statements.

**Axiom VI$_1$**: *In a system of $n$ identical particles, the particles are indistinguishable; they have no individual identity.*

This statement has important physical consequences, which are typical for quantum theory. All have been verified experimentally. Some will be discussed later. It has also consequences for the mathematical description of quantum systems. To these we turn now.

Start for the description of a system of $n$ identical particles, each in its own one particle Hilbert space $\mathcal{H}^{(1)}$, in the $n$-fold tensor product space $\otimes^n \mathcal{H}^{(1)}$. According to Axiom VI$_1$ the unit vector in $\otimes^n \mathcal{H}^{(1)}$

$$\psi = \phi_1 \otimes \cdots \otimes \phi_n,$$

represents the same physical state as the permuted vector

$$U(\sigma)\psi = \phi_{\sigma(1)} \otimes \cdots \otimes \phi_{\sigma(n)},$$

for a permutation $\sigma$ of the indices $1, 2, \ldots, n$. This means that there is an *equivalence relation* between state vectors. Two tensor product unit vectors $\psi$ and $\psi'$ in $\otimes^n \mathcal{H}^{(1)}$ are equivalent, $\psi \sim \psi'$, iff there is a permutation $\sigma$ of the $n$ indices, such that $\psi' = U(\sigma)\psi$, with $U(\sigma)$ the unitary operator in $\mathcal{H} = \otimes^n \mathcal{H}^{(1)}$ which represents $\sigma$.

(9.3,a) **Problem** Prove that this is indeed an equivalence relation, in the sense of Supp. Sect. 19.4.

The group of all permutations of $n$ objects is denoted as $S_n$ and is called the *symmetric group in $n$ variables*. It is a finite group, with $n!$ elements. As such it has a finite number of irreducible representations, which are all finite dimensional, and are—or are equivalent to—unitary representations. An arbitrary representation of $S_n$ is completely reducible, meaning that it can be written as a direct sum—possible infinite—of irreducible ones. The properties of $S_n$ and its representations are discussed in the book by Sagan [3]. See Supp. Chap. 24 for general information on groups and their representations.

The map $\phi_1 \otimes \cdots \otimes \phi_n \mapsto \phi_{\sigma(1)} \otimes \cdots \otimes \phi_{\sigma(n)}$, for each permutation $\sigma$ in $S_n$, defines by linear extension in $\mathcal{H} = \otimes^n \mathcal{H}^{(1)}$, a unitary representation $\{U(\sigma)\}_\sigma$ of $S_n$. The Hilbert space $\otimes^n \mathcal{H}^{(1)}$ can therefore be written as a finite direct sum of Hilbert subspaces

$$\otimes^n \mathcal{H}^{(1)} = \mathcal{H}_{j_1} \oplus \cdots \oplus \mathcal{H}_{j_p}.$$

Each $\mathcal{H}_{j_s}$ carries a—possibly infinite—number of equivalent irreducible representations. The irreducible representations in different $\mathcal{H}_{j_s}$ are mutually inequivalent.

Let us consider the case $n = 2$, which is sufficient for our purpose, as we shall see further on. The group $S_2$, the permutation group relevant for a system of two identical particles, consists, as the group of permutations of two objects, of two elements, the identity element $\sigma_0$, acting in a trivial way as the identity transformation $\sigma_0$ : $\{1, 2\} \to \{1, 2\}$ and the element $\sigma_1$, acting by pair exchange $\sigma_1$ : $\{1, 2\} \to \{2, 1\}$.

Consider as the 'naive' two particle Hilbert space the tensor product space

$$\mathcal{H}^{(2)} = \mathcal{H}^{(1)} \otimes \mathcal{H}^{(1)}$$

Let $\phi_1, \phi_2, \ldots$ be an orthonormal basis of $\mathcal{H}^{(1)}$. Then the tensor products $\phi_j \otimes \phi_k$, for $j, k = 1, 2, \ldots$, form an orthonormal basis for $\mathcal{H}^{(2)}$. To make the action of $S_2$ more transparent, we form a new basis, consisting of the vectors

$$\psi_{jk}^+ = \frac{1}{2}(\phi_j \otimes \phi_k + \phi_k \otimes \phi_j), \quad j, k = 1, 2, \ldots, \quad j \leq k,$$

$$\psi_{jk}^- = \frac{1}{2}(\phi_j \otimes \phi_k - \phi_k \otimes \phi_j), \quad j, k = 1, 2, \ldots, \quad j < k.$$

**(9.3,a) Problem** Show that this second basis is again an orthonormal basis.

For the action of $S_2$ we need only to consider $\sigma_1$, as the action of $\sigma_0$ is trivial.

**(9.3,b) Problem** Show that the action of the permutation $\sigma_1$ on the vectors $\psi_{jk}^+$ and $\psi_{jk}^-$ is

$$\psi_{jk}^+ \mapsto \psi_{jk}^+,$$

$$\psi_{jk}^- \mapsto -\psi_{jk}^-.$$

This result means that the subspaces of $\mathcal{H}^{(2)}$ spanned by the vectors $\psi_{jk}^+$ and $\psi_{jk}^-$ all carry a one dimensional representation of $S_2$, obviously unitary and irreducible. The ones on the spaces of the vectors of the first type are equivalent. We call this the *symmetric representation* $U_s(\cdot)$. The vectors $\psi_{jk}^-$ span the spaces of a second one dimensional representation, clearly inequivalent with $U_s(\cdot)$. It is called the *antisymmetric representation* and is denoted as $U_a(\cdot)$. These are the only irreducible representations of $S_2$. The symmetric groups $S_n$ for $n > 2$ have additional irreducible representations; the group $S_3$, for example, has one more.

The vectors $\psi_{jk}^+$ together span a subspace, which is a direct sum of all the symmetric representations in $\mathcal{H}^{(2)}$. We denote this subspace as $\mathcal{H}_s^{(2)}$. The vectors $\psi_{jk}^-$ span a second subspace, denoted as $\mathcal{H}_a^{(2)}$. The full tensor space $\mathcal{H}^{(2)} = \mathcal{H}^{(1)} \otimes \mathcal{H}^{(1)}$ can be written as a direct sum

$$\mathcal{H}^{(2)} = \mathcal{H}_s^{(2)} \oplus \mathcal{H}_a^{(2)}.$$

In Supp. Sect. 23.3, we show that for a given Hilbert space $\mathcal{H}$ we can define a *symmetric tensor product* $\mathcal{H} \otimes_s \mathcal{H}$ and an *antisymmetric tensorproduct* $\mathcal{H} \otimes_a \mathcal{H}$. We derive there moreover that there exist linear isomorphisms between, on one side $\mathcal{H} \otimes_s \mathcal{H}$ and $\mathcal{H} \otimes_a \mathcal{H}$, and on the other side subspaces of $\mathcal{H} \otimes \mathcal{H}$, obtained by projections $P_s$ and $P_a$, which are defined by linear extension of

$$P_s(\phi_1 \otimes \phi_2) = \frac{1}{2}(\phi_1 \otimes \phi_2 + \phi_2 \otimes \phi_1),$$

and

$$P_s(\phi_1 \otimes \phi_2) = \frac{1}{2}(\phi_1 \otimes \phi_2 - \phi_2 \otimes \phi_1).$$

The ranges of these projections are the subspaces $\mathcal{H}_s^{(2)} = \mathcal{H}^{(1)} \otimes_s \mathcal{H}^{(1)}$ and $\mathcal{H}_a^{(2)} = \mathcal{H}^{(1)} \otimes_a \mathcal{H}^{(1)}$.

The states in $\mathcal{H}_s^{(2)}$ are called *bosons* or *bosonic states*, those in $\mathcal{H}_a^{(2)}$ *fermions* or *fermionic states*.

Let $U(\cdot)$ be the (reducible) representation of $S_2$ in $\mathcal{H}^{(2)} = \mathcal{H}^{(1)} \otimes \mathcal{H}^{(1)}$.

(9.3,c) **Definition** Two non-zero state vectors $\psi'$ and $\psi$ in $\mathcal{H}^{(2)}$ will be called, $\psi \sim \psi'$, iff there is an element $\sigma$ in $S_2$ such that $\psi' = U(\sigma)\psi$.

(9.3,d) **Problem** Show that this is indeed an equivalence relation as defined in Supp. Sect. 19.3.

Note that this definition can be immediately generalized to the situation where $n$ is general. We need only the case $n = 2$, as will be explained further on.

(9.3,e) **Definition** An operator $A$ will be called *physical* iff for every pair of non zero vectors $\psi'$ and $\psi$ with $\psi' \neq \psi$ and $A\psi' \neq A\psi$, the condition $\psi \sim \psi'$ implies $A\psi \sim A\psi'$.

(9.3,f) **Problem** A more obvious definition for an operator $A$ to be physical would be that $\psi' \neq \psi$ implies $A\psi \sim A\psi'$, without further conditions. This, however, does not work. Explain why.

(9.3,g) **Problem** Show that an operator $A$ is physical in the sense of Definition (9.3,e) if and only if it commutes with $U(\sigma_1)$, with $\sigma_1$ the permutation of pair exchange.

With this result we write the relation $AU(\sigma_1) = U(\sigma_1)A$ as a $2 \times 2$ operator-valued matrix relation

$$\begin{pmatrix} A_{ss} & A_{sa} \\ A_{as} & A_{aa} \end{pmatrix} \begin{pmatrix} 1 & 0 \\ 0 & -1 \end{pmatrix} = \begin{pmatrix} 1 & 0 \\ 0 & -1 \end{pmatrix} \begin{pmatrix} A_{ss} & A_{sa} \\ A_{as} & A_{aa} \end{pmatrix}.$$

with the immediate consequence $A_{sa} = A_{as} = 0$. This means that a physical operator cannot connect the boson sector with the fermion sector.

With this we have the situation discussed in the preceding section: a *superselection rule*, with in the Hilbert space $\mathcal{H}^{(2)}$ the subspaces $\mathcal{H}_s^{(1)}$ and $\mathcal{H}_a^{(1)}$ as the superselection subsectors. This is the reason why we called in the beginning of this section the tensor product space $\mathcal{H}^{(2)} = \mathcal{H}^{(1)} \otimes \mathcal{H}^{(1)}$ the 'naive' two particle Hilbert space.

The case of two particles can be easily generalized to $n$ particles. The 'naive' $n$ particle space is the $n$-fold tensor product $\mathcal{H}^{(n)} = \otimes^n \mathcal{H}^{(1)}$. The symmetric group $S_n$ acts on it by a direct sum of irreducible unitary representations $U_1(\cdot), U_2(\cdot), \ldots, U_{j_k}(\cdot)$. In this we have $U_1(\cdot) = U_s(\cdot)$, the symmetric, and $U_{j_k}(\cdot) = U_a(\cdot)$, the antisymmetric representation, both present for every $S_n$ with $n \geq 2$, together with the additional representations that appear for all $n > 2$. The 'naive' $n$-particle space can be written as a direct sum $\mathcal{H}^{(n)} = \mathcal{H}_1 \oplus \cdots \oplus \mathcal{H}_{j_n}$, with $\mathcal{H}_{j_1}, \ldots, \mathcal{H}_{j_k}, \mathcal{H}_{j_1} = \mathcal{H}_s$ and $\mathcal{H}_{j_k} = \mathcal{H}_a$, the ranges of projection operators $P_{j_1}, \ldots, P_{j_k}, P_{j_1} = P_s, P_{j_k} = P_a$. This suggests that other types of states may exist besides boson and fermion states.

Bosons and fermions can be easily distinguished experimentally by their physical properties which follow from the theory as discussed here. It is mainly a matter of a different counting of states. These properties are intrinsic properties of the particles, which do not depend on whether they belong to a two particle system or a to a system of more than two particles. From this we may infer that the mixed symmetry representations of $S_n$ have no physical meaning in the context of the theory of identical particles. So we assume that there exists only bosons and fermions, as is confirmed by the fact that no other types of particles have ever been observed. In the direct sum decomposition for the 'naive' $n$-particle state $\otimes^n \mathcal{H}^{(1)}$, only the subspaces $\mathcal{H}_s$ and $\mathcal{H}_a$ are physically meaningful. We may denote these as $\mathcal{H}_s = \otimes_s^n \mathcal{H}^{(1)}$ and $\mathcal{H}_a = \otimes_a^n \mathcal{H}^{(1)}$.

We state the second part of Axiom VI:

**Axiom VI$_2$:** *There exist two types of particles, bosons and fermions. They are separated by a superselection rule.*

The most important difference between bosons and fermions is based on a simple but deep connection with spin. One has found experimentally that all particles with even spin are bosons; all particles with half integer spin are fermions.

This leads us to the statement of the third part of Axiom VI:

**Axiom VI$_3$:** *Bosons have integer spin quantum numbers*: $0, 1, 2, \ldots$; *fermions half integer quantum numbers*: $\frac{1}{2}, 1\frac{1}{2}, \ldots$

Up until now there is in nonrelativistic quantum mechanics no proof of the connection between spin and statistics, even though the experimental situation is completely unambiguous. However, in relativistic quantum field theory, in particular within the Wightman formulation of axiomatic quantum field, it is possible to prove the connection between spin and statistics. See for a brief discussion of Wightman theory Sect. 16.7.1, and the book by Streater and Wightman (Ref. [4]).

Note that in all this one may read 'states' for 'particles'. An electron has spin $\frac{1}{2}$ and is therefore a fermion. According to the rules for composing spins, discussed in Chap. 8, the state space of a system of two electrons is the direct sum of a spin 0 and a spin 1 space, so this system is a boson state.

All this enables us to formulate Axiom VI as a single statement:

**Axiom VI:** *In quantum theory a composite system made up of n identical particles with each the same Hilbert space $\mathcal{H}$ has as state space either the symmetric tensor product $\otimes_s^n \mathcal{H}$ (a bosonic system) or the antisymmetric tensor product $\otimes_a^n \mathcal{H}$ (a fermionic system). Linear combinations of boson and fermion states are physically meaningless; there is a superselection rule between boson states and fermion states. Moreover, boson states have integer and fermion states half-integer spin.*

Note that in terms of wave functions this means that boson particle systems are described by symmetric and fermion systems by antisymmetric wave functions.

## 9.4 An Example: The Helium Atom

The simplest atom after hydrogen is helium; it is an excellent laboratory in which to study quantum mechanics. A helium atom has two electrons; its nucleus consists of two protons and two neutrons. In a first approximation, with the nucleus fixed in the origin, the Hamiltonian is

$$H = H_0 + V = H_1 + H_2 + V_{12} = \frac{\mathbf{p}_1^2}{2m} - \frac{2e^2}{4\pi\epsilon_0 r_1} + \frac{\mathbf{p}_2^2}{2m} - \frac{2e^2}{4\pi\epsilon_0 r_1} + \frac{2e^2}{4\pi\epsilon_0 |\mathbf{r}_1 - \mathbf{r}_2|}.$$

Without the term $V_{12}$ this would be simply a sum of two noninteracting hydrogen Hamiltonians, for which we can immediately write down the energy eigenvalues and eigenfunctions by using the tensor product structure of the two particle Hilbert space. The energy of the ground state is in this picture just twice that of a separate hydrogen atom. One could try to calculate a more realistic value by looking at $V$ as a perturbation of $H_0$. Not much can be expected from this, however, because the Coulomb interaction between the two electrons is of the same order of magnitude as that of each electron with the nucleus. There are other more sophisticated approximation methods that give better results. They will not be discussed here, because in the context of this chapter, we are not interested in more precise values of the energy eigenvalues, but more in a qualitatively satisfactory description of the system of energy levels, in particular with respect to the consequences of Pauli's exclusion principle. Because electrons are fermions, this principle requires the Hilbert space of a two fermion system like the electron system of the helium atom to be the antisymmetric tensor product of the Hilbert spaces of the separate particles, or more explicitly, the wave function should be antisymmetric under the exchange of the particles.

The Hamiltonian $H_0$ does not act on the spin variables; it also does not involve interaction between the electrons, so for solving the eigenvalue–eigenvector problem of $H_0$ it makes sense to look at the wave functions as products of a spatial wave function for the electrons and a spin function. This corresponds with the tensor product of Hilbert spaces $\mathcal{H} = \mathcal{H}_0 \otimes \mathcal{H}_{\text{spin}}$, with

$$\mathcal{H}_0 = L^2(\mathbb{R}^6, d\mathbf{x}'d\mathbf{x}'') = L^2(\mathbb{R}^3, d\mathbf{x}) \otimes L^2(\mathbb{R}^3, d\mathbf{x}).$$

A product function has the general form

$$\psi(\mathbf{x}', \mathbf{x}''; \rho', \rho'') = \psi_0'(\mathbf{x}')\psi_0''(\mathbf{x}'')\psi_{\text{spin}}(\rho', \rho''),$$

with $\rho'$ and $\rho''$ is $\pm 1$. Note that the spin function is not written as a product. One has for the eigenfunctions

$$\psi_{n',l',m_l',m_s'; n'',l'',m_l'',m_s''}(\mathbf{x}', \mathbf{x}''; \rho', \rho'') = (\psi_0)_{n',l',m_l'}(\mathbf{x}') \, (\psi_0)_{n'',l'',m_l''}(\mathbf{x}'')$$
$$\times (\psi_{\text{spin}})_{m_{s'}; m_s}(\rho', \rho'').$$

According to Pauli's exclusion principle a wave function should satisfy

$$\psi(\mathbf{x}', \mathbf{x}''; \rho', \rho'') = -\psi(\mathbf{x}'', \mathbf{x}'; \rho'', \rho'),$$

i.e. for the eigenfunctions

$$\psi_{n',l',m_l'}(\mathbf{x}') \, \psi_{n'',l'',m_l''}(\mathbf{x}'')\psi_{\text{spin}}(\rho', \rho'') = -\psi_{n',l',m_l'}(\mathbf{x}'') \, \psi_{n'',l'',m_l''}(\mathbf{x}')\psi_{\text{spin}}(\rho'', \rho').$$

In general there are two possibilities. The first is

$$\psi_{n',l',m'_l}(\mathbf{x}')\,\psi_{n'',l'',m''_l}(\mathbf{x}'') = -\,\psi_{n',l',m'_l}(\mathbf{x}'')\,\psi_{n'',l'',m''_l}(\mathbf{x}'),$$

and

$$\psi_{\text{spin}}(\rho',\rho'') = \psi_{\text{spin}}(\rho'',\rho'),$$

and the second

$$\psi_{n',l',m'_l}(\mathbf{x}')\,\psi_{n'',l'',m''_l}(\mathbf{x}'') = \psi_{n',l',m'_l}(\mathbf{x}'')\,\psi_{n'',l'',m''_l}(\mathbf{x}'),$$

and

$$\psi_{\text{spin}}(\rho',\rho'') = -\,\psi_{\text{spin}}(\rho'',\rho').$$

The space $\mathcal{H}_{\text{spin}}$ is a four dimensional Hilbert space. It has an orthonormal basis, consisting of the vectors

$$\chi_+(\rho')\chi_+(\rho''),\;\; \chi_-(\rho')\chi_-(\rho''),\;\; \frac{1}{2}\sqrt{2}\left[\chi_+(\rho')\chi_-(\rho'') + \chi_-(\rho')\chi_+(\rho'')\right],$$

$$\frac{1}{2}\sqrt{2}\left[\chi_+(\rho')\chi_-(\rho'') - \chi_-(\rho')\chi_+(\rho'')\right]$$

with the single particle spin functions defined as

$$\chi_+(+1) = 1,\;\; \chi_+(-1) = 0;\;\;\;\; \chi_-(+1) = 0,\;\; \chi_-(-1) = 1.$$

Within the four dimensional tensor product space $\mathcal{H}_{\text{spin}}$ the first three vectors span the symmetric subspace, which carries the representation of $su(2)$ with total spin quantum number $s = 1$ and $m_s = -1, 0, +1$. The fourth vector spans the one dimensional antisymmetric space, which carries the representation with $s = 0$ and $m_s = 0$.

We apply this to the ground state. This means $n' = n'' = 1$. This gives, in this approximation, that the lowest energy is twice the lowest energy of the hydrogen atom. Then $l$ and $m_l$ are

$$l' = 0,\;\; m'_l = 0;\;\; l'' = 0,\;\; m''_l = 0.$$

The product $\psi_{100}(\mathbf{x}')\psi_{100}(\mathbf{x}'')$ is symmetric. This implies that the function $\psi_{\text{spin}}$ $(\rho', \rho'')$ is antisymmetric. There is therefore only one possibility for the ground state, namely, with the correct normalization,

$$\psi_{1,0,0;\,1,0,0}(\mathbf{x}', \mathbf{x}''; \rho', \rho'') = \frac{1}{2}\sqrt{2}\,\left[\psi_{1,0,0}(\mathbf{x}')\psi''_{1,0,0}(\mathbf{x}'')\,[\chi_+(\rho')\chi_-(\rho'')\right.$$
$$\left. -\chi_-(\rho')\chi_+(\rho'')]\right].$$

In a similar manner the other eigenfunctions, for various values of $n'$ and $n''$, can be studied.

(9.4,a) **Problem** What are the energy eigenfunctions for $n' = 1$ and $n'' = 2$? Show in particular that Pauli's exclusion principle reduces the number of two particle states with these values of the principal quantum from 100 to 45.

*Remark* Because of selection rules which forbid certain transitions, the system of energy eigenvalues consists, at least in low order of perturbation theory, of two isolated subsystems. This means that it looks as if there are two sorts of helium. They are called *orthohelium* (which contains the ground state) and *parahelium*.

## 9.5 Historical Remarks

The historical development of the complex of scientific notions discussed in the preceding sections is that of a piecemeal and fragmentary discovery of at first unrelated facts and ideas, not unlike the earlier development of quantum mechanics itself. One is reminded of an archeological excavation, in which one first discovers fragments of brickwork, than realizes that these are pieces of complete walls. In the end, if one is lucky, one may obtain a single picture of a complete city.

Looking backwards it is easy, with the knowledge and understanding that we have now, to describe the development in the opposite direction, starting with the fundamental fact that quantum particles do not have an individual identity, suggested by Heisenberg's uncertainty relations, and formulated as part of Axiom VI, and then develop from this the full theory, with assuming in addition the spin-statistics relation, up to the full version of Axiom VI. The final picture obtained in this manner is in the end conceptually more satisfying than the sequence of historical events, interesting as they are in themselves. A few words about this history will make this clear.

The context of quantum mechanics in the period, say, from 1925 till 1929, when the elements of the description of systems of identical particles were formulated, was the physics of atoms. In this the picture was that of an atom consisting of a central nucleus with its surrounding electrons, together with the manner in which atoms interact to form molecules. What was experimentally observed were atomic and molecular spectra.

The story starts with Pauli's discovery that looking at atomic spectra one finds that certain specific energy levels in certain atoms are missing, in a very systematic way. This led him in 1925 to the formulation of his *exclusion principle*—for which he later, in 1945, received the Nobel prize—stating, roughly, that two electrons cannot occupy the same state. It was immediately clear that this had important consequences; without it the periodic systems of elements would look very different. Pauli also noted that the electron seemed to have an additional quantum number but he left it at that. The Dutch physicists Goudsmit and Uhlenbeck identified this shortly afterwards as an intrinsic angular momentum; the *spin* of the electron, discussed in Chap. 8.

Up to this point the only elementary particles known were electrons. In nuclear physics as it emerged somewhat later new particles were found which did not satisfy the exclusion principle, for instance, mesons, unstable spin 0 particles found in cosmic radiation and later produced in particle accelerators. From this it became clear that there exist in nature two distinct types of particles, those with integer spin, bosons, and those with half-integer spin, fermions, with important consequences, in particular for quantum statistical mechanics, where Fermi-Dirac statistics takes the place of the statistics of Maxwell-Boltzmann in classical statistical mechanics, as will be seen in Chap. 11.

So, historically speaking, Pauli's exclusion principle is the basic fact in the full description of systems of identical particles—a topic which will be discussed in the next section—from which everything else later followed. But looking backwards it is clear that the theoretical starting point in this is the principle that particles in quantum theory cannot be distinguished. The exclusion principle is a simple consequence of this.

## 9.6 The Fock Space Formalism for Many Particle Systems

### 9.6.1 Introduction

It is convenient, in particular in subjects as solid state and condensed matter physics, to be able to describe within a single framework a system of $n$ identical particles, simultaneously for all values of $n$. Such a framework was first developed in the early thirties by the Russian theoretical physicist Fock [5, 6], and much later put on a proper mathematical footing by Cook [7]. In most physics books the Fock space formalism goes under the name of '*second quantization*', a term which is suggestive but misleading, as will be argued.

The bosonic and the fermionic case are mathematically very similar, even though the corresponding physical properties differ in important aspects. This mathematical structure is explained in Sect. 9.6.2, with both cases discussed simultaneously; after that the discussion will be restricted to the bosonic case, to keep the notation simple. Some remarks on the fermionic case will be made and a few results will be given in the form of problems, results for which the derivation is just a matter of keeping track of additional minus signs. In Sect. 9.6.3 we give a simple explicit realization of the bosonic version of the mathematical structures introduced in Sect. 9.6.2, the model of a system of identical spinless particles of mass $m$, and we end in Sect. 9.6.4 with a brief discussion of quasiparticles, in particular phonons, as an alternative physical model.

In Sect. 9.7 the heuristic 'second quantization' formulation will be discussed. We will briefly remind the reader in Sect. 9.7.2 of a few basic formulas for single particle quantum mechanics, then describe in Sect. 9.7.3 the quantum mechanics of systems of $n$ identical particles, for the values of $n$ separately, and from there in Sect. 9.7.4 set up the full heuristic Fock space formalism in which the information contained in the equations for the separate cases is carried by an single equation for a single Heisenberg picture 'field operator'. Finally the character of the Fock space formalism as an example of quantum field theory is briefly described in Sect. 9.7.5.

### 9.6.2 The Mathematical Framework

The proper mathematical language for the Fock space formalism uses the *symmetric* or *antisymmetric tensor algebras* over a given Hilbert space $\mathcal{H}$ discussed in Supp. Chap. 23, denoted there as

$$T_s(\mathcal{H}) = \oplus_{n=0}^{\infty}(\otimes_s^n \mathcal{H}), \qquad T_a(\mathcal{H}) = \oplus_{n=0}^{\infty}(\otimes_a^n \mathcal{H})$$

and in this section as $\mathcal{H}_s$ (bosons) and $\mathcal{H}_a$ (fermions).

We first discuss the general framework, without specifying a particular type of particle. Let $\mathcal{H}^{(1)}$ be the Hilbert space of a single particle. The Hilbert space of a system of $n$ identical particles is, as we have seen in Sect. 9.3, for bosons the symmetric tensor product $\mathcal{H}_s^{(n)} = \otimes_s^n \mathcal{H}^{(1)}$, and for fermions the antisymmetric tensor product $\mathcal{H}_a^{(n)} = \otimes_a^n \mathcal{H}^{(1)}$. The zero particle space $\mathcal{H}^{(0)}$ is in both cases a one dimensional complex Hilbert space in which a unit vector $\Psi_0$ is specified. The total Hilbert spaces $\mathcal{H}_s$ and $\mathcal{H}_a$ are then the countably infinite direct sums of the tensor product spaces $\mathcal{H}_{s/a}^{(n)}$, i.e.

$$\mathcal{H}_{s/a} = \oplus_{n=0}^{\infty} \mathcal{H}_{s/a}^{(n)} = \oplus_{n=0}^{\infty}(\otimes_{s/a}^n \mathcal{H}^{(1)}),$$

which are in fact *tensor algebras*, the *symmetric and antisymmetric tensor algebras* $T_s(\mathcal{H}^{(1)})$ and $T_a(\mathcal{H}^{(1)})$ over the one particle Hilbert space $\mathcal{H}^{(1)}$ discussed in Supp. Sect. 23.4. To move up and down in these spaces one uses the raising and lowering operators, defined in Supp. Sect. 23.5, but here with a slightly different notation, for which the reason will soon become clear. We have, for all $f$ and $g$ in $\mathcal{H}^{(1)}$, operators $\widehat{\Psi}^*(f)$ and $\widehat{\Psi}(g)$ with the basic commutation relations

$$[\widehat{\Psi}^*(f), \widehat{\Psi}^*(g)]_{\pm} = [\widehat{\Psi}(f), \widehat{\Psi}(g)]_{\pm} = 0, \quad [\widehat{\Psi}(f), \widehat{\Psi}^*(g)]_{\pm} = (f, g),$$

together with the property $\widehat{\Psi}(f)\Psi_0 = 0$, for all $f$ in $\mathcal{H}^{(1)}$. In this $[A, B]$—denotes the commutator $AB - BA$, for the symmetric case and $[A, B]_+$ the anticommutator $AB + BA$ for the antisymmetric case, for operators $A$ and $B$.

All vectors in $\mathcal{H}_{s/a}$ can be obtained as limits of linear combinations of $\Psi_0$ and vectors

$$\widehat{\Psi}^*(f_1)\widehat{\Psi}^*(f_2)\cdots\widehat{\Psi}^*(f_m)\Psi_0,$$

for all possible sequences $f_1, \ldots, f_m$ in $\mathcal{H}^{(1)}$, and all operators in $\mathcal{H}_{s/a}$ as limits of linear combinations of the unit operator and operators

$$\widehat{\Psi}^*(f_1)\widehat{\Psi}^*(f_2)\cdots\widehat{\Psi}^*(f_m)\widehat{\Psi}(g_1)\cdots\widehat{\Psi}(g_n),$$

for all possible sequences $f_1, \ldots, f_m$ and $g_1, \ldots, g_n$ in $\mathcal{H}^{(1)}$. Note that these products of raising and lowering operators have the standard form called 'normal' or 'Wick ordered form', with all raising operators to the left of all lowering operators. By using the commutation—anticommutation relations an arbitrary operator in $\mathcal{H}_{s/a}$ can be written as a limit of linear combinations of Wick ordered operators.

(9.6.2,a) **Problem** Write the operator

$$\widehat{\Psi}(f_1)\widehat{\Psi}(f_2)\widehat{\Psi}^*(f_3)\widehat{\Psi}(f_4)\widehat{\Psi}^*(f_5)$$

as a sum of products in normal-ordered form.

The Hilbert spaces $\mathcal{H}_s$ and $\mathcal{H}_a$ with their tensor algebra structure and with the system of raising and lowering operators

$$\{\widehat{\Psi}^*(f)\}_{f\in\mathcal{H}_{s/a}^{(1)}}, \qquad \{\widehat{\Psi}(g)\}_{g\in\mathcal{H}_{s/a}^{(1)}},$$

are called *Fock spaces*, and the resulting physical formalism the *Fock space formalism*.

It is often convenient to choose an orthonormal basis $\{\phi_j\}_{j=1}^{\infty}$ in $\mathcal{H}^{(1)}$ and use it to define discrete systems of operators $\widehat{\Psi}_j^* = \widehat{\Psi}^*(\phi_j)$ and $\widehat{\Psi}_j = \widehat{\Psi}(\phi_j)$.

(9.6.2,b) **Problem** Prove, for the boson case the commutation relations

$$[\widehat{\Psi}_j^*, \widehat{\Psi}_k^*] = [\widehat{\Psi}_j, \widehat{\Psi}_k] = 0, \qquad [\widehat{\Psi}_j, \widehat{\Psi}_k^*] = \delta_{jk},$$

and for the fermion case the anticommutation relations

$$[\widehat{\Psi}_j^*, \widehat{\Psi}_k^*]_+ = [\widehat{\Psi}_j, \widehat{\Psi}_k]_+ = 0, \qquad [\widehat{\Psi}_j, \widehat{\Psi}_k^*]_+ = \delta_{jk}.$$

With a choice of an orthonormal basis in $\mathcal{H}^{(1)}$, often a basis of energy eigenstates, $\mathcal{H}_{s/a}$ is spanned by the products

$$\left[\prod_{j=1}^{\infty}(\widehat{\Psi}_j^*)^{n_j}\right]\Psi_0 = |n_1, n_2, \ldots >,$$

with $\Psi_0$ the ground state vector, with in the sequence $n_1, n_2, \ldots$ only finitely many $n_j$ different from 0, and with the $n_j$ taking the values $0, 1, 2, \ldots$ for the boson case and 0, 1 for the fermion case.

We use here the so-called 'occupation number representation' in which an $n$-particle state is denoted as $|n_1, n_2, \ldots>$. The notation $|\cdot, \cdot, \ldots>$ comes from Dirac's bra-ket formalism, very popular in physics textbooks. See Supp. Chap. 26 for this.

Operators in $\mathcal{H}^{(1)}$ generate operators in the full Fock space $\mathcal{H}_{s/a}$.

1. *Sum operators.* A given operator $A^{(1)}$ in the one particle space $\mathcal{H}^{(1)}$ gives an operator $\Gamma(A^{(1)})$ in $\mathcal{H}_{s/a}$, obtained by first defining an operator $A^{(n)}$ in each $n$-particle space $\mathcal{H}_s^{(n)}$ by linear extension of

$$A^{(n)}(\psi_1 \otimes_{s/a} \cdots \otimes_{s/a} \psi_n) = (A^{(1)}\psi_1) \otimes_{s/a} \cdots \otimes_{s/a} \psi_n + \psi_2 \otimes_{s/a} (A^{(1)}\psi_2)$$
$$\otimes_{s/a} \cdots \otimes_{s/a} \psi_n + \cdots + \psi_1 \otimes_{s/a} \cdots \otimes_{s/a} (A^{(1)}\psi_n),$$

with $\Gamma(A^{(1)})$ identically zero in $\mathcal{H}^{(0)}$, and then taking the infinite direct sum of the operators $A^{(n)}$ in $\mathcal{H}_{s/a}^{(n)}$. It is shown in Ref. [7], that for $A^{(1)}$ selfadjoint $\Gamma(A^{(1)})$ is also selfadjoint. For two operators $A^{(1)}$ and $B^{(1)}$ in $\mathcal{H}^{(1)}$—bounded, to avoid domain problems—and two complex numbers $\lambda$ and $\mu$ one has the rather obvious linearity property

$$\Gamma(\lambda A^{(1)} + \mu B^{(1)}) = \lambda \Gamma(A^{(1)}) + \mu \Gamma(B^{(1)}).$$

A prime example of an operator $\Gamma(\cdot)$ is the total number operator $\widehat{N} = \Gamma(1^{(1)})$ which has the eigenvalue $n$ in each $\mathcal{H}_{s/a}^{(n)}$.

(9.6.2,c) **Problem** Let $\{\phi_j\}_j$ be an orthonormal basis in $\mathcal{H}^{(1)}$. Consider an operator $A^{(1)}$ in $\mathcal{H}^{(1)}$ with matrix elements $A_{jk}^{(1)}$ with respect to this basis. Show that $\Gamma(A^{(1)})$ is

$$\Gamma(A^{(1)}) = \sum_{j,k=1,2,\ldots} \widehat{\Psi}_j^* A_{jk}^{(1)} \widehat{\Psi}_k,$$

and that

$$\widehat{N} = \sum_{j=1,2,\ldots} \widehat{\Psi}_j^* \widehat{\Psi}_j.$$

2. *Product operators.* There is a second way in which a given operator in $\mathcal{H}^{(1)}$ defines an operator in $\mathcal{H}_{s/a}$. A unitary operator $U^{(1)}$ in $\mathcal{H}^{(1)}$ generates an operator $\Pi(U^{(1)})$ in $\mathcal{H}_{s/a}$, first in $\mathcal{H}^{(n)}$ by linear extension of

$$U^{(n)}(\psi_1 \otimes_{s/a} \cdots \otimes_{s/a} \psi_n) = (U^{(1)}\psi_1) \otimes_{s/a} \cdots \otimes_{s/a} (U^{(1)}\psi_n),$$

with $U^{(0)}\Psi_0 = \Psi_0$, and then as a direct sum operator in $\mathcal{H}_{s/a}$. The unitarity of $U^{(1)}$ implies unitarity of $\Pi(U^{(1)})$. One has the product relation

$$\Pi(U_1^{(1)}U_2^{(1)}) = \Pi(U_1^{(1)})\,\Pi(U_2^{(1)})$$

and the less trivial and important exponential relation, proved in Ref. [7],

$$\Pi(e^{iA^{(1)}}) = e^{i\Gamma(A^{(1)})}.$$

The assignments $A^{(1)} \mapsto \Gamma(A^{(1)})$ and $U^{(1)} \mapsto \Pi(U^{(1)})$ are *functors*, a notion from *category theory*. In fact, the Fock space formalism as developed mathematically by Cook has a strong category theory flavour. This falls however outside the scope of this book.

Using the orthonormal basis $\{\phi_j\}_j$ one can write arbitrary operators in $\mathcal{H}_s$ as (limits of sums of) expressions

$$A^{(m,n)} = \sum_{j_1,\ldots,j_m,k_1,\ldots,k_m=1}^{\infty} \widehat{\Psi}_{j_1}^* \cdots \widehat{\Psi}_{j_m}^* A_{j_1,\ldots,j_m,k_1,\ldots,k_n}^{(m,n)} \widehat{\Psi}_{k_1} \cdots \widehat{\Psi}_{k_n},$$

which map $\mathcal{H}_s^{(j)}$ into $\mathcal{H}_s^{(j-n+m)}$, with, of course, $A^{(m,n)} = 0$ for $n > j$. We denote $A^{(n,n)}$ as $A^{(n)}$ and call it an *n*-particle operator.

### 9.6.3 An Explicit Realization

For simplicity of notation and because of the *mathematical* similarity between bosonic and fermionic structures we shall from this point onwards restrict the discussion to the boson case—apart from a few side remarks. We also choose a more explicit realization.

For this we think of a system of identical bosonic spinless point particles with mass $m$. We consider the $\psi^{(1)}$ in $\mathcal{H}^{(1)}$ not as 'abstract' vectors but as wave functions in $L_2(\mathbb{R}^3, d\mathbf{r})$. Consequently, the Hilbert space $\mathcal{H}^{(n)}$ consists of all square integrable functions $\psi^{(n)}(\mathbf{r}_1, \ldots, \mathbf{r}_n)$, restricted by the requirement of symmetry in the variables $\mathbf{r}_1, \ldots, \mathbf{r}_n$. In this context the raising and lowering $\widehat{\Psi}^*(f)$ and $\widehat{\Psi}(f)$ are called *field operators*. We use the letter $\psi$ to denote wave functions; the symbols $\widehat{\Psi}$ and $\widehat{\Psi}^*$ stand for operators, not functions. The reason for this notation—at first sight somewhat confusing—will shortly become clear. We shall also mention in Sect. 9.6.4 a second realization of $\mathcal{H}_s$, with the vectors in $\mathcal{H}^{(1)}$ 'momentum wave functions' $\phi^{(1)}(\mathbf{p})$ in $L_2(\mathbb{R}^3, d\mathbf{p})$, and with the elements of $\mathcal{H}_s^{(n)}$ symmetric functions $\phi^{(n)}$ in the variables $\mathbf{p}_1, \ldots, \mathbf{p}_n$.

## 9.6.4 Quasiparticles

Raising and lowering operators appeared earlier in this book, in Sect. 6.4, where a pair of such operators, $a^*$ and $a$, were used to solve in a simple algebraic way the eigenvalue–eigenvector problem for the Hamiltonian of the harmonic oscillator. This means that there is a rather trivial Fock space structure over a one dimensional one particle space $\mathcal{H}^{(1)}$ spanned by the ground state, and consequently all the $n$-particle states $\mathcal{H}_s^{(n)}$ also one dimensional. To speak in this case of $n$-particle state does not make much sense, because there is only a single particle involved. One therefore uses in this and related cases the more appropriate term *excitation*. In fact the idea behind it has found broad application.

*An example: phonons.* Consider a piece of solid matter. It is a crystal consisting of a lattice. At each lattice point there is an ion, an atom which has lost one or more of its electrons. These *valence electrons* which in general bind atoms to form molecules, and are much lighter than the ions, move freely through the lattice. The ions sit at fixed points but vibrate, in first approximation as harmonic oscillators, around their equilibrium positions, driven by external forces and through the interaction with the free electrons. This model was first proposed by Einstein, already in 1907, before the advent of quantum mechanics, in its most simple form, without the effects of the moving electrons. He derived important physical consequences from it. The model was refined further by Debye in 1912. In later development the connection was made with the propagation of sound in solids. This is why in this particular case the excitations of this system, the quasiparticles, are called *phonons*. To get an idea of the technical subtlety of modern developments in this area see e.g. the book by Bruus and Flensberg (Ref. [8]).

## 9.6.5 A Remark on Our Notation for Operators

In the many particle context (e.g. in Sect. 9.6.2) $A^{(m,n)}$ denotes an operator that maps $\mathcal{H}^{(n)}$ into $\mathcal{H}^{(m)}$. This notation is also used in the Fock space formalism for normal-ordered operators that have $m$ raising and $n$ lowering operators. Here $A^{(m,n)}$ maps $\mathcal{H}^{(n+j)}$ into $\mathcal{H}^{(m+j)}$, for all $j \geq 0$. $A^{(n,n)}$ is denoted as $A^{(n)}$. Sometimes operators in the Fock Hilbert space $\mathcal{H}_s$ will be denoted just as $A_s$. Finally, the field operators are just $\Psi$ and $\Psi^*$, avoiding thereby the more cumbersome $\Psi^{(0,1)}$ and $(\Psi^*)^{(1,0)}$.

# 9.7 A Heuristic Formulation: 'Second Quantization'

## 9.7.1 Introduction

In the preceding section we described the natural mathematical underpinning of the Fock space description of many particle systems in terms of symmetric and antisymmetric tensor algebras over a given Hilbert space together with a system of raising and lowering operators. The heuristic presentation of the Fock space formalism that one finds in most physics textbooks and that goes under the name 'second quantization' looks very different—certainly at first sight. It is transparent, elegant, and is moreover quite effective as long as one is aware of its limitations. To get there from the rigorous picture presented in the preceding section, one may use the basis-dependent formulation, and take instead of a proper orthonormal basis $\{\phi_j\}_j$ in $\mathcal{H}^{(1)}$ the heuristic basis consisting of three dimensional $\delta$-functions in the position variable $\mathbf{r}$.

To do this we need the idea of a generalized function, a notion which is explained in some detail in Supp. Chap. 25. The basic idea in this is that a function, say $F$ : $\mathbb{R}^1 \to \mathbb{R}^1$, gives rise to a linear functional $F(f) = \int_{-\infty}^{+\infty} F(x)f(x)dx$, defined on a suitable linear space of 'testfunctions' $f$. However, not all linear functionals come from functions. Prime examples are the functional $F(f) = f(0)$, with $F(x) = \delta(x)$, the heuristic Dirac $\delta$-function, and its derivative $F^{(1)}(f) = -f^{(1)}(0)$, with the equally heuristic derivative $\delta^{(1)}(x)$.

In the Fock space formalism the $\widehat{\Psi}^*$ and $\widehat{\Psi}$ are *operator-valued generalized functions*

$$\widehat{\Psi}^*(f) = \int\limits_{-\infty}^{+\infty} \widehat{\Psi}^*(\mathbf{r})f(\mathbf{r})d\mathbf{r}, \quad \widehat{\Psi}(f) = \int\limits_{-\infty}^{+\infty} \widehat{\Psi}(\mathbf{r})\overline{f(\mathbf{r})}d\mathbf{r},$$

with heuristic operators $\widehat{\Psi}^*(\mathbf{r})$ and $\widehat{\Psi}(\mathbf{r})$, the so-called *field operators*. The $\overline{f(\mathbf{r})}$ is the complex conjugate of $f(\mathbf{r})$; the fact that $\widehat{\Psi}$ is conjugate-linear instead of linear in $f$ does not affect the generalized function idea.

The commutation relations for these field operators are, not very surprisingly,

$$[\widehat{\Psi}^*(\mathbf{r}_1), \widehat{\Psi}^*(\mathbf{r}_2)] = [\widehat{\Psi}(\mathbf{r}_1), \widehat{\Psi}(\mathbf{r}_2)] = 0,$$

$$[\widehat{\Psi}(\mathbf{r}_1), \widehat{\Psi}^*(\mathbf{r}_2)] = \delta(\mathbf{r}_1 - \mathbf{r}_2),$$

with again $\widehat{\Psi}(\mathbf{r})\Psi_0 = 0$, for all $\mathbf{r}$ in $R^3$.

The total number operator $\widehat{N}$ is $\int_{-\infty}^{+\infty} \widehat{\Psi}^*(\mathbf{r})\widehat{\Psi}(\mathbf{r})d\mathbf{r}$. Arbitrary operators in $\mathcal{H}_s$ can be written as sums of expressions

$$A^{(m,n)} = \int\limits_{-\infty}^{+\infty} \cdots \int\limits_{-\infty}^{+\infty} d\mathbf{r}_1 \ldots d\mathbf{r}_m \, d\mathbf{r}_1' \ldots d\mathbf{r}_n'$$

$$\widehat{\Psi}^*(\mathbf{r}_1) \ldots \widehat{\Psi}^*(\mathbf{r}_m) \, A^{(m,n)}(\mathbf{r}_1, \ldots, \mathbf{r}_m, \mathbf{r}_1', \ldots, \mathbf{r}_n') \, \widehat{\Psi}(\mathbf{r}_1') \ldots \widehat{\Psi}(\mathbf{r}_n').$$

The $\widehat{\Psi}^*(\mathbf{r})$ and $\widehat{\Psi}(\mathbf{r})$, operators of type $A^{(1,0)}$ and $A^{(0,1)}$, create and annihilate a state with the heuristic non-normalizable wave function

$$\psi_{\mathbf{r}_0}(\mathbf{r}) = \delta(\mathbf{r} - \mathbf{r}_0) = \delta(x - x_0)\delta(y - y_0)\delta(z - z_0),$$

describing a particle with exact position at $\mathbf{r}$, something which is, strictly speaking, not possible in quantum mechanics.

Note that in the context of non relativistic quantum theory in which we find ourselves here, the number of particles in a system is constant, so creation and annihilation does not mean that particles are actually created of annihilated. An important exception was described in Sect. 9.6.4, the case of quasiparticles.

## 9.7.2 A Single Particle: A Reminder

Before starting the construction of the heuristic Fock many-particle formalism we remind the reader of a few basic formulas for the description of a single point particle of mass $m$. The Hamiltonian for such a particle is $H^{(1)} = H_0^{(1)} + H_I^{(1)}$ with the free Hamiltonian

$$H_0^{(1)} = \frac{p^2}{2m} = -\frac{\hbar^2}{2m}\Delta,$$

a second order differential operator in the space $\mathcal{H}^{(1)}$ of one particle wave functions, with $\Delta$ the three dimensional Laplace operator, and $H_I^{(1)}$ describing a possible force on the particle through an external potential $V_{\text{ext}}^{(1)}$.

The action of an arbitrary operator $A^{(1)}$ in $\mathcal{H}^{(1)}$ can be written—at least heuristically, with a possibly symbolic integral kernel for $A^{(1)}$—as

$$(A^{(1)}\psi^{(1)})(\mathbf{r}) = \int_{-\infty}^{+\infty} A^{(1)}(\mathbf{r}, \mathbf{r}')\psi^{(1)}(\mathbf{r}')d\mathbf{r}'.$$

(9.7.2,a) **Problem** Use the information on the derivatives of the Dirac $\delta$-function given in Supp. Chap. 25 to show that $\Delta$ acts in this symbolic manner as

$$(\Delta\psi^{(1)})(\mathbf{r}) = \int_{-\infty}^{+\infty} (\Delta\delta)(\mathbf{r} - \mathbf{r}')\psi^{(1)}(\mathbf{r}')\,d\mathbf{r}',$$

with $(\Delta\delta_{\mathbf{r}})(\mathbf{r}') = (\Delta\delta)(\mathbf{r} - \mathbf{r}')$ the Laplacian of the $\delta$-function concentrated at $\mathbf{r}$.

In the Heisenberg picture (see Sect. 3.6.3) in which the time development of the system is put into the observables instead of in the states, the time evolution of an operator $A^{(1)}$ is described by the general formula

$$A^{(1)}(t) = e^{\frac{it}{\hbar}H^{(1)}} A^{(1)}(0)\, e^{-\frac{it}{\hbar}H^{(1)}},$$

leading to the evolution equation

$$\frac{d}{dt}A^{(1)}(t) = \frac{i}{\hbar}[H^{(1)}, A^{(1)}],$$

an equation which will be used in the next subsection.

### 9.7.3 The Many Particle Situation

We next go to the many particle space context. For each $n \geq 1$ there is an $n$-particle Hilbert space $\mathcal{H}_s^{(n)} = \otimes_s^{(n)}\mathcal{H}^{(1)}$, with $\mathcal{H}^{(0)}$ the one dimensional 0-particle space. A state of the system is represented by an $n$-particle wave function $\psi^{(n)}(\mathbf{r}_1, \ldots, \mathbf{r}_n)$. If the particles are free the Hamiltonian in $\mathcal{H}_s^{(n)}$ is $H_0^{(n)} = -\frac{\hbar^2}{2m}\Delta_n$, with $\Delta_n$ the $n$-dimensional Laplace operator. The $\psi^{(n)}$ satisfy the free $n$-particle Schrödinger equation

$$\frac{\partial}{\partial t}\psi^{(n)} = \frac{i\hbar}{2m}\Delta_n\psi^{(n)}.$$

Particles that are not free either move in an external potential, or interact with each other, or both. An external potential is represented by an operator $V_{\text{ext}}^{(1)}$ in $\mathcal{H}^{(1)}$ which acts by multiplying the wave function with a function $V_{\text{ext}}^{(1)}(\mathbf{r})$.

(9.7.3,a) **Problem** Show that $V_{\text{ext}}^{(1)}$ gives an operator which acts on $n$-particle wave functions $\psi^{(n)}$ as multiplication by the function

$$V_{\text{ext}}^{(n)}(\mathbf{r}_1, \ldots, \mathbf{r}_n) = \sum_{j=1}^{n} V_{\text{ext}}^{(1)}(\mathbf{r}_j),$$

and that therefore the $\psi^{(n)}$ satisfy the $n$-particle Schrödinger equation

$$\frac{\partial}{\partial t}\psi^{(n)} = \frac{i}{\hbar}\left(\frac{\hbar^2}{2m}\Delta_n - V_{\text{ext}}^{(n)}\right)\psi^{(n)}.$$

For interacting particles the simplest situation is that of pair interaction. In that case, to which we restrict the discussion, there is an operator $V_{\text{int}}^{(2)}$ acting in first instance in $\mathcal{H}^{(2)}$ by multiplying the two particle wave functions $\psi^{(2)}$ by a (symmetric) function $V_{\text{int}}^{(2)}(\mathbf{r}_1, \mathbf{r}_2)$.

(9.7.3,b) **Problem** Show that, for $n \geq 2$, $V_{\text{int}}^{(2)}$ defines an operator in each $\mathcal{H}_s^{(n)}$ which acts on $n$-particle wave functions $\psi^{(n)}$ as multiplication by the function

$$V_{\text{int}}^{(n)}(\mathbf{r}_1, \ldots, \mathbf{r}_n) = \sum_{1 \le j < k \le n} V_{\text{int}}^{(2)}(\mathbf{r}_j, \mathbf{r}_k),$$

leading to the same $n$-particle Schrödinger equation as above, but with $V_{\text{int}}^{(n)}$ instead of $V_{\text{ext}}^{(n)}$.

## 9.7.4 The Fock Space Formulation

The information contained in the set of $n$-particle formulas from the preceding subsection—the expressions for the Hamiltonians and the $n$-particle Schrödinger equations, can be subsumed into a single Fock space Hamiltonian $H_s$, leading to a single Schrödinger equation, not for a (numerical) wave function, but for a Heisenberg picture *field operator* in the Fock Hilbert space $\mathcal{H}_s$.

The one parameter group of unitary operators for the time evolution in $\mathcal{H}_s$ is given by a formula with the general form $U_s(t) = e^{-\frac{it}{\hbar} H_s}$. The total number of particles is constant, therefore the $n$-particle evolution operators leave the subspaces $\mathcal{H}_s^{(n)}$ invariant, and consequently $U_s(t)$ is a direct sum of the evolution operators in the separate spaces $\mathcal{H}_s^{(n)}$.

Using the material from Sect. 9.7.1 one obtains the 'second quantized' Hamiltonians for the free case, the cases with an external potential and with interaction between the particles.

(9.7.4,a) **Problem** Show that the 'second quantized' Hamiltonian for free particles is

$$(H_0)_s = \Gamma(H_0^{(1)}) = -\int_{-\infty}^{+\infty} \widehat{\Psi}^*(\mathbf{r}) \frac{\hbar^2}{2m} \Delta \widehat{\Psi}(\mathbf{r}) \, d\mathbf{r},$$

the potential operator

$$(V_{\text{ext}})_s = \Gamma(V_{\text{ext}}^{(1)}) = \int_{-\infty}^{+\infty} \widehat{\Psi}(\mathbf{r}) V^{(1)}(\mathbf{r}) \widehat{\Psi}^*(\mathbf{r}) \, d\mathbf{r},$$

and the pair interaction operator

$$V_{\text{int}}^{(n)} = \int_{-\infty}^{+\infty} \cdots \int_{-\infty}^{+\infty} d\mathbf{r}_1 \ldots d\mathbf{r}_n$$

$$\widehat{\Psi}^*(\mathbf{r}_1) \ldots \widehat{\Psi}^*(\mathbf{r}_n) V_{\text{int}}^{(n)}(\mathbf{r}_1, \ldots, \mathbf{r}_n) \widehat{\Psi}(\mathbf{r}_1) \ldots \widehat{\Psi}(\mathbf{r}_n),$$

with $V_{\text{int}}^{(n)}(\mathbf{r}_1, \ldots, \mathbf{r}_n)$ from Problem (9.6.3,b).

(9.7.4,b) **Problem** Prove, by calculating the commutator between $(H_0)_s$ and $\widehat{\Psi}(\mathbf{r})$, that, for particles in an external field, the Heisenberg picture field operator $\widehat{\Psi}(\mathbf{r}, t)$ satisfies the equation

$$\frac{\partial}{\partial t} \widehat{\Psi} = \frac{i}{\hbar} \left[ \frac{\hbar^2}{2m} \Delta_n - V_{\text{ext}}^{(n)} \right] \widehat{\Psi}.$$

i.e. just the quantum mechanical Schrödinger equation but now *as an operator equation*!

Note that the free particle situation may be usefully described in the momentum representation. This means that one looks at the vectors of the one particle Hilbert space $\mathcal{H}^{(1)}$ as functions of the three dimensional momentum variable $\mathbf{p}$, with the wave functions $\psi(\mathbf{r})$ and the momentum functions $\tilde{\phi}(\mathbf{p})$ connected by Fourier transformation. In the rigorous formulation one has creation and annihilation operators usually denoted as $a^*(\tilde{f})$ and $a(\tilde{g})$, heuristically interpretated as

$$a^*(\tilde{f}) = \int\limits_{-\infty}^{+\infty} a^*(\mathbf{p}) \tilde{f}(\mathbf{p}) \, d\mathbf{p}, \quad a(\tilde{g}) = \int\limits_{-\infty}^{+\infty} a(\mathbf{p}) \overline{\tilde{g}(\mathbf{p})} \, d\mathbf{p},$$

with commutation relations

$$[a^*(\tilde{f}), a^*(\tilde{g})] = [a(\tilde{f}), a(\tilde{g})] = 0, \quad [a(\tilde{f}), a^*(\tilde{g})] = (\tilde{f}, \tilde{g}),$$

and for the heuristic operators

$$[a^*(\mathbf{p}), a^*(\mathbf{p}')] = [a(\mathbf{p}), a(\mathbf{p}')] = 0, \quad [a(\mathbf{p}), a^*(\mathbf{p}')] = \delta(\mathbf{p} - \mathbf{p}').$$

The operators $\widehat{\Psi}(f)$ and $a(\tilde{f})$ are connected by the simple relation $\widehat{\Psi}(f) = a(\tilde{f})$, with $\tilde{f}$ the Fourier transform of $f$, of which the one dimensional version was given in Sect. 4.2.

(9.7.4,c) **Problem** Show that this implies that the heuristic operators $\widehat{\Psi}(\mathbf{r})$ and $a(\mathbf{p})$ are connected by the same Fourier transformation formula *as an operator relation*.

(9.7.4,d) **Problem** Show that the Hamiltonian operator for a free particle in $\mathcal{H}_s$ has the heuristic form

$$(H_0)_s = \int\limits_{-\infty}^{+\infty} a^*(\mathbf{p}) \frac{\mathbf{p}^2}{2m} a(\mathbf{p}) d\mathbf{p}.$$

(9.7.4,e) **Problem** Derive the time evolution formula for the Heisenberg picture operator $a(\mathbf{p}, t)$, with, of course, $a(\mathbf{p}, 0) = a(\mathbf{p})$.

Back to the position representation, to the formulation of time evolution for particles with pair interaction in the heuristic Fock space formalism. We give the main result as a problem which can be solved by using in a standard manner the general time evolution equation for Heisenberg picture operators and the commutation rules for the field operators.

**(9.7.4,f) Problem** Show that for particles with pair interaction the Heisenberg picture field operator $\widehat{\Psi}^*(\mathbf{r}, t)$ satisfies the equation

$$\frac{\partial}{\partial t}\widehat{\Psi}(\mathbf{r}, t) = \frac{i}{\hbar}\left[\frac{\hbar^2}{2m}\Delta + 2\int_{-\infty}^{+\infty}\widehat{\Psi}^*(\mathbf{r}', t)V^{(2)}(\mathbf{r}, \mathbf{r}')\widehat{\Psi}(\mathbf{r}', t)\,d\mathbf{r}'\right]\widehat{\Psi}(\mathbf{r}, t).$$

This operator equation is no longer equal in form to a numerical one particle wave equation. Such a correspondence would in fact be impossible because a pair interaction needs a numerical wave equation in a Hilbert space for at least two particles. Nevertheless, the field operator equation is still very simple and remains illustrative of the great power of the Fock space formalism, namely the possibility of putting the information on all possible $n$-particle wave equations into a single simple operator equation. Note that all this could be formulated rigorously using the concepts discussed in Sect. 9.6.2, but much of the elegance and transparency of the heuristic formulation would be lost.

## 9.7.5 The Fock Space Formalism and Quantum Field Theory

The Fock space formalism presented in the preceding section is an example of a *quantum field theory*. In this chapter quantum field theory was used for the physics of non relativistic particles; the relativistic version, to be discussed in Chap. 16, is the main and in fact indispensable theoretical vehicle for elementary particle or high energy physics.

A quantum field theory in general is characterized by one or more quantum fields, operator valued functions—strictly speaking generalized functions—depending on the position variable $\mathbf{r}$, or $\mathbf{r}$ together with the time variable $t$. They generate the full Hilbert space of the theory from a ground state or 'vacuum state' $\Psi_0$, but do not have a direct physical meaning. They contain nevertheless all physical information of the system, in particular on the time evolution and on symmetries.

In standard nonrelativistic quantum mechanical formalism one describes a state of an $n$-particle system by an *n-particle wave function* $\psi(\mathbf{r}_1, \ldots, \mathbf{r}_n)$, while in the Fock space formalism one has a *field operator* $\widehat{\Psi}(\mathbf{r})$, depending on a single variable $\mathbf{r}$, but which nevertheless contains the quantum mechanical information on the full range of all $n$-particle states. In the Fock many particle formalism the underlying $n$-particle Schrödinger wave functions constitute the quantum part of the theory; there is no 'second quantization', only an auxiliary mathematical structure which is very

convenient but adds nothing physical to what is unfortunately called the 'first quantized theory'. This is contrary to the situation in relativistic quantum field theory.
There the field operators are the primary quantum objects; there are no underlying $n$-particle quantum mechanical wave functions—at least not in a proper sense,
only either classical fields—think of electromagnetism—or auxiliary $\mathbb{R}$ or $\mathbb{C}$-valued
functions.

*In view of this it must be emphasized that in each of the two cases there is only
one 'quantization', in the case discussed in this chapter this is quantum mechanics,
in the second case relativistic quantum field theory. That is why the term 'second
quantization' is misleading.*

The two cases are mathematically similar, but the underlying common mathematical structure has a quite different physical interpretation. In the application of
the mathematics there remains moreover an important difference: the non relativistic Fock space formalism can be stated in a completely rigorous form—as we have
shown—even though most physics textbooks use a convenient heuristic language.
The same language used for relativistic quantum field theory conceals serious mathematical problems, which since the inception of the theory in the 1940s up to the
present day have drawn a lot of attention but have not been solved. This will briefly
be discussed in Chap. 16.

# References

1. Wick, G.C., Wightman, A.S., Wigner, E.P.: The intrinsic parity of elementary particles. Phys.
   Rev. **88**, 101–105 (1952). http://dieumsnh.qfb.umich.mx/archivoshistoricosMQ/ModernaHist/
   Wick.pdf (This is the basic reference for this topic. All later papers refer to it)
2. Fredenhagen, K.: Superselection Sectors. Notes from Lectures held at Hamburg University
   1994/1995. www.desy.de/uni-th/lqp/psfiles/superselect.ps.gz
3. Sagan, B.A.: The Symmetric Group. Representations, Combinatorical Algorithms, and Symmetric Functions. 2nd edn. (Springer, Berlin, 2001)
4. Streater, R.F., Wightman, A.S.: PCT Spin and Statistics and All That. Princeton University Press,
   Princeton (2000)
5. Fock, V.A.: Konfigurationsraum und Zweite Quantelung. Z. Phys. **75**, 622–647 (1932)
6. Fock, V.A.: Zur Quantenelektrodynamik. Physikalisches Zeitschrift der Sowjetunion **6**,
   425–469 (1934) English translation as: On quantum electrodynamics. In: Faddeev, L.D., Khalfin,
   L.A., Komarov, I.V., Fock, V.A. (eds.) Selected Works. 331–368. Chapman and Hall (2004)
7. Cook, J.M.: The mathematics of second quantization.Trans. Amer. Math. Soc. **74**, 222–245
   (1953). http://www.ncbi.nlm.nih.gov/pmc/articles/PMC1063391/pdf/pnas01568-0037.pdf
8. Bruus, H., Flensberg, K.: Many Body Quantum Theory in Condensed Matter Physics. An Introduction. Oxford University Press, New York (2004)

# Chapter 10
# Review of Classical Statistical Physics

## 10.1 Introduction

One of the main themes of this book, emphasized from the beginning, is that quantum theory is in an essential manner a *probabilistic* theory. The word "statistical" in the title of this and the next chapter may therefore look surprising or perhaps unnecessary and superfluous. However, this impression is misleading. In the quantum theoretical description of large systems consisting of very many small subsystems there are *two layers of statistical behaviour*, one using classical notions, and an additional one expressing quantum properties. In other words one has a combination or mixture of classical and nonclassical probabilistic notions. This makes it necessary to understand first the basics of classical statistical physics before going to quantum statistical physics. It is for this reason that, before quantum statistical physics is discussed in the next chapter, a review of classical statistical physics is given in this chapter.

There are very many textbooks that treat both classical and quantum statistical mechanics. Good examples are the books by Balian [1] and Van Vliet [2]. The mathematical foundations are discussed in the books by Ehrenfests [3] and Gallavotti [4]. Historically, the basic reference is the book by Gibbs [5]—still well worth reading.

Classical statistical mechanics has a long history. As a description of physical systems, such as—typically—a large container with gas, it was developed from the second half of the nineteenth century and the beginning of the twentieth century onwards by, among others, Ludwig Boltzmann, James Clerk Maxwell, and above all by Josiah Willard Gibbs, the first American theoretical physicist of real distinction, who can rightly be called the father of the subject. They tried, partially successfully, to derive a statistical description of macroscopic systems from first principles, i.e. from classical mechanics of many particle systems. The result was a system of postulates with certain special probability measures as basis for the theory.

There is one more reason to start with the classical theory. At the time quantum mechanics appeared on the stage, classical statistical physics was already established as a successful theory. One might have tried to start again at the beginning and derive

© Springer International Publishing Switzerland 2015
P. Bongaarts, *Quantum Theory*, DOI 10.1007/978-3-319-09561-5_10

a quantum version of statistical physics in an analogous way as one did in the classical case, i.e. by going back to the underlying mechanics—quantum mechanics in this case—of a system of many particles. Instead the general approach consisted of just introducing a system of postulates for quantum statistical mechanics which was as much as possible analogous to the classical system. Without knowledge of the classical postulates, their origin and the arguments that led to them, these postulates would look arbitrary.

Because classical statistical mechanics is built on classical mechanics and classical probability theory, we will need the material from Chap. 2 and Supp. Chap. 20.

In describing a system consisting of very many small subsystems, for example a gas with roughly $2.5 \times 10^{19}$ molecules in a cubic centimeter, one is not interested in the behaviour of the individual systems—even if a description of this would be technically possible—but in global macroscopic properties, obtained as suitable *averages* over the subsystems. In classical physics this is the domain of *classical statistical physics* and in particular *classical statistical mechanics*. Note that we use the term 'classical statistical mechanics' for a theory for which the underlying 'microscopic' theory is (Hamiltonian) classical mechanics, and 'classical statistical physics' for more general cases. An important example of the second possibility will be given in Sect. 10.12.

The main aim of the founders of classical statistical mechanics was to derive by statistical arguments the laws of thermodynamics, an already existing macroscopic and phenomenological description of bulk systems, from the more fundamental mechanical equations governing the microscopic world. Therefore next a brief intermezzo on thermodynamics.

## 10.2 Thermodynamics

Many textbooks treat thermodynamics and statistical mechanics together as 'statistical thermodynamics'. This can be defended, as far as applications go, on practical grounds. Here we favour a presentation in which it is emphasized that the two theories represent different ways of looking at the same physical situations, each with its own concepts and methods. Thermodynamics is the older theory, a phenomenological macroscopic description and as such—it should be added—in no way outdated; statistical mechanics was developed later from a growing awareness of the existence of an underlying microscopic world of atoms and molecules, which could explain the observed macroscopic data. It should be noted that Planck, who is the father of thermodynamics in its final form, as formulated at the end of the nineteenth century, thought in the beginning of his career that statistical ideas for explaining thermodynamics were irrelevant, an opinion that he later changed drastically, convinced by the work of Boltzmann. See the English translation of Planck's 1897 textbook on thermodynamics [6].

The aim of the founders of statistical mechanics was indeed to derive thermodynamics. In a precise logical sense this means proving that there is an isomorphism between statistical mechanics, in a certain infinite limit—the so-called thermodynamic limit—and thermodynamics, such that in this limit the main quantities of statistical mechanics can be mapped onto the main thermodynamic variables, and in such manner that the equations governing the relations of both systems of variables are similar. This has been done successfully, so far not completely rigorously, but nevertheless in a fairly convincing way. Therefore this chapter discusses three subjects; thermodynamics, statistical mechanics, and the connection between the two; this section gives a short overview of thermodynamics. For an excellent general textbook on thermodynamics, see the book by Gokcen and Reddy [7]. Further useful books are those by Buchdahl [8] and Callen [9].

Thermodynamics deals with general features of the overall behaviour of macroscopic systems, which it describes by a small number of *state variables*. Historically it preceded statistical mechanics; going back to the first half of the the nineteenth century and originating in attempts to understand phenomena like the conversion of heat into mechanical work, as takes place, for instance, in steam engines. From this it evolved further into a general phenomenological theory of matter with as basic notions work, mechanical or otherwise, pressure, heat, temperature, and with derived notions as internal energy, entropy, etc. As a science it kept for a long time a strong applied flavour. The Frenchman Sadi Carnot can be regarded as its founder; other important contributors were Rudolf Clausius, Robert Mayer, William Thomson— later Lord Kelvin, and James Joule, who coined the word 'thermodynamics'.

The final version of the theory, as it is still with us today, is a tightly knit framework of general mathematical rules and relations, based on a few 'laws', axioms or postulates formulating the connections between the main variables from which other relations between observed quantities can be derived. These laws are traditionally stated in terms of 'infinitesimal' changes of the state variables. Thermodynamics is a purely phenomenological macroscopic theory; no assumptions need to be made on the microscopic nature of matter. In fact, during most of its historical development many physicists did not believe in the real existence of atoms and molecules.

In general, an isolated macroscopic physical system will approach equilibrium and will eventually become a stationary state, even though the time needed for this may be very long and in any case different for different systems. Nonequilibrium thermodynamics is not without interest. It remains however a battle field of conflicting opinions, although important generally accepted results have been established. The discussion here will be restricted to equilibrium thermodynamics, for us the heart of the subject.

Let us consider, for the sake of concreteness, the fairly simple standard system of a liquid or a gas of fixed total mass $M$, contained in a container with variable volume $V$. The *Zeroth Law* says that there is an overall quantity, the (absolute) temperature $T$. Two systems in equilibrium and in contact with each other have the same temperature. A system has a total energy, called the *internal energy $U$*. The *First Law* states that this energy is conserved, i.e. $U$ changes 'infinitesimally' in a precisely described manner by the supply of an 'infinitesimal' amount of heat $dQ$ from the external world and by work $pdV$ on the system—due to the displacement of the wall in the direction of the force, with $p$ the external pressure on the container. This is expressed by the infinitesimal formula

$$dQ = dU + pdV.$$

The expression $dQ$ is not in the mathematical sense a differential of a function $Q$; there is in fact no such function. However it turns out that the quotient $(dU + pdV)/T$ is an exact differential form—the inverse of $T$ is a so-called integrating factor, which means that it is the differential of a new in first instance somewhat mysterious state function, the *entropy S*. This is expressed by the infinitesimal relation

$$TdS = dU + pdV,$$

which, in this context, is the *Second Law* of thermodynamics.

This simple system is described by three state variables, for instance $p$ and $V$, two of these independent, the third connected to these by an *equation of state* $F(p, V, T) = 0$. This means that the states of this thermodynamical system can be represented by the points of a two dimensional submanifold of $\mathbb{R}^3$. This geometric picture is due to the Greek mathematician Constant Carathéodory; his role in the formulation of thermodynamics can be compared to the role of von Neumann in quantum theory, with a similar fate: very few physics textbooks mention him or his work. However, Lieb and Yngvason have recently made important contributions to the subject in this spirit. See e.g. [10].

At certain values of the thermodynamical variables different states—different *phases*—may coexist in equilibrium with each other. An obvious example is water and ice at $0\,°C$, at various values of $p$ and $V$. There is even a 'tricritical point' at which three phases, namely water, ice and vapour, coexist in equilibrium. This corresponds with singularities in the manifold of thermodynamic states; which may have critical lines or points. Crossing such a line indicates a *phase transition*. The situation in which different phases coexist can be qualitatively understood by thermodynamics; for a quantative description, e.g. of precise transition temperatures, one needs to understand the underlying microscopic picture.

The differential expression involving $S$ can be written as $dS = \frac{1}{T}dU + \frac{p}{T}dV$, indicating that one may regard $S$ as a function of $U$ and $V$. One may introduce a new variable, the *free energy F*, by performing a so-called Legendre transformation $F = U - TS$, and a further Legendre transformation $G = F + pV$, with $G$ the *Gibbs free energy*. Such transformations give formulations which reflect different points of view for the same system. In the first case $F$ is the basic function depending on the variables $V$ and $T$, in the second case $G$ depending on $T$ and $p$. The functions $F$ and $G$ are called *thermodynamic potentials* and are important in establishing the connection between statistical mechanics and thermodynamics.

(10.2,a) **Problem** Derive from the differential expression for $S$ similar expressions for $F$ and $G$.

Different and more complicated thermodynamical systems have in addition to the internal energy, temperature and entropy, more and other state variables. One may look at systems in containers with permeable walls, i.e. with variable mass $M$, leading to a quantity $\mu$ called the *chemical potential*, and with more components with masses $M_j$ and chemical potentials $\mu_j$. One has magnetic systems in a magnetic field $B$ with a corresponding magnetization of the system, etc.

## 10.3 Classical Statistical Physics (Continued)

In classical statistical mechanics one starts from the phase space $\Gamma$ of a classical mechanical system, in general a symplectic manifold, as was explained in Chap. 2. Let us for simplicity look at a system of $N$ identical point particles enclosed in a large but finite box with volume $V$, so $\Gamma$ is the product of a $3N$-dimensional hypercube with $\mathbb{R}^{3N}$. $\Gamma$ carries a symplectic 2-form $\omega = \sum_{j=1}^{3N} dp_j \wedge dq_j$, written in position and momentum coordinates $(p, q)$, numbered consecutively as $(p_1, \ldots, p_{3N}, q_1, \ldots, q_{3N})$, and a Hamiltonian function $H(p, q)$, a sum $\sum_{j=1}^{3N} p_j^2/2m$ of the kinetic energies, together with a potential energy term which takes care of the interaction between the particles and the interaction between the particles and the wall. In first instance the system is supposed to be *isolated*, with no exchange or interaction of any kind with the world outside the box. There are forces between the particles, but no external forces, except those from the walls, supposed to act only over a very short distance, with a negligible influence on the overall behaviour of the system. The total energy $H(p, q)$ is a constant of the motion.

The central notion of statistical mechanics is that of an *ensemble*, a term coined by Gibbs. Modern physics textbooks describe it in various different but related ways. One book speaks of a "cloud of identical virtual systems moving through phase space", another defines it as "simply a (mental) collection of a very large number of systems, each constructed to be a replica on a thermodynamic (macroscopic) level of the actual thermodynamic system whose properties we are investigating". These colourful characterizations obscure the simple fact that an ensemble is just a *probability distribution in the phase space* $\Gamma$ which embodies our incomplete knowledge of the microscopic state of the system. This is, by the way, an example of how the development of modern mathematics has sometimes passed by physics where archaic notions or at least archaic terminology continue to be in general use, such as in this case, where the original terminology introduced by Gibbs in the beginning of the twentieth century has survived the modern formulation of probability theory of Kolmogorov in the nineteen twenties. See for this the remarks in the Preface of Khinchin's book [11].

Having in this way set the stage for statistical mechanics, the main task is the choice of suitable ensembles which can be used to calculate averages of physical quantities and, finally, after a limit process, to establish the correspondence with thermodynamics. Note that because of the absence of time-dependent external forces these ensembles should be stationary, i.e. time-independent.

For the choice of ensembles we may distinguish historically two approaches, conceptually different but complementary, partially overlapping and leading finally to the same results. One goes back, in a certain not too specific sense, to Boltzmann, the other, in an equally rough sense, to Gibbs. The first is based on the study of the asymptotic behaviour in time of dynamical systems, with ergodicity as a fundamental concept, the second has entropy and its maximalization as its driving principle.

## 10.4 The Three Main Ensembles

Different ensembles are used for different physical situations. The following three are the main ones:

A. *The microcanonical ensemble*

This ensemble describes an isolated system in a container with impermeable walls. It is characterized by the parameter $E$, the total energy of the system, which becomes the internal energy $U$ in the identification with thermodynamics.

B. *The canonical ensemble*

The canonical ensemble describes a system which can exchange energy, but not matter, with its surroundings. The parameter that characterizes it is $\beta$, which in the thermodynamic limit is proportional to the inverse of the absolute temperature $T$.

C. *The grand canonical ensemble*

This ensemble describes a system that is able to exchange both energy *and* matter with the surrounding world. It has two parameters, $\beta$, and what in thermodynamics will become the chemical potential $\mu$.

Ensemble A is the basic ensemble in the Boltzmann approach. B and C are derived from it. The Gibbs approach has B as its point of departure. In the next subsection A will be reviewed in some detail in the spirit of Boltzmann, with the derivation of B and C discussed briefly. After that, in Sect. 10.8, ensemble B will be looked at as the starting point for the Gibbsian approach.

## 10.5 The Microcanonical Ensemble

A fairly convincing argument leading to this ensemble comes from Liouville's theorem: the canonical measure associated with the symplectic structure on $\Gamma$ is invariant under time evolution. For our example of $N$ particles in a box this measure is just the Lebesgue measure $dpdq = dp_1 \ldots dp_{3N} dq_1 \ldots dq_{3N}$.

An obvious suggestion therefore is to assign equal a priori probabilities to sets in $\Gamma$ with equal Lebesgue measure. The total Lebesgue measure of $\Gamma$ is infinite, so Lebesgue measure is not a probability measure on the full $\Gamma$. However the subset $\{(p, q) \in \Gamma \mid E_1 \leq H(p, q) \leq E_2\}$, which is invariant under time development, has finite Lebesgue measure. Therefore, a good candidate for the stationary ensemble for a probabilistic description of an isolated system of $N$ particles in a box with volume $V$, given that one knows that its energy is in a small band between $E$ and $E + \Delta E$, is the *approximate microcanonical ensemble*, that gives the average of a function $f$ on $\Gamma$ as

$$\bar{f}^{mc}_{E,\Delta E} = \frac{1}{\Omega(E, \Delta E, V)} \int\limits_{E \leq H(p,q) \leq E+\Delta E} Cf(p, q)dpdq,$$

with the normalization function

$$\Omega(E, \Delta E, V) = \int\limits_{E \leq H(p,q) \leq E+\Delta E} Cdpdq.$$

$C$ is arbitrary, a constant with dimension $([ML^2T^{-1})^{-3N}$, the dimension of the inverse of an action to the power $3N$. It makes $\Omega$ dimensionless, which will be convenient later. In many physics textbooks $C = h^{-3N}$ or $(N!h)^{-3N}$, with $h$ Planck's constant, the original one without the factor $2\pi$, as in the formulas in the quantum version of statistical physics. Using at this point quantum notions is however conceptually inappropriate or at least premature.

(10.5,a) **Problem**   One may write $\bar{f}^{mc}_{E,\Delta E}$ as $\int_\Gamma f(p, q)\rho^{mc}_{E,\Delta E}(p, q)dpdq$, with a probability density $\rho^{mc}_{E,\Delta E}(p, q)$. What is the expression for this probability density? Hint: use the Heaviside function $H$ defined as $H(x) = 0$, for $x < 0$ and $= 1$ for $x \geq 0$.

   Idealization to a sharp energy value $E$ leads to the (sharp) *microcanonical ensemble*

$$\bar{f}^{mc}_E = \lim_{\Delta E \to 0} \frac{1}{\Delta E} \int\limits_{E \leq H(p,q) \leq E+\Delta E} Cf(p, q)dpdq$$

$$= \frac{1}{\Omega(E, V)} \int\limits_{\Gamma_E} C\frac{f(p, q)\, d\sigma_E}{\mathrm{grad}\, H(p, q)}.$$

In this $\Gamma_E$ is the hypersurface determined by $H(p, q) = E$. The normalization function $\Omega(E, V)$, called the *microcanonical partition function*, important for the connection with thermodynamics, is equal to $\int_{\Gamma_E} C(\mathrm{grad}\, H(p, q))^{-1}d\sigma_E$; $d\sigma_E$ is the measure on $\Gamma_E$, induced by the Lebesgue measure on $\Gamma$, and $\mathrm{grad}\, H(p, q)$ the square root of $\sum_{j=1}^{3N} |\partial H/\partial p_j|^2 + \sum_{j=1}^{3N} |\partial H/\partial q_j|^2$. Note that the surface element $d\sigma_E$ is not invariant under time evolution; the combination with $\mathrm{grad}\, H(p, q)$ however is invariant.

   The limit probability measure is concentrated on the hypersurface $\Gamma_E$; it is—strictly speaking—no longer represented by a probability density on $\Gamma$. Physics textbooks often write it as a singular density on $\Gamma$ by using the Dirac delta function—see for this Supp. Chap. 25.

Early in the history of statistical mechanics Boltzmann and others put also forward arguments, different from, but related to that of equal a priori probabilities for sets of equal Lebesgue measure in $\Gamma$. These were based on general discussions of the behaviour in time of mechanical systems described by first order ordinary differential equations—dynamical systems as they would now be called. It was argued that measurement of a dynamical variable $f(p, q)$ would always mean measurement of a *time average*

$$\lim_{T \to \infty} \frac{1}{T} \int_{t_0}^{t_0+T} f(p(t), q(t)) \, dt,$$

for every trajectory $(p(t), q(t))$, and independent of the initial point $(p(t_0), q(t_0))$. One assumed that trajectories in $\Gamma$ would fill a surface $\Gamma_E$ of constant energy completely, returning to every initial point after some possibly very large time. This together led to the idea that, on $\Gamma_E$, time averages are equal to the phase space average over this surface, which is the microcanonical average. The time evolution of the system was supposed to be *ergodic*. Several mathematical theorems supporting this idea were proved: a pointwise ergodic theorem by Birkhoff in [12], and an $L^2$ ergodic theorem by Neumann in [13]. It became clear, however, that these theorems were only relevant to a very restricted collection of real physical systems. Ergodic theory has since grown into an independent and important mathematical subject, with less and less relevance to statistical mechanics.

## 10.6  The Canonical Ensemble

The microcanonical ensemble may be the basic ensemble of statistical mechanics, at least in the approach of Boltzmann and his successors, but it is not very practical in applications. A second more useful ensemble can be derived from it—or strictly speaking, made plausible.

Consider a very large isolated system of $N_L$ particles in a box with volume $V_L$, described by the (sharp) microcanonical ensemble at energy $E_L$. Suppose that it contains a much smaller, but still quite large subsystem, the system that we want to describe, with $N$ particles in a box with volume $V$. The two systems are separated by a wall, impermeable for matter but with the possibility that energy of the particles can be exchanged. An important consequence of this is that the energy, a constant of the motion for the total system, is no longer conserved for the subsystem. Thermodynamically this corresponds with the situation of a system in contact with a *heat bath*, as will be discussed later.

The microcanonical distribution of the large system induces a probability distribution in the subsystem which one obtains by integrating over the variables not belonging to the subsystem. One then takes in the large system the limit $V_L, N_L \to \infty$, with $N_L/V_L$ remaining fixed, if this limit exists. The result is an ensemble for the subsystem, which is finally the system of interest.

It is a fairly elementary exercise to carry this out, i.e. to do the multiple integral and calculate the limit, for the ideal gas, a system of freely moving point particles with Hamiltonian $H(p, q) = \sum_j p_j^2/2m$. It can be found in most of the standard textbooks on statistical mechanics. The result is the *canonical ensemble* given by the probability density

$$\rho_\beta^{can}(p, q) = \frac{1}{Z(\beta, V)} C e^{-\beta H(p,q)},$$

with the *canonical partition function*

$$Z(\beta, V, N) = \int_\Gamma C e^{-\beta H(p,q)} dp dq$$

as normalization. It is characterized by a parameter $\beta$, determined by the given environment, i.e. by the fixed value $N_L/V_L$, and connected to the inverse of the absolute temperature $T$ in thermodynamics as $\beta = 1/k_B T$. It will reappear in Sect. 10.9. Note that the energy $E = H(p, q)$ of the resulting system is now a random variable.

For the case of an ideal gas this derivation of the canonical ensemble is mathematically rigorous. The result is a legitimate probability distribution, but physically meaningless, because the idea behind this energy reservoir set up is that a small system is placed in a large reservoir with which it can exchange energy and that because of this it reaches after some time an equilibrium state as a result of this exchange. However, in this derivation there is no interaction between the given system and its larger environment.

The important contribution of Khinchin was to show, using limit theorems of probability theory, that the ideal gas result remains valid for a large class of systems in which there is a nontrivial interaction between the particles over sufficiently short distances. See [11]. It should be noted that almost none of the many physics textbooks on statistical mechanics mention Khinchin's work. There are, to the best of the author's knowledge, two exceptions: one is the book by Münster [14], the other a book on statistical mechanics and elasticity theory, surprisingly, which explains in a clear manner the main elements of Khinchin's ideas. It is a book by Weiner [15, Chap. 2].

## 10.7 The Grand Canonical Ensemble

One may finally consider a system that can exchange energy *and* matter with its surroundings. Arguments similar to those used in the preceding case lead to an ensemble, the *grand canonical ensemble*, in which both the energy and the number of particles are stochastic variables. The space on which it is defined is the disjoint union $\Gamma = \sqcup_{N=0}^{\infty} \Gamma_N$, with $\Gamma_N$ the phase space for the system having $N$ particles—$\Gamma_0$ consisting of a single point. Each element of $\Gamma$ is a point from one particular space $\Gamma_N$; a function on $\Gamma$ is represented by a sequence $f = \{f^N\}_{N=0}^{\infty}$ of functions $f^N(p^N, q^N)$ on $\Gamma_N$. An example is the Hamiltonian, a sequence $H = \{H_N\}_N$, consisting of functions $H_N(p^N, q^N)$, and, trivially, $H_0 = 0$, with $(p^N, q^N) = (p_1, \ldots, p_N, q_1, \ldots, q_N)$, for $N = 1, 2, \ldots$

The grand canonical ensemble is described by an infinite sequence of probability densities on the $\Gamma_N$, in fact a sequence of weighted canonical densities, characterized by $\beta$ and a new parameter, the chemical potential $\mu$, at this point a purely statistical mechanical parameter, later to be identified with the chemical potential from thermodynamics. The grand canonical average of a stochastic variable $f = \{f^N\}_{N=0}^{\infty}$ is, in this spirit,

$$\overline{f}^{\text{gc}} = \frac{1}{\Xi(\beta, \mu, V)} \sum_{N=0}^{\infty} e^{\beta \mu N} (\overline{f^N})^{\text{can}} .$$

with

$$(\overline{f^N})^{\text{can}} = \int_{\Gamma_N} f^N(p^N, q^N) C e^{-\beta H_N(p^N, q^N)} dp^N dq^N ,$$

with the *grand canonical partition function*, a weighted sum

$$\Xi(\beta, \mu, V) = \sum_{N=0}^{\infty} e^{\beta \mu N} Z(\beta, V, N),$$

and $Z(\beta, V, N)$ the canonical partition function, for $N = 0, 1, 2, \ldots$

## 10.8 The Canonical Ensemble in the Approach of Gibbs

For Gibbs the basic ensemble was the canonical ensemble. He characterized it by the requirement of maximality of what he called the "average index of probability"

$$S[\rho] = - \int_{\Gamma} \rho(p, q) \ln \rho(p, q) dp dq,$$

—or what now would be called the statistical entropy—with as constraint a given average of the total energy $\overline{E}$. He had no proof for this assumption; in fact it is still not possible to derive it directly from the microscopic dynamics of the system; nevertheless it has turned out to be very powerful as a general postulate. Recently it has been given a general interpretation in the context of information theory, developed by Claude Shannon in the late forties, and applied later to statistical mechanics by E.T. Jaynes. See Shannon's basic paper [16] and two papers by Jaynes [17]. Jaynes's general position is that uncertainty in a physical situation is caused by information deficiency; he calls the entropy the *missing information*. His work can be seen as a natural continuation and clarification of that of Gibbs. It is strange therefore that it has had not much influence on the teaching of statistical mechanics. Even in modern textbooks his name, let alone his work, is usually not mentioned. Exceptions are three recent books, two by Grandy [18, 19] and one by Ben-Naim [20], that both make a strong case for basing statistical mechanics on Jaynes's maximal entropy principle.

## 10.9 From Statistical Mechanics to Thermodynamics

Thermodynamics was developed in the early nineteenth century as a general theory of heat and work in matter. When it became gradually clear that matter consisted of atoms and molecules, a need arose for a more fundamental microscopic mechanical explanation. The aim of statistical mechanics was to provide such an explanation. In this subsection the relation between thermodynamics and statistical mechanics will be discussed.

One cannot simply identify statistical mechanics with thermodynamics, at least not in the form in which it was discussed in the preceding sections. An intermediate step is needed to make the connection; one has to take the *thermodynamic limit* of statistical mechanics. That the existence of such a limit could not be taken for granted was first realized by van Hove in 1949, who wrote pioneering articles on this subject. It later became a major field of research in which further important mathematically rigorous contributions were made by Lieb, Ruelle, Lanford and many others. See in particular [21, 22].

In studying the thermodynamic limit there are two options:

1. One may try to prove that the principal quantities, like, for instance, the energy per volume or per particle, have a finite limit when the volume or particle number goes to infinity. The existence and the properties of the thermodynamic limit in this sense has been successfully studied for particular classes of concrete systems, with various types of interactions between particles. A good impression of the flavour of the sort of rigorous functional analysis needed for this type of work is given in the book by Catto et al. [23]. General theorems are however still lacking.
2. Taking limits over different sequences may lead to different results. This phenomenon is easier to handle by studying directly an infinite system. Of course, a real system is finite, with a finite number of particles in a finite volume. In statistical physics it is nevertheless so large that an infinite system is a useful mathematical idealization. For such an approach new mathematical methods from operator algebra theory have been developed. A bit more on this will be said in Chap. 12.

Let us finally give the explicit formulas for the relation between the main notions of statistical mechanics and thermodynamics, the principal goal of statistical mechanics right from its beginning. In each of the three main ensembles there is a central notion, the *partition function*. Each of these can be identified, using an appropriate version of the thermodynamic limit—as we discussed, with a basic thermodynamical quantity, in two possible varieties, as explained in the foregoing.

a. The *microcanonical ensemble*. The central notion is the microcanonical partition function, given in Sect. 10.5 as

$$\Omega(E, V) = \int_{\Gamma_E} C(\operatorname{grad} H(p, q))^{-1} d\sigma_E$$

Behind this is the well-known relation $S = k_B \ln \Omega$, with $S$ the *(thermodynamic) entropy* $S$, and $k_B$ Boltzmann's constant, equal to

$$k_B = 1.3806503 \times 10^{-23} \,\mathrm{J\,K^{-1}}.$$

As $S = k \ln W$ it can be found on Boltzmann's tombstone in the Zentralfriedhof (Central Cemetery) in Vienna. It was central to Boltzmann's ideas on statistical mechanics, even though the formula in this explicit form is nowhere to be found in Boltzmann's work. In any case the relation statistical mechanics to thermodynamics for this ensemble is $\Omega \Longrightarrow S$.

b. The *canonical ensemble*. The canonical partition function is given in Sect. 10.6 as

$$Z(\beta, V, N) = \int_{\Gamma} C e^{-\beta H(p,q)} dp\, dq$$

The thermodynamical quantity that is related to this partition function is the *(Helmholtz) free energy* $F$ through the formula $F = -k_B \ln Z$, with $\beta = 1/k_B T$, so one has $Z \Longrightarrow F$.

c. The *grand canonical ensemble*. The partition function in this case is given in Sect. 10.7 as

$$\Xi(\beta, \mu, V) = \sum_{N=0}^{\infty} e^{\beta \mu N} Z(\beta, V, N),$$

with $Z(\beta, V, N)$ the canonical partition function, for $N = 0, 1, 2, \ldots$. This is again related with a thermodynamical quantity, the *Gibbs free energy* $G$ through the formula $G = -k_B \ln \Xi$. The relation between statistical mechanics and thermodynamics is in this case $\Xi \Longrightarrow G$.

## 10.10 Summary

Let us in conclusion formulate the basics of classical statistical mechanics in a small set of postulates, valid, for instance, for a mechanical system of $N$ particles.

The first postulate is:

1. The basis of equilibrium statistical mechanics is the choice of appropriate ensembles, i.e. probability measures on the phase space of the classical system, to be used for the calculation of averages of physical quantities, leading to satisfactory macroscopic, i.e. thermodynamic, descriptions of the system.

Then there are postulates on the choice of such ensembles for various physical situations. The three most common are:

2. An isolated system with given total energy is described by the microcanonical ensemble.
3. A system in contact with an energy reservoir is described by the canonical ensemble.
4. A system in contact with a heat reservoir with which it can exchange matter is described by the grand canonical ensemble.

These last postulates are suggested either by plausible arguments and partial proofs in the Boltzmann-Khinchin approach, or follow from the Gibbs-Jaynes maximal entropy principle, which cannot be derived from the microscopic properties of the system, but may be regarded as a methodological choice for statistical inference. In addition to these axioms there are rules for the correspondence with the macroscopic theory of thermodynamics.

## 10.11 Kinetic Gas Theory

The kinetic theory of gases was a statistical theory of bulk matter, mainly gases, that preceded statistical mechanics proper. It was developed by Maxwell and somewhat later by Boltzmann, roughly between 1860 and 1880, in order to understand the macroscopic properties of gases, such as the temperature and pressure, from the underlying dynamics of the particles.

By borrowing statistical ideas from the social sciences, Maxwell calculated averages, such as the average velocity of the particles of a system of $N$ point particles in a fixed volume $V$. This was an average over the points of what was called $\mu$-space, the 6-dimensional space of the position and momentum coordinates of a single particle, not the $6N$-dimensional phase space, later introduced by Gibbs. Within statistical physics, kinetic gas theory is therefore a distinct subject, even though by now most of its results can be derived from more general statistical mechanics in the sense of Gibbs.

One of the main results of kinetic gas theory, in the description of ideal or near-ideal gases, was a derivation, from the general principles of free particle motion, of the probability distribution for the velocity of the molecules. It is given by the formula

$$f(v) = 4\pi v^2 \left(\frac{m}{2\pi k_B T}\right)^{3/2} e^{-mv^2/2k_B T},$$

and is now known as the *Maxwell-Boltzmann distribution*. Looking back from the modern view point of statistical mechanics, it is easy to derive this formula from ensemble theory, in particular from the canonical ensemble. In the following problem the reader is asked to do this.

(10.11,a) **Problem**  Derive the Maxwell-Boltzmann distribution $f(v)$ for the case of the ideal gas, i.e. for a system of $N$ point particles with Hamiltonian $H(p, q) = \sum_{j=1}^{3N} p_j^2/2m$. Hint: use the fact that in this case the canonical distribution function $\rho_\beta^{can}(p, q)$ is the $N$-fold product of $N$ identical one particle distribution functions $\rho_{\beta,1}^{can}(\vec{p}, \vec{q})$, and use the standard Gaussian integrals, for $a > 0$,

$$\int_0^\infty e^{-ax^2} dx = \frac{1}{2}\sqrt{\frac{\pi}{a}}, \quad \int_0^\infty x^2 e^{-ax^2} dx = \frac{1}{4}\sqrt{\frac{\pi}{a^3}}.$$

In quantum statistical physics we shall meet the Bose-Einstein and Fermi-Dirac distributions, interesting deviations from the classical Maxwell-Boltzmann distribution discussed here.

## 10.12  General Statistical Physics: The Ising Model

In this book we make a distinction between *quantum mechanics* and *quantum theory*. Quantum mechanics is the quantum version of classical (Hamiltonian) mechanics; quantum theory is more general. We similarly distinguish between *classical statistical mechanics* and the more general *classical statistical physics*, the title of this chapter. To illustrate this distinction we discuss an example of a classical statistical model, not based on an underlying Hamiltonian particle system: the *Ising model*.

The Ising model is a simple model in statistical physics. In its one dimensional form it is due to the German physicist Wilhelm Lenz, who suggested it as a problem to his student Ernst Ising, who gave in 1924 a complete solution, i.e. the canonical partition function in closed form [24].

The model in its general form has $N$ discrete variables $s_1, \ldots, s_N$, 'spins', each taking the values $-1$ or $+1$. There is an energy function which has the general form

$$E(s_1, \ldots, s_N) = -\frac{1}{2} \sum_{j,k=1}^{N} J_{jk} s_j s_k,$$

with the $J_{ij} = J_{ji}$ real constants that determine the strength of the interaction between the $i$ and $j$. Studying the model means studying the canonical partition function

$$Z(\beta, N) = \sum_{s_1=\pm 1} \cdots \sum_{s_N=\pm 1}^{2} e^{-\beta E(s_1,\ldots,s_N)},$$

and, if possible, determining its behaviour in the thermodynamic limit.

The variables $s_i$ can be thought of as spins attached to a crystal lattice in $n$-dimensional space. The physical interpretation of the model is that it gives a drastically simplified picture of magnetism. Think of a solid piece of iron, a crystal with atoms fixed at regular positions, having electrons, spin-$\frac{1}{2}$ particles with a magnetic moment. Of course, this is a quantum mechanical situation; nevertheless, by neglecting the spatial properties of atoms and electrons, and finally, by allowing the spins to have only two values, one gets a classical model, because all the observables of the system commute—in the point of view of this book, the defining property of what 'classical' means.

Ising studied the one dimensional case, a chain of magnets, with the energy function

$$E(s_1, \ldots, s_N) = -\frac{1}{2} J \sum_{j=1}^{N-1} s_j s_{j+1},$$

i.e. with a constant $J > 0$ and nearest neighbour interaction. His result was in a certain sense disappointing. The partition function, or rather the partition function divided by $N$, remains in the thermodynamic limit $N \to \infty$, with $E/N$ remaining constant, a smooth function of $\beta$. This means that the model does not have a phase transition. Ising believed this to be a general result and did no further work on the model.

After many attempts by others, Lars Onsager, with methods of great mathematical sophistication, succeeded in finding an exact solution for the two dimensional square lattice model in 1944. He did not publish the complete result; this was later done by Yang [25]. In this dimension there is indeed a phase transition. The two dimensional case with an external magnetic field $H$ in the direction of the spins, i.e. with an additional energy term $-\sum_{i=1}^{N} H s_i$ has so far not been solved. Onsager received the Nobel Prize in 1968, not for his solution of the Ising model, but for a different important achievement, his reciprocal relations.

The model of Lenz and Ising, simple as it is, has given rise to a large and fruitful area of research—one might say an ever growing industry. The number of papers having the word 'Ising' in the title up till the present has been estimated to be between 10.000 and 20.000.

# References

1. Balian, R.: From Microphysics to Macrophysics, Methods and Applications to Macrophysics. I, II. Springer, Berlin (2007)
2. Van Vliet, C.: Equilibrium and Non-equilibrium Statistical Mechanics. World Scientific, New Jersey (2008) (Two of the many excellent and comprehensive modern textbooks on thermodynamics and statistical mechanics, classical as well as quantum.)
3. Ehrenfest, P., Ehrenfest, T.: The Conceptual Foundations of the Statistical Approach in Mechanics. Dover, New York (1990) (An English translation of Begriffliche Grundlagen der statistischen Auffassung in der Mechanik, published in 1912 in one of the volumes of the Enzyclopädie der mathematischen Wissenschaften. A classic.)
4. Gallavotti, G.: Statistical Mechanics. A Short Treatise. Springer, Berlin (1999). (A mathematically rigorous exposition of classical and quantum statistical mechanics. An internet version can be downloaded at: http://ipparco.roma1.infn.it/pagine/deposito/1998/libro.pdf.)
5. Willard Gibbs, J.: Elementary Principles in Statistical Mechanics: Developed with Special Reference to the Rational Foundation of Thermodynamics. Dover, New York (1960) (The original edition was published in 1902 and can be found at: https://ia601200.us.archive.org/18/items/ElementaryPrinciplesInStatisticalMechanics/Gibbs-ElementaryPrinciplesInStatisticalMechanics.pdf. A classic text, still worth reading.)
6. Planck, M.: Treatise on Thermodynamics. Dover, New York (2010) (Translated from the German. First edition 1897.)
7. Gokcen, N.A., Reddy, R.G.: Thermodynamics., 2nd edn. Plenum Press, New York (1996)
8. Buchdahl, H.A.: The Concepts of Classical Thermodynamics. Cambridge University Press, Cambridge (1966) (Reissued as paperback 2009.)
9. Callen, H.A.: Thermodynamics and an Introduction to Thermostatistics, 2nd edn. Wiley, New York (1985) (Available at: http://keszei.chem.elte.hu/1alapFizkem/H.B.Callen-Thermodynamics.pdf. The first of these two books discusses thermodynamics in an axiomatic manner, the second is an excellent textbook stressing general principles.)
10. Lieb, E.H., Yngvason, J.: The Mathematical Structure of the Second Law of Thermodynamics. In: Current Developments in Mathematics, 2001 pp. 89–130 International Press 2002
11. Khinchin, A.I.: Mathematical Foundations of Statistical Mechanics. Original Russian edition 1943. English translation. Dover (1949) (The first serious discussion of the mathematical foundations of classical statistical mechanics and still the best in what we call the Boltzmann approach. The probabilist J.L. Doob wrote: "[Khinchin] shows how to make classical statistical mechanics a respectable rigorous discipline, with a consistent mathematical content". Note Khinchin's severe criticism of the lack of mathematical rigour in discussions on statistical mechanics by physicists in the preface and the first introductory chapter. These can be found separately at: http://www.gap-system.org/~history/Extras/Khinchin_preface.html, http://www.gap-system.org/~history/Extras/Khinchin_introduction.html.)
12. Birkhoff, G.D.: Proof of the ergodic theorem. Proc. Natl. Acad. Sci. USA **17**, 656–660 (1931)
13. von Neumann, J.: Proof of the quasi-ergodic hypothesis. Proc. Natl. Acad. Sci. U.S.A. **18**, 70–82 (1932)
14. Münster, A.: Statistical Thermodynamics. Springer, Berlin (1969)
15. Weiner, J.H.: Statistical Mechanics of Elasticity, 2nd edn. Dover, New York (2003)
16. Shannon, C.E.: A mathematical theory of communication. The Bell Syst. Tech. J. **27**, 379–423 (1948) (The founding paper of information theory. For anyone new to information theory, this clearly written paper will be a great introduction. To be found at: http://cm.bell-labs.com/cm/ms/what/shannonday/shannon1948.pdf.)
17. Jaynes, E.T.: Information theory and statistical physics. Phys. Rev. **106**, 620–630, and **108**, 171–180 (1957) (Jaynes' two basic papers on his information theoretical approach to statistical mechanics. To be found at: http://bayes.wustl.edu/etj/articles/theory.1.pdf, http://bayes.wustl.edu/etj/articles/theory.2.pdf.)
18. Grandy, W.T.: Foundations of Statistical Mechanics Volume I: Equilibrium Theory. Springer, New York (1987)

19. Grandy Jr, W.T.: Entropy and the Time Evolution of Macroscopic Systems. Oxford University Press, Oxford (2008)
20. Ben-Naim, A.: A Farewell to Entropy: Statistical Thermodynamics Based on Information. World Scientific, Singapore (2008) (Three recent textbooks that make a strong case for Jaynes' approach to statistical mechanics. There is a second volume of [18] on nonequilibrium processes, outrageously priced. The same is true for [19], a book of 209 pages.)
21. Ruelle, D.: Statistical Mechanics. Rigorous Results, 2nd edn. World Scientific, Singapore (1999). First edition Benjamin 1969
22. Ruelle, D.: 10 Thermodynamic Formalism, 2nd edn. Cambridge University Press, Cambridge (2004). First edition Addison Wesley 1978
23. Catto, I., Le Bris, C., Lions, P.-L.: Mathematical Theory of Thermodynamical Limits: Thomas-Fermi Type Models. Oxford University Press, Oxford (1998)
24. Ising, E.: Beitrage zur Theorie des Ferromagnetismus. Z. Phys. **31**, 253–258 (1925)
25. Yang, C.N.: The spontaneous magnetization of a two dimensional Ising model. Phys. Rev. **85**, 808–815 (1952)

# Chapter 11
# Quantum Statistical Physics

## 11.1 Introduction

What we know about the fundamentals of gases—kinetic theory, the relations between pressure and temperature, van der Waals' theory of molecular interactions, is based on classical statistical mechanics. For more general, and in particular more recent subjects we need a quantum version of statistical mechanics. Note that Planck's calculation of the energy spectrum of black-body radiation, which stood right at the beginning of quantum theory, can—with hindsight—be seen as an application of quantum statistical mechanics 'avant la lettre'. See Sect. 11.7 for remarks on this.

More modern developments in physics such as solid-state and condensed matter physics, the theory of metals, semiconductors, phase transitions, and, for instance, a large part of molecular physics and astrophysics, can only be understood properly by quantum statistical physics. This is in particular true for such a recent topic as Bose-Einstein condensation—to be discussed in Sect. 11.8, superfluidity, and superconductivity, all purely quantum phenomena.

Classical statistical mechanics, as we have seen in the preceding chapter, is a combination of classical mechanics and classical probability theory. Its conceptual framework is in principle simple; its dynamical equations are completely known. All relevant physical quantities can be calculated as averages over probability measures on the phase space of the mechanical systems.

The dynamical equations for quantum systems of many particles, of more recent vintage, because built on quantum mechanics, are equally well known. Quantum statistical mechanics provides a formally analogous but nevertheless different and typically nonclassical prescription for calculating averages.

The aim of the founders of classical statistical mechanics was to derive the appropriate probability measures, or 'ensembles' as they called them, from first principles—successfully, in a not too rigorous sense. This led to the various ensembles discussed in the preceding chapter.

Quantum statistical mechanics might have started with analogous discussions of what can be called 'quantum ergodic theory', a subject which by now indeed exists

© Springer International Publishing Switzerland 2015
P. Bongaarts, *Quantum Theory*, DOI 10.1007/978-3-319-09561-5_11

but does not seem to be of great relevance to the foundations of quantum statistical mechanics. What happened instead was that quantum analogues of classical statistical mechanical concepts at the level of ensemble theory were just postulated. In most presentations of the subject—with the exception of a laudable attempt in a book by Jancel [1], the arguments for these postulates are neither transparent nor very convincing, but fortunately they have been amply justified by experimental results.

Most standard textbooks on statistical physics discuss also quantum statistical physics. See for these the list of references at the end of 10.

## 11.2  What is an Ensemble in Quantum Statistical Physics?

We start by giving a short answer to this question, an answer that will be explained and made more precise in the next sections:

*An ensemble in quantum statistical physics is represented by, as is now generally accepted, a density operator, i.e. a positive selfadjoint trace class operator D in the Hilbert space $\mathcal{H}$ of the quantum system, with trace equal to one.*

We may also ask how such a quantum ensemble is used for calculation of averages. The answer, again short and to be discussed more fully later, is:

*The operator D gives the average of an observable represented by a bounded selfadjoint operator A by the formula*

$$\overline{A} = \mathrm{Tr}\,(DA).$$

*More generally, for a not necessarily bounded selfadjoint operator A, the distribution function $F(\alpha)$ for the corresponding observable is given by*

$$F(\alpha) = \mathrm{Tr}\,(DE_\alpha),$$

*for $E_\alpha$ a spectral projection from the spectral resolution of A.*

This prescription can already be found in von Neumann's 1932 book on the foundations of quantum mechanics (See Supp. Chap. 21, Ref. [1], English translation, Chap. IV), not quite in this form and with using the term 'statistical operator' instead of 'density operator'.

Note that physicists tend to use the term *density matrix*, which is natural because they often have a fixed orthonormal basis of $\mathcal{H}$ in the back of their mind.

(11.2,a) **Problem** Suppose that D is a one dimensional projection P. Let $\psi_P$ be the unit vector on which P projects, which is defined up to a phase factor. Show that the expression for the average for a bounded operator A is $\overline{A} = (\psi, A\psi)$, and that the distribution function for A not necessarily bounded is equal to $F(\alpha) = (\psi, E_\alpha\psi)$, with the $E_\alpha$ the spectral projections of A.

The above result means that 'ordinary' quantum physics can be seen as a special case of quantum statistical physics. This will be put in a general perspective in Sect. 11.5.

## 11.3 An Intermezzo—Is there a Quantum Phase Space?

In setting up quantum statistical mechanics one may be tempted to look for a *quantum phase space*. There are encouraging facts pointing in this direction. Let $\mathcal{H}$ be the Hilbert space of a quantum system. The states are represented by unit vectors $\psi$ in $\mathcal{H}$. Multiplication of such a state vector by a phase factor gives the same physical state, so it is better to say that the state space is the space of equivalence classes $[\psi]_\sim$ of unit vectors with respect to the equivalence relation $\psi_1 \sim \psi_2 \Leftrightarrow \psi_1 = e^{i\phi}\psi_2$, i.e. what amounts to what is called the *projective Hilbert space* $P(\mathcal{H})$. Linear operators in $\mathcal{H}$ descend to maps in the quotient space $P(\mathcal{H})$, not, however, to linear maps, as $P(\mathcal{H})$ is not a vector space. Some other physically useful properties also descend to $P(\mathcal{H})$. The norm $||\psi||$ of a vector $\psi$ in $\mathcal{H}$ obviously gives a meaningful quantity in $P(\mathcal{H})$. This means that unitary operators, in particular the one parameter group $U_t$ of time evolution operators, preserve the image of this quantity. The absolute value of an inner product $(\psi_1, A\psi_2)$, an important quantum mechanical expression, also remains meaningful.

The next step in this approach would be to construct a probability space with $P(\mathcal{H})$ as sample space, and with the observables as stochastic variables, i.e. measurable functions on $P(\mathcal{H})$. But all such variables commute, meaning that one would be setting up a classical physical theory, not a quantum one. This is clearly the wrong direction in which to proceed.

## 11.4 An Approach in Terms of Linear Functionals

Short of deriving quantum statistical physics from first principles, i.e. from the quantum mechanics of systems of many particles, which so far has not made much headway, as far as it has been tried, the best thing is to try to develop a rigorous and natural quantum analogue of classical statistical mechanics. A good way to do this is to look in a somewhat different and special manner at the classical situation.

Consider observables represented by bounded—or, to be more precise, by essentially bounded measurable functions (in the sense of measure theory, see Supp. Chap. 19); arbitrary measurable functions can be obtained as pointwise limits of these. All bounded observables $f$ have finite means $\overline{f}$. This means that an ensemble gives the collection of all such averages; it is a linear functional on the linear space of the bounded observables. Note that such an *expectation functional E* is *positive*, i.e. $f \geq 0$ implies $E(f) \geq 0$, and *normalized*, i.e. with $E(1) = 1$. It is moreover continuous, i.e. an element of the (topological) dual of the vector space of bounded observables. The linear span of expectation functionals to be used is smaller than this dual, see below.

We complexify this classical picture, which does not add any information, but is more convenient for the correspondence with what follows.

From this point of view on the classical situation an obvious analogue for the quantum situation immediately suggests itself: a definition of a *quantum ensemble* as a linear functional on the linear space of bounded quantum observables, i.e. the space of bounded operators $B(\mathcal{H})$ in the Hilbert space of the quantum system. For the same reason as in the classical case, a reason that will be explained in Sect. 12.4.3 of the next chapter, we do not use the full (topological) dual $(B(\mathcal{H}))^*$ of $B(\mathcal{H})$, but a linear subspace, which, as it turns out, is in one-to-one correspondence with the linear space of *trace class operators* in $\mathcal{H}$.

(11.4,a) **Definition** A bounded operator $A$ in $\mathcal{H}$ is called trace class iff for an orthonormal basis $\{\psi_i\}_i$ the sum $\sum_{i=i}^{\infty} |(\psi_i, A\psi_i)|$ is convergent.

It is not hard to prove that if there is convergence for one particular orthonormal basis, then it holds for all other such bases. In this case the expression

$$\mathrm{Tr}\,(A) := \sum_{i=i}^{\infty} |(\psi_i, A\psi_i)|$$

is finite and independent of the choice of $\{\psi_i\}_i$. It is called the *trace* of $A$, and is clearly a generalization of the notion of trace of a finite matrix to infinite dimensional Hilbert space operators. The expression $\sum_{i=1}^{\infty} ((A^*A)^{1/2}\psi_i, \psi_i)$ is denoted as $||A||_1$ and is called the *trace norm* of $A$. It is a true norm, in the sense of Supp. Sect. 21.2. The relation with the usual operator norm is $||A|| \leq ||A||_1$.

A selfadjoint trace class operator has a purely discrete real spectrum; all eigenvalues except zero have finite multiplicity; zero is also the only point of accumulation of the spectrum. A positive selfadjoint trace class operator $D$ with $\mathrm{Tr}\,(A) = 1$ defines a positive, real, normalized linear functional on $B(\mathcal{H})$ by the formula $\overline{A} = \mathrm{Tr}\,(DA)$, which makes sense for all bounded selfadjoint operators $A$, because the trace class operators form a *two-sided-ideal* in $B(\mathcal{H})$, meaning that the product of a trace class operator with an arbitrary bounded operator is again trace class. The trace satisfies $\mathrm{Tr}\,(AB) = \mathrm{Tr}\,(BA)$, for a pair of bounded operators $A$ and $B$, of which at least one is trace class. See for this and other information on trace class operators Reed and Simon [3].

Positive selfadjoint trace class operators $D$ with $\mathrm{Tr}\,(1) = 1$ will be called *density operators*.

*The simple analogy arguments given in this section provide sufficient justification for the choice of density operators as "quantum ensembles".*

The density operator formalism that we have introduced for quantum statistical physics clearly no longer fits in the system of quantum theory axioms discussed in Chap. 3. A new generalized system is needed that includes the old one as a special case, as was already clear from the answer of Problem (11.2,a). Such a generalized axiom system will be set up in the next section.

## 11.5  An Extended System of Axioms for Quantum Theory

In Sect. 3.3.4, we made a distinction between a first level of quantum theory axiomatics, what we called 'pure state quantum theory', and a second more general level, 'mixed state quantum theory'. It is in this chapter that we are going to formulate the axioms for this second level for which we have made the preparations in the preceding section.

The main concepts from the system as described in Chap. 3 were *state*, *observable*, the *rule for physical interpretation which connects these two*; further *dynamics* and *symmetries*. All this reappears in the new set up as will be explained in this section.

In Sect. 3.5, we first gave a preview of the axiom system, before separately discussing in Sects. 3.6–3.8 the five axioms, numbered from I to V, together with details of their properties, relations and physical meaning. Much of this remains relevant for the new generalized system and does not need to be repeated. Axiom numbers will now have primes, i.e. $I'$, $II'$, etc.

- **The Hilbert space** $\mathcal{H}$

We start with an ambient Hilbert space $\mathcal{H}$, as a fixed background object, not necessarily as a space of quantum states.

- **States**

**Axiom $I'$**: *A state of a quantum system is represented by a selfadjoint positive trace class operator with trace one.*

The space of states will be denoted $S$. It is a selfadjoint subset of $B(\mathcal{H})$, the complex linear space of all bounded operators in $\mathcal{H}$.

(11.5,a) **Problem** Let $D_1$ and $D_2$ be two different elements of $S$, and let $\lambda$ be a real number with $0 < \lambda < 1$. Show that the sum $\lambda D_1 + (1 - \lambda)D_2$ is again an element of $S$. (Hint: use the properties of the trace norm, as given in Sect. 11.4).

This result means that $S$ is a *convex* space, defined as a subspace of a real vector space with this property of the addition. Two different points in a convex space are connected by a line segment. Some points are endpoints of lines, *extremal points*, in the sense that they cannot be written as a convex sum of two different points, with a parameter $\lambda$, $0 < \lambda < 1$. These points form a kind of boundary of the convex space.

This means that our convex space of density operators $S$ is a disjunct union of an 'interior set' and a 'boundary set', to be denoted as $S_m$ and $S_p$ respectively. The elements of $S_m$ are called *mixed states* and those of $S_p$ *pure states*. We have the following important theorem:

(11.5,b) **Theorem** *A density operator $D$ is a pure state if and only if $D$ is a one dimensional projection.*

(11.5,c) **Problem** Prove this theorem. (Hint: Show first that an arbitrary density operator $D$ can be written, uniquely, in fact, as

$$D = \sum_{n=1}^{\infty} d_n P_n,$$

with $\{P_n\}_n$ a sequence of nonzero, mutually orthogonal projections, with as sum the unit operator).

Note that the $\{d_n\}_n$ form a sequence of decreasing nonnegative numbers, which converge to zero; their sum is one. All $P_n$ are finite dimensional.

One dimensional projection operators are in one-to-one correspondence with the unit vectors in $\mathcal{H}$, or rather with the elements of the projective Hilbert space $P(\mathcal{H})$, as defined in Sect. 11.3, i.e. with the state space of 'ordinary' quantum theory. That this is the right physical interpretation, which exhibits the fact that the new axiomatic set up is indeed a generalization of the axiom system discussed in Chap. 3, and in fact includes it, will be clear from what follows.

• **Observables**

**Axiom II'**: *The observables of a quantum system are represented by the selfadjoint operators in $\mathcal{H}$.*

This is the same as in Chap. 3. No further comments are required.

• **The relation between I' and II'. Physical interpretation**

**Axiom III'**: *If the system is in a state represented by a density operator $D$, then the probability distribution function for the outcome of the measurement of an observable $A$ is $F_D(\alpha) = \mathrm{Tr}(DE_\alpha)$, with the $E_\alpha$ the spectral projections of $A$. For $N$ commensurable observables $A_1, \ldots, A_N$ the joint distribution function is $F_D(\alpha_1, \ldots, \alpha_N) = \mathrm{Tr}(DE_{\alpha_1,\ldots,\alpha_N})$, with $E_{\alpha_1,\ldots,\alpha_N}$ the projections from the $N$-parameter spectral resolution of the sequence of commuting selfadjoint operators $\{A_N\}_N$.*

• **Time evolution**

**Axiom IV'**: *The time evolution of a system, in the general case of a time-dependent dynamics, is described by a two parameter system of unitary operators $\{U(t_2, t_1)\}_{t_1,t_2}$ which acts on the density operator $D$ as $D$ going to $U(t_2, t_1)D\,U(t_2, t_1)^{-1}$. In the case of a time-independent dynamics—the most common case, there is a one parameter group $\{U(t)\}_t$ acting on $D$ as $D$ going to $U(t)D\,U(t)^{-1}$.*

(11.5,d) **Problem** Let $D$ be a pure state, i.e. a projection operator on a unit vector $\psi$. Show that for a time-independent dynamics, given by the group of unitary operators $\{U(t)\}_t$, one recovers the standard Schrödinger picture dynamics $\psi_t = U(t)\psi$.

• **Symmetries**

**Axiom V'**: *A symmetry of the system is given by a unitary operator $U$ acting on a density operator $D$ as $D$ going to $UD\,U^{-1}$.*

(11.5,e) **Problem** Let $D$ be pure state, i.e. a projection operator on a unit vector $\psi$. Show that for a symmetry of the system, given by a unitary operator $U$, one recovers the formula $\psi' = U\psi$.

The results from Problems (11.2,a), (11.5,d) and (11.5,e) show that standard quantum theory is indeed included in extended quantum theory as a special case.

Both in standard and in extended quantum theory one starts with a Hilbert space $\mathcal{H}$, either as state space or as ambient space. The basic concept is in any case that of *state*. After that comes the concept of *observables*, selfadjoint operators in $\mathcal{H}$. In recent years a further generalization has been developed in which one starts from an abstract algebraic notion of observable, then proceeds in the next step to the notion of state as a linear functional on the algebra of observables. The combination of the algebra of observables and the choice of a state as a functional then gives in a natural manner rise to a Hilbert space. This algebraic picture is useful for systems with an infinite number of degrees of freedom, such as in quantum field theory and in the statistical physics of actually infinite systems. We shall briefly discuss the axiom system for what is usually called 'algebraic quantum theory' in the next chapter.

## 11.6 The Explicit Form of the Main Quantum Ensembles

Again, the standard quantum ensembles are not derived from first principles, as one does—more or less—in classical statistical statistical mechanics, but postulated by analogy.

The classical ensembles that we discussed in the preceding chapter are given by probability densities, in fact functions of the Hamiltonian, with the exception of the microcanonical ensemble—which can be formulated in terms of a heuristic $\delta$-function, or more rigorously as a measure concentrated on the hypersurface $H(p, q) = E$. Averages $\overline{f}$ of functions $f$ are calculated as integrals over the product of such a density and $f$. A trace of an operator can be seen as a 'noncommutative' integral, a point of view that will be discussed in more detail in the next chapter.

It is therefore natural to define the quantum average of an observable, represented by a selfadjoint operator $A$, as the trace of the product of a density operator with $A$. The quantum microcanonical ensemble stands again somewhat apart. Classically one uses the 'sharp' microcanonical ensemble at a precise value $E$ of the energy. This is an idealization, but it makes sense because at each $E$ there is a manifold of 'microstates'. In the quantum case the Hamiltonian has a discrete spectrum of energy eigenvalues $\{E_n\}_n$, with each $E_n$ having finite multiplicity, so there are only a finite number of 'microstates' or none at all at a given energy $E$. Therefore an 'approximate' quantum microcanonical ensemble for a small interval $[E, E + \Delta E]$ is more appropriate. Note that we count the eigenvalues $E_n$ in such way that $E_n < E_{n+1}$, for all $n$. The discrete eigenvalues of the energy of a many particle system are extremely close to each other, so even a small energy interval will give very many quantum 'microstates'.

For the ensembles discussed in the preceding chapter, one postulates quantum density operators which are operator functions of the Hamilton operator, analogous to the classical expressions. The Hamilton operator for a system in a finite box has a purely discrete spectrum. For the density operators that will be defined, and which have to be trace class operators, one has to require somewhat stronger properties of the Hamiltonians, requirements that in realistic cases are always satisfied. To emphasize the analogies we use the notation $\hat{\rho}$ instead of $D$.

a. The *approximate quantum microcanonical ensemble* for a system with energy between $E$ and $E + \Delta E$, consisting of $N$ particles in a box with volume $V$, is given by the density operator:

$$\hat{\rho}_{E,\Delta E,V,N}^{mc} = \frac{1}{\Omega(E, \Delta E, V, N)} \sum_{\{n \,|\, E \leq E_n < E+\Delta E\}} P_n \,,$$

with $P_n$ the projection on the eigenspace of the eigenvalue $E_n$ of the Hamiltonian operator $H$—we assume for these eigenvalues $E_n < E_{n+1}$, for all $n$. The normalization function $\Omega$, called the *microcanonical partition sum*, is equal to

$$\Omega(E, \Delta E, V, N) = \sum_{\{n \,|\, E \leq E_n < E+\Delta E\}} Tr\,(P_n),$$

with $Tr\,(P_n)$, the dimension of the eigenspace of $E_n$, finite, because otherwise $\hat{\rho}_{E,\Delta E,V,N}^{mc}$ cannot be trace class.

b. The *canonical ensemble* for a system characterized by a temperature parameter $\beta = (k_B T)^{-1}$ in a volume $V$ is given by

$$\hat{\rho}_{\beta,V,N}^{can} = \frac{1}{Z(\beta, V, N)} e^{-\beta H_N} \,,$$

with $H_N$ the Hamiltonian operator of the $N$-particle quantum system, and with the normalization function, the *canonical partition sum* $Z$, equal to

$$Z(\beta, V, N) = Tr\,(e^{-\beta H_N}).$$

c. *The grand canonical ensemble* for a system with a variable number $N$ of particles, characterized by $\beta$ and the chemical potential $\mu$ is described by a density operator

$$\hat{\rho}_{\beta,V,\mu}^{gc} = \frac{1}{\Xi(\beta, V, \mu)} e^{-\beta(H_N - \mu \hat{N})} \,,$$

with $\hat{N}$ the operator that counts the (now variable) number of particles in the box $V$. This density operator acts, of course, in a many particle Hilbert space, the symmetric or antisymmetric tensor algebra over a given one particle Hilbert space, discussed in Chap. 9, contrary to the cases a and b where it acts in the Hilbert space of a fixed number of particles $N$. The normalization function $\Xi$, the *grand canonical partition sum*, is equal to

$$\Xi(\beta, V, \mu) = \text{Tr}\,(\hat{\rho}_{\beta,V,\mu}^{\text{gc}}) = \sum_{N=0}^{\infty} e^{\beta\mu N} Z(\beta, V, N) = \text{Tr}\,(e^{-\beta H_N}).$$

Note that all these partition sums are dimensionless, so insertion of a factor $C$ as in the classical case is not needed.

Various thermodynamic quantities are obtained from $\Omega$, $Z$ and $\Xi$ in the same manner, involving the same problems, as in the classical situation, which was discussed in Sect. 10.9.

## 11.7 Planck's Formula for Black-Body Radiation

It is sometimes said that quantum theory was born on December 14, 1900, the day on which Max Planck gave a talk in Berlin to the *Deutsche Physikalische Gesellschaft* in which he presented his derivation of a new formula for the frequency distribution of black-body radiation. Later in 1901 he published his result in the *Annalen der Physik* Ref. [4].

Consider a cavity with perfectly absorbing (black) walls, kept at a fixed temperature $T$. Its interior will be filled with radiation having all possible wavelengths. Suppose that this radiation is in equilibrium with the walls of the cavity. It was already known in the middle of the nineteenth century that the distribution over the various wavelengths, or equivalently, over the frequencies, was independent of the material of the walls of the cavity. Toward the end of that century the form of this distribution was experimentally known with great precision. There were several attempts to find a theoretical underpinning of these experimental results. One was the radiation law of Rayleigh and Jeans (1900), expressing the intensity of the radiation as

$$\rho_{\text{RJ}}(\nu, T) = \frac{2\nu^2 k_B T}{c^3},$$

with $\nu$ the frequency of the radiation, $T$ the absolute temperature, $c$ the velocity of light, and $k_B$ Boltzmann's constant. It is a good approximation only for low frequencies. A second law was that of Wien (1896),

$$\rho_{\text{W}}(\nu, T) = b\,\nu^3 e^{-\frac{a\nu}{T}},$$

with two constants $a$ and $b$ to be determined. This formula works quite well for high frequencies.

The formula presented by Planck in his 1900 talk and in subsequent papers gave an interpolation between the formulas of Rayleigh-Jeans and Wien; it agreed with experiment over the full range of frequencies. It reads, in slightly modernized form,

$$\rho(\nu, T) = \frac{2h}{c^3} \frac{\nu^3}{e^{h\nu/k_B T} - 1},$$

with $h$ a new universal physical constant with the dimension of an action, now typical for quantum theory and rightly called Planck's constant.

(11.7,a) **Problem** Show that Planck's formula agrees for $\nu \to 0$ with the formula of Rayleigh-Jeans, and for $\nu \to \infty$ with that of Wien. Determine the constants $a$ and $b$ in the second case.

Planck's early work was in thermodynamics; the final form it reached in the second half of the nineteenth century is to a large extent due to him. At first he did not see much in trying to find a microscopic statistical basis for thermodynamics. However, the work of Boltzmann convinced him of the merits of a statistical approach, and in particular of the value of the notion of entropy, as was already noted in Sect. 10.2. He also was familiar with Maxwell's theory of electromagnetic radiation. He used the familiarity with these three topics in his work on the problem of black-body radiation. The model that he considered was based on the interaction between (classical) electromagnetic radiation in the cavity and resonators in the walls. His great discovery, an extremely unwelcome fact for him, "An act of desperation", as he seems to have called it, was that he could obtain the experimentally known frequency distribution only by assuming that energy was exchanged between the walls and the radiation in the cavity in discrete quantities $h\nu$, with $\nu$ the frequency of the radiation and $h$ a new universal physical constant. Planck certainly did not 'quantize' the energy in the sense of modern quantum theory. What he exactly quantized remains a matter of dispute among historians of science. His proof is very hard reading for a modern physicist; in any case parts of it are not completely convincing. But his law stands, as one of the basic formulas of modern physics. It took Planck many years to really accept quantum theory in its full sense, as it grew from his work. It was Einstein who somewhat later drew the full consequences of Planck's idea, by showing, in the context of his work on the photo-electric effect, that light, and electromagnetic radiation in general, can manifest itself as a stream of massless particles, *photons*, each with energy $h\nu$. See Ref. [5].

Planck was very much a classical physicist. Reading about him one is reminded of "Jacob", a fictional German professor of theoretical physics, the main character of an entertaining and erudite novel written by Russell MacCormmack, a well-known historian of nineteenth century theoretical physics Ref. [6].

The history of the birth of Planck's radiation law, what happened before, and what came later, is brilliantly described in a book by Kuhn Ref. [7].

The Indian physicist S.N. Bose gave in 1924 in a four-page article a simple and straightforward but in a certain sense rather curious proof of Planck's law [8]. His set up is very simple. He treats the radiation in the cavity as a system of $N$ noninteracting particles, an ideal gas of photons, the particles representing electromagnetic radiation, introduced in 1905 by Einstein in his groundbreaking paper on the photo-electric effect. Interaction with the wall of the cavity played no role. He applies statistical ideas on this system, and divides for this, in the spirit of Boltzmann, the microscopic state space of a single particle in small cells of volume $h^3$. A 'microstate' of the system is then an assignment of the $N$ particles to these cells. They are supposed to have equal probability, essentially the microcanonical ensemble, not in the sense of Gibbs in the full phase space of the $N$-particle system, but in the one particle phase space of Boltzmann. In a straightforward calculation the entropy $S$ is obtained from the formula $S = k \log W$, with $W$ the total probability summed over the microstates. Then Planck's radiation law is finally found by using the thermodynamic relation $T^{-1} = \frac{\partial S}{\partial E}$, in an appropriately chosen thermodynamic limit.

It is worth noting that Bose's simple proof is based, and solely based, on a very particular way of counting the microstates contributing to $W$. This can be best understood by looking at a simple example. Suppose we have two identical particles, $p_1$ and $p_2$, which can be placed in two cells, $c_1$ and $c_2$. We enumerate the possible 'microstates' as

(1) $p_1$ in $c_1$, and $p_2$ in $c_2$,
(2) $p_1$ in $c_2$, and $p_2$ in $c_1$,
(3) both $p_1$ and $p_2$ in $c_1$,
(4) both $p_1$ and $p_2$ in $c_2$.

Each of these four possibilities has probability $1/4$. For Bose (1) and (2) are the same states. This means three different microstates, each with probability $1/3$. With this manner of counting microstates, used in the full formalism for a system of a very large number of particles, with the number $h$ a small but finite constant, Bose was able to derive Planck's radiation law.

Bose in his paper made no remark on his special way of counting microstates; it is therefore hard to resist the impression that he just made an error in elementary statistics. After his paper was rejected by the *Philosophical Magazine*, he sent the manuscript to Einstein, who found it of great interest, translated it into German and had it published in the *Zeitschrift für Physik*, with a postscript in which he praised it as an important piece of research. At that moment Einstein also did not see anything strange or remarkable in Bose's way of counting states, but later, when he himself did further work in this direction, he realized that it was just this special sort of statistical argument that enabled him to obtain various interesting quantum results. Eventually this way of counting quantum states of systems of identical particles became part of a very important general principle, namely that particles in quantum theory are indistinguishable, do not have, strictly speaking, an identity. We discussed this already fully in Chap. 9. For bosonic particles this means that one has to use *Bose-Einstein statistics*, as it is now called.

Whether Bose just made an elementary error or whether he worked from a vague but correct intuition is hard to tell. The psychological aspect of this question is discussed in a short but interesting paper by the Nobel laureate Max Delbrück Ref. [2].

## 11.8 Bose-Einstein Condensation

To a large extent the effects of quantum theory only appear at the submicroscopical level. An important exception to this is Bose-Einstein condensation, BEC for short.

In 1925 Einstein published a paper in which ideas put forward earlier by Bose were developed further [9]. (See the preceding section for more information on the paper of Bose). Einstein studied an ideal gas of particles following Bose's statistics and found that at very low temperatures the particles could be expected to coalesce in the lowest energy level to form a sort of new phase. He was not sure that this idea should be taken seriously.

It took 70 years before this idea became physical reality, at least in the laboratory. In the meantime other macroscopic quantum phenomena were observed, *superconductivity* (1911) and *superfluidity* (1937), with hindsight phenomena of a very similar nature. Finally in 1995, with the progress made in the physics of very low temperatures, Eric Cornell and Carl Wieman [10], and independently Wolfgang Ketterle [11], succeeded to observe with a microscope a true Bose-Einstein condensate, at something like a millionth degree Kelvin above absolute zero, the first on a system of rubidium atoms, the second with sodium atoms. The three of them received the Nobel prize in 2001. Further progress in low temperature physics led later to the discovery of BEC at room temperature.

It is not very hard to calculate the BEC effect for a system of noninteracting bosons, in a finite box with volume $V$ and with periodic boundary conditions. Using 'second quantization', i.e. the Fock space formalism discussed in Sect. 9.6. one finds that the average occupation number in an energy eigenstate $E_j$ in the grand canonical ensemble is equal to

$$<n_k>_{gc} = \frac{\mathrm{Tr}\,(e^{-\beta(H-\mu N)}n_k)}{\mathrm{Tr}\,(e^{-\beta(H-\mu N)})},$$

with $H$ the Hamiltonian, $N$ the total number operator, $n_k$ the number operator for the energy eigenstate $k$ and $\mu$ the chemical potential. By taking the thermodynamic limit $V \to \infty$ and $N \to \infty$, with $\mu$ and the density, the quotient $N/V$, constant, one proves that there is a critical temperature below which all the particles occupy the lowest energy state, forming a Bose-Einstein condensate. For this derivation and for further details, see the book of Verbeure [12].

Most of the physics textbooks on statistical physics listed in the reference section of 10 treat both the classical and the quantum case.

## References

1. Jancel, R.: Foundations of Classical and Quantum Statistical Mechanics. Pergamon Press, New York (1969) (Translated from the 1963 French edition.)
2. Delbrück, M.: Was Bose-Einstein statistics arrived at by serendipity? J. Chem. Edu. **57**, 467–470 (1980)

3. Reed, M., Simon, B.: Methods of Mathematical Physics. I: Functional Analysis. Academic Press, New York (1972) (This book is because of its readability our main reference for this chapter. A text book on functional analysis, with the theory of operators in Hilbert space as its central topic, especially written for applications in mathematical physics. It is the first of a series of four books by the same authors. The other volumes are more specialized, but contain nevertheless useful material, in particular II and IV.)

4. Planck, M.; Über das Gesetz der Energieverteilung im Normalspectrum. Ann. Phys. **309**, 553–563 (1901). Available at: http://www.physik.uni-augsburg.de/annalen/history/historic-papers/1901_309_553-563.pdf (An English translation at: http://web.ihep.su/owa/dbserv/hw.part2?s_c=PLANCK+1901+.)

5. Einstein, A.: Über einen die Erzeugung und Verwandlung des Lichtes betreffenden heuristischen Gesichtspunkt. Ann. Phys. **17**, 132–148 (1905). Available at: http://www.physik.uni-augsburg.de/annalen/history/einstein-papers/1905_17_132-148.pdf (One of the five groundbreaking papers Einstein published in 1905, his 'Annus Mirabilis'. An English translation appeared in the American Journal of Physics, 33, May 1965)

6. McCormmach, R.: Night Thoughts of a Classical Physicist. Harvard University Press, Harvard (1982) (Russell MacCormmack is a historian of nineteenth century theoretical physics. His book is a novel, set in Germany, during the First World War, with as main character "Jacob", a professor of theoretical physics. He has had a thorough education in classical physics, and is now confronted with new ideas such as those of Planck, Einstein and others, which threaten his picture of physics. This professor is fictional, but he is a composite of various historic persons; everything he says or does or is engaged in comes from historical reality, and is as such scrupulously documented in an appendix. A wonderful book.)

7. Kuhn, T.S.: Black-body Theory and the Quantum Discontinuity, 1894–1912. Oxford University Press, Oxford (1978) (Reprinted with a new afterword by the Chicago University Press, 1987.)

8. Bose, S.N.: Plancks Gesetz und Lichtquantenhypothe. Eng. Z. Phys. **26**, 178–181 (1924)

9. Einstein, A.: Quantentheorie des einatomigen idealen Gases. English: Quantum theory of ideal monatomic Gases. Sitzungsbericht Preuss. Akad. Wiss. **23**, 3–14 (1925). Available at http://web.physik.rwth-aachen.de/~meden/boseeinstein/einstein1925.pdf

10. Anderson, M.H., Ensler, J.R., Matthews, M.R., Wieman, C.E., Cornell, E.A.: Observation of Bose-Einstein condensation in a dilute atomic vapor. Nature **269**, 198–201 (1995)

11. Davis, K.B., Mewes, M.O., Andrews, M.R., van Druten, N.J., Durfee, D.S., Kurn, D.M., Ketterle, W.: Bose-Einstein condensation in a gas of sodium atoms. Phys. Rev. Lett. **75**, 3969–3979 (1995)

12. Verbeure, A.F.: Many-Body Boson Systems. Half a Century Later. Springer, London (2011)

# Chapter 12
# Physical Theories as Algebraic Systems

## 12.1 Introduction

It is in principle possible to describe both quantum and classical physics within a single algebraic framework. In such a formulation physical systems appear as 'algebraic dynamical systems'.

It was in quantum theory that an algebraic point of view of the sort that we have in mind first appeared. The mathematical basis of quantum theory is the use of operators in Hilbert space, as we have explained in detail in Chap. 3; from studying such operators it is only one step to the study of *algebras of operators*.

Quantum theory is assumed to be valid for all physical phenomena—at least as long as new experimental evidence will not teach us otherwise—but classical theories provide good approximations in many situations. Therefore it makes sense to try to develop an algebraic framework also for classical physics—not obvious as this may look at first sight. One of the advantages of such a general formulation is that it would narrow the conceptual gap between classical and quantum physics. In any case, a desire for unification has always been a guiding thought in theoretical physics.

This chapter will present such an algebraic framework, in which classical physics is obtained as a special case by requiring the relevant algebras to be *commutative*, and in turn quantum physics appears as a generalization—or perhaps a deformation—of classical physics. It is true that this does not add much, concretely and explicitly, to known theories and it is certainly not helpful in performing calculations. It must also be admitted that it still has loose ends. Considerable white areas remain; many details have not been worked out, but all this does not diminish its conceptual attraction.

There is a recent mathematical development that connects algebra and geometry in a new way, which may be seen as a background for this. It goes under the name "noncommutative geometry", has as its origin the work of Gelfand and Naimark on the one-to-one correspondence between compact topological spaces and commutative $C^*$-algebras—see the next section and for more details Supp. Chap. 27—and has been developed into a broad mathematical discipline by Alain Connes. See for this his book [1] and also the book of Gracia-Bondía et al. [2].

© Springer International Publishing Switzerland 2015
P. Bongaarts, *Quantum Theory*, DOI 10.1007/978-3-319-09561-5_12

The general idea of this is—very naively formulated, first the observation that many mathematical theories start from a 'space', i.e. a point set with additional structure, for instance a measure space, a topological space or a differentiable manifold, and secondly that such a 'space' can be characterized by a *commutative algebra*, namely the algebra of the appropriate functions on the underlying set, measurable, continuous and smooth functions. The next step, the truly interesting one, is then that by looking at similar *noncommutative algebras* new mathematical theories are obtained based on 'virtual spaces', which have no meaning as point sets, but are intuitive ideas that suggest interesting generalizations of properties of the commutative case.

In this spirit the general theme of this chapter will be:

**Classical physics**   ⟷   **Commutative algebras**
**Quantum physics**   ⟷   **Noncommutative algebras**

In this scheme the algebras are of the same type, e.g. commutative and noncommutative $C^*$-algebras, or commutative and noncommutative von Neumann algebras, etc., notions that will be used in the next sections of this chapter and on which extensive mathematical details are given in Supp. Chap. 27.

Ideas of this type have been around already for some time and are discussed at various places in the literature; see for instance the recent books by Faddeev and Yakubovskiĭ (Preface, Ref. [3]) and Strocchi (Preface, Ref. [5], pp. 10–23). These ideas will be developed fully and in a more systematic way in this chapter.

A proposal of formulating quantum physics in terms of abstract algebras of observables with averages given by positive linear functionals on these algebras was first put forward in 1947 by Segal in a classic paper [3] and applied later by others to specific topics in quantum physics, in particular to attempts to give a rigorous mathematical basis to quantum systems with an infinite number of degrees of freedom such as quantum field theory and quantum statistical mechanics in the thermodynamic limit. See the seminal paper by Haag and Kastler [5].

This chapter differs somewhat from the rest of this book. It does not simply reproduce known standard material, but presents and develops instead a general scheme, new, particularly in many details, but based on a complex of ideas to be found in recent literature. The content is, apart from generalities, independent from the remainder of the main text of this book, but does depend, especially for the mathematics, on two supplementary chapters, Supp. Chap. 21 and in particular Supp. Chap. 27, which should be read in parallel with this chapter.

*Remark* In a dynamical system time is seen as an external parameter, which describes maps of the system onto it self. This is typical for most of nonrelativistic physics, although not, for example, for the quantum theory of lattices, higher dimensional quantum analogues of the Ising model discussed in Sect. 10.12, in which one is interested in calculating the properties of equilibrium states by maximizing the entropy and where time evolution does not play a role. In relativistic physics one has time evolution, but time is there no longer a separate notion, but is part of 'spacetime'. One might in that case speak of 'algebraic covariance system' instead of 'algebraic dynamical system'. Relativistic physics will be discussed in Chaps. 15 and 16.

## 12.2 'Spaces': Commutative and Noncommutative

### 12.2.1 Introduction

The central idea of this chapter is that of 'space'. First as a precise notion: a point set $X$, carrying an additional structure characterized by the commutative algebra $C(X)$ of appropriate functions on $X$, which characterizes this structure. Then, in a more suggestive and heuristic way, in the quantum situation, in which there is no longer a point set $X$ but in which we do have the noncommutative version of the algebra $C(X)$ from the classical case, no longer an algebra of functions.

### 12.2.2 A List of 'Spaces'

In this section we present an annotated list of 'spaces' and their virtual noncommutative analogues. For more mathematical details see Supp. Chap. 27.

(1) *Topological spaces.*

The history of the subject begins with the theorem of Gelfand and Naimark, which states that there is a one-to-one correspondence between commutative $C^*$-algebras and locally compact topological spaces, and more special, between commutative $C^*$-algebra and compact topological spaces. See Supp. Sect. 27.4.

This leads to a suggestion for noncommutative $C^*$-algebras:

*'compact quantum topological space'* means " *noncommutative $C^*$-algebra".*

(2) *Measure spaces* or *probability spaces.*

There is a similar relation between commutative von Neumann algebras and algebras of measurable functions on measure spaces or, more in particular, a probability space, which is a bit more complicated because it is a correspondence between equivalence classes. See Supp. Sect. 27.5.

This leads to a suggestion for noncommutative von Neumann algebras:

*'quantum measure'* or *'probability space'* means *"noncommutative von Neumann algebra".*

(3) *Smooth manifolds.*

Smooth, or differentiable, manifolds are topological manifolds with a differentiable structure (See Supp. Chap. 20). The associated algebra of smooth functions $C^\infty(\mathcal{M})$ has for a compact manifold $\mathcal{M}$ a Fréchet space as its underlying vector space, and for a noncompact $\mathcal{M}$ an LF space, i.e. a direct limit of a sequence of Fréchet spaces. See Supp. Sect. 27.6. No full analogue of the Gelfand-Naimark theorem for commutative $C^*$-algebras is known for the algebras of smooth functions on manifolds. One has however Theorem (27.6.a).

This leads to an admittedly rather vague suggestion for noncommutative smooth ∗-algebras:

'*quantum manifold*' means "*noncommutative smooth ∗-algebra*".

Note that if we restrict to bounded smooth functions, the corresponding noncommutative smooth ∗-algebra will be a subalgebra of a $C^*$-algebra.

(4) *Symplectic manifolds.*

A symplectic manifold is a smooth manifold provided with a symplectic form, i.e. a closed nondegenerate 2-form $\omega$, which defines a *Poisson bracket* $\{\cdot, \cdot\}$ on the algebra of smooth functions. This makes $C^\infty(\mathcal{M})$ into a *Poisson algebra*. A manifold with a Poisson bracket is called a *Poisson manifold*, and is slightly more general than a symplectic manifold. Poisson geometry, as a special topic in differential geometry, has been thoroughly investigated. Less attention has been paid to the properties of $C^\infty(\mathcal{M})$ as a special case of a locally convex topological algebra.

This nevertheless leads to a suggestion for noncommutative Poisson ∗-algebras:

'*Quantum Poisson manifold*' means "*noncommutative smooth Poisson ∗-algebra*".

## 12.3 An Explicit Description of Physical Systems I

### 12.3.1 Introduction

In this section and in the next one we shall apply the general concepts formulated in the preceding section to explicit physical systems, both classical and quantum, in the context of an algebraic axiom system. In Chap. 3 we formulated a first system of axioms for what we called the first level of quantum theory, with as basic notions unit vectors in a Hilbert space as states and selfadjoint operators in that space as observables. A second level of axiomatics was introduced in Chap. 11, still with a Hilbert space, but now as an ambient background space, and with density operators describing states. This generalization was necessary for the description of quantum statistical physics. The use of these two levels is standard in mainstream quantum physics.

This chapter presents a general algebraic framework for quantum theory as a noncommutative version of a similar framework for classical physics, suggesting a third level, in which one starts from an abstract algebra, much in the spirit of Ref. [3]. This third formalism is not—or, maybe, not yet—part of the standard curriculum of physics. It is of great conceptual interest as a general theoretical framework, but has in a practical sense not much to offer in ordinary classical and quantum mechanics. It does however play an important role in the description of systems with an infinite number of degrees of freedom, the thermodynamic limit in statistical physics, both classical and quantum, and relativistic quantum field theory. Over the last forty years algebraic approaches have been developed in these areas, along slightly different lines, by a broad phalanx of mathematicians and mathematical physicists—we already mentioned the names of pioneers such as Irving Segal, Rudolph Haag, Daniel Kastler and Huzihiro Araki.

## 12.3.2 A First Sketch of an Algebraic Axiom System

Here is a simple system of axioms for the algebraic description of both classical and quantum systems in physics in the spirit of this chapter. In the following section we shall develop this system in more detail and apply it to concrete physics. Precise definitions and mathematical properties of all notions occurring in this chapter can be found in Supp. Chap. 27.

- **Axiom I″**: Observables: *a complex ∗-algebra $\mathcal{A}$.*

  We will use $C^*$-, von Neumann and (LF) Fréchet-Poisson ∗-algebras $\mathcal{A}$.
- **Axiom II″**: States: *positive normalized real linear functionals $\omega$ on $\mathcal{A}$.*

  A state on a ∗-algebra $\mathcal{A}$ is a positive normalized linear functional $\omega$ on $\mathcal{A}$. It is called 'real' iff $\overline{\omega(a)} = \omega(a^*)$, for all $a$ in $\mathcal{A}$.
- **Axiom III″**: Axiom I″ and Axiom II″ are combined to give probabilistic predictions for measurements, along the same lines as earlier.
- **Axiom IV″**: Time evolution: *a one parameter group $\{\phi(t)\}_{t \in \mathbb{R}}$, or more general, a two parameter system*

$$\{\phi(t_2, t_1)\}_{t_1, t_2 \in \mathbb{R}}$$

  *of ∗-automorphisms of $\mathcal{A}$.*
- **Axiom V″**: Symmetries: *various ∗-automorphisms and groups of ∗-automorphisms of $\mathcal{A}$, commuting with the time evolution automorphisms.*

## 12.4 An Explicit Description of Physical Systems II

### 12.4.1 Introduction

In this and the next sections we give a more precise formulation of the five axioms in Sect. 12.3.2, together with remarks on their meaning for concrete physical systems.

## 12.4.2 Observables

On this third level the starting point is not a Hilbert space, but the *algebra of observables*. Nevertheless a state dependent Hilbert space will automatically emerge from the data specified according to the first two axioms.

**Axiom I″**: *There is an abstract $C^*$-algebra $\mathcal{A}$ of what may be called preobservables, the 'skeleton' of a von Neumann operator algebra of physical observables.*

It is convenient to start with statistical physics and statistical quantum physics—more particular statistical mechanics and statistical quantum mechanics, as the general situation, regarding classical mechanics and quantum mechanics as a special degenerate case. For the observables this does not make any difference; for the states and for the dynamics the two cases have to be discussed separately, as will be done in the next sections.

Classical statistical mechanics is essentially probability theory; 'ensembles' in physics parlance are probability distributions. This means that we use as basic function algebra for the observables the measurable functions—the random variables—on a probability space. Even though these functions form an algebra, it is convenient, in view of the correspondence with the quantum case, to restrict the algebraic discussion to the essentially bounded functions, and look at the unbounded functions as pointwise limits of bounded ones. As we have argued, this gives us a commutative von Neumann algebra. The quantum case is then described by a noncommutative von Neumann algebra of bounded operators; with the unbounded operators, not an algebra because of obvious domain problems, as limits of bounded operators.

Measurable functions form the most general background. These include the smaller system of bounded continuous functions, forming a commutative $C^*$-algebra, according to the Gelfand-Naimark correspondence, leading for the quantum case to a noncommutative $C^*$-algebra. Classical statistical mechanics has a more specific mathematical structure than general probability theory. The measure space, the phase space, is a manifold, in fact a symplectic manifold. The algebra of observables form what we have called a smooth *-algebra, with the bounded smooth functions a subalgebra of a commutative $C^*$-algebra, which is in turn a subalgebra of the commutative von Neumann algebra of bounded measurable functions on the phase space of the system.

An important feature of classical mechanics as a Hamiltonian theory is the Poisson bracket $\{\cdot, \cdot\}_c$ on the algebra of smooth functions on phase space. It is a second order differential expression in the canonical coordinates for position and momentum. See for its definition and properties Supp. Sect. 27.7 and also Supp. Chap. 20. Every *-algebra of operators in quantum theory is in an obvious manner a Poisson *-algebra. We define the quantum Poisson bracket simply as

$$\{A_q, B_q\}_q = i[A_q, B_q],$$

for each pair of bounded operators $A_q$ and $B_q$, and with some caution also for suitable unbounded operators. Note that the imaginary $i$ is necessary because of the reality condition for the quantum Poisson bracket.

(12.4.2,a) **Problem** Show that $\{\cdot, \cdot\}_q$ indeed satisfies the requirements for a Poisson bracket as formulated in Supp. Sect. 27.7.

For quantum mechanics we have the Poisson $*$-subalgebra consisting of bounded smooth quantum observables, an algebra that we are not able to specify in a precise and explicit manner, except that we know that it is a noncommutative version of a Poisson-Fréchet or LF $*$-algebra. In quantum mechanics unbounded observables are ubiquitous; the Poisson bracket between such operators has to be used with caution, because of domain problems.

There is a suggestive relation between the Poisson bracket for classical variables and the corresponding quantum observables, first noticed by Dirac in one of his sudden brilliant insights (See [4], p. 86–87). We have the following well-known formulas on a classical phase space $\mathcal{M} = \mathbb{R}^{3N}$ with coordinates $p_j$ and $q_k$ and quantum mechanics with operators $P_j$ and $Q_k$:

- *Poisson bracket between the classical variables*

$$\{p_j, p_k\}_c = \{q_j, q_k\}_c = 0, \qquad \{p_j, q_k\}_c = \delta_{jk},$$

- *Heisenberg commutation relations between the quantum operators*

$$[P_j, P_k] = [Q_j, Q_k] = 0, \qquad [P_j, Q_k] = \frac{\hbar}{i}\delta_{jk},$$

leading to the quantum Poisson brackets

$$\{P_j, P_k\}_q = \{Q_j, Q_k\}_q = 0 \qquad \{P_j, Q_k\}_q = \hbar\{p_j, q_k\}_c.$$

One has similar relations for the time evolution equations using a Poisson bracket form of the Liouville equations, as we shall show further on. This suggests a notion of 'quantization' as a general linear map $\mathcal{Q}$ from the classical observables to the quantum observables, generated by

$$P_j = \mathcal{Q}(p_j) \qquad Q_k = \mathcal{Q}(q_k),$$

with the properties

$$\mathcal{Q}(f)^* = \mathcal{Q}(f),$$

for each (real) classical observable $f$, and

$$\{\mathcal{Q}(f), \mathcal{Q}(g)\}_q = \hbar\,\mathcal{Q}(\{f, g\}_c),$$

for all pairs of classical observables $f$ and $g$. This idea is however too naive; the Groenewold-van Hove theorem (1946, 1951) states that for $\mathcal{M} = \mathbb{R}^{3N}$ such a map does not exist. See [6, 7].

There is a fairly obvious and direct reason why the above prescription for a general quantization formula can not be meaningful. Consider as an example the simple expression $pq^4p$, for simplicity in dimension one, so with a single pair of canonical variables $(p, q)$. This is a real variable, in principle a classical observable, although in practice not very useful. The corresponding quantum expression $PQ^4P$ is selfadjoint (on some suitable domain), so does represent a possible physical observable. The classical expression $q^2p^2q^2$ is of course equal to $pq^4p$. The corresponding quantum expression $Q^2P^2Q^2$ is also selfadjoint but different from $PQ^4P$. This example shows that the naive quantization, defined as a linear map $\mathcal{Q}$ on all classical phase space functions, as proposed above, is not well defined.

(12.4.2,b) **Problem** Prove, by using the Heisenberg canonical commutation relations, that one has

$$PQ^4P = Q^2P^2Q^2 + 2\hbar^2Q^2.$$

The fact that a second order term in $\hbar$ appears in the difference between $PQ^4P$ and $Q^2P^2Q^2$ points to a better way of looking at the relation between the classical and the quantum algebras of observables.

We should first note again, as we did in the preface, that the idea of "quantizing a classical theory" has a historical origin, because during a long period after the beginning of quantum theory all quantum theories were constructed by 'quantizing' existing classical theories. This is nevertheless in principle incorrect as a way of seeing the relation between classical and quantum physics. Quantum theory is the truly fundamental theory; classical theories should arise as approximations, by taking a 'classical limit', i.e. letting the typically quantum parameter $\hbar$ go to zero in an appropriate sense.

From this point of view a quantum theory may be seen as a *deformation* of a classical theory. Its operators can be represented as power series in $\hbar$, with the lowest order term the classical part. The classical limit gives this lowest order term, which may coincide for different quantum operators, as in the example in Problem (12.4.2a). *Deformation quantization* as a special topic will be reviewed in the next chapter.

(12.4.2,c) *Remark* In the case of more general types of classical and quantum statistical physics, such as discrete spin lattices, discussed in Sect. 10.12, the situation is much simpler, as we do not need smooth functions algebras and their noncommutative quantum analogues for the observables. $C^*$- and von Neumann algebras are sufficient.

## 12.4.3 States

**Axiom II″**: *There is a convex set of states, the positive normalized linear functionals on $\mathcal{A}$.*

A state $\omega$ determines a representation $\pi_\omega$ of $\mathcal{A}$ in a Hilbert space $\mathcal{H}_\omega$, the GNS-representation, discussed in detail in Supp. Sect. 27.9. The weak closure $\widehat{\mathcal{A}}$ of the operator algebra $\pi_\omega(\mathcal{A})$ in $\mathcal{H}_\omega$ is the representation dependent von Neumann algebra of physical observables.

In ordinary quantum mechanics and quantum statistical mechanics $\mathcal{A}$ is just the algebra of bounded operators in a separable Hilbert space $\mathcal{H}$, which is already a von Neumann operator algebra, so both the distinction between 'pre-observables' and true 'physical observables' and the use of a GNS-representation is superfluous, i.e. $\widehat{\mathcal{A}} = \mathcal{A}$.

We again discuss first the general case of classical and quantum statistical mechanics. Classical statistical mechanics is, as noted in the preceding section, an example of a probability theory, so a state is an *ensemble*, i.e. a probability distribution on the phase space $\Gamma$ of a mechanical system. The functions that serve as observables are the smooth functions on $\Gamma$, in first instance the bounded smooth functions, or even, for technical convenience, bounded with compact support, in which case the ensemble is a state in the precise mathematical sense, i.e. with finite expectations for all variables. For the quantum case the state is analogously defined on a not yet precisely specified noncommutative smooth $*$-algebra, or rather as a state on a $C^*$- or von Neumann algebra which contains this algebra as a subalgebra.

We restrict the states to *normal states*, which were defined by an additional continuity requirement in Supp. Chap. 27, Definition (27.8b). There is an equivalent definition which brings out the physical relevance of this property:

(12.4.3,a) **Theorem** *A state $\omega$ on a von Neumann algebra $\mathcal{A}$ is normal iff for every countable system $\{E_j\}_j$ of mutually orthogonal projections one has $\omega(\sum_j E_j) = \sum_j \omega(E_j)$.*

In the classical case this means that the states are probability measures that are $\sigma$-additive, a standard assumption in ordinary measure theory (See Supp. Chap. 19). For the quantum case this is the noncommutative version of $\sigma$-additivity. It is equivalent to the fact that a normal state is given by a density operator. This justifies the choice of such linear functionals as general physical quantum states, a choice presented in Sect. 11.4, and originally due to von Neumann.

In classical and quantum mechanics, regarded as degenerate cases of the above situation, we have for a classical mechanical state a point measure, and in quantum mechanics a density operator which is a pure state, just a one dimensional projection operator, which defines, up to a phase factor, a unit state vector.

## 12.4.4 Physical Interpretation

**Axiom III''**: *The physical interpretation in terms of the results of the measurement of an observable A in a state $\omega$ combines Axioms I'' and II''' and follows the same line as in the two earlier versions of the axiom system in Chaps. 3 and 11.*

This means that we are again in a Hilbert space situation, with the GNS-representation space $\mathcal{H}_\omega$ and selfadjoint operators $A$ in the operator algebra $\widehat{\mathcal{A}}$, the von Neumann algebra of physical observables, generated by the represented $C^*$-operator algebra $\pi_\omega(\mathcal{A})$ in $\mathcal{H}_\omega$, which leads to statements about expectations for the measurement of observables $A$ in the a given state $\omega$, just like in Chaps. 3 and 11.

### 12.4.5 Time Evolution

**Axiom IV″**: *The time evolution of a system is described by a one parameter group* $\{\phi_t\}_{t\in\mathbb{R}}$ *of* ∗-*automorphisms of* $\mathcal{A}$ *acting on the state* $\omega$ *as*

$$\omega_t(A) = \omega(\phi_{-t}(A)), \qquad \forall A \in \mathcal{A}.$$

*In addition to the 'autonomous' time evolution by a one parameter group of automorphisms one also has the possibility of a 'nonautonomous' dynamics for systems which evolve under a two parameter system* $\{\phi_{t_2 t_1}\}_{t_1, t_2 \in \mathbb{R}}$ *of* ∗-*automorphisms, which satisfies the relation*

$$\phi_{t_3 t_2} \phi_{t_2 t_1} = \phi_{t_3 t_1}, \qquad \forall t_1, t_2 \in \mathbb{R}.$$

Non-autonomous time evolution automorphisms describe, for instance, systems that evolve under the influence of a time-dependent external force. The general features of this for quantum systems have been briefly discussed in Sect. 3.6.5. Most of what was said in the foregoing about autonomous time evolution can be generalized to the non-autonomous case, by applying some fairly obvious techniques. This will not be discussed further.

Time evolution automorphisms should be unitarily implementable in $\mathcal{H}_\omega$, i.e. there should exist a one parameter group of unitary operators $U_t$ acting in the GNS-representation Hilbert space $\mathcal{H}_\omega$, such that

$$\omega_t(A) = \omega(U_{-t}\, \pi_\omega(A)\, U_t), \qquad \forall t \in R^1.$$

A state $\omega$ on $\mathcal{A}$ is called *invariant* with respect to the time evolution automorphisms iff

$$\omega_0(A) = \omega_0(\phi_{-t}(A)), \qquad \forall A \in \mathcal{A}, \qquad \forall t \in R^1.$$

It is convenient to requires continuity for the one parameter group $\{\phi_t\}_{t\in R^1}$, such that there is strong continuity for the group $\{U_t\}_{t\in R^1}$.

Most quantum physical models have a 'ground state' $\omega_0$ which is invariant with respect to the $\phi_t$.

(12.4.5,a) **Problem** Suppose that a state $\omega$ is invariant with respect to the automorphisms $\phi_t$. Show that the $\phi_t$ are unitarily implementable in $\mathcal{H}_\omega$, with the unitary operators $U_t$ defined by the formula

$$U_t \, \pi_\omega(A)\psi_\omega = \pi_\omega(\phi_t(A))\psi_\omega \,, \qquad \forall A \in \mathcal{A}.$$

Show that this implies that the cyclic vector $\psi_\omega$ in $\mathcal{H}_\omega$ is invariant under the operators $U_t$, i.e. that $U_t\psi_\omega = \psi_\omega$, for all $t$ in $R^1$.

The vector $\psi_{\omega_0}$, corresponding with the ground state $\omega_0$, is, for instance in the hydrogen atom (in Chap. 7) and in many other similar quantum mechanical systems, the state of lowest (discrete) energy. It appears in the Fock many particle formulation of quantum mechanics (in Chap. 9) and in quantum field theory (in Chap. 15) as the *no-particle state*.

In systems based on the Hamiltonian formalism the time evolution automorphisms are connected with symplectic transformations that leave both classical and quantum Poisson brackets invariant, another reminder of the remarkable relation between classical and quantum systems through the Poisson bracket.

### 12.4.6 Symmetries

**Axiom V″**: *Symmetries, (groups of symmetries) of physical systems are described by ∗-automorphisms (groups of ∗-automorphisms) of $\mathcal{A}$, that commute with the time evolution automorphism.*

Symmetry automorphisms are always required to leave the basic state $\omega_0$ invariant, are therefore, according to the answer to Problem (12.4.5,a), unitarily implemented by a unitary operator in the GNS-representation space $\mathcal{H}_{\omega_0}$, which leaves the state vector $\psi_{\omega_0}$ invariant.

In Hamiltonian classical mechanics symmetries, like time evolution, are represented by symplectic transformations; in both classical and quantum theories the Poisson bracket is left invariant. This again reinforces the suggestive idea—useful although somewhat naive—of canonical quantization discussed in Sect. 12.4.2.

For systems for which time development is not relevant, equilibrium ensembles in statistical physics, lattice systems such as the Ising spin system and its generalizations (in Sect. 10.12) a symmetry still determines the properties of the classical or quantum Hamiltonian and is therefore also important in these cases.

In relativistic theory time is a coordinate dependent concept. What matters in this respect is spacetime. Time evolution is therefore part of symmetry with respect to Lorentz transformations. In this case the $C^*$-algebra $\mathcal{A}$ is of a special form; it is an inductive limit, the closure in a certain sense of a system of 'local' $C^*$-algebras $\mathcal{A}(O)$, for all open sets $O$ in four dimensional spacetime (See Chap. 15). See for the notion of inductive limit Supp. Sect. 26.5. Lorentz transformations $\Lambda$, together with spacetime translations over spacetime vectors $a$, map these open sets onto each other and induce on $\mathcal{A}$ a group of $*$-automorphisms $\phi_{(\Lambda,a)}$. One says that a relativistic quantum field theory is *Lorentz covariant*, meaning that its basic state $\omega_0$ is invariant with respect to these automorphisms, with as consequence unitary implementability, as was to be proved in Problem (12.4.5,a).

*Final remark*: The third level of axioms, discussed in this section, is not of great use in ordinary classical or quantum theory but comes into its own for systems with an infinite number of degrees of freedom, the thermodynamic limit of finite systems in classical and quantum statistical physics and relativistic quantum field theory. This last application, *algebraic axiomatic quantum field theory* will briefly appear in Chap. 15.

## 12.5 Quantum Theory: von Neumann Versus Birkhoff

### 12.5.1 Introduction

In the early years of quantum theory von Neumann was not the only mathematician interested in this exciting new physics. There was also George D. Birkhoff—father of Garrett Birkhoff, also an important mathematician—who developed an alternative algebraic formalism, discussing it with von Neumann, working together with him (See Ref. [8]). It is of interest to compare and relate the two approaches, even though they have a considerable overlap. Von Neumann's approach may be simply characterized by the phrases "selfadjoint operators in Hilbert space as observables" and "linear functionals on the algebra of operators as states". In his 1932 book (Ref. [9]) he gave as the mathematical description of quantum states—general, because including quantum statistical states—linear functionals on the algebra of operators in a Hilbert space, with these functionals evaluated on selfadjoint operators, giving the expected value of the corresponding observables. He then derived—more or less—that such a functional is represented by a density operator. This we discussed in Sect. 11.2. In fact, the presentation of the mathematical framework of quantum theory in this book, in particular in Chap. 3 and Supp. Chap. 21, is based on the ideas of von Neumann. There is therefore no need to say more here about his approach to quantum theory, so in the next sections we shall focus on Birkhoff's ideas.

### 12.5.2 Birkhoff's Approach: Introduction

For Birkhoff quantum theory was a *nonclassical proposition calculus* with as basic elements *yes-no questions* about the properties of a physical system, to be answered by measurements. These yes-no, or 1–0 questions are represented by (orthogonal) projection operators in Hilbert space, the same as that of von Neumann. Together they form a nonclassical version of a $\sigma$-algebra, a generalization of what we know from standard probability theory. A quantum $\sigma$-algebra is an example of a well-known general mathematical object called a *complete orthomodular lattice*. In this particular context one often uses the term *quantum logic*. We give the main properties in the next section.

### 12.5.3 Lattices

We take as basic object in the mathematics of lattices a *poset*—short for *partially ordered set*. For the definition and some properties of this see Supp. Sect. 18.6.

(12.5.3,a) **Definition** A poset $X$ is called a *lattice* iff each pair of elements $x$ and $y$ has an infimum and a supremum. A *bounded lattice* has a minimum element, denoted as 0 and a maximum element denoted as 1.

In a lattice we denote the infimum of a pair of elements $x$ and $y$ as $x \wedge y$ (*meet*) and the supremum as $x \vee y$ (*join*) of $x$ and $y$.

(12.5.3,b) **Problem** Derive, for all $x$, $y$ and $z$ in $X$, the following relations:

$$x \wedge x = x, \quad x \vee x = x, \qquad\qquad\qquad\qquad\qquad\qquad (idempotence)$$
$$x \wedge y = y \wedge x, \quad x \vee y = y \vee x, \qquad\qquad\qquad\qquad (commutativity)$$
$$x \wedge (y \wedge z) = (x \wedge y) \wedge z, \quad x \vee (y \vee z) = (x \vee y) \vee z. \quad (associativity)$$
$$x \vee (x \wedge y) = x, \quad x \wedge (x \vee y) = x \qquad\qquad\qquad (absorption)$$

Note that associativity means that we may write $x_1 \wedge \ldots \wedge x_n$ and $x_1 \vee \ldots \vee x_n$; implying that each finite set of elements has an infimum and a supremum.

(12.5.3,c) **Definition** A bounded lattice $X$ is called *orthocomplemented* iff there is a map $x \mapsto x^c$ with the properties $(x^c)^c = x$, $0^c = 1$, $(x \vee y)^c = x^c \wedge y^c$, for all $x$ and $y$ in $X$.

(12.5.3,d) **Definition** A lattice $X$ is called *complete* ($\sigma$-*complete*), iff every $A \subset X$ (every countable $A \subset X$) has an infimum and a supremum.

(12.5.3,e) **Definition** An orthocomplemented lattice $X$ is called a *Boolean algebra* iff it has the additional property of *distributivity* : $x \vee (y \wedge z) = (x \vee y) \wedge (x \vee z)$, for all $x$, $y$ and $z$ in $X$. One has Boolean algebras, but also complete Boolean algebras, and in particular $\sigma$-complete Boolean algebras, or *Boolean $\sigma$-algebras*, which play a fundamental role in measure theory.

(12.5.3,f) *Example* Show that the power set $P(X)$ of a set $X$, the system of all subsets of $X$, is a complete Boolean algebra, the $\sigma$-algebra of measurable sets of a measure space a Boolean $\sigma$-algebra.

The Boolean algebra is the mother of all lattice systems. Its idea was proposed for use in a logical calculus of truth values by the British logician George Boole in 1847. The term dates from 1913.

An orthocomplemented lattice without the distributivity property is called a *quantum logic*, because of its role in the formulation of quantum theory, as will be explained in the next section.

### 12.5.4  Birkhoff's Approach: Continuation

We have seen that a classical probability theory is equivalent to a commutative von Neumann algebra. Such an algebra is completely determined by its projections.

(12.5.4,a) **Problem** Let $A$ be a commutative von Neumann algebra in a Hilbert space $\mathcal{H}$, with $\mathcal{E}(A) \subset B(\mathcal{H})$ the system of its projections. Show that $\mathcal{E}(A)$ is a $\sigma$-complete orthocomplemented distributive lattice, i.e. a Boolean $\sigma$-algebra, with respect to the definitions

$$E_1 \prec E_2 \iff E_1 E_2 = E_1, \quad \text{or} \quad \inf\{E_1, E_2\} = E_1 \wedge E_2 = E_1 E_2,$$

with

$$0 = 0_{\mathcal{H}}, \quad 1 = 1_{\mathcal{H}}, \quad E^c = 1_{\mathcal{H}} - E,$$

and verify in particular the distributivity condition.

Note that there is a one-to-one correspondence $E \leftrightarrow \mathcal{M}_E$ between projections and closed subspaces of $\mathcal{H}$ through $Ex = x \iff x \in \mathcal{M}_E$, or $\mathcal{M}_E = E(\mathcal{H})$.

Birkhoff defines for quantum theory a *quantum proposition calculus*.

(12.5.4,b) **Problem** Show that the system $\mathcal{E}(A)$ of projections in a noncommutative von Neumann algebra $A$ is a *quantum $\sigma$-logic*. Verify that in this case the distributivity property does not hold.

Birkhoff goes on to introduce functions $\mu : \mathcal{E}(A) \to R^1$, *quantum probability measures*, with the properties

(1) $0 \le \mu(E) \le 1$, for all projections $E$ in $\mathcal{E}(A)$,
(2) $\mu(0_{\mathcal{H}}) = 0$ and $\mu(1_{\mathcal{H}}) = 1$,
(3) $\mu(\sum_j E_j) = \sum_j \mu(E_j)$, for every at most countable set $\{E_j\}_j$ of mutually orthogonal projections in $\mathcal{E}(A)$.

This leads to *nonclassical quantum probability theories* $(A, \mathcal{E}(A), \mu)$.

(12.5.4,c) **Problem** Show that a sum of projections as in (Chap. 3) converges strongly to a projection. (Use material from Supp. Sect. 21.8.)

### 12.5.5  The Relation von Neumann—Birkhoff: Gleason's Theorem

In classical measure theory (and probability theory) one goes from measure to integration. There is a one-to-one relation between measurable sets $F$ and characteristic functions $f_F$ defined as $f_F(t) = 1$ for $1 \in F$, and 0 elsewhere. An integral of a measurable function, an average of a random variable, is defined as a limit of linear combinations of characteristic function.

In quantum theory we have instead of the $\sigma$-algebra $\mathcal{F}$ of measurable sets (events) the nondistributive quantum $\sigma$-logic $\mathcal{E}(A)$, with the projections $E$ taking the place of measurable sets $F$.

It is clear that a quantum system in the sense of von Neumann, with $\omega$ a state functional on $B(\mathcal{H})$, immediately leads to a quantum probability system in the sense of Birkhoff. The implication in the other direction seems also fairly obvious, but this is less simple than in the classical case. An arbitrary selfadjoint operator $A$ in $\mathcal{A}$ can be obtained as a limit of operators $A_n$ which are sums of linear combinations $\sum_k \lambda_k E_k$ of mutually orthogonal finite dimensional spectral projections $E_k$ in $\mathcal{A}$. The expectation value of $A$ (the 'integral over the quantum random variable') should then be $\lim_{n \to \infty} \sum_k \lambda_k \mu(E_k)$. However the proof that this limit is independent of the choice of the sequence $\{E_k\}_k$ is highly nontrivial. For this we have a celebrated theorem by Gleason. See [10]:

(12.5.5,a) **Theorem** (Gleason) *Let $\mathcal{H}$ be a (separable) Hilbert space of dimension at least three, with $\mathcal{A} = B(\mathcal{H})$ the von Neumann algebra of all bounded operators, and $\mathcal{E}(\mathcal{A})$ the system of projections in $\mathcal{A}$. Let $\mu$ be a function $\mu : \mathcal{E}(\mathcal{A}) \to \mathbb{R}^1$, a quantum probability measure, with the properties (1), (2) and (3). Then $\mu$ has a unique extension to a state $\omega_\mu$ on $\mathcal{A}$. It has the form $\omega_\mu(A) = \mathrm{Trace}\,(DA)$, for all $A$ in $\mathcal{A}$, with $D$ a density operator, i.e. a positive trace class operator with trace equal to one.*

The proof is highly nontrivial. It has been simplified considerably by, among others, Cooke et al. [11] but remains also in this version far from elementary. For a very readable introduction to Gleason's proof, see [12].

That Gleason's theorem is nontrivial is illustrated by the fact that it does not hold for the simple case of a two dimensional Hilbert space.

(12.5.5,b) **Problem** Consider the algebra $\mathcal{A} = B(\mathcal{H})$ in the two dimensional Hilbert space $\mathcal{H}$ of a single spin-$\frac{1}{2}$ quantum system. The system of projections $\mathcal{E}(\mathcal{A})$ in this case consists of the trivial projections $0_{\mathcal{A}}$, $1_{\mathcal{A}}$ and all one dimensional projections $E$. Let $D$ be a density operator in $\mathcal{H}$ and $\rho_D$ the quantum probability function on $\mathcal{E}(\mathcal{A})$ determined by $\rho_D(E) = \mathrm{Tr}\,(DE)$. Each $E$ determines, up to phase factors, an orthonormal basis $\{\psi_1, \psi_2\}$ in $\mathcal{H}$. This allows us to associate with $\rho_D$ a function $\hat{\rho}_D : \mathcal{H} \longrightarrow [0, 1]$. Show that $\hat{\rho}_D$ is continuous.

(12.5.5,c) **Problem** In the case described in Problem (12.5.5,b) the system $\mathcal{E}(\mathcal{A})$ consists, apart from the two trivial projections, of pairs of mutually orthogonal projections $E$ and $F = 1 - E$. Choose one such pair, say $\{E_1, F_1\}$. Define $\sigma(E_1) = \sigma(F_1) = \frac{1}{2}$, and assign to $\sigma$ for all other pairs the values 0 or 1. Show that this $\sigma$ satisfies the requirements of a quantum probability function in the sense of Birkhoff, but that it cannot be extended to a linear functional $\omega_D$ on the algebra $\mathcal{A}$.

Note that Gleason's theorem holds only for $\mathcal{A} = B(\mathcal{H})$, i.e. for a factor of type $I_n$, for all finite $n$, $n = 2$ excepted, and $I_\infty$. It does not seem to be known whether an analogue of the theorem for general factors exists, even though for this purpose various appropriately generalized lattices have been studied.

*Conclusion*: Birkhoff's lattice formulation can be derived from the formulation of von Neumann in term of operators in a Hilbert space. So far it is not known whether the two formulations are completely equivalent in all physical situations. The approach of Birkhoff stresses the character of quantum theory as a noncommutative version of classical measure (probability) theory, and has as such a definite conceptual attraction. Von Neumann's formulation has the great technical advantage of the linear context of functional analysis.

# References

1. Connes, A.: Noncommutative Geometry. Academic Press, Boston (1994). http://www. alainconnes.org/docs/book94bigpdf.pdf. (One of the great mathematics books written in the last twenty years. "The reader of the book should not expect proofs of theorems. This is much more a tapestry of beautiful mathematics and physics which contains material to intrigue readers ..." (Vaughan Jones). "... a long discourse or letter to friends" (Segal, I.E.: in a review. Bull. Amer. Math. Soc. **33**, 459–465 (1996). The author was not amused). The serious student will get much help from the next book)
2. Gracia-Bondía, J.M., Várilly, J.C., Figueroa, H.: Elements of Noncommutative Geometry. Birkhauser, Boston (2001) (The best general textbook on noncommutative geometry in the spirit of Connes. The 'authorized version' at the time this book was published)
3. Segal, I.E.: Postulates for general quantum mechanics. Ann. of Math. **48**, 930–948 (1947) (The founding paper on the subject of this chapter)
4. Farmelo, G.: The Strangest Man. The Hidden Life of Paul Dirac. Quantum Genius. Faber and Faber, London (2009) (A biography which not only gives a fascinating picture of the person of Dirac, but also of the environment in which quantum theory emerged and in which Dirac was one of the main actors)
5. Haag, R., Kastler, D.: An algebraic approach to quantum field theory. J. Math. Phys. **5**, 848–861 (1964) (An important paper applying the ideas of [3] to quantum field theory)
6. Groenewold, H.J.: On the principles of elementary quantum mechanics. Physica **12**, 405–460 (1946)
7. van Hove L.: Sur certaines représentations unitaires d'un groupe infini des transformations. Mémoire de l'Académie Royales des Sciences de Belgique **26**, 1–102 (1951)
8. Birkhoff, G., von Neumann, J.: The logic of quantum mechanics. Ann. Math. **37**, 823–843 (1936). http://www.fulviofrisone.com/attachments/article/451/the%20logic%20of% 20quantum%20mechanics%201936.pdf
9. von Neumann, J.: Mathematical Foundations of Quantum Mechanics. Princeton University Press, Princeton (1996). https://app.box.com/s/fsp81filfebei98umd6u (Translated from the original 1932 German Springer edition. A great classic, written by the creator of the mathematical formalism of quantum mechanics. The part on the theory of measurement may have been overtaken by more recent developments, but the general discussion of quantum theory, in particular of selfadjoint operators and the spectral theorem remains instructive and fresh)
10. Gleason, A.M.: Measures on the closed subspaces of a Hilbert space. J. Math. Mech. **6**, 885–893 (1957). http://www.iap.tu-darmstadt.de/tqp/uebungen/qinfo11/Gleason.pdf
11. Cooke, R., Keane, M., Moran, W.: An elementary proof of Gleason's theorem. Math. Proc. Cambridge. Philos. Soc. **98**, 117–128 (1985)
12. Granström, H.: Gleason's Theorem. Thesis University of Stockholm, Stockholm (2006). http:// kof.physto.se/theses/helena-master.pdf

# Chapter 13
# Quantization

## 13.1 Introduction

In this chapter we discuss the general notion of quantization. It has a range of meanings; the first and most obvious among those is the historic one. In the beginning of the twentieth century theoretical physicists struggled with serious fundamental problems in what we now call classical physics. In his search for a new model of atomic physics Heisenberg discovered that by employing noncommuting canonical variables $p_j$ and $q_k$ he could correctly calculate the energy levels of various systems such as the harmonic oscillator. Immediately after this Max Born observed that these variables were infinite matrices, which later, in von Neumann's approach, meant operators in Hilbert space.

In time a more formal definition of quantization was developed, as a procedure which assigns in a systematic way to each classical observable a quantum observable. There is standard *canonical quantization*, which starts from classical mechanics based on a symplectic manifold as phase space with its Poisson bracket for the classical observables and results in selfadjoint operators in the quantum mechanical Hilbert space. We discuss two different examples of this, Born–Jordan quantization in Sect. 13.2, and, more extensively, Weyl quantization in Sects. 13.3 and 13.4. Born–Jordan and Weyl quantization are different, which shows that quantization is not a unique procedure. Note that both procedures are up till now restricted to linear phase spaces. In Sect. 13.5 we discuss a general scheme for canonical quantization.

*Deformation quantization* has the same point of departure. However, it avoids Hilbert space altogether. Quantum results are formulated in terms of the algebra of classical observables, provided with a new deformed noncommutative product. One has strict and formal deformation quantization, discussed in Sects. 13.6 and 13.7.

*Geometric quantization* goes, in principle, all the way from Hamiltonian classical mechanics to quantum mechanics in Hilbert space, using techniques from differential geometry. This approach is meaningful—but not very successful—for arbitrary symplectic manifolds. It is one of these subjects—like for instance ergodic theory, or formal deformation quantization—that originated as an attempt to solve or elucidate

© Springer International Publishing Switzerland 2015
P. Bongaarts, *Quantum Theory*, DOI 10.1007/978-3-319-09561-5_13

a general problem in physics, but after a while developed into a purely mathematical theory. The result is a lot of interesting and beautiful mathematics; its contribution to physics has been rather disappointing. Because of this and of the fact that a lot of very technical material from differential geometry is required—far outside what is given in Supp. Chap. 20—we shall not discuss it.

Finally, there is *Feynman's path integral quantization*, which stands apart from all other approaches to quantization. It does not use the Hamiltonian for the classical system nor Hilbert space notions for the quantum theory, but is a prescription for a direct calculation of quantum transition amplitudes from classical data. It is discussed in Sect. 13.8.

Originally quantization meant starting with classical physical theory and obtaining from this a quantum physical theory. The classical electromagnetic field is a valid—and as such very successful—theoretical model, which can be used to construct a quantized version, as was first done by Dirac [1]. However, this is an exception among field theories. All other quantum field theories, such as the scalar and the Dirac quantum field, are constructed by starting from classical fields, which as such have no physical meaning, but play a role as auxiliary mathematical objects used to obtain quantum field theories. Note in this respect that, contrary to ordinary classical mechanics, field theories have an infinite number of degrees of freedom and their quantization is therefore mathematically much more difficult. Quantum field theory will be discussed in Chap. 16.

There is a further complication: so far we have only discussed the quantization of boson systems. For fermion systems one has to use a different quantization prescription. A simple quantum system like a nonrelativistic free particle with spin 1/2 has no underlying classical phase space description; there is no symplectic manifold in the usual sense.

For fermion theories we introduce a notion of *pseudo-classical systems*, again auxiliary, purely mathematical models, with no physical meaning. We shall apply this to a toy model, the fermionic harmonic oscillator, in Sects. 13.9.2 and 13.9.4. The precise mathematical nature of the quantization of fermionic systems drew in the beginning relatively little attention from mathematicians or mathematical physicists. Later, however it turned out that a new branch of mathematics, 'super mathematics', i.e. the theory of *superalgebras*, 'superdifferentiable geometry', with *supermanifolds*, could be developed for this, and, as has happened in other cases, became a flourishing field of pure mathematics.

The basic notions of all this are given in Sect. 13.9, with a brief introduction to 'super mathematics' in Sect. 13.9.4.

## 13.2 Born–Jordan Quantization

Max Born and Pascual Jordan published in 1925 an article [2], in which they extended the results of a paper on quantum mechanics, just published at that time by Heisenberg [3]—in fact what might be called the founding paper of the subject. They observed in particular that Heisenberg's algebraic noncommuting variables $p_j$ and $q_k$ were infinite matrices. From then onwards Heisenberg's formulation of quantum mechanics became known as *matrix mechanics*. What interests us here is that they put forward a quantization prescription for arbitrary monomials $p^m q^n$

$$p^m q^n \;\;\rightarrow\;\; \frac{1}{m+1} \sum_{k=0}^{m} P^{m-k} Q^n P^k.$$

Their argument for this is part of long discussion and reformulation of Heisenberg's results in which they used in particular the similarity between the Hamilton time evolution equations in the classical and quantum case. It is not easy to summarize.

## 13.3 Weyl Quantization 1

Another prescription which eventually surpassed that of Born and Jordan in popularity was put forward in 1927 by Hermann Weyl in a paper [4], and in his famous book on the application of group theory to quantum mechanics [5]. Weyl gave a general Fourier integral formula, to be discussed in the next section, from which a prescription for polynomials can be derived. This integral formula emerged later in mathematics as a particular way of defining *pseudodifferential operators*, a topic of great interest, although it has not much to do with quantum theory or quantization and will therefore not be discussed here. Weyl's prescription for the quantization of arbitrary monomials, a result of his integral formula, reads

$$q^m p^n \;\;\rightarrow\;\; \frac{1}{2^n} \sum_{k=0}^{n} \frac{n!}{(n-k)!k!} P^{n-k} Q^m P^k.$$

The Born–Jordan and the Weyl quantizations are different, but only for $m > 1$, $n > 1$.

(13.5,a) **Problem** Verify that one has, for both Born–Jordan and Weyl,

$$qp \;\;\rightarrow\;\; \frac{1}{2}(PQ + QP),$$

but for Born–Jordan

$$q^2 p^2 \;\;\rightarrow\;\; \frac{1}{3}(P^2 Q^2 + PQ^2 P + Q^2 P^2),$$

and for Weyl

$$q^2 p^2 \;\;\rightarrow\;\; \frac{1}{4}(P^2 Q^2 + 2PQ^2 P + Q^2 P^2).$$

The only physically realistic case in which products of position and moment variables occur is that of a particle with electric charge $e$ moving in a given magnetic potential $A(x)$, in which the Hamiltonian is

$$H = \frac{1}{2me} \left( \mathbf{p} + \frac{e}{c}\mathbf{A} \right) \cdot \left( \mathbf{p} + \frac{e}{c}\mathbf{A} \right),$$

with the vector potential $\mathbf{A}$. This means that for all practical purposes both quantizations can considered to be physically equivalent. In any case, when there are different quantizations for some classical expression, physical experiments should decide which is correct.

## 13.4 Weyl Quantization 2: An Integral Formula

Weyl's quantization formula, a particular kind of integral formula, first presented in a paper [4], and in his book [5],was clearly suggested by his interest in applying group theory to quantum physics, a subject that he pioneered. See for his active interest in the new quantum mechanics [6] (Note in passing that he also wrote one of the first books on general relativity [7], so here we have again a mathematician with great interest in and knowledge of physics).

His arguments for choosing his quantization formula are not easy to follow, but the resulting formula's are elegant and clear. One of the things that bothered him was the sloppy way his physics colleagues treated the canonical operators $p_j$ and $Q_k$, unbounded operators which cannot simply be added and multiplied with each other because of domain problems. For this reason he proposed to use instead the exponentiated forms, the one parameter unitary groups $U(\alpha) = e^{\frac{i}{\hbar}\alpha P}$ and $V(\beta) = e^{i\beta Q}$. They satisfy the commutation relation

$$U(\alpha)V(\beta) = e^{i\alpha\beta}V(\beta)U(\alpha).$$

or

$$e^{\frac{i}{\hbar}\alpha P}e^{i\beta Q} = e^{i\alpha\beta}e^{i\beta Q}e^{\frac{i}{\hbar}\alpha P}.$$

which is the rigorous form of the usual commutation relation

$$[P, Q] = -i\hbar 1_{\mathcal{H}}.$$

The operators $U(\alpha) = e^{\frac{i}{\hbar}\alpha P}$ and $V(\beta) = e^{i\beta Q}$ can be combined to form what is nowadays called the *Weyl operator*

$$W_\hbar(\alpha, \beta) = e^{-\frac{i}{2}\alpha\beta}e^{\frac{i}{\hbar}\alpha P}e^{i\beta Q},$$

which can also be written as

$$W_\hbar(\alpha, \beta) = e^{i(\frac{\alpha}{\hbar}P + \beta Q)},$$

which is in fact a definition of the right hand side, because a linear combination of the two noncommuting selfadjoint operators $P$ and $Q$ is a priori not well defined.

The notion of *Weyl operator* and in particular 'Weyl system' has been championed by Irving Segal who used it as basis for his work on the mathematical foundations of quantum field theory [8]. It is for the canonical operators formulated in this exponential Weyl form that the Stone–von Neumann uniqueness theorem holds.

The argument from Weyl's [4] paper on quantum theory and also from his book, leading to his formula for (relatively) arbitrary functions of the operators $P$ and $Q$, can be summarized as follows, again for the sake of simplicity, for the one dimensional case:

Let $f(p, q)$ be a classical function on the (two dimensional) phase space $\mathbb{R}^2$. The aim of Weyl's procedure is to assign to this an operator function $f(P, Q)$ in the Hilbert space $L^2(\mathbb{R}^1, dx)$. The usual Fourier transformation in quantum mechanics connects the 'position representation' with the 'momentum representation', i.e. $\mathcal{H} = L^2(\mathbb{R}^1, dx)$ with $\widehat{\mathcal{H}} = L^2(\mathbb{R}^1, dp)$, according to the formulas

$$\hat{\psi}(p) = \frac{1}{(2\pi\hbar)^{1/2}} \int_{-\infty}^{+\infty} \psi(x)e^{-\frac{i}{\hbar}px}\, dx,$$

and its inverse

$$\psi(x) = \frac{1}{(2\pi\hbar)^{1/2}} \int_{-\infty}^{+\infty} \hat{\psi}(p)e^{+\frac{i}{\hbar}px}\, dp.$$

Here we do something different. We start with square integrable functions $f(p, q)$ on the phase space and transform these to a space $\mathcal{H}' = L^2(R^2, du\, dv)$ according to

$$\hat{f}(u, v) = \frac{1}{2\pi\hbar} \int\int f(p, q)e^{-i(\frac{u}{\hbar}p + vq)}dp\, dq,$$

with inverse

$$f(p, q) = \frac{1}{2\pi\hbar} \int\int \hat{f}(u, v)e^{i(\frac{u}{\hbar}p + vq)}du\, dv.$$

Weyl's definition of the operator $f(P, Q)$ simply employs the last formula, rewrites it as an operator formula, uses the second form of the formula for $W_\hbar(\cdot, \cdot)$, which gives

$$f(P, Q) = \frac{1}{2\pi\hbar} \int\!\!\int \hat{f}(u, v) e^{i(\frac{u}{\hbar}P + vQ)} du\, dv$$

$$= \frac{1}{2\pi\hbar} \int\!\!\int \hat{f}(u, v)\, e^{-\frac{i}{2}uv} e^{\frac{i}{\hbar}uP} e^{ivQ} du\, dv.$$

Of course, whether this operator $f(P, Q)$ is well-defined depends on the properties of the numerical function $f(p, q)$. A nontrivial matter which has to be investigated carefully. Weyl's integral formula defines a *pseudo differential operator* in $L^2(R^1, dx)$.

By choosing $f(p, q) = q^m p^n$ one finds Weyl's prescription for the quantization of monomials.

A question that remains is whether Weyl's quantization idea can be applied to a classical system for which the phase space is not $\mathbb{R}^{2n}$ but an arbitrary symplectic manifold $\mathcal{M}$. A possible approach might be to try to cover $\mathcal{M}$ by local Darboux coordinate systems.

## 13.5 A General Scheme for Canonical Quantization

For classical mechanics where the observables are in first instance the real elements of the ∗-algebra of complex-valued functions on phase space, a $2n$-dimensional symplectic manifold $\mathcal{M}$, this assignment

$$C^\infty(\mathcal{M}) \rightarrow B(\mathcal{H}), \quad f \mapsto Q(f),$$

should satisfy the following requirements:

(1) linearity,
(2) $Q(1) = 1_{\mathcal{H}}$,
(3) reality, i.e. $Q(\overline{f}) \rightarrow Q(f)^*$, i.e. real functions map to selfadjoint operators.
(4) irreducibility, i.e. the operators $Q(f)$, for all $f$ in $C^\infty(\mathcal{M})$, form an irreducible system in $\mathcal{H}$.

A quantization in the above sense is in principle completely determined by the quantization of a general polynomial, in fact of linear combinations of monomials, i.e. of all the operators $Q(p_j^m q_k^n)$, for $m, n = 0, 1, \ldots$ This is the case for Born–Jordan and Weyl quantization.

There is a suggestive relation between the Poisson bracket for classical variables $f$ and $g$ in classical mechanics and the commutator of the corresponding quantum variables, the operators $Q(f)$ and $Q(g)$. It was first noticed by Dirac in [9], and is given as

$$\{f, g\}_{\text{classical}} \quad \mapsto \quad i\hbar[Q(f), Q(g)].$$

One has, in this manner, for instance, for the canonical variables $p_j$ and $q_k$ the Poisson bracket

$$\{p_j, q_k\} = \delta_{jk}$$

and for the canonical operators $Q(p_j) = P_j$ and $Q(q_k) = Q_k$; the commutator

$$[Q(p_j), Q(q_k)] = -i\hbar\delta_{jk},$$

Note also the relation between the formulas for the time evolution of an observable. Define for a moment, for a pair of classical variables $a$ and $b$ and a pair of quantum observables $A$ and $B$

$$\{a, b\}_{\mathrm{cl}} = \{a, b\}, \qquad \{A, B\}_{\mathrm{qu}} = \frac{i}{\hbar}[A, B].$$

Then there is clearly a general suggestion for a correspondence

$$\{\cdot, \cdot\}_{\mathrm{cl}} \mapsto \{\cdot, \cdot\}_{\mathrm{qu}}.$$

It is true for the time evolution for a classical function

$$f = f(p(t), q(t)),$$

given by

$$\frac{df}{dt} = \{H, f\}_{\mathrm{cl}},$$

and

$$\frac{dF}{dt} = \{H, F\}_{\mathrm{qu}}$$

for a quantum observable in the Heisenberg picture,

$$F(t) = e^{\frac{i}{\hbar}tH} H F(0) e^{-\frac{i}{\hbar}tH}.$$

Because of all this it is tempting to add a fifth property to the definition of the quantization map, namely (5)

$$[Q(f), Q(g)] = -i\hbar Q(\{f, g\}), \quad \forall f, g \in C_{\mathbb{C}}^{\infty}(\mathbb{R}^{2n}),$$

which we can write, by using the definition

$$i[Q(f), Q(g)] = \{Q(f), Q(g)\}_{\mathrm{qu}},$$

as

$$\{Q(f), Q(g)\}_{qu} = \hbar Q(\{f, g\}_{cl}), \quad \forall f, g \in C_{\mathbb{C}}^{\infty}(\mathbb{R}^{2n}).$$

However, this assumption is too naive; a theorem of Groenewold and van Hove states that even for the simple case of $\mathcal{M} = \mathbb{R}^{2n}$ a quantization map $Q$ with these five properties for all variables $f$ and $q$ does not exist [10, 11]. What can be expected instead is a relation $(5')$ of the form

$$\{Q(f), Q(g)\}_{qu} = \hbar Q(\{f, g\}_{cl}) + \cdots,$$

with higher order terms in $\hbar$ of the form

$$\hbar^{n} m_{n}(f, g), \quad n = 1, 2, \ldots, \quad m_{0}(f, g) = Q(\{f, g\}_{cl}),$$

in which the $m_{n}$ are bilinear maps

$$m_{n} : C_{\mathbb{C}}^{\infty}(\mathbb{R}^{2n}) \times C_{\mathbb{C}}^{\infty}(\mathbb{R}^{2n}) \rightarrow B(\mathcal{H}),$$

in fact bidifferential operators on $\mathbb{R}^{2n}$, like the Poisson bracket.

The idea that there is a correspondence between Poisson brackets in classical mechanics and operator commutators in quantum mechanics underlies both Born–Jordan and Weyl quantization although it is not explicitly expressed.

## 13.6 Strict Deformation Quantization

Planck's constant $\hbar$ appeared in all the quantum formulas in the preceding discussions. This was to emphasize the idea that quantum theories are *deformations* of classical theories, with $\hbar$ as deformation parameter. Note that $\hbar$ is a constant of nature. Choosing units for physical variables will change the *numerical value* of $\hbar$. If we use the kilometer as unit of length this numerical value becomes vanishing small; we are approaching the *classical limit* of the quantum system. In this and the next section we shall discuss explicit ways in which this idea of deformation quantization has been expressed.

In the first place there is *strict deformation quantization*, an attempt to construct for a given symplectic manifold a quantized system, analytically dependent on $\hbar$. This has been undertaken by Rieffel [12, 13]. He starts from a commutative $C^{*}$-algebra $\mathcal{A}$ of continuous functions on a given symplectic manifold $\mathcal{M}$. In $\mathcal{A}$ he defines a deformation, depending on $\hbar$, of the given algebra multiplication, together with a deformation both of the norm and of the $*$-operation on $\mathcal{A}$. In this manner he obtains a *field* of noncommutative $C^{*}$-algebras $\{\mathcal{A}_{\hbar}\}_{\hbar \in \mathbb{R}^{1}}$. He has obtained interesting rigorous results for several manifolds $\mathcal{M}$, none of which is however of particular physical interest.

## 13.7 Formal Deformation Quantization

### 13.7.1 Introduction

The difficulties of obtaining rigorous results in the deformation approach, as is obvious from the work of Rieffel, has led to a purely algebraic approximative approach based on formal power series. If one goes only up to first order this is called *first order* or *infinitesimal* deformation quantization. The general case, to be discussed in this section is called *formal deformation quantization*.

Formal deformation quantization as a subject was initiated in 1977 in two long and fairly complicated papers by what might be called the 'Dijon School', consisting of the physicists Moshé Flato (†1998)—its de facto leader, François Bayen, Christian Fronsdal, Daniel Sternheimer, all from Dijon, together with the differential geometer André Lichnerowicz (†1998) from the Collège de France in Paris [14]. This work went unnoticed for several years, but then started to draw attention; more and more papers were written, developing the ideas of what then indeed became to be known as 'formal deformation quantization', a true avalanche, culminating in the work of Fedosov [15] and finally Kontsevich [16] with his formality theorem, for which he, among other things, obtained a Fields Medal in 1998.

The original intention of the Dijon school was to find a description of quantum mechanics which was to be an alternative to the standard Hilbert space formalism, incorporating Dirac's idea of the connection between the classical Poisson bracket and the quantum commutator, dependent on Planck's constant $\hbar$ as a deformation parameter. However this approach developed gradually into an independent mathematical field, of great interest, but with less and less relevance to physics, something that may remind us of what happened to ergodic theory, that started as an attempt to lay a basis for classical statistical mechanics, and later independently developed into a general theory of the asymptotic behaviour of groups of measure preserving transformations.

The main idea of formal deformation quantization is to avoid difficult problems of functional analysis by working with formal power series, leaving aside the questions of their convergence. A certain justification for this can be found in the fact that it is not unusual in physics to work with asymptotic series. In any case, this makes it mathematically into a problem of algebra.

So instead of trying to find a strict deformation of the algebra of classical observables, one constructs a larger auxiliary algebra of formal power series in the deformation parameter $\hbar$, with coefficients in the classical algebra, on which a deformed (noncommutative) product, a so-called *star product* is defined. The mathematical background for this will be discussed in the next subsection. For a clear introduction to formal deformation quantization and in particular to the formality theorem of Kontsevich, see the book of Chiara Esposito [47].

### 13.7.2  Formal Deformations of an Associative Algebra

Let $\mathcal{A}$ be a complex associative $*$-algebra, with unit element.

The multiplication on $\mathcal{A}$ is a symmetric bilinear map

$$\mathcal{A} \times \mathcal{A} \rightarrow \mathcal{A}, \quad (a, b) \mapsto ab, \quad \forall a, b \in \mathcal{A}.$$

It is associative i.e. satisfies the relation $(ab)c = a(bc)$, for all $a, b, c$ in $\mathcal{A}$. One has, of course, $m(1_\mathcal{A}, a) = m(a, 1_\mathcal{A}) = a$, for all $a \in \mathcal{A}$.

We may consider a different—not necessarily commutative—multiplication in $\mathcal{A}$, i.e. a second bilinear map

$$m : \mathcal{A} \times \mathcal{A} \rightarrow \mathcal{A}, \quad (a, b) \mapsto m(a, b), \quad \forall a, b \in \mathcal{A},$$

not necessarily with $m(a, b) = m(b, a)$, with $m(1, a) = m(a, 1) = a$, for all $a$ in $\mathcal{A}$, i.e. with the same unit, and with the associativity of $m$ expressed as

$$m(a, m(b, c)) = m(m(a, b), c), \quad \forall a, b, c \in \mathcal{A}.$$

Because of the possibility of noncommutativity of this new multiplication one has to impose the reality condition $m(a, b)^* = m(b^*, a^*)$, instead of $(ab)^* = a^* b^*$ for the original multiplication in the commutative algebra $\mathcal{A}$.

Two such products, $m_1$ and $m_2$, are *equivalent*, i.e. essentially the same, iff there exists an invertible linear map $T : \mathcal{A} \rightarrow \mathcal{A}$ such that

$$T(m_1(a, b)) = m_2(T(a), T(b)), \quad \forall a, b \in \mathcal{A}.$$

This is sometimes called 'gauge equivalence'.

(13.7.2,a) **Problem** Show that this is indeed an equivalence relation, as defined in Supp. Sect. 19.3.

Of particular interest are products depending on a deformation parameter $\lambda$, according to the power series

$$m_\lambda(a, b) = m_0(a, b) + m_1(a, b)\lambda + m_2(a, b)\lambda^2 + \cdots,$$

with, of course, $m_0(a, b) = ab$, meaning that the new multiplication is in order 0 the same as the original one.

(13.7.2,b) **Problem** Show that one has

$$m_j(1, a) = m_j(a, 1) = 0, \quad \forall j = 1, 2, \ldots, \quad \forall a \in \mathcal{A}.$$

One may ask whether such a power series is convergent, i.e. if the deformation is analytic in $\lambda$. The central point of the approach of Bayen et al. is that one does not try to find an answer to this question, which in any case will usually be negative, but to use instead a formulation in terms of *formal* power series. These form an associative algebra, an extension of $\mathcal{A}$, and is denoted as $\mathcal{A}[[\lambda]]$. There is a simple down to earth description of this algebra in which a power series

$$a(\lambda) = a_0 + a_1\lambda + a_2\lambda^2 + \cdots$$

is represented by an infinite sequence

$$\hat{a} = (a_0, a_1, \ldots), \quad a_j \in \mathcal{A}, \quad j = 0, 1, 2, \ldots.$$

with obvious addition and scalar multiplication. The pointwise multiplication of the power series in $\lambda$ leads to the multiplication in terms of the sequences.

(13.7.2,c) **Problem** Show that this product is given by the formula

$$\hat{a}\hat{b} = (a_0b_0, a_0b_1 + a_1b_0, a_0b_2 + a_1b_1 + a_2b_0, \ldots).$$

There is no need to use these infinite sequences of elements of $\mathcal{A}$ explicitly, nor the algebra $\mathcal{A}[[\lambda]]$ itself. We just write everything in terms of power series in the indeterminate variable $\lambda$ and use this as a bookkeeping device. Things will then be taken care of automatically.

Consider a vector space $\mathcal{A}$ which has two associative multiplications $m_1$ and $m_2$. These multiplications are *equivalent* iff there exists an invertible linear map $T$ from $\mathcal{A}$ in itself such that, for all $a$ and $b$ in $\mathcal{A}$, one has $m_2(a, b) = T(m_1(T^{-1}(a), T^{-1}(b)))$.

In line with this we define equivalence of two formal deformations by an invertible linear map $T_\lambda : \mathcal{A}[[\lambda]] \to \mathcal{A}[[\lambda]]$, defined by a formal power series

$$T_\lambda(a) = T_0(a) + T_1(a)\lambda + T_2(a)\lambda^2 + \cdots, \quad \forall a \in \mathcal{A},$$

with $T_0(a) = a$, for all $a$ in $\mathcal{A}$. Note that $T_\lambda$ does not necessarily restrict to a map $\mathcal{A} \to \mathcal{A}$.

(13.7.2,d) **Problem** Show that the power series

$$T_\lambda(a) = a + T_1(a)\lambda + T_2(a)\lambda^2 + \cdots$$

defines an inverse of the operator $T_\lambda$, given by a power series

$$(T^{-1})_\lambda = a + (T^{-1})_1(a)\lambda + (T^{-1})_2(a)\lambda^2 + \cdots,$$

with, for $n \geq 1$, the coefficients successively determined by

$$(T^{-1})_n(a) = -T_n(a) - (T^{-1})_1(T_{n-1}(a)) + \cdots - (T^{-1})_{n-1}(T_1(a)),$$

for all $a$ in $\mathcal{A}$. Show that this implies that every $T_\lambda$ with $T_0(a) = a$, is invertible in the sense of operators in $\mathcal{A}[[\lambda]]$.

In this context there are two questions to be answered:

[1]. **Existence**: For a given sequence $\{m_j\}_{j=0,1,\dots}$ one has to establish *associativity*.

(13.7.2,e) **Problem** Show that the condition for this is

$$\sum_{j+k=n} m_j(m_k(a,b),c))\lambda^n = \sum_{j+k=n} m_j(a,m_k(b,c))\lambda^n,$$

for $n = 0, 1, 2, \dots$ and for all $a, b, c$ in $\mathcal{A}$. Work out the first few terms.

[2]. **Equivalence**

(13.7.2,f) **Problem** Show that the equivalence of two formally deformed products $(m_\lambda)_1$ and $(m_\lambda)_2$ can be written as a sequence of relations for each order in $\lambda$,

$$\sum_{j+k=n} T_j((m_1)_k(a,b)) = \sum_{j+k+l=n} (m_2)_j(T_k(a), T_l(b)),$$

for all $a, b$ in $\mathcal{A}$. Work out a few lowest order terms.

Cohomology theory is a part of mathematics that has many applications in a broad field of other mathematical fields. A particular branch, *Hochschild cohomology*, was invoked by Gerstenhaber [17, 18] to reformulate in an elegant manner the above conditions for existence and equivalence of formal deformations of associative algebras. This lies outside the scope of this book and will therefore not be discussed here.

## 13.7.3 Formal Deformation Quantization

All this can be applied to the commutative algebra $\mathcal{A} = C_{\mathbb{C}}^\infty(\mathcal{M})$ of complex-valued smooth functions on the classical phase space of a physical system, a symplectic manifold, with a Poisson bracket $\{\cdot, \cdot\}$, which makes $C_{\mathbb{C}}^\infty(\mathcal{M})$ into a Poisson algebra. Formal deformation quantization then means finding a formal deformation of the point wise multiplication of functions from $C_{\mathbb{C}}^\infty(\mathcal{M})$, in terms of a sequence $\{m_j(\cdot, \cdot)\}_{j=0,1,2\dots}$, satisfying the associativity condition given above. The deformation should be 'in the direction of the classical description', i.e. one should start with an $m_1(\cdot, \cdot)$ such that $\{f, g\} = \frac{1}{2}(m_1(f, g) - m_1(g, f))$, for all $f, g$ in $C_{\mathbb{C}}^\infty(\mathcal{M})$ and with $\{\cdot, \cdot\}$ the classical Poisson bracket. The bilinear expressions $m_j(\cdot, \cdot)$ should be bidifferential operators, like the Poisson bracket. The maps $T_j(\cdot)$ that describe equivalences should likewise be differential operators.

This was the central idea of the seminal papers of Bayen et al. [14], followed by many papers by these authors and others, culminating in the work of Fedosov [15] and finally that of Kontsevich [16], who proved the existence of a formal deformation for symplectic, respectively Poisson manifolds, as was already observed in the beginning of Sect. 13.7.1.

## 13.8 Feynman's Path Integral Quantization

### 13.8.1 Introduction

Path integral quantization is completely different from all the other quantization methods discussed in this chapter. Like these it starts from the description of a physical system by classical mechanics, however, not in terms of the Hamiltonian but of the Lagrangian formulation (see for this Sect. 2.4). Its results are not expressed in terms of Hilbert space and its operators, but give directly the quantum mechanical transition amplitude for the physical process at hand.

The transition amplitude from an initial time $t_1$ to a later time $t_2$ is the (singular) integral kernel of the time evolution operator $e^{-\frac{i}{\hbar}(t_2-t_1)H}$ in the position representation. It is often called the *propagator* and is denoted, in the case of a single nonrelativistic particle, as $K(\mathbf{x_2}, t_2; \mathbf{x_1}, t_1)$, or in the more heuristic Dirac bra-ket notation as $\langle \mathbf{x_2} | e^{-\frac{i}{\hbar}(t_2-t_1)H} | \mathbf{x_1} \rangle$. (The bra-ket notation is explained in Supp. Chap. 26).

As physicist Feynman was a genius, but he never cared for mathematical rigour and was driven in his work by a very direct and creative physical intuition. As he explains in his paper [19], and also in his later book (Feynman and Hibbs [20]), the path that a quantum mechanical particle will follow in going from $\mathbf{x_1}$ at $t = t_1$ to $\mathbf{x_2}$ at $t = t_2$, will be the result of a probability that is the sum—in fact the integral— of contributions from the different paths, for each of which Feynman postulates an exponential with an imaginary phase containing the classical action with a factor $1/\hbar$ in front. The true classical path, for which the action is minimal, will give the most important contribution, corrected by higher order terms in $\hbar$, coming from the other paths. This idea leads to the central formula of his approach,

$$K(\mathbf{x_2}, t_2; \mathbf{x_1}, t_1) = N \int e^{\frac{i}{\hbar} S[\gamma]} \mathcal{D}[\gamma].$$

The problem with this formula is that the expression $\mathcal{D}[\gamma]$, a kind of infinite dimensional Lebegue measure on the space of all possible paths, with $\gamma_1(t_1) = x_1$ and $\gamma_2(t_1) = x_2$, does not exist, as was shown, for example, by Cameron [21]. Moreover, the normalization constant $N$ is infinite, so as an integral in the accepted sense the path integral formula does not make sense. As a historical note it may be observed that the idea of using the classical action in determining the time evolution of a physical particle had been put forward in 1933 by Dirac in a rather obscure journal and that Feynman knew this paper and was inspired by it. See [22].

There is no doubt about the great physical importance of Feynman's approach; many fruitful applications have been developed, but its mathematical basis, or at least a great part of it, has remained to this day unclear. This was realized already quite in the beginning, after publication of Feynman's paper. Generations of mathematicians and mathematical physicists have tried to get a grip on it, up until the present time, with only partial success.

There are several excellent modern books on path integrals, e.g. [44, 45] and [46].

There are two rigorous approaches to the Feynman path integral. One is by means of the Trotter product formula, which gives meaning to Feynman's formula as a limit of finite dimensional integrals. In the other one continues the formula for real time to a similar formula for imaginary time, leading to a stochastic process, which can be described in terms of a path integral over classical stochastic paths, indeed an integral in the accepted mathematical sense. The connection between the two types of path integrals is expressed by the Feynman–Kac formula.

### 13.8.2 The Trotter Product Formula

Divide the interval $[t_a, t_b]$ in $n$ equal parts, i.e.

$$[t_0 = t_a, t_1], \ [t_1, t_2], \ \ldots, \ [t_{n-1}, t_n = t_b],$$

with $t_{j+1} - t_j = \Delta t$, for $j = 0, 1, \ldots, n-1$, so $t_b - t_a = n\Delta t$. Write

$$e^{-\frac{i}{\hbar}(t_b-t_a)H} = e^{-\frac{i}{\hbar}(t_n-t_{n-1})H} \ldots e^{-\frac{i}{\hbar}(t_1-t_0)H} = (e^{-\frac{i}{\hbar}\Delta t H})^n$$

Since $H_0$ and $V$ do not commute, we have

$$e^{-\frac{i}{\hbar}\Delta t(H_0+V)} \neq e^{-\frac{i}{\hbar}\Delta t H_0} e^{-\frac{i}{\hbar}\Delta t V},$$

nevertheless, the difference obviously becomes small for $\Delta t \to 0$, i.e. one has $\lim_{\Delta t \to 0} \left( e^{-\frac{i}{\hbar}\Delta t(H_0+V)} - e^{-\frac{i}{\hbar}\Delta t H_0} e^{-\frac{i}{\hbar}\Delta t V} \right) = 0$. For $n \to \infty$, so $\Delta t \to 0$, one may ask which effect will win in the product $(e^{-\frac{i}{\hbar}\Delta t H})^n$, the vanishing of the difference between $e^{-\frac{i}{\hbar}\Delta t(H_0+V)}$ and $e^{-\frac{i}{\hbar}\Delta t H_0} e^{-\frac{i}{\hbar}\Delta t V}$ or the effect of the number of factors $(e^{\frac{i}{\hbar}\Delta t H})^n$. The answer is given by the *Trotter formula* (or *Lie-Trotter formula*)

$$\lim_{n\to\infty} \left( e^{i\frac{t}{n}A} e^{i\frac{t}{n}B} \right)^n = e^{it(A+B)}$$

This equality is true for arbitrary bounded operators $A$ and $B$, with the limit in the sense of operator norm convergence. The proof for this is straightforward. We are interested however in the case in which both $A$ and $B$ are unbounded selfadjoint operators. A useful condition is then that $A+B$ is selfadjoint with its domain equal to the intersection of the domains of $A$ and $B$. The convergence is in the strong operator topology. A somewhat stronger result is that for which $A$ and $B$ are bounded from above or from below and $A + B$ is only essentially selfadjoint on the intersection of the domains of $A$ and $B$. For a proof see the first volume of the book by Reed and Simon (Ref. [23], pp. 295–297, and for the stronger case the references on p. 308) (The various types of operator convergences are explained in Supp. Sect. 21.6.2, and essential selfadjointness in Supp. Sect. 21.9). For Trotter's original paper, see [24].

The relevance of Trotter's formula to the Feynman path integral was first noted by Nelson [25].

### 13.8.3 Application of Trotter's Formula

We divide again, as in the preceding subsection, the interval $[t_a, t_b]$ in $n$ equal subintervals, with the length of each interval $(t_b - t_a)/n = \Delta t$. For the time evolution of a quantum mechanical system from $t_0 = t_a$ to $t_n = t_b$ Trotter's formula gives

$$e^{-\frac{i}{\hbar}(t_b-t_a)(H_0+V)} = \lim_{n\to\infty} \left( e^{-\frac{i}{\hbar}\Delta t\, H_0} e^{-\frac{i}{\hbar}\Delta t\, V} \right)^n$$

$$= \lim_{n\to\infty} e^{-\frac{i}{\hbar}(t_n-t_{n-1})H_0} e^{-\frac{i}{\hbar}(t_n-t_{n-1})V} \cdots e^{-\frac{i}{\hbar}(t_1-t_0)H_0} e^{-\frac{i}{\hbar}(t_1-t_0)V}.$$

We translate this result into a formula for the integral kernels

$$K(\mathbf{x_b}, t_b; \mathbf{x_a}, t_a) = \int_{R^1} \cdots \int_{R^1} K_0(\mathbf{x_n}, t_n; \mathbf{x_{n-1}}, t_{n-1})\, e^{-\frac{i}{\hbar}(t_n-t_{n-1})V(\mathbf{x_{n-1}})}$$

$$\cdots K_0(\mathbf{x_1}, t_1; \mathbf{x_0}, t_0)\, e^{-\frac{i}{\hbar}(t_1-t_0)V(\mathbf{x_0})} d\mathbf{x_{n-1}} \ldots d\mathbf{x_1}.$$

The free evolution operator is diagonal in the momentum representation, i.e. it is there the multiplication by $e^{-\frac{i}{\hbar}(t_j-t_{j-1})\frac{p^2}{2m}}$. By using the Fourier transformation and its inverse, together with the standard integral

$$\int_{-\infty}^{+\infty} e^{iu^2} du = \sqrt{\pi}\, e^{\frac{i\pi}{4}},$$

one obtains

$$K_0(\mathbf{x_j}, t_j; \mathbf{x_{j-1}}, t_{j-1}) = \left( \frac{m}{2\pi i\hbar(t_j - t_{j-1})} \right)^{3/2} e^{-\frac{i}{2}\frac{m}{\hbar}\frac{(\mathbf{x_j}-\mathbf{x_{j-1}})^2}{(t_j-t_{j-1})}}.$$

This gives finally

$$K(\mathbf{x_b}, t_b; \mathbf{x_a}, t_a) = \lim_{n\to\infty} \left( \frac{nm}{2\pi i\hbar(t_b - t_a)} \right)^{3n/2}$$

$$\int_{R^1} \cdots \int_{R^1} e^{\frac{i}{\hbar}\sum_{j=1}^{n}\left[ \frac{1}{2}m\left(\frac{\mathbf{x_j}-\mathbf{x_{j-1}}}{t_j-t_{j-1}}\right)^2 - V(\mathbf{x_{j-1}})(t_j-t_{j-1}) \right]} d\mathbf{x_{n-1}} \ldots d\mathbf{x_1}.$$

So far, all this is, in principle, completely rigorous. The integral kernel $K(x_b, t_b; x_a, t_a)$ is not itself an infinite dimensional integral as Feynman's path integral formula suggests, but a limit, for $n \to \infty$, of a sequence of $n$-dimensional integrals.

Our next step is to give this formula a heuristic interpretation, going from a discrete picture to a continuous one. We write the normalization factor in front of the $n$-fold integral as

$$\lim_{n \to \infty} \left( \frac{nm}{2\pi i \hbar (t_b - t_a)} \right)^{3n/2} = N$$

This number $N$ is infinite. We also write the $n$-fold integration element as

$$\lim_{n \to \infty} dx_{n-1} \ldots dx_1 = \mathcal{D}[\gamma].$$

This infinite-dimensional Lebesgue volume element is mathematically meaningless. We finally look at the sum in the exponent in the integrand as a discrete approximation of an integral over paths $\gamma(t)$. This gives

$$\lim_{n \to \infty} \sum_{j=1}^{n} \left[ \frac{1}{2} m \left( \frac{x_j - x_{j-1}}{t_j - t_{j-1}} \right)^2 - V(x_{j-1}) \right] (t_j - t_{j-1})$$

$$= \int_{t_a}^{t_b} \left[ \frac{1}{2} m \left( \frac{d\gamma(t)}{dt} \right)^2 - V(\gamma(t)) \right] dt,$$

the classical action of the system. The result is Feynman's formula, a heuristic version of a rigorous result, namely the limit of a sequence of well-defined finite dimensional integrals, but not an integral it self.

### 13.8.4 Connection with Brownian Motion

Compare the following two cases:
1. The free *Schrödinger equation* in one space dimension

$$i\hbar \frac{\partial \psi(x, t)}{\partial t} = -\frac{\hbar^2}{2m} \Delta \psi(x, t)$$

Its solution is given by a system of unitary time evolution operators

$$\psi(t_2) = e^{-\frac{i}{\hbar}(t_2 - t_1)H_0} \psi(t_1), \quad \text{for} \ -\infty < t_1 < t_2 < +\infty$$

with integral kernel $K_0(x_2, t_2; x_1, t_1)$, for $-\infty < t_1 < t_2 < +\infty$. Note that one has, for $t_1 < t_2 < t_3$, an obvious multiplication relation

$$\int\limits_{x_2=-\infty}^{+\infty} K_0(x_3, t_3; x_2, t_2) K_0(x_2, t_2; x_1, t_1) \, dx_2 = K_0(x_3, t_3; x_1, t_1),$$

Because of time and space translation invariance the $K_0(\ldots)$ may be written as $K_0(x_2 - x_1; t_2 - t_1)$. For this kernel $K_0(\ldots)$ Feynman introduced his general, mathematically heuristic, path integral formula, given in Sect. 13.8.1.

2. The free *diffusion equation* in one space dimension

$$\frac{\partial \rho(x, t)}{\partial t} = D \frac{\partial^2}{\partial x^2} \rho(x, t)$$

with $D$ the diffusion constant. Its solution is given by a system of time evolution operators

$$\rho(t_2) = e^{-(t_2-t_1)A} \rho(t_1), \quad \text{for} \quad 0 < t_1 < t_2 < \cdots < +\infty.$$

with $A$ a positive selfadjoint operator. Note that the evolution operators are not unitary but selfadjoint and moreover not invertable. Observe that one has, also in this case, but now for $0 < t_1 < t_2 < t_3$, a multiplication relation

$$\int\limits_{x_2=-\infty}^{+\infty} \rho(x_3, t_3; x_2, t_2) \rho(x_2, t_2; x_1, t_1) \, dx_2 = \rho(x_3, t_3; x_1, t_1),$$

which in the context to be discussed below is usually called the *Chapman-Kolmogorov equation*. Of the two cases to be compared, the first, quantum mechanics does not need further explanation here; we need to say something about the second.

The term diffusion covers a wide range of physical processes. The type that we have in mind here is the one that gives a macroscopical description of an underlying microscopical phenomenon, Brownian motion, the random motion of pollen particles suspended in a liquid, seen under a microscope, in 1827 by the British botanist Robert Brown. This motion is caused by the continuous bombardment of the particles by the multitude of much smaller molecules of the liquid. It was for the first time mathematically described by Einstein in one of the five groundbreaking papers that he wrote in 1905 [26]. His description was experimentally confirmed by Perrin in 1909 [27].

Brownian motion is a process that is the continuous form of an extremely simple discrete stochastic process, the *random walk*. It is isotropic in space, i.e. invariant under space translation; the possible positions $x$ of the particles are an equal distance $\Delta x$ apart. The process evolves in equal time steps $\Delta t$; they are statistically independent. Like Brownian motion it is a *Markov process*, implying that it has a state space, the set of integer numbers, and conditional transition probabilities that connect a discrete time point with later time points. The transition probability $\rho((i - j)\Delta x; \Delta t)$ for movement over a single time interval $\Delta t$ is

$$\rho((i-j)\Delta x; \Delta t) = \frac{1}{2}, \quad \text{for } |i-j| = 1, \quad \text{otherwise } 0, \; i, j \in \mathbb{N}.$$

**(13.8.4,a) Problem** Use Pascal's triangle to show from this that the transition probability for $n$ discrete time intervals is

$$\rho(s\Delta x; n\Delta t) = \frac{1}{2^n} \binom{n}{k} = \frac{n!}{(n-k)!k!},$$

with $s = 2k - 1$ and $0 \le k \le n$.

**(13.8.4,b) Problem** Use the formula for the one-step probability transition function to derive the difference equation

$$\frac{\rho(x; t + \Delta t) - \rho(x, t)}{\Delta t} = \frac{(\Delta x)^2}{2\Delta t} \frac{\rho(x + \Delta x; t) - 2\rho(x, t) + \rho(x - \Delta x, t)}{(\Delta x)^2}.$$

Taking the limit $\Delta x, \Delta t \to 0$ while $D = \frac{(\Delta x)^2}{2\Delta t}$ is kept constant gives immediately the differential equation

$$\frac{\partial}{\partial t}\rho(x, t) = D \frac{\partial^2}{\partial x^2} \rho(x, t),$$

the diffusion equation, which now appears as the equation for the probability transition function for Brownian motion, as a continuous stochastic process, obtained from the discrete random walk process. This was Einstein's 1905 discovery.

**(13.8.4,c) Problem** Check that the solution of the one dimensional diffusion equation is

$$\rho(x_2, x_1; t) = \frac{1}{\sqrt{4\pi Dt}} \exp\left[-\frac{(x_2 - x_1)^2}{4Dt}\right].$$

A proper mathematical definition, in the context of probability theory, was given by Wiener in [28]. Brownian motion, or the Wiener process as it now often called, is a collection random variables $\{X_t\}_{t \in \mathbb{R}^1_+}$, which are, as all random variables, measurable functions on a certain probability measure space $(\Omega, \mathcal{F}, \mu)$. Wiener has shown that $\mu$ is concentrated on a space of 'paths', real-valued functions on $\mathbb{R}^1_+$. These paths are continuous but nowhere differentiable. This corresponds with the extreme irregularity of the motion of Brownian particles as seen under the microscope. The mathematical reason for this is that $\frac{(\Delta x)^2}{2\Delta t}$ has to remain constant in the limit $\Delta x, \Delta t \to 0$, implying that the velocities have to go to infinity.

Mark Kac knew Feynman's work at an early stage [29]. He found a general mathematical formula, which among many other interesting things, gave a formula connecting Feynman's path integral with a similar formula in diffusion theory.

*Two final remarks:*

1. The discussion as given here can be extended to the more physical three dimensional situation. This implies considerable but not insurmountable technical problems, the treatment of which might obscure the basic simplicity of the connection between the (heuristic) Feynman path integral and the (rigorous) Wiener path integral.

2. The same is true for the restriction to the free equations in both cases. Adding a potential energy term to the Schrödinger equation and a suitable term describing some sort of external influence on the Brownian particle would not destroy the analogy.

For a few remarks on stochastic processes, see Supp. Sect. 22.4.

## 13.9  Fermion Quantization

### 13.9.1  Introduction

So far we have discussed quantization only for boson systems. In this section we shall explain how fermionic quantum systems are obtained from—or at least related to—what one may call *pseudo-classical systems*. These systems have no physical meaning as such, but can be used as auxiliary models that lead to corresponding physical quantum systems, in a way which is different but nevertheless in a certain interesting way analogous to what is done in the boson case. This requires a new branch of mathematics, 'super mathematics', originally a complex of heuristic ideas developed by physicists, but later put on a firm footing by mathematicians.

We shall in this context consider a simple model, the fermionic oscillator (in Sects. 13.9.2 and 13.9.4). Finally we shall say a few words on the fermionic path integral (in Sect. 13.9.5); Sect. 13.9.3 will give a brief overview of the main elements of 'super mathematics'.

### 13.9.2  The Fermionic Oscillator: The Quantum Picture

First as a reminder a few remarks on the classical one dimensional harmonic oscillator, treated extensively in Chap. 6. Canonical quantization leads there from the Hamiltonian picture, with variables $p$ and $q$, to the quantum model, with operators $P$ and $Q$, and with classical Poisson brackets corresponding (more or less) with commutators. The operator expression for the energy is similar to that of the classical energy. The physical results are however very different: the allowed values for the energy, in particular, are not continuous, but discrete, with eigenvalues $E_n = (n + 1/2)\hbar\omega$, for $n = 0, 1, 2, \ldots$ Together with the eigenfunctions they are found by power series methods from classical mathematical physics. We have shown in Sect. 6.4, that there is a second, much shorter and much more elegant algebraic method, using a pair of annihilation and creation operators, satisfying simple commutation relations. This formalism is our starting point for the fermion oscillator. Remember that in this second approach the description of the quantum boson oscillator is based on the *commutation relations* for the operators $\mathbf{a}$ and $\mathbf{a}^*$

$$[\mathbf{a}, \mathbf{a}] = [A^*, A^*] = 0, \qquad [\mathbf{a}, \mathbf{a}^*] = 1.$$

These operators are obtained from $P$, $Q$ by the linear expressions

$$\mathbf{a} = \alpha P + i\beta Q, \qquad \mathbf{a}^* = \alpha P - i\beta Q,$$

with $\alpha = (2\hbar)^{-1/2}(m\omega)^{-1/2}$ and $\beta = (2\hbar)^{-1/2}(m\omega)^{+1/2}$.

The fermion oscillator model is an extremely simple quantum system with a two dimensional Hilbert space. Whatever its physical interpretation or merits, it is a bona fide quantum model and it is moreover very instructive for the problems to be discussed here. Its formulation may be based on an annihilation operator $A$ and a creation operator $A^*$, which satisfy the *anticommutation relations*

$$[\mathbf{a}, \mathbf{a}]_+ = [\mathbf{a}^*, \mathbf{a}^*]_+ = 0, \qquad [\mathbf{a}, \mathbf{a}^*]_+ = 1.$$

Note that this implies that $\mathbf{a}$ and $\mathbf{a}^*$ are nilpotent, i.e. that $\mathbf{a}^2 = (\mathbf{a}^*)^2 = 0$. Their standard representation space is a two dimensional Hilbert space $\mathcal{H}$, spanned by a unit vector $\psi_0$, with $\mathbf{a}\psi_0 = 0$, and a second vector $\psi_1 = \mathbf{a}^*\psi_0$.

(13.9.2,a) **Problem** Show that $\mathbf{a}\psi_1 = \psi_0$ and $\mathbf{a}^*\psi_1 = 0$ and that the system $\{\psi_0, \psi_1\}$ is an orthonormal basis.

The boson oscillator suggests that the Hamiltonian for the fermion case is

$$H = (\mathbf{a}^*\mathbf{a} + 1/2)\hbar\omega,$$

with $\omega$ the frequency. The eigenvalues of $H$ obviously are

$$E_n = (n + 1/2)\hbar\omega, \qquad n = 0, 1,$$

with the corresponding eigenvectors $\psi_0$ and $\psi_1$.

(13.9.2,b) **Problem** Use the general time evolution formula for a Heisenberg picture operator $F(t) = e^{\frac{i}{\hbar}Ht} F(0) e^{-\frac{i}{\hbar}Ht}$

$$\frac{dF}{dt} = \frac{i}{\hbar}[H, F]$$

to show that the evolution of $\mathbf{a}(t)$ and $\mathbf{a}^*(t)$ is given by

$$\mathbf{a}(t) = e^{-i\omega t}\mathbf{a}(0), \qquad \mathbf{a}^*(t) = e^{i\omega t}\mathbf{a}^*(0).$$

This is, of course, similar to the boson case.

Following the boson case, we write the annihilation and creation operators as fairly obvious linear expressions in canonical operators $P$ and $Q$.

$$\mathbf{a} = \alpha' P + i\beta' Q, \qquad \mathbf{a}^* = \alpha' P - i\beta' Q,$$

with the inverse relations

$$P = (2\alpha')^{-1}(\mathbf{a} + \mathbf{a}^*), \qquad Q = -i(2\beta')^{-1}(\mathbf{a} - \mathbf{a}^*),$$

and with $\alpha' = \beta' = (2\hbar)^{-1/2}$.

(13.9.2,c) **Problem** Show that these relations imply that the following somewhat surprising but not unreasonable anticommutation relations hold:

$$[P, P]_+ = [Q, Q]_+ = \hbar, \qquad [P, Q]_+ = 0,$$

implying that $P$ and $Q$ are not nilpotent but have $P^2 = Q^2 = \frac{1}{2}\hbar$.

(13.9.2,d) **Problem** Show that the Hamiltonian in terms of the canonical $P$ and $Q$ is equal to

$$H = \left(\frac{i}{\hbar}PQ + 1\right)\hbar\omega.$$

(13.9.2,e) **Problem** Derive from this Hamiltonian the time evolution equation for the Heisenberg picture operators $P(t)$ and $Q(t)$, and show that the general solution $Q(t)$ has the same form as the general solution of the classical oscillator equation, in the classical variable $q(t)$.

This result may look a bit surprising because the expression for the Hamiltonian in $P$ and $Q$ is rather different from that of the boson harmonic oscillator. That there is nevertheless a great similarity is due to the fact that all relations, both for the classical as well as the boson and fermion oscillator, are linear.

The generalization of all this to an arbitrary finite or countably infinite system of noninteracting fermion oscillators, with different frequencies $\omega_1, \omega_2, \omega_3, \ldots$, is obvious. Infinite systems of fermion oscillators play an important role in quantum field theory; as will be briefly mentioned in Sect. 13.9.5. In order to avoid complications due to Pauli's exclusion principle, we shall assume that as particles these oscillators are all different.

## 13.9.3 Intermezzo: 'Supermathematics'

### 13.9.3.1 Introduction

Unlike the boson oscillator, the quantum fermion oscillator cannot be obtained from a description of classical mechanics in terms of symplectic manifolds. In this case a proper classical model does not exist; however one can think of auxiliary 'pseudo-classical' models which serve the same purpose. For the formulation of these models, and also for *supersymmetry*, a speculative theory in particle physics, which was developed in the nineteen seventies, physicists have developed a new branch of heuristically formulated mathematics, 'super mathematics', later made rigorous by mathematicians, in fact in two different ways. One, closely following the heuristic ideas of the physicists, but nevertheless strictly rigorous, may be called 'analytic'; the other, more intrinsically mathematical, is purely algebraic. In this book the algebraic interpretation will in principle be used. This is in accordance with our general emphasis on—and preference for—algebraic formulations, which we have expressed at various places. Nevertheless, heuristic formulations in the spirit of the 'analytic' interpretation will sometimes be useful. It is a general feature of physical theory that heuristic formulations give often a more direct intuitive insight than the corresponding rigorous formulations, even though we need to have the latter at hand in order to avoid serious mistakes. Other examples of this are Dirac's $\delta$-function (in Supp. Chap. 26) and 'second quantization' (in Sect. 9.6).

### 13.9.3.2 Superalgebra

'Super' notions are generalizations of well-known standard notions obtained by adding *grading* and minus signs in multiplication formulas, by which commutativity becomes 'super commutativity'. The main ingredient is what is heuristically called *anticommuting c-numbers* or *Grassmann variables*. Much of the basic mathematical ideas go back to the work of Berezin.

(13.9.3.2,a) **Definition** A vector space $V$ is called *graded* iff it is written as a direct sum $V = V^{(0)} \oplus V^{(1)}$; $V^{(0)}$ is called the *even* part of $V$ and $V^{(1)}$ the *odd* part. The numbers 0 and 1 are called *degrees*.

(13.9.3.2,b) **Definition** An algebra $\mathcal{A}$, associative, with unit element, is called a *graded algebra* iff it is graded as vector space and iff moreover the product of two homogeneous elements $a^{(a)}$, $b^{(b)}$ has degree $a + b$.

(13.9.3.2,c) **Problem** Show that the unit element of $\mathcal{A}$ is even.

(13.9.3.2,d) **Definition** An algebra $\mathcal{A}$ is called *super commutative* or *almost commutative* iff one has, for all homogenous elements $a^{(a)}$ and $b^{(b)}$ in $\mathcal{A}$,

$$a^{(a)}b^{(b)} = (-1)^{ab}b^{(b)}a^{(a)}.$$

(13.9.3.2,e) *Remarks*

1. The grading here is sometimes called $\mathbb{Z}_2$-*grading* to distinguish it from other types of grading that we do not use.
2. Addition and multiplication of degrees is modulo 2.
3. We use the term *almost commutative* instead of, or besides, super commutative to indicate that it differs very little from commutativity. Almost anything that exists in commutative algebra can be given meaning in the almost commutative case, with the addition of minus signs at appropriate places. See as an example [30].

(13.9.3.2,f) **Problem** Show that the linear operators in a graded vector space form a graded algebra, but not an almost commutative (graded) algebra.

A *super Lie algebra* is a 'super' generalization of the notion of Lie algebra. It is briefly discussed in Supp. Sect. 24.10. In physics it is applied to *supersymmetry*. See Sect. 17.2.4.

### 13.9.3.3 Supermanifolds

Almost all objects that appear in ordinary differential geometry have generalizations to the super case: vector fields, differential forms, tensor fields, etc. There is an extension of differential geometry, called *super differential geometry* and in particular a super generalization of ordinary symplectic geometry, as discussed in Supp. Sect. 20.7.

The basic notion in this context is that of a *supermanifold*. A supermanifold has dimension $(m, n)$, with an even part of dimension $m$ an ordinary manifold. In the 'analytic' approach it is defined in terms of coordinate patches that have to be joined together, like in the case of an ordinary manifold, but now with as coordinates $m$-tuples of elements of an indeterminate commutative algebra and $n$-tuples of elements of an equally indeterminate Grassmann algebra, both possibly infinite dimensional. In the 'algebraic' interpretation such a supermanifold is an $m$-dimensional ordinary manifold, provided with a vector bundle with as fibre the Grassmann algebra over $n$ generators. See for the definition and some elementary properties of vector bundles (see Supp. Sect. 20.2.4). Discussion of the pseudo-classical description of general fermion systems will require super symplectic manifolds. For $m = 0$ we are back in the 'linear phase space' of, for instance, a system of fermionic oscillators or spins.

According to the algebraic point of view, stressed in this book, a symplectic manifold is characterized by the (commutative) algebra of its $C^\infty$-functions, an infinite-dimensional Lie algebra, with the Poisson bracket as Lie bracket. This is the tool for formulating all mathematical properties of a classical Hamiltonian system, which is—as we have seen in Sect. 13.5—the starting point for quantization. For general fermion systems there is a 'virtual phase manifold', a *super symplectic manifold*, or in the more simple case here, a 'virtual linear phase space', characterized by an algebra of what may be called 'pseudo-classical variables'.

### 13.9.3.4  Super Analysis

The smooth functions on a $2n$-dimensional classical phase space $C_{\mathbb{C}}^{\infty}(\mathcal{M})$ form a commutative algebra; the polynomials an infinite dimensional subalgebra. A polynomial can be written in the case of $\mathcal{M} = \mathbb{R}^{2n}$ as a finite linear combination of monomials

$$f(x_1, \ldots, x_{2n}) = \alpha_0 1 + \sum_{s=1}^{2n} \sum_{1 \le j_1 < \ldots < j_s \ldots \le j_{2n}} \alpha_{j_1, \ldots, j_s}\, e_{j_1}^{k_{j_1}} \cdots e_{j_s}^{k_{j_s}},$$

with $j_l = 0, 1, \ldots$, with the $\alpha_0$ and the $\alpha_{j_1, \ldots, j_{2n}}$ complex coefficients, and with $e_{j_k}$ the linear functions $e_{j_k}(x_1, \ldots, x_{2n}) = x_{j_k}$.

The 'functions' that represent the pseudo-classical observables on the $2n$-dimensional 'linear pseudo-classical phase space' of a system of $n$ fermion oscillators are not proper functions, not mappings, but elements of an algebra $\mathcal{A}_{2n}$, generated by the unit $1_{\mathcal{A}_{2n}}$ and $2n$ anticommuting elements $e_1, \ldots, e_{2n}$. Any such 'function' has the unique form

$$f = \alpha_0 1_{\mathcal{A}_{2n}} + \sum_{s=1}^{2n} \sum_{1 \le j_1 \le \ldots \le j_s \ldots \le j_{2n}} \alpha_{j_1, \ldots, j_s}\, e_{j_1} \cdots e_{j_s},$$

with again $\alpha_0$ and $\alpha_{j_1, \ldots, j_s}$ complex coefficients. The separate terms in this sum are scalar multiples of 'anticommuting monomials'.

The algebra of commuting polynomials is the symmetric tensor algebra $T_s(V)$; the anticommuting polynomials form the antisymmetric algebra $T_a(V)$, as they are denoted and discussed in Supp. Sect. 23.4, but with here $T_s(V) = \mathbb{R}^{2n}$. Both play also an important role in 'second quantization', with $V$ a general Hilbert space, as discussed in Sect. 9.6.2. The antisymmetric algebra as it is used here is often called the *Grassmann algebra* over $2n$ generators $\{e_j\}_j$.

Pseudo-classical 'functions' can be differentiated. There is, for each $i = 1, \ldots, 2n$, a partial derivative 'with respect to the $i$th variable', denoted as $\partial_i$ and defined algebraically by linear extension of

$$\partial_i 1 = 0, \qquad \partial_i(e_{j_1} \cdots e_{j_s}) = -\sum_{k=1}^{s}(-1)^k e_{j_1} \cdots e_{j_{k-1}} \delta_{ik} e_{j_{k+1}} \cdots e_{j_s}.$$

(13.9.3.4,a) **Problem** Show that $[\partial_i, \partial_j]_+ = 0$, for all $i, j = 1, 2, \ldots$

The above partial derivatives act from the left; they may be denoted as $\partial_i^L$. There are also partial derivatives from the right, denoted as $\partial_i^R$ and defined, by linear extension of

$$1 \partial_i^R = 0, \qquad (e_{j_1} \cdots e_{j_s})\partial_i^R = -\sum_{k=1}^{s}(-1)^k e_{j_1} \cdots e_{j_{k-1}} \delta_{ik} e_{j_{k+1}} \cdots e_{j_s}.$$

A more heuristic notation is suggested by the analogy with the boson situation. It is

$$\partial_i^L = \frac{\partial^L}{\partial x_i}, \qquad \partial_i^R = \frac{\partial^R}{\partial x_i}.$$

### 13.9.3.5 The Berezin Integral

Pseudo-classical 'functions' can also be integrated. The Berezin integral is not a measure-theoretical but an algebraic notion; it is a linear map of the algebra of almost commutative functions into itself. It again exemplifies the idea that many seemingly different mathematical notions have a unified representation in algebraic terms.

*First the heuristics.* The integral of a function $f$ in a single Grassmann variable $\theta$ is defined by linear extension of

$$\int d\theta = 0, \qquad \int \theta d\theta = 1.$$

Note that the Berezin integral cannot be defined for a finite interval. This is in accordance with the fact that the variable $\theta$ does not take numerical values; it is a symbolic algebraic object. For $n$ variables $\theta_1 \ldots \theta_n$ the integral becomes

$$\int d\theta_i = 0, \qquad \int \theta_j d\theta_i = \delta_{ji},$$

and

$$\int \theta_{j_1} \cdots \theta_{j_s} d\theta_i = -\sum_{k=1}^{s} (-1)^k \theta_{j_k} \cdots \theta_{j_{k-1}} \delta_{ik} \theta_{j_{k-1}} \cdots \theta_{j_s}.$$

These formulas determine the general integral $\int f(\theta_1, \ldots, \theta_n) d\theta_i$.

*Next the rigorous definition.* For a single variable the Berezin integral is a linear map $\int : \mathcal{A}_1 \to \mathcal{A}_1$, with $\mathcal{A}_1$ the Grassmann algebra over a single generator $e$, given by linear extension of

$$\int 1 = 0, \qquad \int e = 1.$$

For $n$ variables the $i$th integral is a linear map $\int^i : \mathcal{A}_n \to \mathcal{A}_n$, with $\mathcal{A}_n$ the Grassmann algebra over $n$ generators $e_1, \ldots, e_n$, given by linear extension of

$$\int^i 1 = 0, \qquad \int^i e_j = \delta_{ji},$$

and

$$\int^i e_{j_1} \cdots e_{js} = -\sum_{k=1}^{s}(-1)^k e_{j_k} \cdots e_{j_{k-1}} \delta_{ik} \cdots e_{j_{k+1}} \cdots e_{js}.$$

Note the surprising fact that the Berezin integral, heuristically written as $\int \theta_{j_1} \cdots$ $\theta_{js} d\theta_i$, is identical to the partial derivative, equally heuristically written as $\frac{\partial}{\partial \theta_i}$. Even more surprising is the fact that even though the notions of differentiation are identical, the Berezin integral is nevertheless the point of departure of a wide range of super versions of formulas from ordinary integration theory, such as the chain rule, change of coordinates, partial integration, etc.

### 13.9.3.6  Literature

There is by now a good choice of books and papers on the mathematics of super-analyis, superalgebras and supermanifolds. For the 'analytic' approach, for instance, Rogers [31], DeWitt [32] and Tuynman [33], for the 'algebraic' interpretation Leites [34], Berezin [35] and Kostant [36], and for the relation between the two Batchelor [37, 38].

## 13.9.4  Application to the Pseudo-classical Fermion Oscillator

After this intermezzo we turn to the pseudo-classical description of a fermion oscillator. We apply all the material of the proceeding subsection—or part of it—on a fermionic oscillator.

The almost commutative algebra of pseudo-classical observables is the complex Grassmann algebra $\mathcal{A}_2$ over 2 generators, complex generators $a$ and $a^*$ or selfadjoint generators $p$ and $q$. For a system of $n$ fermion oscillators one has $\mathcal{A}_{2n}$, over generators $a_j, a_j^*$ or $p_j, q_j$, for $j = 1, \ldots, n$.

We have the obvious linear relations

$$p = (2\hbar)^{1/2}(\mathbf{a} + \mathbf{a}^*), \quad q = -i(2\hbar)^{1/2}(\mathbf{a} - \mathbf{a}^*)$$

with inverse

$$\mathbf{a} = (2\hbar)^{-1/2}(p + iq), \quad \mathbf{a}^* = (2\hbar)^{-1/2}(p - iq).$$

An important notion in this framework would be a super version of the Poisson bracket, which can be connected with super commutators of the quantum picture. With this we would have a complete conceptually satisfying symmetry between boson and fermion canonical quantization. However a satisfactory super Poisson bracket $\{\cdot, \cdot\}^S$ that gives the brackets that correspond with the anticommutators of the canonical operators $P$ and $Q$, namely

$$\{p, p\}^S = \{q, q\}^S = 1, \qquad \{p, q\}^S = 0,$$

has so far not been found. There are various approaches to fermion quantization in terms of Grassmann variables in the literature, for example, that by Martin [39], who already in 1957, long before supersymmetry, came up with the idea of pseudo-classical mechanics, and later by Casalbuoni [40] and Berezin and Marinov [41]. However none of this is truly satisfactory; mainly because they do not have pairs of fermionic canonical variables, and therefore also no satisfactory super Poisson bracket.

### 13.9.5 The Fermionic Path Integral

Much of what was put forward in the preceding subsection concerning fermion quantization, constructing a fermion quantum model from a pseudo-classical system, had to do with trying to obtain an elegant general picture for the relation between classical and quantum physics, with not much consequences for physical practice.

There is one area of physics where the Berezin integral is of great practical value and is in fact generally used: fermionic quantum field theory. The path integral there is infinite dimensional, totally heuristic, but is nevertheless used with great effectiveness in supplying the formulas and diagrams for perturbative calculations that are typical for the applications of quantum field theory to elementary particle physics. Strictly speaking, the integration is not over a space of paths but of *functionals* of *field configurations*. These field configurations, for example Dirac spinor fields $\psi(\mathbf{x}, t)$, are not real- or complex-valued, but, speaking heuristically, Grassmann number valued, or, more rigorously, elements of an infinite dimensional Grassmann algebra $\mathcal{A}_\infty$ of pseudo-classical fields.

A free quantum field can be seen as an infinite system of harmonic oscillators, with a continuum of frequencies, which becomes a discrete set when the field is put in a finite box in space. In the calculations of the behavior of interacting quantum fields one treats these fields as perturbations of free fields. For fermion fields, such as the Dirac spinor fields, this involves infinite systems of the pseudo-classical fermion oscillator that we discussed in Sects. 13.9.2 and 13.9.4.

The idea of 'anticommuting functionals' was developed by Berezin in his book [42]. Elements of this went back to Schwinger [43].

# References

1. Dirac, P.A.M.: The quantum theory of the emission and absorption of radiation. Proc. R. Soc. Lond. Ser. A. **A114**, 243–265 (1927). Available at: http://rspa.royalsocietypublishing.org/content/114/767/243.full.pdf or http://hermes.ffn.ub.es/luisnavarro/nuevo_maletin/Dirac_QED_1927.pdf
2. Born, M., Jordan, P.: Zur Quantenmechanik. Z. Phys. **34**, 858–888 (1925)
3. Heisenberg, W.: Über die quantentheoretische Umdeutung kinematischer und mechanischer Beziehungen. Z. für Phys. **33**, 879–893 (1925) (An English translation of these two fundamental articles is in Van der Waerden, see Chap. 3, Ref. [4]. The 1967 North-Holland edition is available at: https://ia601208.us.archive.org/14/items/SourcesOfQuantumMechanics/VanDerWaerden-SourcesOfQuantumMechanics.pdf.)
4. Weyl, H.: Quantenmechanik und Gruppentheorie. Z. Phys. **46**, 1–46 (1927)
5. Weyl, H.: The Theory of Groups and Quantum Mechanics (Translated from the 2nd revised German edition 1931 of Gruppentheorie und Quantenmechanik) Dover 2003. Translated from the second revised edition as: Theory of Groups and Quantum Mechanics. Dover 1950 (The English translation is available at: https://ia600807.us.archive.org/20/items/ost-chemistry-quantumtheoryofa029235mbp/quantumtheoryofa029235mbp.pdf.)
6. Scholz, E.: Weyl entering the 'new' quantum mechanics discourse. Contribution to the Conference "History of Quantum Physics", Berlin, July 2–6, 2007. Accessible at: http://www2.math.uni-wuppertal.de/~scholz/preprints/HQ_1_ES.pdf
7. Weyl, H.: Raum, Zeit, Materie. Springer, various editions. Translated as Space, Time, Matter. Dover 1952 (Weyl's collected works were published by Springer in 1968)
8. Segal, I.E.: Representations of the canonical commutation relations. Lectures given at the 1965 Cargèse Summer School Notes by P.J.M. Bongaarts and Th. Niemeijer. In "Applications of Mathematics to Problems in Theoretical Physics", pp. 107–170. Maurice Lévy, F. Lurcat, Editors Gordon and Breach 1967
9. Dirac, P.A.M.: [13] The Fundamental Equations of Quantum Mechanics. Proceedings of the Royal Society London **A109**, 642–653 (1925). Available at: http://rspa.royalsocietypublishing.org/content/109/752/642.full.pdf+html
10. Groenewold, H.J.: On the principles of elementary quantum mechanics. Physica **12**, 405–460 (1946)
11. van Hove, L.: Sur le problème des relations entre les transformations unitaires de la mécanique quantique et les transformations canoniques de la mécanique classique. Acad. Roy. Belgique, Bull. Cl. Sci. **37**, 610–620 (1951)
12. Rieffel, M.A.: Deformation Quantization and Operator Algebras. Proceedings of Symposia in Pure Mathematics **51**, 411–423 (1990). Available at: http://math.berkeley.edu/~rieffel/papers/deformation.pdf
13. Rieffel, M.A.: Deformation Quantization and $C^*$-Algebras. Contemporary Mathematics **167**, 67–97 (1994). Available at: http://math.berkeley.edu/~rieffel/papers/quantization.pdf
14. Bayen, F., Flato, M., Fronsdal, C., Lichnerowicz, A.,Sternheimer, D.: Deformation theory and quantization I. Deformation of symplectic structures. Ann. Physics **111**, 61–110 (1978) II. Physical applications. Ann. Physics **111**, 111–151 (1978)
15. Fedosov, B.V.: A simple geometrical construction of deformation quantization. J. Differ. Geom. **40**, 213–238 (1994)
16. Kontsevich, M.: Deformation quantization of Poisson manifolds. Lett. Math. Phys. **66**, 157–216 (2003). Available at: http://label2.ist.utl.pt/vilela/Papers/kontsevich.pdf The preprint version is available at: http://arxiv.org/pdf/q-alg/9709040v1.pdf
17. Gerstenhaber, M.: On the deformation of rings and algebras I. Ann. Math. **79**, 59–103 (1963)
18. Gerstenhaber, M.: On the deformation of rings and algebras II. Ann. Math. **84**, 1–19 (1966)
19. Feynman, R.P.: Space-Time Approach to Non-Relativistic Quantum Mechanics. Rev. Modern Phys. **20**, 367–387 (1948). Available at: http://web.ihep.su/dbserv/compas/src/feynman48c/eng.pdf

20. Feynman, R.P., Hibbs, A.R.: Quantum Mechanics and Path Integrals. McGraw-Hill 1965, Dover 2012
21. Cameron, R.H.: A family of integrals serving to connect the Wiener and Feynman integrals. J. Math. Phys. **39**, 126–140 (1960)
22. Dirac, P.A.M.: The Lagrangian in quantum mechanics. Phys. Zeitschr. Sowjetunion **3**, 64–72 (1933). Reprinted in Quantum Electrodynamics. J. Schwinger (ed.) Dover 1958. Available at http://www.ifi.unicamp.br/~cabrera/teaching/aula%2015%202010s1.pdf
23. Reed, M., Simon, B.: Methods of Mathematical Physics. I: Functional Analysis Academic Press 1972 (This book is because of its readability our main reference for this chapter. A text book on functional analysis, with the theory of operators in Hilbert space as its central topic, especially written for applications in mathematical physics. It is the first of a series of four books by the same authors. The other volumes are more specialized, but contain nevertheless useful material, in particular II and IV.)
24. Trotter, H.F.: On the product of Semi-Groups of Operators. Proc. Amer. Math. Soc. **10**, 545–551 (1959). Available at: http://www.ams.org/journals/proc/1959-010-04/S0002-9939-1959-0108732-6/S0002-9939-1959-0108732-6.pdf
25. Nelson, E.: Feynman integral and the Schrödinger equation. J. Math. Phys. **5**, 332–343 (1964)
26. Einstein, A.: Über die von der molekularkinetischen Theorie der Wärme geforderte Bewegung von in ruhenden Flüssigkeiten suspendierten Teilchen (English translation: On the Motion Required by the Molecular Kinetic Theory of Heat of Small Particles Suspended in a Stationary Liquid. Ann. Phys. **17**, 549–560 (1905). Available at: http://www.pitt.edu/~jdnorton/lectures/Rotman_Summer_School_2013/Einstein_1905_docs/Einstein_Brownian_English.pdf.)
27. Perrin, J.: Mouvement brownien et realité moléculaire. Annales de Chimie et de Physique **18**, 5–114 (1909). English translation: Brownian Movement and Molecular Reality. Taylor and Francis 1910
28. Wiener, N.: Differential space. J. Math. Phys. **2**, 132–174 (1923). Available at: http://math.iisc.ernet.in/~manju/Brownian/Differential%20Space.PDF
29. Kac, M.: On Distributions of Certain Wiener Functionals. Trans. Amer. Math. Soc. **65**, 1–13 (1949). Available at: http://www.ams.org/journals/tran/1949-065-01/S0002-9947-1949-0027960-X/S0002-99471949-0027960-X.pdf. Mark Kac did mention that he had been influenced by Feynman (Besides the book by Feynman and Hibbs [20], which has at now mainly historical value, there are many competent modern books on Feynman's path integral, for example.)
30. Bongaarts, P.J.M., Pijls, H.G.J.: Almost commutative algebra and differential calculus on the quantum hyperplane. J. Math. Phys. **35**, 959–970 (1994). Available at: http://dare.uva.nl/document/30673
31. Rogers, A.: Supermanifolds. Theory and Application. World Scientific (2007)
32. DeWitt, B.: Supermanifolds. 2nd Edition. Cambridge (1992)
33. Tuynman, G.M.: Supermanifolds and Supergroups. Basic Theory. Kluwer, Boston (2004). There is a recent paperback edition published by Springer in 2012
34. Leites, D.A.: Introduction to the theory of super manifolds. Russ. Math. Surveys **35**, 1–64 (1980)
35. Berezin, F.A.: Introduction to Superanalysis. Reidel, New York (1987)
36. Kostant, B.: Graded manifolds, graded Lie theory and pre quantization. In: Bleuler K., Reetz A. (eds.) Differential Geometric Methods in Mathematical Physics. Lecture Notes in Mathematics vol. **570**, pp. 177–306. Springer (1977)
37. Batchelor, M.: Two approaches to supermanifolds. Trans. Amer. Math. Soc. **258**, 257–270 (1980). Available at: http://www.ams.org/journals/tran/1980-258-01/S0002-9947-1980-0554332-9/S0002-9947-1980-0554332-9.pdf
38. Batchelor, M.: Graded manifolds and supermanifolds. In: Clarke C.J.S., Rosenblum A., Seifert H.J. (eds.) Mathematical Aspects of Superspace. Reidel (1984)
39. Martin, J.L.: Generalized Classical Dynamics, and the "Classical Analogue" of a Fermi Oscillator. Proc. R. Soc. Lond. **A251**, 536–542 (1959)

40. Casalbuoni, R.: On the quantization of systems with anticommuting variables. Nuovo Cim. **33A**, 115–125 (1976)

41. Berezin, F.A., Marinov, M.S.: Particle spin dynamics as the Grassmann variant of classical mechanics. Ann. Phys. **104**, 336–362 (1977)

42. Berezin, F.A.: The Method of Second Quantization. Academic Press, New York (1966)

43. Schwinger, J.: On the Green's Functions of Quantized Fields. I. Proc. Natl. Acad. SCI. USA **37**, 452–455 (1951). Available at: http://www.pnas.org/content/37/7/452.long

44. Schulman, L.S.: Techniques and Applications of Path Integration. Wiley (1981), Dover 2005

45. Roepstorff, G.: Path Integral Approach to Quantum Physics. An Introduction. Springer, original edition 1994, paperback edition 1996

46. Chaichian, M., Demichev, A.: Path Integrals in Physics I. Stochastic Processes and Quantum Mechanics. Institute of Physics (2001)

47. Esposito, C.: Formality Theory: From Poisson Structures to Deformation Quantization. Springer, Heidelberg (2014)

# Chapter 14
# Scattering Theory

*Scattering theory is the study of an interacting system on a scale of time and/or distance which is large compared to the scale of the interaction itself. As such, it is the most effective means, sometimes the only means, to study microscopic nature.* (Reed-Simon III, p. ix. Ref. [1])

## 14.1 Introduction

Scattering is a phenomenon that appears in many areas of physics. In nonrelativistic quantum mechanics the basic example is the collision process between two particles, as described by the Schrödinger equation. When there is translation invariance, this can be reduced, by introducing centre of mass coordinates, to the case of a single particle scattered by a potential.

There are two essentially equivalent approaches to scattering theory: time-dependent and time-independent. The first is more rigorous; it uses 'abstract' operator language, and has as its central notions wave operators, the S-operator, and the interaction picture; the second is in terms of 'explicit' scattering wave functions, with as important notion Green's functions and the Lippmann-Schwinger equation. We shall emphasize the time-dependent approach, as it fits in best with the general operator language used in this book. Time-independent scattering theory, which we shall also discuss, is widely used in physical practice; it is closer to actual experiments and more suitable for explicit calculations, but its formulation is less rigorous. We shall explain the relation between the two approaches.

Scattering theory also has two distinct faces. One part is mathematically rigorous; it is almost—or has in fact become—pure mathematics; as such it mainly applies to scattering of a nonrelativistic quantum mechanical particle by a potential, and, to a certain extent, to the scattering in systems of more than two particles, situations for which many rigorous results are known. On the other hand its general formulas can be used in a formal, i.e. heuristic, manner, with no proof of the existence of integrals, limits, and no care being taken of the domains of unbounded operators. This is the

© Springer International Publishing Switzerland 2015
P. Bongaarts, *Quantum Theory*, DOI 10.1007/978-3-319-09561-5_14

standard practice in quantum field theory, as used in elementary particle physics. There the formulas generate a perturbation series for the S-operator, with all terms undefined or just infinite, but from which nevertheless by so-called renormalization methods very precise numerical results can be extracted, that are fully confirmed by experiments.

The description of a scattering process is *asymptotic in time*, which is particularly visible in the time-dependent formalism. The study and description at all (finite) times of a scattering process would be quite difficult. Moreover, it is not of interest. What is physically relevant is the outcome of a scattering experiment, i.e. what the final result is, ideally at $t = +\infty$, for given initial data, ideally at $t = -\infty$. The best possible theoretical result is a unitary operator which connects initial and final data, and which is usually denoted as S, and called the *S-operator* or *scattering operator*.

For a brief review of the history of scattering theory, see Ref. [2].

## 14.2  Time-Dependent Scattering: The General Formalism

### 14.2.1  *Wave Operators I*

For the discussion of a general quantum scattering process, leading to an S-operator, one considers two systems, given by two Hamiltonians. There is a "free" system, the reference system, with a Hamiltonian $H_0$, and an "interacting" system, the system of actual interest, with a Hamiltonian $H = H_0 + \lambda H_I$, with $H_I$ the interacting part. The free system is supposed to be solved, i.e. the eigenvalues and eigenfunctions of $H_0$ are known, while those of the interacting Hamiltonian $H$ have to be determined, at least approximatively and so far as they determine the asymptotic behaviour in time with respect to the free system. The presence of the coupling parameter $\lambda$ underlines that there is a perturbation theoretical aspect in this. As this chapter deals exclusively with quantum theory, without the need to compare the results with classical scattering theory, we may simplify the notation by using so-called natural units, i.e. with $\hbar = 1$.

Let $\psi_I$ be a given state vector. Because of the interaction it will evolve in time according to $\psi_I(t) = e^{-iHt}\psi_I$. Suppose that there is another vector $\phi_{\text{in}}$ evolving as $\phi_{\text{in}}(t) = e^{-iH_0t}\phi_{\text{in}}$, such that it is equal to $\psi_I(t)$ for $t \to -\infty$, i.e. such that

$$\lim_{t \to -\infty} ||\phi_{\text{in}}(t) - \psi_I(t)|| = 0,$$

or

$$\lim_{t \to -\infty} ||e^{-iH_0t}\phi_{\text{in}} - e^{-iHt}\psi_I|| = 0,$$

which is the same as

$$\lim_{t \to -\infty} || e^{iHt} e^{-iH_0 t} \phi_{in} - \psi_I || = 0.$$

This means that there is a limit operator

$$W_- = \lim_{t \to -\infty} e^{iHt} e^{-iH_0 t},$$

defined in the strong sense (For this notion, see Supp. Sect. 21.6.2). The operator $W_-$ is called a *wave operator* or, sometimes, *Møller wave operator*. The strong limit of a unitary operator is not necessarily unitary; it is in general an *isometry*.

### 14.2.2 Intermezzo: Isometries and Partial Isometries

Let $W$ be a bounded operator in the Hilbert space $\mathcal{H}$. It defines a surjective and injective map from $(\text{Ker}(W))^\perp$, the orthogonal complement of its null space onto $\text{Ran}(W)$, its range. Due to the boundedness, i.e. the continuity of $W$, $(\text{Ker}(W))^\perp$ is a closed subspace of $\mathcal{H}$. Note that in general $\text{Ran}(W)$ need not to be closed.

(14.2.2,a) **Definition** The operator $W$ is called a *partial isometry* if and only if

$$(W\psi_1, W\psi_2) = (\psi_1, \psi_2), \quad \forall \psi_1, \psi_2 \in (\text{Ker}(W))^\perp.$$

This definition is geometric; at the end of this intermezzo we shall give a simple but slightly less intuitive algebraic definition. A partial isometry is called an isometry when its null space is trivial, i.e. when it is an injective map from $\mathcal{H}$ onto its range. When its hermitian conjugate is also an isometry, then it is unitary.

(14.2.2,b) **Problem** Show that the strong limit of a system of isometries $\{U(t)\}$ is an isometry. Use for this and the next problems the Hilbert space methods discussed in Supp. Chap. 21.

(14.2.2,c) **Problem** Show that for a partial isometry $W$ the subspace $\text{Ran}(W)$ is closed.

Given this result it is clear that there is an inverse map, for the moment called $W_1^+$, from $\text{Ran}(W)$ onto $(\text{Ker}(W))^\perp$, which can be extended to a linear map $W^+$ from the full Hilbert space $\mathcal{H}$ onto $(\text{Ker}(W))^\perp$ by defining

$$W^+ \psi_0 = 0, \quad \forall \psi_0 \in (\text{Ran}(W))^\perp.$$

(14.2.2,c) **Theorem** $W^+$ *is also a partial isometry. Moreover* $W^+ = W^*$, *the hermitian conjugate of* $W$.

*Proof* We have to show that $W_1^+$ preserves the inner product. One checks easily that $W^+ W$ is the projection on $(\text{Ker}(W))^\perp$ and similarly $W W^+$ the projection on $\text{Ran}(W)$. Let $\phi, \phi'$ be arbitrary vectors in $\text{Ran}(W)$, with $\psi = W^+ \phi$ and $\psi' = W^+ \phi'$. One has $(W^+ \phi, W^+ \phi') = (\psi, \psi')$, which, because $W$ is an isometry, is equal to $(W\psi, W\psi')$. This together gives, $\forall \phi, \phi' \in \text{Ran}(W)$,

$$(W^+\phi, W^+\phi') = (W\psi, W\psi') = (WW^+\phi, WW^+\phi') = (\phi, \phi'),$$

which proves that $W^+$ is a partial isometry.

Next the proof of $W^+ = W^*$. Arbitrary $\psi$ and $\phi$ in $\mathcal{H}$ can be written—by using the direct sum decompositions—as $\psi = \psi_0 + \psi_1$ and $\phi = \phi_0 + \phi_1$. We have, because of the various orthogonality relations,

$$(W\psi, \phi) = (W\psi_1, \phi_1), \quad (\psi, W^+\phi) = (\psi_1, W^+\phi_1)$$

The partial isometry of $W$ gives

$$(\psi_1, W^+\phi_1) = (W\psi_1, WW^+\phi_1) = (W\psi_1, \phi_1).$$

The combination of these three relations lead to

$$(W\psi, \phi) = (\psi, W^+\phi), \quad \forall \psi, \phi \in \mathcal{H},$$

which proves $W^+ = W^*$.

(14.2.2,d) **Problem** Show that a bounded operator $W$ is a partial isometry iff either $W^*W$ or $WW^*$ is a projection operator.

This equivalent algebraic definition of the notion of partial isometry is useful in the context of $C^*$-algebras, and in the von Neumann algebra framework, discussed in Chap. 12 and Supp. Chap. 27.

### 14.2.3 Wave Operators II: The S-Operator

After this intermezzo we return to the time-asymptotic behavior of a scattering system, now for $t \to +\infty$.

Let $\psi_I$ be again a state vector evolving in time according to $\psi_I(t) = e^{-i(H_0 + \lambda H_I)t}\psi_I$, and suppose now that there is a freely evolving vector $\psi_{\text{out}}$ such that

$$\lim_{t \to +\infty} ||\psi_{\text{out}}(t) - \psi_I(t)|| = 0.$$

If such an $\psi_{\text{out}}$ exists for every $\psi_I$, then there is a limit operator, an isometry,

$$W_+ = \lim_{t \to +\infty} e^{iHt}e^{-iH_0t}.$$

This second operator $W_+$ is also called a wave operator. For given $H_0$ and $H$ the pair of isometries $(W_-, W_+)$ is called a *scattering system*. It is called *complete* iff $\text{Ran}(W_+) = \text{Ran}(W_-)$. We denote the common range of $W_-$ and $W_+$ as $\mathcal{H}_{\text{sc}}$, the space of *scattering states*.

(14.2.3,a) **Problem** Show that $\mathcal{H}_{sc}$ is a subspace of the space associated with the continuous spectrum of $H$, if $H_0$ is the free Hamiltonian of a three dimensional nonrelativistic particle.

In this case we may add to the definition of a complete scattering system the condition that $\mathcal{H}_{sc}$ is equal to this space.

A complete scattering system has a *S-operator* or *scattering operator* defined as

$$S = W_+^* W_-.$$

This is a unitary operator in $\mathcal{H}$. We use the notation $\mathcal{H}_{in}$ for incoming free states and $\mathcal{H}_{out}$ for outgoing free states. Both spaces are just $\mathcal{H}$; using this notation means that we may look at an arbitrary free state as asymptotically equal to an interacting state, for $t \to -\infty$ or $t \to +\infty$. In this picture $S$ is a unitary operator from $\mathcal{H}_{in}$ onto $\mathcal{H}_{out}$.

(14.2.3,b) **Theorem** $e^{iHt} W_\pm = W_\pm e^{iH_0 t}$, $H W_\pm = W_\pm H_0$

*Proof*

$$e^{iHt} W_\pm = e^{iHt} \left[ \lim_{t' \to \pm\infty} e^{iHt'} e^{-iH_0 t'} \right]$$

$$= \lim_{t' \to \pm\infty} \left[ e^{iH(t+t')} e^{-iH_0 t'} \right]$$

$$= \lim_{t' \to \pm\infty} \left[ e^{iH(t+t')} e^{-iH_0(t+t')} \right] e^{iH_0 t}$$

$$= \left[ \lim_{t'' \to \pm\infty} \left[ e^{iHt''} e^{-iH_0 t''} \right] \right] e^{iH_0 t} = W_\pm e^{iH_0 t}.$$

So

$$e^{iHt} W_\pm = W_\pm e^{iH_0 t}, \qquad \forall t \in \mathbb{R}^1.$$

Taking the derivative in $t = 0$, on appropriate vectors, gives

$$H W_\pm = W_\pm H_0.$$

(14.2.3,c) **Problem** Show that because $W_\pm^* W_\pm = 1_\mathcal{H}$, one has the *intertwining relation*

$$W_\pm^* H W_\pm = H_0.$$

This means that the restriction of $e^{iHt}$ and $H$ to $\mathcal{H}_{sc}$ are unitarily equivalent with $e^{iH_0 t}$ and $H_0$. For a repulsive potential $H$ has no bound state, so $\mathcal{H}_{sc} = \mathcal{H}$ and both $W_\pm$ and $W_\pm^*$ are unitary.

(14.2.3,d) **Corollary** $SH_0 = H_0 S$.

(14.2.3,e) **Problem** Prove this.

This means that the scattering process conserves the energy of the free system.


## 14.2.4 Multi-channel Scattering

Scattering in a system of more than two particles is a very complicated process; a theoretical description has to make drastic approximations in order to obtain useful numerical results. The first simplification is to assume that there is only interaction between pairs of particles. One of the complicating facts is that many particle scattering can be inelastic, which means that the number and sort of the incoming particles may differ from that of the outgoing particles. In principle the creation and annihilation of particles is a relativistic effect; nevertheless approximative nonrelativistic Schrödinger equation methods can be used in this situation.

To understand the situation one may consider the following simple, somewhat artificial but very instructive example which we take from the book of Taylor [3, Chap. 16]: three different (spinless) particles $a$, $b$ and $c$, interacting through short-range two body potentials. We assume that particles $b$ and $c$ have two bound states, denoted as $(bc)$, the ground state, $(bc)'$, the first excited state, and the bound state $(ac)$ of the particles $a$ and $c$. Note that because the energy is conserved under scattering we can look independently at a fixed energy interval, in this case the lowest part, starting with the ground state of the total system, and including the ground state $(bc)$ of $b$ and $c$, and finally the bound states $(bc)'$ and $(ac)$.

In the scattering process we have *channels* which remain stable at $\pm\infty$. To the free states $\{a, b, c\}$, which form the 0-channel we add channels $\{a, (bc)\}$, $\{a, (bc)'\}$ and $\{b, (ac)\}$. These together form additional in and out Hilbert spaces of free states, which are in fact, just as in the single channel case the same Hilbert space. Due to the interaction (at finite time) there will be transition from one channel to another. So for instance, $\{a, b, c\} \to$, or $\{a, (bc)\} \to \{a, (bc)\}$ is elastic scattering, $\{a, (bc)\} \to \{a, b, c\}$ is a break up process. For each channel there is a *threshold energy*, the energy above which the channel can be reached from other channels. There are also scattering states for each channel, for energies starting at their thresholds, incoming and outgoing, to be paired with the free incoming and outgoing states in the in and out Hilbert spaces, which by this pairing are distinctive.

The scattering is not described by a single set of wave operators $W_\pm$ but instead by separate wave operators for each channel. For the $\{a, b, c\}$ channel the wave operators are just the limits for $t \to \pm\infty$ for $e^{iHt} e^{-iH_0 t}$, which should exist in the usual free state Hilbert space. In the $\{a, (bc)\}$ channel the free states are not the usual free states, because the bound state $(bc)$ is stable and remains present in the limit of infinite time. The interacting states evolve under the influence of all three two-particle potentials $V_{ab}(\mathbf{x_a} - \mathbf{x_b})$, $V_{ac}(\mathbf{x_a} - \mathbf{x_c})$ and $V_{bc}(\mathbf{x_b} - \mathbf{x_c})$. For the free states the potentials $V_{ab}$ and $V_{ac}$ become negligible for $t \to \pm\infty$; the potential $V_{bc}$ remains present, so is part of a modified 'channel Hamiltonian'

$$H_0' = \frac{\mathbf{p}_a^2}{2m_a} + \frac{\mathbf{p}_b^2}{2m_b} + \frac{\mathbf{p}_c^2}{2m_c} + V_{bc}.$$

The wave operators for this channel are therefore the limits for $t \to \pm\infty$ of $e^{iHt}e^{-iH_0't}$. This limit should exist in addition to the usual free state Hilbert spaces. The same holds for the other channels, each leading to a pair of channel wave operators, acting in its specific extended Hilbert space part. It does not make sense to combine the in and the hermitian conjugate of the out wave operator to get a channel S-operator, because the essential idea of multichannel scattering is that there are transitions between the channels. We should instead form a direct sum wave operator for the direct sum of the usual free Hilbert space together with the additional free spaces, this separately for the incoming and the outgoing case. The resulting total incoming wave operator $W_{\text{total}-}$ should then be combined with the outgoing operator $W_{\text{total}+}^*$ to give a total S-operator $S_{\text{total}}^*$, acting in the extended free state Hilbert space. If the limits exist and the ranges of the total wave $W_{\text{total}-}$ and $W_{\text{total}+}^*$ form together a unitary operator, we have what might be called a 'good scattering system'. See for all of this, [2–4].

### 14.2.5 The Interaction Picture

There is an alternative way to describe a scattering system within the time-dependent approach, namely, by means of the interaction picture, a picture halfway between the Schrödinger and the Heisenberg picture. As such it was introduced as a third dynamical picture alongside these two dynamical pictures in Sect. 3.8.7. It finds its most useful application in this chapter. Here is again the definition for the time evolution operator in this picture

$$U_I(t_2, t_1) = e^{iH_0 t_2} e^{-iH(t_2 - t_1)} e^{-iH_0 t_1}.$$

These operators form a two parameter system of unitary operators as defined in Sect. 3.8.5, describing the dynamics of a non-autonomous quantum system, here at least formally, because the underlying free and interacting dynamical systems are autonomous. This system satisfies the operator-valued ordinary differential equation

$$\frac{\partial}{\partial t_2} U_I(t_2, t_1) = -i\lambda H_I(t_2) U_I(t_2, t_1),$$

with $H_I(t) = e^{iH_0 t} H_I e^{-iH_0 t}$ the *interaction picture Hamiltonian*. The S-operator is obtained from this as

$$S = \lim_{t_1 \to -\infty, t_2 \to +\infty} U_I(t_2, t_1).$$

Note that, strictly speaking, this is not the definition of the S-operator, because as a statement it is slightly weaker than the definition in term of the wave operators.

(14.2.5,a) **Problem** Why?

As observed in Sect. 3.8.7, this Schrödinger differential equation for $U_I(t_2, t_1)$, in the variable $t_2$, together with the initial condition $U_I(t_1, t_1) = 1$, is equivalent to the integral equation

$$U_I(t_2, t_1) = 1 - \lambda i \int_{t_1}^{t_2} dt\, H_I(t) U_I(t, t_1).$$

### 14.2.6 Perturbation Theory: The Dyson Series

An approximation of the solution of the integral equation for the system $U_I(t_2, t_1)$ by an infinite series is obtained by iteration, according to

$$U_I(t_2, t_1) = 1 - \lambda i \int_{t_1}^{t_2} dt'\, H_I(t') \left( 1 - \lambda i \int_{t_1}^{t'} dt''\, H_I(t'') U_I((t'', t_1) \right)$$

$$= 1 - \lambda i \int_{t_1}^{t_2} dt'\, H_I(t') + \lambda^2 \left[ (-i)^2 \int_{t_1}^{t_2} dt' \int_{t_1}^{t'} dt''\, H_I(t') H_I(t'') \right] + \cdots$$

This is a perturbation series in the coupling parameter $\lambda$; it is called the *Dyson series*. The terms of this series can be rewritten by the use of the so-called *time ordered product*. The meaning of this is clear from looking at the second order term

$$\int_{t_1}^{t_2} dt' \int_{t_1}^{t'} dt''\, H_I(t') H_I(t'') = \frac{1}{2} \int_{t_1}^{t_2} \int_{t_1}^{t_2} dt' dt''\, T(H_I(t') H_I(t'')),$$

with

$$T(H_I(t') H_I(t'')) = H_I(t') H_I(t''), \quad t'' \geq t',$$

and

$$T(H_I(t') H_I(t'')) = H_I(t'') H_I(t'), \quad t'' \leq t'.$$

(14.2.5,a) **Problem** Write down the formula for the time ordered product or the term of general order $n$.

Carrying this procedure a step further results in the symbolic exponential

$$U_I(t_2, t_1) = T\left(e^{-\lambda i \int_{t_1}^{t_2} dt' H_I(t')}\right).$$

Taking the limits $t_1 \to -\infty$, $t_2 \to +\infty$ gives the equally symbolic exponential expression

$$S = T\left(e^{-\lambda i \int_{-\infty}^{+\infty} dt' H_I(t')}\right).$$

*Important Remark* Whether the formulas in this and the preceding section have a mathematically rigorous meaning depends on the precise properties of the interaction part $H_I$ of the total Hamiltonian $H$. At this point the two distinct faces of scattering theory, mentioned in the introduction, show up. The free Hamiltonian $H_0$ is mathematically all right, both in Schrödinger quantum mechanics as well as in the second quantization formalism of relativistic quantum field theory. The possible meaning and properties of $H_I$ are however very different in the two cases. For the first case $H_I$ is a potential $V(\mathbf{x})$; for a large class of such potentials it can be shown that good scattering systems rigorously exist. This will be discussed in the next section. In the quantum field theory case the interaction operator $H_I$ has in general no satisfactory mathematical meaning. The S-operator formalism, with in particular the perturbation series that it generates, ill-defined as it is, provides nevertheless the basis for many important calculations in elementary particle physics, by systematically generating all terms that need to be calculated.

## 14.3 Potential Scattering

In this section we briefly discuss aspects of the mathematically rigorous scattering theory of a three dimensional nonrelativistic particle in a given central potential. In this $H_0$ is the standard free Hamiltonian

$$H_0 = -\frac{\hbar^2}{2m}\Delta,$$

and $H = H_0 + V$ the total Hamiltonian, with a potential $V(|\mathbf{x}|)$.

In obtaining a mathematically satisfactory scattering theory for this situation there are two problems that we have to deal with.

1. $H_0$ is a selfadjoint operator, with domain the space of functions that are twice differentiable, almost everywhere, such that the second derivatives are square integrable. It is essentially selfadjoint on the space of Schwartz functions. This follows easily from the selfadjointness properties of the position and momentum operators mentioned in Supp. Sect. 21.9.2. Adding a bounded potential like that of a finite square well gives an operator $H = H_0 + V$ with the same properties as $H_0$. However, most physical potentials are infinite at $\mathbf{x} = 0$, which poses the

problem of the selfadjointness of the resulting total Hamiltonians $H$. We did not worry too much about this problem in the explicit cases discussed so far, the harmonic oscillator (in Chap. 6) and the hydrogen atom (in Chap. 7), because we were in both cases able to give explicit solutions of the eigenvalue–eigenfunction problem.

The general mathematical problem is however quite difficult. It was first discussed by Tosio Kato, who showed, among other things, that $H$ is selfadjoint for potentials that are square integrable. See his fundamental paper [5]. His book [6] gives an extensive discussion of all this. Later much more refined results have been obtained by many others. This is reviewed in the book of Reed and Simon [1].

2. The next question is whether, given the existence of $H$ as selfadjoint operator, the pair $(H_0, H)$ is a good scattering system, i.e. whether the wave operators and eventually the S-operator exists.

   Scattering as we presented it in the preceding sections has in the first place to do with asymptotics in time. Conceptually this is however coupled with notions of asymptotic behaviour in space. The general idea in this is that an incoming particle behaves like a free particle at an early time, when it is very far from the origin of the potential and behaves again like a free particle, at a later time, again at great distance from the origin. Only potentials that go sufficiently fast to zero for $|\mathbf{x}|$ going to infinity can have this effect, so this will be an obvious requirement for a well-behaved potential scattering theory.

   The work of Kato, Simon, and many others, has established limits on possible solutions of both problems. They obtained results for a whole series of sometimes arcane, subtly defined but physically not very relevant classes of potentials, making this subject to a large extent into an interesting field of pure mathematics, like ergodic theory.

From a direct physical point of view there are however general conditions, that potentials $V(\mathbf{x})$ should meet, in particular for $|\mathbf{x}| \to \infty$ and $\mathbf{x} = 0$, to get a physically reasonable scattering theory. Following the book of Taylor [3, Chap. 2, pp. 26–27], we give the three conditions that are sufficient for this.

1. $V(\mathbf{x}) = O(|\mathbf{x}|^{-3-\epsilon})$ as $|\mathbf{x}| \to \infty$, for some $\epsilon > 0$. Note that $V(\mathbf{x}) = O(|\mathbf{x}|^p)$ means that $|V(|\mathbf{x}|)| \leq C|\mathbf{x}|^p$ for some constant $C$.
2. $V(\mathbf{x}) = O(|\mathbf{x}|^{-2+\epsilon})$ as $|\mathbf{x}| \to 0$, for some $\epsilon > 0$.
3. $V(|\mathbf{x}|)$ is continuous on $0 < |\mathbf{x}| < +\infty$, except possibly at a finite number of discontinuities.

The Coulomb potential has a long range action; it falls off at infinity more slowly than $|\mathbf{x}|^{-3}$, so it does not satisfy the first requirement. The operators $e^{iHt}e^{-iH_0t}$ do not converge for $t \to \pm\infty$; the wave operators and the S-operator do not exist. A generalized scattering formalism that covers the Coulomb potential is due to Dollard [7, 8]. Its main idea is the use of a 'distorted' free time evolution $e^{-iH_0^c t}$, for which modified wave operators exist, leading to a scattering theory that is in agreement with experiments.

## 14.4 Perturbation Theory: General Remarks

Interesting nontrivial quantum theoretical models for which we have exact solutions, i.e. eigenvalues and eigenfunctions of the Hamiltonian in closed form, are rare. One has the elementary examples of the harmonic oscillator (in Chap. 6) and the hydrogen atom (in Chap. 7) and not much else. These are the exceptions; the general rule is that physical results are obtained by means of approximation. So much of concrete work in theoretical physics consists therefore of approximative calculations.

Very often such calculations use perturbation theory. In this we consider a given 'free' system of which the solution is known, say with a Hamiltonian $H_0$, together with a Hamiltonian with interaction, in general, $H(\lambda) = H_0 + \lambda H_1 + \lambda^2 H_2 + \cdots$, with $\lambda$ a real perturbation parameter, the 'coupling constant'. The operator-valued function $H(\lambda)$ is developed in to a power series in $\lambda$, which hopefully is convergent for a $\lambda$-interval around 0. The results should be corrections in first and higher order of eigenvalues and eigenvectors (time-independent perturbation theory) and of the time development of the system, i.e. transition amplitudes (time-dependent perturbation theory).

Perturbation theory is an important part of physical practice; it also has interesting mathematical aspects. In Sect. 14.2.6 we have seen an important example, the Dyson scattering series. The size of this book does not permit a full treatment of perturbation theory; this section provides some general remarks. The reason to discuss it here in this section is that there are clear similarities between perturbation and scattering theory.

Like scattering theory, perturbation theory has two distinct faces. On one hand it is a rigorous mathematical subject, which deals primarily with the perturbation of Schrödinger operators, again almost pure mathematics. As such it mainly applies to scattering of a nonrelativistic quantum mechanical particle by a potential, and, to a certain extent, to the scattering in systems of more than two particles, situations for which many rigorous results are known. See for this the book of Reed and Simon [1].

On the other hand, it is a practical method widely used in physics, often in a mathematically careless but practically effective way. This is true for quantum mechanical problems, where one will be satisfied when the results of the first few terms of the power series are small compared to the 'free' results, without worrying whether the series is convergent. This is even more so for its application in quantum field theory as used in elementary particle physics (Chap. 16). There the formulas generate a Dyson perturbation series for the S-operator, with all terms undefined or just infinite, but from which nevertheless by so-called renormalization methods valid and very precise numerical results can be extracted.

There are also similarities in the mathematical problems that have to be resolved, in particular the question whether a given total Hamiltonian $H(\lambda) = H_0 + \lambda V$ in quantum mechanics is a well-defined selfadjoint operator with a physically acceptable spectrum, and for which values of $\lambda$ this is the case.

The basic problem in the mathematical theory of perturbation theory in quantum mechanics is in how far the operator valued function $H(\lambda)$ is analytic in the coupling parameter $\lambda$. The mathematical basis of this is also due to Tosio Kato; much further work has been done by others. This is reviewed in the book by Reed and Simon [1]. An excellent book for this and the preceding section is also the recent book by Teschl [9].

## 14.5  Time-Independent Scattering Theory

### 14.5.1  Stationary Scattering States

The time-dependent scattering formalism can be used for very general quantum systems, scattering of nonrelativistic particles, relativistic quantum field theory, while the time-independent method applies mainly to the scattering of particles. Here we discuss the scattering of two particles, reduced to the scattering of a single particle in a central potential. Multichannel scattering will not be discussed.

The central notion of the time-independent approach is that of *stationary scattering states*, which we denote as $\psi_{\mathbf{p}}^-$ and $\psi_{\mathbf{p}}^+$. They are eigenfunctions of $H$, for the continuous spectrum—and therefore nonnormalizable. They are paired with eigenfunctions $\phi_{\mathbf{p}}^{\pm}$ of $H_0$. Because of this pairing we use the momentum $\mathbf{p}$ as index for the $\psi_{\mathbf{p}}^{\pm}$, even though they are not eigenfunctions of momentum like the $\phi_{\mathbf{p}}^{\pm}$.

The two eigenvalue–eigenfunction equations

$$H_0\phi = E\phi, \qquad H\psi = (H_0 + \lambda H_I)\psi = E\psi,$$

for the same eigenvalue $E$, imply the single equation

$$\psi = \lambda \frac{1}{E - H_0} H_I \psi + \phi.$$

Of course, the formal inverse $1/(E - H_0)$ is not well-defined; it can be given a more precise meaning by adding an "infinitesimal" term $\pm i\epsilon$, which gives, for the relation between the $\psi_{\mathbf{p}}^{\pm}$ and the $\phi_{\mathbf{p}}^{\pm}$

$$\psi_{\mathbf{p}}^{\pm} = \lambda \frac{1}{E \mp i\epsilon - H_0} H_I \psi_{\mathbf{p}}^{\pm} + \phi_{\mathbf{p}}^{\pm}.$$

This is the *Lippmann-Schwinger equation*. It is an implicit equation for the $\psi_{\mathbf{p}}^{\pm}$, given the $\phi_{\mathbf{p}}^{\pm}$. It can be solved approximately by iteration, according to

$$\psi_{\mathbf{p}}^{\pm} = \phi_{\mathbf{p}}^{\pm} + \frac{\lambda}{E \mp i\epsilon - H_0} H_I \phi_{\mathbf{p}}^{\pm}$$
$$+ \frac{\lambda}{E \mp i\epsilon - H_0} H_I \frac{\lambda}{E \mp i\epsilon - H_0} H_I \phi_{\mathbf{p}}^{\pm} \cdots$$

This series is called the *Born series*; its first order term describes the Born approximation. The relation of this series with the Dyson series in Sect. 15.2.5 is obvious.

The expression $1/(E \mp i\epsilon - H_0)$ is often called a *retarded—advanced Green's function* and written as $G_0^{\pm}(E)$. This is a general notion from the theory of inhomogeneous linear differential equations. If $L$ is a linear differential operator acting on some space $V$ of functions or distributions, a Green's function is a right-inverse, i.e. an operator $G$ such that $GL = 1_V$. For a given function $f$ the equation $L\psi = f$ is solved by $\psi = Gf$. $G$ is usually given as a (possibly singular) integral kernel.

## 14.5.2 From the Time-Dependent to the Time-Independent Formulation

Instead of following the direct but heuristic road to time-independent scattering theory that we sketched in the preceding subsection, we derive it in this subsection rigorously from the time-dependent formalism.

The general idea is that we consider again pairs of states associated with each other asymptotically, for $t \to \pm\infty$, but these states should now be proper Hilbert space states, i.e. normalizable wave packets. For the case of a single particle in a potential the free eigenfunctions are the usual plane wave functions $\phi_{\mathbf{p}}(\mathbf{x}) = 1/(2\pi)^{3/2}e^{i\mathbf{p}\cdot\mathbf{x}}$, each of which can be an incoming state $\phi_{\mathbf{p}}^-$ or an outgoing state $\phi_{\mathbf{p}}^+$, with energy eigenvalues $E = \mathbf{p}^2/2m$, defining ingoing and outgoing scattering states, special eigenfunctions with the same energy.

Of course a plane wave $\phi_{\mathbf{p}}(\mathbf{x})$ describes an incoming or an outgoing particle with momentum $\mathbf{p}$ in a heuristic manner. Such a particle can be described rigorously by a function

$$\phi_{\mathbf{p}_0}^{\pm}(\mathbf{x}) = \int_{\mathbb{R}^3} f_{\mathbf{p}_0}(\mathbf{p})\phi_{\mathbf{p}}(\mathbf{x})\,d\mathbf{p},$$

with an $L_2$-function $f_{\mathbf{p}_0}(\mathbf{p})$ sharply peaked at $\mathbf{p} = \mathbf{p}_0$. If our system is a complete scattering system, in the sense of Sect. 15.2, then there are, according to time-dependent scattering theory, unique scattering states $\psi_{\mathbf{p}_0}^{\pm} = W_{\pm}\phi_{\mathbf{p}_0}^{\pm}$, with the same (approximative) energy $E = \mathbf{p}_0^2/2m$.

We shall consider the case with incoming particles, with wave operator $W_-$, in detail. The case with outgoing states, i.e. with $W_+$, then follows immediately by substituting everywhere in the formulas plus signs for minus signs.

We write

$$W_-^* \psi_{\mathbf{p}_0}^- = \lim_{t \to -\infty} e^{iH_0 t} e^{-iHt} \psi_{\mathbf{p}_0}^-.$$

Define the vector-valued functions

$$A(t) = e^{iH_0 t} e^{-iH t} \psi_{\mathbf{p}_0}^-,$$

$$B(t) = \left(1 - \lambda i \int\limits_{t'=0}^{t} e^{iH_0 t'} H_I e^{-iH t'} dt'\right) \psi_{\mathbf{p}_0}^-.$$

One has

$$A(0) = B(0) = \psi_{\mathbf{p}_0}^-$$

and

$$\frac{d}{dt} A(t) = \frac{d}{dt} B(t) = -\lambda i \, e^{iH_0 t} H_I \, e^{-iH t} \psi_{\mathbf{p}_0}^-,$$

which shows that $A(t) = B(t)$, for all $t \geq 0$, and therefore that

$$W_-^* \psi_{\mathbf{p}_0}^- = \phi_{\mathbf{p}_0}^- = \lim_{t \to -\infty} e^{iH_0 t} e^{-iH t} \psi_{\mathbf{p}_0}^-$$

$$= \psi_{\mathbf{p}_0}^- - \lambda i \int\limits_{t=0}^{+\infty} e^{iH_0 t} H_I e^{-iH t} dt \; \psi_{\mathbf{p}_0}^-.$$

The left hand side is meaningful for all $\psi_{\mathbf{p}_0}^-$ in $\mathcal{H}_{\mathrm{sc}}$; the operator $H_I$ is in general unbounded, so for the right hand side a reasonable assumption about the domain of the operator expression in the integral should be made. Similar assumptions have to be made for the remainder of this subsection.

The integral above is absolutely convergent, which allows us to insert a factor $e^{-\epsilon t}$, leading to the formula

$$\phi_{\mathbf{p}_0}^- = \psi_{\mathbf{p}_0}^- - \lambda \lim_{\epsilon \downarrow 0} i \int\limits_{t=0}^{+\infty} e^{-\epsilon t} \, e^{iH_0 t} H_I e^{-iH t} dt \; \psi_{\mathbf{p}_0}^-.$$

We may write this as

$$\int\limits_{\mathbf{p} \in \mathbb{R}^3} f_{\mathbf{p}_0}(\mathbf{p}) \phi_{\mathbf{p}}^- \, d\mathbf{p} = \int\limits_{\mathbf{p} \in \mathbb{R}^3} f_{\mathbf{p}_0}(\mathbf{p}) \psi_{\mathbf{p}}^- \, d\mathbf{p}$$

$$- \lambda \lim_{\epsilon \downarrow 0} i \int\limits_{t=0}^{+\infty} \left( \int\limits_{\mathbf{p} \in \mathbb{R}^3} e^{-\epsilon t} \, e^{iH_0 t} H_I e^{-iH t} f_{\mathbf{p}_0}(\mathbf{p}) \psi_{\mathbf{p}}^- \, d\mathbf{p} \right) dt$$

$$= \int\limits_{\mathbf{p} \in \mathbb{R}^3} f_{\mathbf{p}_0}(\mathbf{p}) \psi_{\mathbf{p}}^- \, d\mathbf{p}$$

$$-\lambda \lim_{\epsilon \downarrow 0} i \int_{t=0}^{+\infty} \left( \int_{\mathbf{p} \in \mathbb{R}^3} e^{-\epsilon t - i E_{\mathbf{p}} t} \, e^{i H_0 t} \, H_I \, f_{\mathbf{p}_0}(\mathbf{p}) \psi_{\mathbf{p}}^- \, d\mathbf{p} \right) dt.$$

Changing the order of integration gives

$$\int_{\mathbf{p} \in \mathbb{R}^3} f_{\mathbf{p}_0}(\mathbf{p}) \, \phi_{\mathbf{p}}^- \, d\mathbf{p} = \int_{\mathbf{p} \in \mathbb{R}^3} f_{\mathbf{p}_0}(\mathbf{p}) \, \psi_{\mathbf{p}}^- \, d\mathbf{p}$$

$$-\lambda \lim_{\epsilon \downarrow 0} i \int_{\mathbf{p} \in \mathbb{R}^3} \left( \int_{t=0}^{+\infty} e^{i(H_0 - E_{\mathbf{p}} + i\epsilon)t} \, dt \right) H_I \, f_{\mathbf{p}_0}(\mathbf{p}) \psi_{\mathbf{p}}^- \, d\mathbf{p}.$$

Using the general formula

$$\int_{t=0}^{+\infty} e^{iAt} \, dt = \int_{t=0}^{+\infty} \frac{1}{iA} \, de^{iAt} = \left[ \frac{1}{iA} e^{iAt} \right]_{t=0}^{t=+\infty},$$

one obtains

$$\int_{\mathbf{p} \in \mathbb{R}^3} f_{\mathbf{p}_0}(\mathbf{p}) \, \phi_{\mathbf{p}}^- \, d\mathbf{p} = \int_{\mathbf{p} \in \mathbb{R}^3} f_{\mathbf{p}_0}(\mathbf{p}) \, \psi_{\mathbf{p}}^- \, d\mathbf{p}$$

$$+ \lambda \lim_{\epsilon \downarrow 0} \int_{\mathbf{p}_0 \in \mathbb{R}^3} \frac{1}{H_0 - E_{\mathbf{p}} + i\epsilon} H_I \, f_{\mathbf{p}_0}(\mathbf{p}) \psi_{\mathbf{p}}^- \, d\mathbf{p},$$

or, with a change in the order of the terms,

$$\psi_{\mathbf{p}_0}^- = \phi_{\mathbf{p}_0}^- - \lambda \lim_{\epsilon \downarrow 0} \frac{1}{H_0 - E_{\mathbf{p}} + i\epsilon} H_I \psi_{\mathbf{p}_0}^-,$$

the *Lippmann-Schwinger equation for normalizable wavefunctions*. The expression $\lim_{\epsilon \downarrow 0} [1/(H_0 - E_{\mathbf{p}} + i\epsilon) H_I \psi_{\mathbf{p}_0}^-]$ is the *retarded Green function* $G_0^-(E)$, the limit $\epsilon \downarrow 0$ of the *resolvent* $R(H_0 - z)$, with

$$R(A - z) = \frac{1}{A - z},$$

a bounded Hilbert space operator, for every selfadjoint operator $A$ and every complex $z$ with Im $z \neq 0$. It is in fact an operator-valued holomorphic function in $z$, a very important property which is widely used in further developments, but will not be used here.

Choosing $f_{\mathbf{p}_0}(\mathbf{p}) = \delta(\mathbf{p} - \mathbf{p}_0)$ gives immediately the Lippmann-Schwinger equation for the (nonnormalizable) eigenfunctions $\phi_{\mathbf{p}}^-$ and $\psi_{\mathbf{p}}^-$ of $H_0$ and $H$, in the form given in the preceding subsection. A more rigorous relation can be obtained by looking at the $\phi_{\mathbf{p}}^-$ and $\psi_{\mathbf{p}}^-$ not as eigenfunctions but as (possibly singular) integral kernels of operators. In this manner $\phi_{\mathbf{p}}^-(\mathbf{x})$ is nothing but the integral kernel $\mathcal{F}_0(\mathbf{p}, \mathbf{x})$ of the unitary Fourier transformation from the $\mathbf{x}$-representation of $\mathcal{H}$ to its $\mathbf{p}$-representation. Similarly, the eigenfunction $\psi_{\mathbf{p}}^-(\mathbf{x})$ is the integral kernel $\mathcal{F}(\mathbf{p}, \mathbf{x})$ of the isometric operator from the $\mathbf{x}$-representation of $\mathcal{H}$ to the (asymptotic) $\mathbf{k}$-representation of $\mathcal{H}_{sc}$. We leave it to the reader to distill from the above equation for normalizable wave functions a rigorous Lippmann-Schwinger equation for the integral kernels $\mathcal{F}_0$ and $\mathcal{F}$.

All this is for the incoming states. By interchanging plus signs and minus signs at the appropriate places, one gets the same results for the outgoing states. Note that we use the mathematical sign convention for the $W_\pm$, $\psi^-$, etc., i.e. with the minus sign for the incoming wave operators, scattering states, etc., and the plus sign for the outgoing objects. Most physics texts use the opposite convention, probably because of the $+i\epsilon$ in the Lippmann-Schwinger equation for $\psi_{\mathbf{p}}^-$.

There is an abundant and ever growing literature on scattering theory, on the time-independent formalism, much of which is rather heuristic, and on the time-dependent approach, rigorous, but often extremely technical. An excellent textbook for this is the book by Newton [10].

# References

1. Reed, M., Simon, B.: Methods of Modern Mathematical Physics III: Scattering Theory. Academic Press, New York (1979) (The best comprehensive modern textbook on the mathematics of single-channel scattering theory, in the time-dependent operator approach emphasized in this book. It is essentially a book inspired by a physical problem, but nevertheless a pure mathematics book. Note : The quote at the front of this chapter is from page ix of this book and is reprinted here with kind permission of Elsevier.)
2. Hunziker, W., Sigal, I.M.: Time-dependent scattering theory of N-body quantum systems. Rev. Math. Phys. **12**, 1033–1084 (2000)
3. Taylor, J.R.: Scattering Theory: The Quantum Theory of Nonrelativistic Collisions. Dover (2006) (This book gives a full discussion both of time-dependent and time-independent scattering theory; it explains in particular the precise relation between the two. It also has a chapter in which multi-channel scattering is clearly explained. In a certain way it is the opposite of the book of Reed and Simon: a physics book with great attention for the mathematical background.)
4. Faddeev, L.D., Merkuriev, S.P.: Quantum Scattering Theory for Several Particle Systems. Translated from the 1985 Russian Edition. Springer 1993, paperback edition 2010 (A basic reference for the scattering of many particle systems. Faddeev's work laid the basis of the mathematical theory of multi-channel scattering.)
5. Kato, T.: Fundamental properties of hamiltonian operators of: Schrödinger Type. Trans. Amer. Math. Soc. **70**, 195–211 (1951)
6. Kato, T.: Perturbation Theory for Linear Operators. Springer (1966), Reprint Edition 1995 (After almost 50 years still the basic reference for the subject.)
7. Dollard, J.D.: Quantum-mechanical scattering theory for short-range and Coulomb interactions. Rocky Mt. J. Math. **1**, 5–88 (1971). http://projecteuclid.org/download/pdf_1/euclid.rmjm/1250131999, An errata at: http://projecteuclid.org/download/pdf_1/euclid.rmjm/1250187234

(In addition to its treatment of the generalized scattering formalism needed for the Coulomb potential, this paper can be recommended as a general introduction to time-dependent scattering theory for a nonrelativistic particle in a given potential, which is both clear and rigorous.)

8. Dollard, J.D.: Asymptotic convergence and Coulomb interactions. J. Math. Phys. **5**, 729–738 (1964)
9. Teschl, G.: Mathematical Methods in Quantum Mechanics With Applications to Schrödinger Operators. American Mathematical Society, Providence (2009)
10. Newton, R.G.: Scattering Theory of Waves and Particles, McGraw-Hill, 1966. 2nd edition, Dover 2013 (A very comprehensive (768 pages) textbook on all aspects of classical and quantum scattering theory, but mainly in the time-independent formulation.)

# Chapter 15
# Towards Relativistic Quantum Theory

## 15.1 Introduction

Two new physical theories changed twentieth century physics in a dramatic way: quantum theory and the theory of relativity. The relation between the two is still far from clear; in their present form they are in a fundamental way incompatible with each other. To understand this is one of the main challenges for fundamental physics today.

This book explains the basics of nonrelativistic quantum theory. In this chapter the road will be sketched to a relativistic version of the theory; in the next chapter we shall give a very brief introduction to what is the preliminary endpoint of this road, relativistic quantum field theory, a theory full of imperfections, not to say grave problems, in particular in its mathematical structure, but nevertheless very successful as a physical theory with an ability to predict experimental results in elementary particle physics with great accuracy.

In the twenties, when quantum mechanics, as we know it now, emerged over a short period of a few years, Einstein's theory of relativity, in particular his theory of special relativity, was already generally accepted as a universal theoretical framework for the physical world, even though it was realized that its consequences in terrestrial physics could only be expected to show up in situations where particles with very high velocities, approximately the velocity of light, would be involved, velocities that could at that time not be realized in experiments. The theoretical need for a relativistic version of quantum theory was nevertheless almost immediately felt and work began on it.

## 15.2 Einstein's Special Theory of Relativity

Einstein wrote the fundamental paper in which he introduced the theory of special relativity in 1905 [1], his 'Annus Mirabilis', the year in which he published five groundbreaking papers. His aim was to solve a problem related to motion

© Springer International Publishing Switzerland 2015
P. Bongaarts, *Quantum Theory*, DOI 10.1007/978-3-319-09561-5_15

and Maxwell's theory of electromagnetism, as is clear from the title of the paper (in English): "On the Electrodynamics of Moving Bodies". This was connected with the problem of the *aether*, a substance which was supposed to fill the cosmos, serving as the carrier of electromagnetic waves and which seemed to have very contradictory properties. This question had been discussed by various nineteenth century physicists, such as Poincaré, Lorentz and FitzGerald.

It was a problem that was connected with the propagation of light as an electromagnetic wave. One would expect that its speed would depend on the motion of the earth through the aether. In 1889 Michelson and Morley showed by a very ingenious and precise interferometer experiment that there was no such dependence. The speed of light was always the same. Special relativity, in fact, abolished the concept of aether altogether. Einstein later developed his general theory of relativity, which gave a new description of gravitation. In 1921 he wrote a short book on both the special and the general theory [2].

The basic notion of classical Newtonian mechanics is that of an *inertial system*, a system of Euclidean coordinates in which a body not subjected to forces of any kind is either in rest or in rectilinear motion with constant velocity. This is connected with the following important principle, discussed in 2, and assumed to be generally valid:

*All inertial systems are physically completely equivalent; there is no preferred inertial system.*

Let $L$ be a given inertial system with spatial coordinates $x$, $y$, $z$. Consider a second inertial system $L'$ with coordinates $x'$, $y'$, $z'$, moving with respect to $L$ with constant velocity $v$ in the positive $x$-direction. This means that there is a coordinate transformation

$$x' = x - vt, \quad y' = y, \quad z' = z.$$

This is called a *Galilei transformation*. All such transformations for all possible velocities and in all possible directions, together with all rotations in space, form a group, the *homogeneous Galilei group*. If we also allow shifts of the origin of both position coordinates and time, it becomes the *inhomogeneous Galilei group*, the symmetry group of classical mechanics. We say that classical mechanics is *Galilei covariant* and that its equations are *Galilei invariant*.

(15.2,a) **Problem** Write down the general Galilei transformation from a given system $L$ to a system $L'$ moving with a velocity $\mathbf{v} = (v_x, v_y, v_z)$ with respect to the system $L$.

According to Einstein the principle of inertia still holds, but it has to be interpreted in a new way. In the classical picture space is a three dimensional Euclidean vector space. Strictly speaking, it is an affine space, roughly a vector space in which the origin is irrelevant and can be chosen at will, a fine point that we shall not worry about. Time is an external parameter which describes the motion of a system in this space.

In Einstein's theory the starting point is a different relation between the coordinates of two inertial systems. In the above example of an inertial system $L'$ moving with respect to a given system $L$ the correct transformation formulas are now

$$x' = \gamma(x - vt), \quad y' = y, \quad z' = z, \quad t' = \gamma(t - vx/c^2),$$

with $\gamma = (1 - v^2/c^2)^{-1/2}$ and $c$ the velocity of light.

Two things should be noted in these formulas.

(1) This is not a three dimensional transformation, with $t$ as an external parameter, but a truly four dimensional one. It means that we live in a four dimensional spacetime, with no absolute separation between space and time. What is space and what is time depends on the observer.
(2) The velocity of light $c$ is roughly equal to 300,000 km/s. Almost no moving terrestrial objects, with the exception of electrons in an atom or subatomic particles in accelerators have a speed that comes near to this. The quotient $v/c$ is very small and the factor $\gamma$ is to a high precision equal to one, which means that it is difficult to detect the difference between classical and relativistic mechanics.

(15.2,b) **Problem** If we perform two successive Galilei transformations in the $x$-direction, with the velocity $v_1$ going from $L$ to $L'$ and $v_2$ from $L'$ to $L''$, the resulting velocity $v$ from $L$ to $L''$ will be $v = v_1 + v_2$. Show that this is different in Einstein's theory and calculate $v$ there as $v = (1 + (v_1 v_2)/c^2)^{-1/2}(v_1 + v_2)$.

Note that for velocities much smaller than that of light this formula is also very close to the pre-Einstein formula.

(15.2,c) **Problem** Suppose that an object moves with a velocity $u$ in the $x$ direction of the inertial system $L$. Show that its velocity $u'$ in $L'$ is

$$u' = \frac{u - v}{1 - uv/c^2}.$$

By substituting $u = c$ one gets $u' = c$, i.e. the velocity of light is the same in different inertial systems, a fundamental feature of special relativity, in agreement with the results of the Michelson and Morley experiment.

One can show that the bilinear form

$$B(x, y, z, t, x', y', z', t') = c^2 tt' - (xx' + yy' + zz'),$$

together with the associated quadratic form

$$Q(x, y, z, t, x, y, z, t) = c^2 t^2 - (x^2 + y^2 + z^2),$$

is invariant under Lorentz transformations. In fact Lorentz transformations can be defined as the linear transformations that leave this form invariant. They form a group sometimes denoted as $SO(1, 3)$ (pseudo-orthogonal group in 4 dimensions). For $t = t' = 0$ it restricts to the standard positive definite Euclidean inner product of space—apart from an overall minus sign, but on spacetime as a whole it is indefinite; the corresponding four dimensional spacetime 'length' $(ct)^2 - (x^2 + y^2 + z^2)$ can be positive, negative or zero. Spacetime as a four dimensional vector space provided with this inner product is called *Minkowski space*, for reasons that will become clear in the next section.

Recommended reading for all this is a collection of the basic papers on relativity theory by, among others, Einstein, Minkowski and Lorentz, in English translation [3].

## 15.3 Minkowski Spacetime Diagrams

Strangely enough the explicit idea of a four dimensional spacetime is not due to Einstein but to the mathematician Hermann Minkowski, a number theorist, who gave a talk on this in 1908 and subsequently published it in a paper in 1909.

> The views of space and time which I wish to lay before you have sprung from the soil of experimental physics, and therein lies their strength. They are radical. Henceforth space by itself, and time by itself, are doomed to fade away into mere shadows, and only a kind of union of the two will preserve an independent reality.

See [4], and also for an interesting paper on the mathematical and physical context of Minkowski's work [5].

It took some time before Einstein saw the merit of this geometric point of view.

The essential difference between the nonrelativistic and relativistic idea about spacetime lies in the relation between time and space in the two cases. In the first case it is just the combination of a linear time variable with the three dimensional space plane which describes space. To understand the difference with the second case it is enough to look at the relation between the time variable $t$ and a single space variable, any of the three, say $x$. The situation can then nicely be drawn in a two dimensional picture, called a *Minkowski diagram*. Diagrams of this sort form a tool of great pedagogical value; they are used everywhere in undergraduate teaching of special relativity.

Figure 15.1 shows empty spacetime. The $x$-axis is horizontal; the $t$-axis vertical, with the time running in upward direction. In Fig. 15.2 we see 'world lines', the histories of moving point particles, and also *events*, particles at certain spacetime points A and B. In this line (1) shows a particle in rest, while line (2) a particle moving with positive constant velocity. Line (3) shows a particle at rest till a certain time, at which it starts an accelerated motion; after a while it is again at rest and is finally annihilated in some way at a certain spacetime point. Of course, all this is with respect to the given coordinate system.

Figure 15.3 shows a Galilei transformation. Notice that the $x$-axis remains in place; the new $t'$-axis is slanted with respect to the old $t$-axis. Figure 15.4 shows a Lorentz transformation. Mathematically a sequence of $n$ numbers can always be regarded as coordinates for an $n$-dimensional space, but in physics more is required,

**Fig. 15.1** Empty spacetime

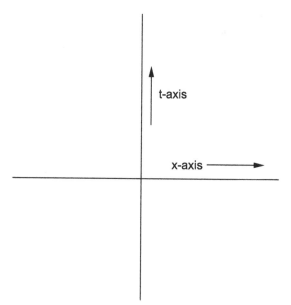

**Fig. 15.2** World lines and spacetime events

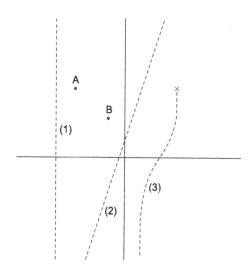

for instance, that objects in this 'space' can be 'rotated' in a meaningful way. This is the case for Lorentzian spacetime, but not in the non relativistic situation. Note that for the pictures we assume that we use units such that $c = 1$. For units second and meter the difference between Figs. 15.3 and 15.4 would be impossible to see.

We could have drawn in all four pictures the lines $x = \pm t$, representing the restriction to two dimensional spacetime of a three dimensional hyper surface, the *light cone*. Mathematically this is set of zeros of the fundamental bilinear form of

**Fig. 15.3** Galilei
transformation

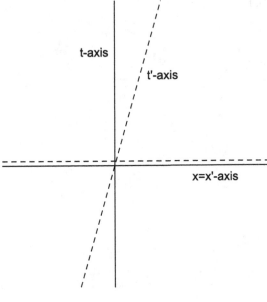

**Fig. 15.4** Lorentz
transformation

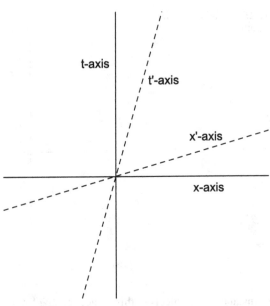

special relativity. It has important physical meanings. The world lines on it represent
particles that move with the velocity of light (and pass through the origin at $t = 0$).
In quantum theory these would be *photons*, the particles connected with the electromagnetic field. World lines inside the cone, which consist of a *forward* part ($t > 0$),
and a *backward* part ($t < 0$), describe particles moving with a speed less than that

of light. World lines passing through the origin and moving outside of the light cone describe particles moving faster than light, *tachyons*. They would have very weird properties. So far they have not been observed.

An introductory book on special relativity is that of French [8]. A general textbook is that of Rindler [10].

## 15.4 The Klein-Gordon Equation

One of the basic features of special relativity is that space and time have to be treated on the same footing. The Schrödinger equation is clearly not relativistic; as a differential equation it is of first order in time and of second order in the space variables. Immediately after the appearance of the first papers on quantum mechanics a search for a relativistic analogue started. Schrödinger himself found such an equation but did not publish it because he was aware of its problematic aspects. Several others found the same equation and did publish it, among them Oskar Klein and Walter Gordon, after whom the equation was eventually named. It is not hard to prove that there exists no Lorentz invariant first order differential equation. The *Klein-Gordon equation* is a second order equation. For the description of a single free particle of mass $m$ it reads

$$\left( \left( \frac{1}{c^2} \frac{\partial^2}{\partial t^2} - \frac{\partial^2}{\partial x^2} - \frac{\partial^2}{\partial y^2} - \frac{\partial^2}{\partial z^2} \right) + \frac{m^2 c^2}{\hbar^2} \right) \phi(x, y, z, t) = 0.$$

Before studying this equation further we introduce notation which is convenient for all discussions in relativity, among other things because it brings out the four dimensional spacetime character of the theory, and is therefore used in most books on relativity and especially in books on relativistic quantum field theory, the subject of the next chapter. It is also important and even indispensable in general relativity, which involves differential geometry on manifolds. In this theory spacetime is no longer a flat space, but a four dimensional differentiable manifold. It has a symmetric indefinite contravariant 2-tensor field $g(\cdot, \cdot)$ which defines locally, in the tangent space of each point of spacetime, the Minkowski metric of special relativity.

We shall use the following conventions:

(1) We shall have spacetime coordinates $x^\mu$, with $x^1 = x$, $x^2 = y$, $x^3 = z$ and $x^0 = ct$. The symbol $x$ will from now on mean the quadruple $(x^0, x^1, x^2, x^3)$. Latin indices $j, k$, etc. are used for the spatial coordinates with $\mathbf{r} = (x^1, x^2, x^3)$.
(2) We use lower (covariant) indices and upper (contravariant) indices, as introduced in Supp. Chap. 20.
(3) We employ the Einstein convention already introduced in Supp. Chap. 20: summation signs are omitted when summing over the same index, one covariant, the other contravariant. Example : $x^\mu y_\mu$ for $\sum_\mu x^\mu y_\mu$.

(4)  A Lorentz transformation $\Lambda$ acts on the spacetime coordinates as a $4 \times 4$ matrix: $(x')^\mu = \Lambda^\mu{}_\nu x^\nu$.

(5)  The Minkowski inner product is written with the help of a covariant 'metric 2-tensor' $g$, with elements $g_{00} = 1$, $g_{jk} = -\delta_{jk}$, and with the other $g_{\mu\nu}$ being 0. So $\langle x, y \rangle = g_{\mu\nu} x^\mu x^\nu$. Sometimes the inverse (contravariant) 'metric 2-tensor' is used with the same elements $g^{\mu\nu} = g_{\mu\nu}$. This seems a bit trivial, but it facilitates the simple writing of various formulas. It also reflects the fact that special relativity can be seen as a locally linearized form of general relativity.

(15.4,a) **Problem** Show that these two tensors are each others inverses, by calculating $g^{\mu\rho} g_{\rho\nu}$.

(6)  The partial derivative $\frac{\partial}{\partial x^\mu}$ is written as $\partial_\mu$, with also $\partial^\mu = g^{\mu\nu} \partial_\nu$.

(15.4,b) **Problem** Use these conventions to write the Klein-Gordon equation simply as $(\partial^\mu \partial_\mu + m^2 c^2 / \hbar^2) \phi(x) = 0$.

Note that in elementary particle theory one generally uses units such that $c = 1$ and $\hbar = 1$. In that context the Klein-Gordon equation becomes just $(\partial^\mu \partial_\mu + m^2) \phi(x) = 0$.

(15.4,c) **Problem** Show that the Klein-Gordon equation is indeed Lorentz invariant, i.e. that if the function $\phi(x)$ is a solution than also the Lorentz transformed function $\phi'(x) = \phi(\Lambda^{-1} x)$.

The Klein-Gordon equation is a relativistic equation. However, can we use it as a quantum mechanical wave equation? It has two serious problems:

(1)  The solutions of the Schrödinger equation have a positive definite inner product. For a unit vector $\psi$ one has $(\psi, \psi) = \int \rho(\mathbf{r}) \, d\mathbf{r} = 1$ with $\rho(\mathbf{r}) = |\psi(\mathbf{r})|^2$ the positive normalized probability density for finding the particle at position $\mathbf{r}$, one of the basic facts of quantum mechanics. The normalization of the probability density is conserved in time. A similar expression $(\phi, \phi) = \int \phi(\mathbf{r})^* \phi(\mathbf{r}) \, d\mathbf{r}$ for solutions for the Klein-Gordon equation is useless in this respect. The equation is second order in time, so for a solution one has to give as Cauchy data both $\phi$ and its time derivative $\dot{\phi}$. There is in this case only one mathematically reasonable sesquilinear expression, namely

$$(\{\phi_1, \dot{\phi}_1\}, \{\phi_2, \dot{\phi}_2\}) = i \int (\overline{\phi}_1 \dot{\phi}_2 - \overline{\dot{\phi}}_1 \phi_2) \, d\mathbf{r}.$$

This is however indefinite; the expression $(\{\phi, \dot{\phi}\}, \{\phi, \dot{\phi}\})$ can be negative, as can be checked easily, and this excludes a probabilistic interpretation.

(2)  The Fourier transform of a solution of the Klein-Gordon equations contains both positive and negative frequencies. This means that if we still want to use it as a quantum mechanical equation there will be particles with negative energy, which is unacceptable physically.

For these two reasons the Klein-Gordon equation is *physically* unacceptable as a Schrödinger type wave equation and has as such to be dismissed. As a *mathematical* equation it will reappear in the next chapter, however with a very different physical interpretation.

## 15.5 The Dirac Equation

There are no Lorentz invariant first order differential equations, that is to say not for *scalar* functions. It was Dirac's idea to look for first order equations for functions with more components. In 1928 [6] he proposed the following equation for a function $\psi : \mathbb{R}^4 \to \mathbb{C}^4$, written in his notation as

$$i\hbar\frac{\partial}{\partial t}\psi(\mathbf{r}, t) = \left( \beta mc^2 + i\hbar c \sum_{j=1}^{3} \alpha_j \frac{\partial}{\partial x^j} \right) \psi(\mathbf{r}, t).$$

In this there are three $4 \times 4$ matrices $\alpha_j$ and a single matrix $\beta$ with the properties $\alpha_j^2 = \beta^2 = 1$, and with all pairs of different matrices anticommuting.

For the modern formulation we use the conventions introduced in the preceding section and moreover the so-called $\gamma$-*matrices* defined as

$$\gamma^0 = \beta, \quad \gamma^k = \beta\alpha_k, \quad k = 1, 2, 3.$$

(15.5,a) **Problem** Show that these matrices satisfy the anticommutation relations

$$[\gamma^\mu, \gamma^\nu]_+ = 2g^{\mu\nu}.$$

(15.5,b) **Problem** Show that this leads to a compact and elegant form of the Dirac equation, namely

$$(i\gamma^\mu\partial_\mu - mc/\hbar)\psi(x) = 0.$$

For units with $c = \hbar = 1$ it becomes $(i\gamma^\mu\partial_\mu - m)\psi(x) = 0$.

The Dirac equation is a Lorentz invariant equation, which is a bit more difficult to see than in the case of the Klein-Gordon equation. We have the following situation:

The connected part of the Lorentz group $O(1, 3)$, denoted as $L_+^\uparrow$, consists of the Lorentz transformations that do not reverse space or time. It has the group $SL(2, \mathbb{C})$ as its universal covering group; there is a two-to-one group homomorphism $g \mapsto \Lambda(g)$ from $SL(2, \mathbb{C})$ onto $L_+^\uparrow$. $SL(2, \mathbb{C})$ acts on the four dimensional complex vector space spanned by the four components of $\psi$ as $g \mapsto S(g)$, a nonunitary representation, while $O(1, 3)$ itself acts on the spacetime coordinates according to $(x')^\mu = \Lambda^\mu{}_\nu x^\nu$. Together this gives the Lorentz transformation—or rather $SL(2, \mathbb{C})$

transformation—of $\psi$ as $\psi'(x) = S(g)\psi(\Lambda^{-1}(g)x)$ (See for the notion of covering group Supp. Sect. 24.5).

(15.5,c) **Problem** Explain why one has to use $\Lambda(g^{-1})$ instead of $\Lambda(g)$ in this transformation formula.

(15.5,d) **Problem** Show that the Dirac equation is indeed invariant under these transformations, i.e. that solutions are transformed into solutions.

As with the Klein-Gordon equation we have to ask whether the Dirac equation is acceptable as a quantum mechanical wave equation. We shall see that it is much better in this respect but still not perfect.

(1) The solutions of the Dirac equation do have a time-independent positive inner product

$$(\psi, \varphi) = \int \sum_{j=1}^{4} \overline{\psi}_j(\mathbf{r}) \, \varphi_j(\mathbf{r}) \, d\mathbf{r}.$$

By substituting in this expression a unit vector $\psi = \varphi$, one gets

$$\rho(\mathbf{r}) = \sum_j \overline{\varphi}_j(\mathbf{r}) \, \varphi_j(\mathbf{r}),$$

which can be interpreted as a probability density for the position $\mathbf{r}$. There remain problems with the definition of a position operator for relativistic particles. Solutions have been proposed, none of these totally convincing.

(2) The Fourier transform of a solution contains positive and negative frequencies, just as for the Klein-Gordon equation, so we have the same problem with unphysical negative energies. In a 1930 paper in which Dirac discussed this problem, he proposed an ingenious and imaginative heuristic solution for this problem. See [7]. According to Dirac the vacuum is filled with infinitely many negative energy particles. There may be a few 'holes' in this 'sea'. These should be regarded as positive energy particles with positive charge.

At the time only three elementary particles were known, the electron, with negative electric charge, the proton, with positive charge and the neutron, electrically neutral—as the name suggests. Because of this Dirac believed that he had found an equation that described electrons and protons. In 1932 a new positively charged particle was discovered in cosmic radiation by Anderson. It had the same mass as an electron and was called a *positron*. Dirac now realized that he had found an equation for the electron and the positron as its *antiparticle*. For this he received the Nobel prize in 1933, sharing it with Schrödinger.

The Dirac equation has more strong points:

(1) For a particle in an electromagnetic field an interaction term is added, which gives, again with $c = \hbar = 1$,

$$(i\gamma^{\mu}(\partial_{\mu} + ieA_{\mu}(x)) - m)\psi(x) = 0$$

in which $e$ is the electric charge and $A_{\mu}(x)$ the electromagnetic 4-potential, appearing in the relativistic formulation of Maxwell's theory of the electromagnetic field. For the electron in the simplest model for the hydrogen atom, only $A_0(x)$ is nonzero; it is the Coulomb potential $-e^2/4\pi\epsilon_0 r$ (see Chap. 7). The other three components describe a possible magnetic field. The eigenvalue–eigenfunction equation for the hydrogen atom can be solved, just like in the case of the Schrödinger equation. Forgetting the unphysical negative energy eigenvalues one gets the hydrogen spectrum with relativistic corrections. Lines are split; one obtains the so-called *fine structure* of the spectrum, with values in good agreement with experiments. In this way the Dirac equation is a definite improvement over the nonrelativistic Schrödinger equation.

(2) The electron spin is automatically included in the Dirac equation; there is no need to add it 'by hand' to the wave equation as was necessary with the nonrelativistic Pauli spin in Chap. 8.

However, notwithstanding these successes, the Dirac equation—and notwithstanding Dirac's ingenious 'hole theory' interpretation of it—eludes all attempts to give it a consistent and clear mathematical meaning as a single particle wave equation. In fact, Dirac's interpretation does hint at a many particle situation. That is the direction in which we have to go to give a mathematically and physically satisfactory description of the behaviour of relativistic particles. For this we need the mathematical apparatus of 'second quantization' that we have developed in Sect. 9.5.

The Dirac equation, but also the Klein-Gordon equation and other relativistic equations will acquire a new life as equations for *operator fields*, in the context of *relativistic quantum field theory*. This will be discussed in the next chapter.

A general quantum mechanics text book that also treats relativistic quantum mechanic is that of Messiah [9].

# References

1. Einstein, A.: Zur Elektrodynamik bewegter Körper. Ann. Phys. **17**, 891–921 (1905) Einstein's first paper on the theory of relativity. An English translation of this paper (On the Electrodynamics of Moving Bodies). http://www.fourmilab.ch/etexts/einstein/specrel/specrel.pdf. See also [5]
2. Einstein, A.: Relativity. The Special and the General Theory. Henry Holt and Company 1921. https://ia700305.us.archive.org/2/items/cu31924011804774/cu31924011804774.pdf
3. Einstein, A., Lorentz, H.A., Weyl, H., Minkowski, H.: The Principle of Relativity. A collection of original papers on the special and general theory of relativity. Translated from the 1913 German edition, Dover 1952. All these pioneering papers are very readable. A treasure for the history of relativity theory
4. Minkowski, H.: Raum und Zeit. Teubner Verlag 1909 This is the text of the talk Minkowski gave in 1908. He died suddenly shortly before it was published. For an English translation, see [5]

5. Walter, S.: Minkowski, Mathematicians, and the Mathematical Theory of Relativity. In: The Expanding World of General Relativity. H. Goenner, J, Renn, J. Ritter, T. Sauer (eds.) Birkhäuser 1999. http://www.univ-nancy2.fr/DepPhilo/walter/papers/einstd7.pdf
6. Dirac, P.A.M.: The quantum theory of the electron. Proc. R. Soc. Lond. **A 117**, 610–624 (1928). http://rspa.royalsocietypublishing.org/content/117/778/610.full.pdf
7. Dirac, P.A.M.: A theory of electrons and protons. R. Soc. Lond. **A 126**, 360–365 (1930). http://rspa.royalsocietypublishing.org/content/126/801/360.full.pdf (These two papers are among Dirac's greatest contributions to theoretical physics.)
8. French, A.P.: Special Relativity. Norton 1968. Note finally that a few physics textbooks contain useful reviews of relativistic quantum mechanics, for instance the book by Messiah
9. Messiah, A.: Quantum Mechanics. Volume I and II. Translated from the French. North-Holland 1961. Reissued in one volume in 1999 by Dover. The North-Holland edition is available at: https://ia601206.us.archive.org/23/items/QuantumMechanicsVolumeI/Messiah-QuantumMechanicsVolumeI.pdf, and https://ia601605.us.archive.org/0/items/QuantumMechanicsVolumeIi/Messiah-QuantumMechanicsVolumeIi.pdf
10. Rindler, W.: Introduction to Special Relativity, 2nd edn. Oxford University Press, Oxford (1991)

# Chapter 16
# Quantum Field Theory and Particle Physics: An Introduction

## 16.1 Introductory Remarks: Some History

The preceding chapter ended with the Dirac equation as the culmination of the search for relativistic quantum mechanics. As a relativistic wave equation it was fairly successful, in particular in Dirac's brilliant but heuristic 'hole theory' picture, although it defied all attempts to formulate it rigorously. See Refs. [15, 16] it suggested, however, the way to the proper description of relativistic particles, namely within a many particle context. This means quantum field theory.

Quantum field theory is an important tool in different areas of physics. Its most important role is in elementary particle physics, which could not have existed without it. This role has a long history, from nuclear physics in the nineteen fifties down to the more recent development of the Standard Model, an all-encompassing theoretical framework that underlies the modern view of the subatomic world. Because of this we shall give a brief overview of its historical development and sketch its present form in Sect. 16.9.

In general fields are physical quantities with one or more components depending on space and time variables, and with an evolution in time prescribed by partial differential equations. We know many such fields, the electromagnetic field, a combination of electric and magnetic fields, evolving in time according to Maxwell's equations, and Einstein's gravitational field, for example. Phenomena like hydrodynamics, aerodynamics, elasticity and many others are also described by fields.

Historically the main field for fundamental physics was the electromagnetic field. It was understood that matter consisting of atoms and molecules was kept together by electric and magnetic forces transmitted by fields. It is therefore not surprising that the founders of quantum theory tried to find a quantized version of this field. One of the more fruitful ideas was to see an electromagnetic field as a collection of infinite many harmonic oscillators of all possible frequencies. As one knew how to quantize systems of oscillators this made sense. For a free field, i.e. a field in the vacuum, not interacting with charges and current this was not hard, even though this meant dealing with systems with an infinite number of degrees of freedom, a situation in

© Springer International Publishing Switzerland 2015
P. Bongaarts, *Quantum Theory*, DOI 10.1007/978-3-319-09561-5_16

which von Neumann's uniqueness theorem for the operators that satisfy the canonical commutator relations no longer holds. Attempts at constructing a quantized version of interacting systems started already in the early nineteen thirties with work by Paul Dirac. Work in further development of this ran into serious problems: persistent infinities at each level of proposed theories. From a practical point of view these problems were overcome by a procedure called *renormalization*, invented partly independently by Feynman, Dyson, Schwinger and Tomonaga in the nineteen fifties. Renormalization theory is a method, ad hoc but effective, in which most infinities, divergent integrals in fact, can be made to cancel. All this without really understanding why this is so. However, as a way to test experimentally the predictions of quantum field theories it has turned out to be very successful. This is in particular true for *quantum electrodynamics*, the theory that describes the electromagnetic interactions between particles, where certain quantities can be calculated with a precision up to six or seven decimals. So renormalization theory, a very complicated technical procedure, has become a standard tool for elementary particle theory. Nevertheless, the true mathematical nature of the theory remains mysterious. Fifty years of hard work by competent mathematical physicists has not resulted in remedying this. There have been various attempts to clarify the situation, each of them with interesting results. Some of these attempts will be briefly described later in this chapter.

The standard work on quantum field theory is now the three volume book by Weinberg [1]; An introductory book is that of Zee [2].

## 16.2 Quantum Field Theory as a Many Particle Theory

Because of the high velocities involved, elementary particle physics is necessarily relativistic. The main reason that relativistic quantum mechanical equations fail is that they describe single particle states. However, a characteristic aspect of elementary particle physics, or as it is usually called, high energy physics, is that particles can be created and annihilated.

In the preceding chapter we discussed the Klein-Gordon equation as a first candidate for a quantum mechanical relativistic wave equation. As such it had to be abandoned. The second candidate, the Dirac equation was better, but still not good enough. It took some time before the situation was properly understood and that it was realized that the same equations could be used, with a completely different mathematical meaning and different physical understanding, to describe in a relativistic way, as equations for quantum field operators, the many particle situations appearing in elementary particle physics.

In Chap. 9 we discussed many particle systems in quantum mechanics. It was explained there that the Hilbert spaces of systems of $n$ non-identical particles are $n$-fold tensor products of the Hilbert spaces of the separate particles. For identical particles these tensor products had to be symmetrized (bosons) or antisymmetrized (fermions), with as results symmetric tensor products $\mathcal{H}_s^{(n)} = \otimes_s^n \mathcal{H}^{(1)}$ and antisymmetric tensor products $\mathcal{H}_a^{(n)} = \otimes_a^n \mathcal{H}^{(1)}$.

Quantum field theory describes processes in which particles can be annihilated or created, the possibility of which is a typical feature of elementary particle physics. This happens, for instance, in high energy accelerators, when an electron and its antiparticle, a positron, collide, are annihilated, giving rise in this process to the creation of a photon, the particle that carries the quantized version of electromagnetic radiation. This means that we need a many particle Hilbert space, the countably infinite direct sum of all the tensor product spaces. Such a space is called a Fock space after the Russian physicist V.A. Fock who in 1932 first introduced them. We have

$$\mathcal{H}_s^{\text{Fock}} = \oplus_{n=0}^{\infty} (\otimes_s^n \mathcal{H}^{(1)}),$$

for bosonic particles, and

$$\mathcal{H}_a^{\text{Fock}} = \oplus_{n=0}^{\infty} (\otimes_a^n \mathcal{H}^{(1)}),$$

for fermionic ones. The properties of Fock spaces were described in detail in Chap. 9. In the next section some of this will be repeated, not in the context of the 'second quantization' of the Schrödinger equation, but for the construction of the simplest example of a relativistic quantum field theory.

## 16.3  Fock Space and Its Operators

Fock spaces have a characteristic system of operators, *annihilation and creation operators*, names that suggest the physical role that they play, namely that of creating and annihilating particles. Mathematically they are used to move up and down in the many particle Fock space. As we shall discuss as example only a field theory for bosonic particles, we restrict ourself to the bosonic case.

Let $\mathcal{H}^{(1)}$ be a given one particle Hilbert space, with elements $f$, $g$, etc. We define *creation operators*, for every $f$ in $\mathcal{H}^{(1)}$ and $n = 0, 1, \ldots$, as operators

$$a^*(f) : \mathcal{H}_s^{(n)} \rightarrow \mathcal{H}_s^{(n+1)}$$

by linear extension of

$$a^*(f)(f_1 \otimes \cdots \otimes f_n)_s = (f \otimes f_1 \otimes \cdots \otimes f_n)_s,$$

and *annihilation operators*, for every $g$ in $\mathcal{H}^{(1)}$ and for $n = 1, 2, \ldots$, as operators

$$a(g) : \mathcal{H}_s^{(n)} \rightarrow \mathcal{H}_s^{(n-1)}$$

by linear extension of

$$a(g)(f_1 \otimes \cdots \otimes f_n)_s = [(g, f_1)(f_2 \otimes \cdots \otimes f_n)_s + \cdots + (g, f_n)(f_1 \otimes \cdots \otimes f_{n-1})_s].$$

There is one dimensional zero particle space $\mathcal{H}^{(0)}$, with a preferred state $\Omega_0$, the ground state or *vacuum state*. Each annihilation operator acts on $\Omega_0$ as $a(f)\Omega_0 = 0$. It is not hard to derive the following commutation relations, for all $f$ and $g$ in $\mathcal{H}^{(1)}$,

$$[a^*(f_1), a^*(f_2)] = [a(g_1), a(g_2)] = 0, \qquad [a(g), a^*(f)] = (g, f).$$

The operators $a(f)$ and $a^*(g)$ defined, respectively, as maps from $\mathcal{H}_s^{(n)}$ to $\mathcal{H}_s^{(n-1)}$, and from $\mathcal{H}_s^{(n)}$ to $\mathcal{H}_s^{(n+1)}$, are obviously operators in the direct sum space $\oplus_{n=0}^{\infty}\mathcal{H}_s^{(n)}$, the boson Fock space $\mathcal{H}_s^{\text{Fock}}$.

For the Fermion case one has similarly defined creation and annihilation operators, with a few additional minus signs in the action of the annihilation operators, and anticommutation instead of commutation relations.

## 16.4 The Scalar Quantum Field

The simplest quantum field theory is a theory with a single real scalar field operator, in which the Klein-Gordon equation acquires a second life. It describes a system of spin zero particles, electrical neutral, for example $\pi^0$-mesons, produced from other particles in accelerators, but also found in cosmic radiation. The really interesting situation in physics is one in which there are several different fields, describing different particles, interacting with each other, e.g. electrons and positrons, protons and neutrons. The free scalar field on its own is only a pedagogical example. Free means in this context that there are no attracting or repulsive forces between the particles, and that they do not collide with each other. Its great advantage is that it can be formulated with complete mathematical rigor, unlike interacting quantum field theories which all suffer from very serious mathematical problems, notwithstanding their great success as physical theories. As we cannot go into detail on the subject of interacting quantum fields and their difficulties, the scalar field is an excellent model to understand at least the first steps in setting up quantum field theory.

For this the one particle Hilbert space $\mathcal{H}^{(1)}$ consists of functions $f(p)$, with the variable $p = (p^0, p^1, p^2, p^3)$ a 4-vector in momentum space, restricted to a *forward mass shell*, a hypersurface in this space defined by the relation

$$p^0 = \sqrt{(p^1)^2 + (p^2)^2 + (p^3)^2 + m^2},$$

which is invariant under Lorentz transformations $p^\mu = \Lambda^\mu_\nu p^\nu$. In this $p^0$ is the energy, $m$ the mass, sometimes called the rest mass, and $\mathbf{p} = (p^1, p^2, p^3)$ the 3-momentum. Note that we have chosen units such that $c = \hbar = 1$, as is usual in elementary particle physics. The operator $a^*(f)$ creates a particle with a momentum wave packet $f(p)$. One would like to create particles with a sharp momentum $p$ with a creation operator $a^*(p)$. This is, strictly speaking, impossible, just as it is, again strictly speaking, impossible to have particles with an exact position. Nevertheless, we do this, for both cases, heuristically.

With this idea in mind we write, with $\mathbf{p} = (p^1, p^2, p^3)$,

$$a^*(f) = \int a^*(p) f(p) \frac{d\mathbf{p}}{p^0},$$

and

$$a(g) = \int a(p) \overline{g(p)} \frac{d\mathbf{p}}{p^0}.$$

(16.4,a) **Problem** Show that the heuristic operators $a^*(p)$ and $a(p)$ must satisfy the commutation relations

$$[a^*(p), a^*(p')] = [a(p), a(p')] = 0, \qquad [a(p), a^*(p')] = p^0 \delta(p - p').$$

Sets of heuristic formulas of this type, for all sorts of particles, bosonic, fermionic, with different masses and spins, form the basis of all further developments of relativistic quantum field theory that one can find in standard textbooks.

## 16.5  The Scalar Quantum Field: The Field Operators

The scalar quantum field itself, a system $\{\hat{\phi}(x)\}_{x \in R^4}$, of (heuristic) operators depending on points of spacetime, is defined by performing a Fourier transformation on a combination of the (heuristic) annihilation operators and (heuristic) creation operators

$$\hat{\phi}(x) = \frac{1}{(2(2\pi)^3)^{1/2}} \int \left( a(p) e^{-ipx} + a^*(p) e^{+ipx} \right) \frac{d\mathbf{p}}{p^0},$$

with again $\mathbf{p} = (p^1, p^2, p^3)$ and $x = (x^0 = t, x^1, x^2, x^3)$ and the $p^0$ in the numerator under the integral to make the formula manifestly Lorentz invariant. From this one derives the field equation

$$(\partial^\mu \partial_\mu + m^2) \hat{\phi}(x) = 0,$$

just the Klein-Gordon equation, but now as an operator equation describing a many particle system. There are canonical equal-time commutation relations

$$[\hat{\phi}(\mathbf{x_1}, t), \hat{\phi}((\mathbf{x_2}, t)] = [\hat{\pi}(\mathbf{x_1}, t), \hat{\pi}((\mathbf{x_2}, t)] = 0,$$

$$[\hat{\pi}(\mathbf{x_1}, t), \hat{\phi}(\mathbf{x_2}, t)] = -i\delta(\mathbf{x_1} - \mathbf{x_2}),$$

with $x_j = ((x_j)^1, (x_j)^2, (x_j)^3)$, $(x_j)^0 = t$ and $\hat{\pi}(\mathbf{x}, t) = \frac{\partial}{\partial t}\hat{\phi}(\mathbf{x}, t)$, the 'canonical momentum' conjugate to $\hat{\phi}(\mathbf{x}, t)$. Note that we are obviously in the Heisenberg picture, in which the observables depend on time while the state vectors in $\mathcal{H}_s^{\text{Fock}}$ remain constant. The Hamiltonian operator for this quantum system is

$$H_0 = \int \frac{1}{2}\left[\hat{\pi}^2 + (\nabla\hat{\phi} \cdot \nabla\hat{\phi}) + m^2\hat{\phi}^2\right]d\mathbf{x},$$

with

$$\nabla = \left(\frac{\partial}{\partial x^1}, \frac{\partial}{\partial x^2}, \frac{\partial}{\partial x^3}\right).$$

The operator $H_0$ as the energy operator is a constant of the motion, so it is time-independent, even in the Heisenberg picture.

This heuristic formulation can be made rigorous by starting from creation and annihilation operators depending on square integrable functions on momentum space, in the spirit of Sect. 15.3, then rewriting the Fourier transformation and finally obtaining the field operator as a functional $\hat{\phi}(f)$ on a linear space of test functions $f$ on spacetime. This means that the free scalar quantum field is a mathematically well-defined model. However, by doing this the elegance and transparency of the heuristic description is lost. We would also no longer see the similarity and the differences between the Klein-Gordon equation as a one-particle quantum mechanical wave equation with numerical solutions, not very successful in this role, and as a many particle operator equation, completely acceptable in a mathematical and physical sense.

(16.5,a) **Problem** Express the field operator $\hat{\phi}(f)$ in the rigorous creation and annihilation operators.

The free scalar quantum field is, together with other similarly constructed free field theories, such as the Dirac quantum field theory, and quantum electrodynamics, the point of departure for general field theories for interacting particles. Note however that for the very important class of gauge fields, to be briefly mentioned in Sect. 16.9.4, this is not true, since they do not have free linear models to start with.

## 16.6 The Scalar Field with Self-Interaction

The next step is a quantum field theory for one type of scalar particles that interact with each other. A possible sort of interaction is given by the interaction Hamiltonian

$$H_I = \frac{1}{4}\int \hat{\phi}^4 d\mathbf{x}.$$

Together with the Hamiltonian of the free theory this gives a total Hamiltonian

$$H = H_0 + \lambda H_I,$$

with $\lambda$ a so-called coupling constant. This leads to the nonlinear field equation

$$(\partial^\mu \partial_\mu + m^2)\hat{\phi}(x) + \lambda \hat{\phi}^3(x) = 0$$

By using the Fourier formula for $\hat{\phi}(x)$ one can write $H_0$ as an expression in annihilation and creation operators in the Fock Hilbert space. This leads to a well-defined selfadjoint operator. It is unbounded, but has a perfectly acceptable dense domain of definition. This is not the case for the interaction Hamiltonian $H_I$, which can also be written as an expression in creation and annihilation operators. However, it carries all the vectors of the Fock Hilbert space into 'non-normalized' vectors, meaning that the domain of definition of this expression as an operator only contains the zero vector.

Elementary particle physicists do not worry about the mathematical existence of this and other similar formal operator expressions. They just use them as point of departure for perturbative calculations of processes between various particles. The terms of the resulting power series expansions all lead to divergent integrals. These divergencies are removed by the systematic prescriptions of *renormalization theory*. The final results are finite and give excellent predictions for the outcomes of experiments. A rigorous mathematical basis for this still does not exist, even though much work has been done to clarify this situation. In the next section the most important attempts will be briefly discussed.

## 16.7  Towards a Rigorous Quantum Field Theory

### 16.7.1  Wightman Axiomatic Field Theory

In the nineteen sixties Arthur Wightman began, first with Lars Gårding, a series of papers in which he studied vacuum expectation values of products of field operators, the so-called $n$-point functions [3]. If we restrict ourself again to the simple example of a single real scalar field, as we do in most of this chapter, this means studying the functions

$$W_n(x_1, \ldots, x_n) = (\Omega_0, \hat{\phi}(x_1) \ldots \hat{\phi}(x_n)\Omega_0),$$

with the $\hat{\phi}(x_j)$ the field operators at spacetime points $x_j$ and $\Omega_0$ the vacuum state. Strictly speaking, the $W_n$ are not functions, but tempered distributions, so the proper way of writing the $W_n$ is in terms of 'smeared field operators' as

$$W_n(f_1, \ldots, f_n) = (\Omega_0, \hat{\phi}(f_1) \ldots \hat{\phi}(f_n)\Omega_0),$$

with the $f_j$ complex-valued test functions in $\mathcal{S}^{\mathbb{C}}(\mathbb{R}^4)$. See for the notion of (tempered) distributions Supp. Chap. 25. All relevant properties of the quantum field are reflected in properties of these Wightman functions.

(16.7.1,a) **Problem** Show that the fact that $\hat{\phi}$ is a real scalar field corresponds with the property

$$\overline{W_n(x_1, \ldots, x_n)} = W_n(x_n, \ldots, x_1)$$

or more rigorously

$$\overline{W_n(f_1, \ldots, f_n)} = W(\overline{f_n}, \ldots, \overline{f_1}),$$

for all $f_j$ in $\mathcal{S}^{\mathbb{C}}(\mathbb{R}^4)$.

Relativistic covariance of the theory means that there is a unitary representation $\{U(\Lambda, a)\}_{(\Lambda, a) \in \mathcal{P}}$ of the Poincaré group $\mathcal{P}$ in the Hilbert space of the quantum field such that

$$U(\Lambda, a)\, \hat{\phi}(x)\, U(\Lambda, a)^{-1} = \hat{\phi}(\Lambda x + a),$$

for all $(\Lambda, a) \in \mathcal{P}$ and all spacetime points $x$, which moreover leaves the vacuum state $\Omega_0$ invariant. (See for the properties of groups and their representations Supp. Chap. 24). To write this in a rigorous way we note that the Poincaré group acts on $\mathcal{S}^{\mathbb{C}}(\mathbb{R}^4)$ by linear maps $T(\Lambda, a)$ as

$$(T(\Lambda, a)f)(x) = f(\Lambda x + a).$$

(16.7.1,b) **Problem** Show that the transformation formula for the rigorous field operator is

$$U(\Lambda, a)\, \hat{\phi}(f)\, U(\Lambda, a)^{-1} = \hat{\phi}(T(\Lambda, a)^{-1}f),$$

for all $f$ in $\mathcal{S}^{\mathbb{C}}(\mathbb{R}^4)$.

(16.7.1,c) **Problem** Show that the natural manner to express the Poincaré covariance of the theory in terms of the Wightman functions is

$$W_n(\Lambda x_1 + a, \ldots, \Lambda x_n + a) = W_n(x_1, \ldots, x_n),$$

and more rigorously

$$W_n(f_1, \ldots, f_n) = W_n(T(\Lambda, a)^{-1}f_1, \ldots, T(\Lambda, a)^{-1}),$$

for all Poincaré transformations $(\Lambda, a)$, and all $f_j$ from $\mathcal{S}^{\mathbb{C}}(\mathbb{R}^4)$.

All other properties of the scalar quantum field can in a similar way be characterized by properties of the Wightman functions, for example the definite positivity of the inner product of the Hilbert space, the positivity of the energy, and certain locality and causality properties. One can put all these properties, those of the quantum field in an operator context and—equivalently—those of the Wightman functions, in a system of axioms, the *Wightman axioms*. We shall not spell out this system here. See [4, 5].

The equivalence between the two pictures is indeed very strong, as is stated in the *reconstruction theorem*, the high point of Wightman theory: given a system of functions $\{W_n\}_{n=0,1,\ldots}$, satisfying the Wightman axioms, one can construct from these a Hilbert space, with a unique system of quantum field operators $\{\hat{\phi}(x)\}_{x \in \mathbb{R}^4}$, a vacuum state $\Omega_0$, together with a unitary representation of the Poincaré group, such that the vacuum expectation values of the products of field operators are equal to the given functions $W_n$.

Note that even though the Wightman axioms are supposed to describe all quantum field models, with or without interaction, they have nevertheless nothing to say about dynamics or interaction between fields. This is not surprising as they all are linear statements, while the essence of interaction in field theory, as in elementary particle physics in general, is nonlinearity. The simplest interaction, in our simple, physically not very realistic model, the real scalar field is a self interaction term $\int \hat{\phi}^4(x)\mathrm{d}x$, on which the Wightman theory cannot say anything.

Wightman theory is technically complicated, but its main results, in particular the *reconstruction theorem* and the *spin-statistic theorem* (in Sect. 9.3), are conceptually simple. There is a point of view, developed by Borchers and Wyss [6, 7], and relatively unknown or not emphasized in the vast literature on Wightman theory, that provides a transparent picture of it. It fits in the algebraic scheme developed in Chap. 12.

It starts with the observation that the Wightman function $W_n$, in its rigorous version, is an $n$-linear functional

$$\times^n \mathcal{S}^{\mathbb{C}}(\mathbb{R}^4) \mapsto \mathcal{S}^{\mathbb{C}}(\mathbb{R}^4), \qquad (f_1, \ldots, f_n) \to W_n(f_1, \ldots, f_n),$$

with the $f_j$ test functions from the Schwartz space $\mathcal{S}^{\mathbb{C}}(\mathbb{R}^4)$. It defines therefore a linear functional on the $n$-fold tensor product $\otimes^n \mathcal{S}^{\mathbb{C}}(\mathbb{R}^4)$; this means that the sequence $\{W_n\}_n$ defines a linear functional on the tensor algebra $T(\mathcal{S}^{\mathbb{C}}(\mathbb{R}^4))$.

We have the following pair of objects:

- a $*$-algebra $\mathcal{A}_B = T(\mathcal{S}^{\mathbb{C}}(\mathbb{R}^4))$, the Borchers algebra, the algebra of observables of this particular quantum system,
- a positive normalized linear functional $\omega_B$ on $\mathcal{A}_B$, the sequence of Wightman functions, the state of the system.

These are just what is needed as prerequisites for the GNS construction explained in Supp. Sect. 27.9: a $*$-algebra $\mathcal{A}$, together with a positive normalized linear functional $\omega$ on $\mathcal{A}$. The GNS construction then gives a Hilbert space $\mathcal{H}_\omega$, a representation $\pi_\omega$ of $\mathcal{A}$ in $\mathcal{H}_\omega$ with a cyclic vector $\Omega_\omega$, such that the value $\omega(a)$ is equal to the expectation value $(\Omega_\omega, \pi_\omega(a)\Omega_\omega)$. This means that the GNS theorem is in this case just Wightman's reconstruction theorem. The fact that $\mathcal{A}_B$ is not a $C^*$-algebra, but a more general topological algebra has no consequences for the proof.

Note that the Poincaré group acts linearly on the test function space, as we have seen, which induces an action by $*$-automorphisms of the Borchers algebra $\mathcal{A}_B$. Lorentz or more general Poincaré covariance of the field translates into invariance of the Wightman functions, which in turn gives invariance of the state functional $\omega_B$. An addition to the GNS construction procedure shows then that the Poincaré automorphisms of $\mathcal{A}_B$ are implemented by unitary operators in $\mathcal{H}_\omega$.

There is a generalization of the Borchers-Wyss formulation of Wightman theory, useful for the description of the Maxwell quantum field which has complications which the other free fields do not have. The Maxwell field has two different fields, the tensor field, an antisymmetric covariant 2-tensor $\widehat{F}$, and a potential field $\widehat{A}$, a covariant vector field. The first is needed for the physical interpretation, the second for the formulation of interaction with matter fields. $\widehat{F}$ can be obtained by a simple differentiation from $\widehat{A}$, in components $\widehat{F}_{\mu\nu}(x) = \partial_\mu \widehat{A}_\nu(x)$. Different fields $\widehat{A}$, different *gauges*, describe the same (physical) $\widehat{F}$. There are two Borchers algebras; the relation between the two fields induces an algebra homomorphism between the two. The unpleasant fact about the field $\widehat{A}$ is that there is no *positive* Lorentz invariant functional. One has the choice between using either a not manifestly Lorentz covariant gauge or to let the $\widehat{A}$ act in a space with an indefinite metric. This situation, two fields, two spaces of test functions, relations between these, is hard to understand in the original Wightman approach. With the algebraic reformulation of Borchers and Wyss, this is much easier. Positivity of the state functional for $\widehat{A}$ can be replaced by continuity; the main idea of the GNS construction still works. For more details, see [8, 9].

The merit of the Wightman approach is that it provides a general framework, in which much interesting work has been done, in particular on general aspects of locality and causality aspects of quantum theory in relativistic spacetime, even though no other explicit and physical meaningful examples are available than systems of free fields, describing non-interacting particles.

### 16.7.2 Constructive Quantum Field Theory

In the nineteen sixties James Glimm and Arthur Jaffe took up the challenge of the Wightman approach and started a program to construct nontrivial examples of relativistic quantum field theory, which would fit in the Wightman scheme, by forceful methods of approximation, in the first place by spatial cut-offs, usually in models in lower dimensional spacetime, in which the divergencies are less virulent. Much work along these lines was done by many mathematical physicists; interesting nonlinear models of interacting quantum fields were constructed, scalar fields with quartic self interaction, scalar and spinor fields with so-called Yukawa interaction, with or without space cut-offs, but always in lower spacetime dimensions, never one of the standard quantum field theories in four dimensional spacetime. See for an overview of the final results [10].

### 16.7.3 Algebraic Quantum Field Theory

Algebraic quantum field theory is in its origin based on work of I.E. Segal, who first suggested the idea of an algebraic formulation of quantum theory, with the observables as point of departure. See [11]. The first paper in which an explicit system of axioms for algebraic quantum field theory was presented was written by Haag and Kastler [12]. In this scheme fields are derivative objects; the primary notion is that of local and quasi-local observables. One assumes that to all open subsets $O$ of four dimensional spacetime a $C^*$-algebra $\mathcal{A}(O)$ is assigned with its selfadjoint elements the physical observables localized in $O$. All these *local algebras* can be put together with the help of a certain limit procedure which gives after closure the *$C^*$-algebra of quasi-local observables* of the system. The set of Haag-Kastler axioms contains a number of assumptions on this system $\{\mathcal{A}(O)\}_{O \subset \mathbb{R}^4}$: consistency with inclusion and taking intersections and unions of sets, covariance under Poincaré transformations, with respect to the action of a representation of the Poincaré group by ∗-automorphisms of the algebraic system as a whole. Furthermore locality conditions in the sense that regions of spacetime the points of which are space-like to each other—see for this notion the preceding chapter—cannot influence each other, etc. See [13, 14]. For the underlying mathematics, see Supp. Chap. 27.

With respect to the success of this approach in mathematically understanding or justifying quantum field theory as it is practised in elementary particle physics, the same can be said as in the case of Wightman theory. The free fields satisfy the axioms, but up till now it has not been shown that any of the interacting theories of particle physics can be described in this manner. But beautiful theorems about the basic properties of spacetime quantum theory have been proved. Great contributions have also been made to pure mathematics, in particular to the theory of operator algebras.

## 16.8 Quantum Field Theory: Concluding Remarks

Quantum field theory is a curious theory. It began with Dirac in the nineteen twenties, ran into the difficulties of infinities, was revived in the nineteen fifties in a practical sense as renormalized quantum electrodynamics through the work of Feynman and others, and reached in the nineteen sixties its greatest success in the so-called gauge field theories, that provided the theoretical basis for the *Standard Model* of elementary particles (Weinberg, Salam, Glashow, Veltman, 't Hooft), which encompasses our present knowledge of the fundamental constituents of matter. This will be discussed in more detail in the next section. At the same time the mathematical basis of the theory is at present still not understood, notwithstanding much hard work by mathematicians and mathematical physicists during more than fifty years.

Various approaches have been tried for developing such a basis. *Axiomatic quantum field theory* in the late nineteen fifties (Wightman) has given a beautiful set of general axioms, with important theorems proved, but with only the free fields as examples. In *constructive field theory* (Glimm and Jaffe) one has later worked on nontrivial examples, with a certain success for a limited class of simplified models. *Algebraic quantum field theory*, using $C^*$-algebras, with Haag and Araki as pioneers, a beautiful theory which again has lead to interesting theorems, but with very few results on explicit nontrivial examples. For these reasons the mathematical investigation of quantum field theory remains completely open.

## 16.9 Elementary Particle Physics: A Brief Review

### 16.9.1 Introduction

This chapter is devoted to quantum field theory, or at least to basic aspects and elementary examples. In its advanced fully developed form it is the main tool for present day elementary particle physics.

Elementary particle physics, or 'high energy physics' as its practioners like to call it, is still the most prestigious and, with astronomy and space science, most expensive area of modern physics. Its flagship is at present the 'Standard Model', an inclusive theoretical scheme in which all particles that are known so far and are considered to be truly elementary, have a logical place. It is hard to understand or to appreciate it without knowing something of the way it developed from much simpler models in atomic and nuclear physics. That is the reason why this section is included in this chapter.

### 16.9.2 Atomic Physics

Quantum mechanics when it first appeared was able to explain the structure and behaviour of atoms and molecules, something in which classical physics had failed.

In this early form of submicroscopic physics the following particles, considered to be elementary, were known.

- The *electron*, already discovered and studied by J.J. Thomson and various other physicists in the second half of the nineteenth century. Its antiparticle, the *positron*, theoretically predicted in 1930 by Dirac in his groundbreaking paper (see Refs. [15, 16]), and observed in experiment by Carl Anderson in 1932 [17].
- The *photon*, identified by Einstein as the particle of electromagnetic radiation, in 1905, in his work on the photoelectric effect (see Ref. [18]). In the quantum theory

of atomic structure photons played a role in understanding transitions between different energy levels of atoms.

- *Atomic nuclei*, of various type, depending on the sort of atom. They were not seen as elementary particles, even though nothing was known of their structure.

The force acting in atomic physics, inside atoms, and between different atoms and molecules, was the *electromagnetic force*, after the gravitational force, the second of the four fundamental forces that are known at present.

The general quantum field theory describing the electromagnetic interactions between electrons and their antiparticles, photons, and charged particles in general is called *quantum electrodynamics*. It is plagued by divergent integrals, that can be circumvented however by *renormalization*, a somewhat ad hoc procedure, invented, just after the end of the Second World War, by Richard Feynman, Julian Schwinger and Freeman Dyson in the USA, and independently by Sin-Itiro Tomonaga in Japan. Although even now quantum electrodynamics and renormalization are not understood in a rigorous mathematical way, the results of its calculations agree in an extremely precise manner with experiment. There is a very worthwhile collection of historical papers on quantum electrodynamics, edited by Schwinger [19].

### 16.9.3 Nuclear Physics

The nuclei of all atoms consist of positively charged *protons* and neutral *neutrons*, of almost the same mass, much heavier than the electron, in fact roughly 1,836 times the electron mass. The hydrogen nucleus is just a singe proton. The number of protons in a nucleus is called the *atomic number* and is denoted as Z. The chemical properties of an atom are determined by Z. The number of neutrons is denoted as N. The atomic weight of an element is $Z+N$. Atoms with the same Z and different N are called *isotopes*. They have the same chemical properties.

At first sight it seems surprising that atomic nuclei can hold together, because of the repulsive force between the positively charged protons. The solution to this problem is a new fundamental force, the *strong nuclear force*. It was predicted in 1935 by Hideki Yukawa [20] as a force which would be transmitted by hypothetical massive bosons. They were found in 1947 and later, as $\pi$-*mesons*, *pions* for short, $\pi^0$ (neutral) and $\pi^+$, $\pi^-$ (charged).

There was a remaining problem on what can happen inside an atomic nucleus. This has to do with radioactive decay. After the early discovery of radioactivity by Henri Becquerel in 1896, and Marie Curie in 1898, it became clear in the nineteen thirties that atomic nuclei were not stable, but could decay spontaneously into nuclei of other elements. Later ways of transforming elements artificially were found.

In a typical such process, $\beta$-decay, already observed by Rutherford in 1903, a neutron is converted into a proton, with emission of a fast electron ($\beta$-radiation). It is a phenomenon in the nucleus which cannot be explained by the strong interaction. An curious feature of this process is that energy-momentum does not seem to be conserved. This led Pauli in 1930 to suggest that the reason for this might be an unknown particle created in the process, a neutral massless particle, with almost no interaction with other particles, not susceptible to electromagnetic or strong forces. It was observed in 1956 and was called the *neutrino*.

In this way it turned out that there was another new fundamental force: the *weak nuclear force*. It is much weaker than the electromagnetic force: the strong force is hundred times stronger than the electromagnetic force, and this is in turn $10^{12}$ times stronger than the weak force.

The result was a system of two main sets of particles.

1. *Hadrons*: consist of *baryons* (heavy particles): protons and neutrons, and *mesons*: pions and kaons
2. *Leptons* (light particles): the electron, the muon, a kind of heavier electron discovered in 1936, and the neutrino. Leptons do not feel the strong force.

### 16.9.4 The Particle Zoo: Towards the Standard Model

Much further progress was made by the development of particle accelerators, machines in which charged particles such as electrons and protons, were given high velocities by means of electric and magnetic fields, in order that other particles might be created in collisions. There were linear accelerators, but circular machines later dominated the field. An early example was the cyclotron, invented by Ernest Lawrence in 1931.

Its modern descendant is the gigantic Large Hadron Collider (LHC) at CERN (Genève), finished in 2008 an now the largest machine of its sort. In it protons are accelerated to 99.9999964 % of the velocity of light. It is built in a tunnel with a circumference of 16 miles, running deep under the surface, partially in Switzerland and partially in France.

Another useful device was the *cloud chamber*, invented already in 1911 by Charles Wilson, and later the much larger *bubble chamber* invented by Donald Glaser in 1952, both able to make the paths of particles in collision processes visible.

Also useful in this respect was the investigation of cosmic radiation, which might bring in particles with extremely high velocities.

As a consequence many new particles made their appearance, were detected experimentally and precisely measured for the first time.

- The *antiproton*, the antiparticle of the proton, the nucleus of the hydrogen atom, with positive charge equal to minus that of the electron. It was observed in an accelerator experiment in 1955, although much earlier detected in cosmic radiation.
- The *neutron*, with charge zero, with a mass slightly larger than that of a proton. It was first observed by Chadwick in 1932. There is also an *antineutron* with the same charge zero and mass. It was detected in 1956 by using a particle accelerator. Neutrons and antineutrons are unstable.

- Various types of *mesons*, massive particles, bosons, that mediate in the interaction between nucleons (protons and neutrons). There are $\pi$-mesons or pions: the $\pi^+$-*meson*, positively charged, its antiparticle, the $\pi^-$-*meson*, negatively charged, and the neutral $\pi_0$-meson, its own antiparticle. There are also $\rho$-mesons, $\eta$-mesons, etc.
- A host of other particles, all unstable and very short lived, more than 150. A well-known physicist is known to have said that if he knew all the names of these particles he would be a botanist. There was, not surprisingly, a general feeling that in this chaotic situation some kind of systematization would be desirable.

The first important step toward more order was the unification of the weak and electromagnetic forces into a single field theory. An *electro-weak theory* was formulated in the early 1960s by Sheldon Glashow, Steven Weinberg and Abdus Salam [21]. This implied the existence of a new force transmitted by bosonic particles; the charged $W^+$ and $W^-$, and the neutral $Z^0$, all three observed in accelerator experiments in 1983. The theory has a group theoretical basis, with a (partial) symmetry with respect to the product group $SU(3) \times U(1)$. It is a so-called *gauge field theory*, a notion that went back to a short paper by Yang and Mills [22], published in 1954, at that time not getting much attention, but later recognized as an important and original paper.

Gauge theories have interesting mathematical aspects, in which a combination of group theory and differential geometry plays a role. This lies outside the material that we have assembled in Supp. Chaps. 20 and 24, so we shall not discuss it. Very important was the proof of the renormalizability of such theories by Veltman and 't Hooft in 1972.

The next step came with the growing suspicion that particles as the proton and the neutron were not truly elementary, but were composed of smaller not yet discovered particles. The work of Murray Gellmann, Yuval Ne'eman in the early nineteen sixties contributed to this. In this symmetry with respect to the group $SU(3)$, the 'Eightfold Way', played an important role. A problem in their approach was that the new hypothetical particles that they needed would have non-integral charge, namely $\pm \frac{1}{3}$ and $\pm \frac{2}{3}$. The new scheme was in many ways so compelling that the fractional charge was gradually accepted. The particles in question were called 'quarks' by Gell-Mann. All this led to *quantum chromodynamics*, a new fundamental quantum field theory treating the strong interaction, explaining the way hadrons and mesons are made up from quarks and antiquarks, introducing new force carrying particles, the *gluons*, and new quantum numbers, *charm* and *color*. Over a period of years all these predictions were experimentally verified.

There were two problems for which eventually solutions were found.

1. Experiments suggested that there were inside a proton, for instance, much smaller point like particles, which could be identified as quarks. But they were 'confined'; they could not be observed outside the proton. Why not? The answer given by David Gross and Frank Wilczek in 1973 was '*asymptotic freedom*', the idea that the force between two quarks inside a proton is described by a potential which becomes weak when the quarks are close together [23].

2. Some particles in the scheme have masses which they should not have. The solution to this problem was suggested, already in 1964, by François Englert, Robert Brout and independently by Peter Higgs: a mechanism that could generate these masses. It involved a new particle. This *Higgs particle* was in 2012 finally observed in the Large Hadron Collider at Genève, a result which completed the *Standard Model*.

### 16.9.5  The Standard Model

The Standard Model is a framework in which all the known elementary particles and their interactions are described. It can be pictured by the following scheme.

```
| u  c  t | g | H |
| d  s  b | γ |
| – – – – – |
| e  μ  τ | Z |
| νe νμ ντ | W |
```

Explanation: Name (notation) (charge, spin, mass)
1. *Quarks.*

Two upper left rows of three particles; columns of two; total of six.

- *Up* (u) $(+\frac{2}{3}, +\frac{1}{2}, 2.3 \text{ MeV/c}^2)$
- *Down* (d) $(-\frac{1}{3}, +\frac{1}{2}, 4.8 \text{ MeV/c}^2)$
- *Charm* (c) $(+\frac{2}{3}, +\frac{1}{2}, 1.275 \text{ GeV/c}^2)$
- *Strange* (s) $(-\frac{1}{3}, +\frac{1}{2}, 95 \text{ MeV/c}^2)$
- *Top* (t) $(+\frac{2}{3}, +\frac{1}{2}, 173,07 \text{ GeV/c}^2)$
- *Bottom* (b) $(-\frac{1}{3}, +\frac{1}{2}, 4.18 \text{ GeV/c}^2)$

2. *Leptons.*

Two lower left rows of three particles; columns of two, six in total.

- *Electron* (e) $(-1, +\frac{1}{2}, 0.511 \text{ MeV/c}^2)$
- *Electron* **neutrino** $(\nu_e)$ $(0, +\frac{1}{2}, <2.2.\text{eV/c}^2)$
- *Muon* $(\mu)$ $(-1, +\frac{1}{2}, 105.7 \text{ MeV/c}^2)$
- *Muon neutrino* $(\nu_\mu)$ $(0, +\frac{1}{2}, <0.17 \text{ MeV/c}^2)$
- *Tau particle* $(\tau)$ $(-1, +\frac{1}{2}, 1.777 \text{ GeV/c}^2)$
- *Tau neutrino* $(\nu_\tau)$ $(0, +\frac{1}{2}, <15.5 \text{ MeV/c}^2)$

3. *Gauge bosons.*

A column of four particles on the right.

- *Gluon* (g) (0, 1, 0)
- **Photon** ($\gamma$) (0, 1, 0)
- *Z-boson* (Z) (0, 1, 91.2 GeV/c$^2$)
- *W-boson* (W) (0, probably 1, 80.4 GeV/c$^2$)

4. *A separate particle*—extreme upper right.

- *Higgs particle* (H) (0, 0, about 126 GeV/c$^2$)

The particles under 1 and 2 are fermions; those under 3 and 4 bosons.

Units of mass in particle physics: eV means electronVolt; MeV means Mega-electronVolt; GeV means Giga-electronvolt.

1 MeV = $10^3$ eV; 1 GeV = $10^6$ eV.

The **boldface** particles are the particles that were traditionally considered to be elementary and remain so in the standard model. However, particles such as the proton, neutron, the mesons, are no longer elementary; they are build up from the various types of quarks and antiquarks. A proton is made from two Up and one Down quark, a neutron from two Down and one Up quark, and the $\pi^+$ meson from one Up and one anti-Down quark, etc.

The standard model is far from perfect:

1. Neutrinos have after all a small mass (as was shown in experiments from 1998 onwards). The different types of neutrinos can be converted into each other ('neutrino oscillation').
2. It is not understood why very similar particles have masses that are different in many orders of magnitude.
3. Gravity, the fourth fundamental force in the universe is not included in the standard model.
4. Finally, the standard model has 17–23 free parameters whose values cannot be derived from theory but have to be found by experiments.

The standard model is a *phenomenological model*. An underlying more basic theory does not yet exist. More will be said on this in the next chapter.

For a very readable and entertaining overview of elementary particle physics in general, see the book by Veltman [24]. See also [24] of Chap. 17.

# References

1. Weinberg, S.: The Quantum Theory of Fields. 1. Foundations. 2. Modern Applications. 3. Supersymmetry. Cambridge University Press, Cambridge (2005). (This very comprehensive three volume book is now the standard textbook on quantum field theory)
2. Zee, A.: Quantum Theory in a Nutshell, 2nd edn. Princeton University Press, Princeton (2010). (A more accessible introductory book on the subject)
3. Gårding, L., Wightman, A.S.: Fields as operator-valued distributions in relativistic quantum field theory. Arkiv f. Fysik, Kungl. Svenska Vetenskapsak. **28**, 129–189 (1964)
4. Bogoliubov, N.N., Logunov, A.A., Oksak, A.I., Todorov, I.T.: General Principles of Quantum Field Theory. Kluwer, Boston (1989)

5. Streater, R.F., Wightman, A.S.: PCT, Spin and Statistics, and All That. Princeton University Press, Princeton (2000)
6. Borchers, H.J.: On the Structure of the Algebra of Field Operators, pp. 214–236. Nuovo Cimento, US (1962)
7. Wyss, W.: The field algebra and its positive linear functionals. Commun. Math. Phys. **27**, 223–234 (1972). https://projecteuclid.org/download/pdf_1/euclid.cmp/1103858251
8. Bongaarts, P.J.M.: Maxwell's equations in axiomatic quantum field theory I. Field tensor and potentials. J. Math. Phys. **18**, 1510–1516 (1977)
9. Bongaarts, P.J.M.: Maxwell's equations in axiomatic quantum field theory II. Covariant and noncovariant gauges. J. Math. Phys. **23**, 1881–1898 (1982)
10. Glimm, J., Jaffe, A.: Quantum Physics : A Functional Integral Point of View, 2nd edn. Springer, US (1987)
11. Segal, I.E.: Postulates for general quantum mechanics. Ann. Math. **48**, 930–948 (1947)
12. Haag, R., Kastler, D.: An algebraic approach to quantum field theory. J. Math. Phys. **5**, 848–861 (1964)
13. Araki, H.: Mathematical Theory of Quantum Fields. Oxford University Press, Oxford (1999) (Translated from the Japanese)
14. Roberts, J.E., Roepstorff, G.: Some basic concepts of algebraic quantum theory. Commun. Math. Phys. **11**, 321–338 (1969). http://projecteuclid.org/download/pdf_1/euclid.cmp/1103841262
15. Dirac, P.A.M.: A theory of electrons and protons. Proc. R. Soc. Lond. **A 126**, 360–365 (1930) http://rspa.royalsocietypublishing.org/content/126/801/360.full.pdf
16. Dirac, P.A.M.: The quantum theory of the electron. Proc. R. Soc. Lond. **A 117**, 610–624 (1928) http://rspa.royalsocietypublishing.org/content/117/778/610.full.pdf
17. Anderson, C.D.: The positive electron. Phys. Rev. **43**, 491–494 (1933) http://authors.library.caltech.edu/7189/1/ANDpr33b.pdf
18. Einstein, A.: Über einen die Erzeugung und Verwandlung des Lichtes betreffenden heuristischen Gesichtspunkt. Ann. der Phys. **17**, 132–148 (1905)
19. Schwinger, J. (ed.): Selected Papers on Quantum Electrodynamics, Revised edition 2012. Dover, New York (1958)
20. Yukawa, H.: On the interaction of elementary particles. I. Progr. Theoret. Phys. **1**, 1–10 (1935). http://web.ihep.su/dbserv/compas/src/yukawa35/eng.pdf
21. Glashow, S.L.: Partial-symmetries of weak interactions. Nucl. Phys. **22**, 579–588 (1961) http://www.physics.princeton.edu/mcdonald/examples/EP/glashow_np_22_579_61.pdf
22. Yang, C.N., Mills, R.L.: Conservation of isotopic spin and isotopic gauge invariance. Phys. Rev. 96, 191–195 (1954) http://fisicafundamental.net/relicario/doc/yang-mills.pdf
23. Gross, D.J., Wilczek, F.: Asymptotically free gauge theories. I, II. Phys. Rev. D **8**, 3633–3652 (1973), Phys. Rev. D **9**, 980–993 (1974) http://www.aps.org/about/pressreleases/upload/Asymptotically_Free_Gauge_Theories_I.pdf, http://www.aps.org/about/pressreleases/upload/Asymptotically_Free_Gauge_Theories_II.pdf (Two important papers highlighting the road to the final Standard Model, very technical, quoted here just to show how work in this area looks like)
24. Veltman, M.: Facts and Mysteries in Elementary Particle Physics. World Scientific, Singapore (2003). The first chapter is available at http://www.worldscientific.com/doi/suppl/10.1142/5088/suppl_file/5088_chap1.pdf

# Chapter 17
# Concluding Remarks

## 17.1 Introduction

In this concluding chapter I shall venture a few remarks on what I believe to be the present status of physics and its relations with the rest of exact science. These remarks are highly personal; many of my colleagues will strongly disagree with me. For this reason I shall use in this chapter the more personal "I" instead of "we". The reader of this book may consider what I have written here as belonging to the sort of provocations that sometimes lead to fruitful discussions.

The 20th century was the century of physics. It became dominated by two great new theories, first the theory of relativity, then quantum theory. This book is devoted to the second. The applications of quantum theory have changed our live in many respects. This is less so with relativity, but both theories have led to deep changes in our philosophical picture of the world, relativity in our ideas on space and time, quantum theory on matters to do with causality and what we should consider reality.

The 21th century will be the century of the biological sciences, in particular cell biology, the science of the brain, DNA, sciences which have just began, but are making rapid progress. Their practitioners have the same feeling of excitement as the pioneers of quantum mechanics in the 1920s. Their results will have a greater impact on human life than either quantum theory or relativity ever had.

What remains of the importance of physics? In the first place, the life sciences have physics as an absolutely essential basis, together with the applications of computer science, the importance of which is growing every day. But what about physics itself? The most prestigious and most fundamental part of physics was and is still elementary particle physics, with its flagship the Standard Model, discussed in the preceding chapter. It cannot be denied, however, that elementary particle physics is stagnant, has in fact been essentially stagnating for the last 40 years. Yes, the Higgs particle was found last year. A fantastic technical achievement. Scientifically speaking, it more or less completed the standard model.

To put things in a historical perspective, one should compare the standard model with the periodic table of the elements, formulated in the 19th century.

© Springer International Publishing Switzerland 2015
P. Bongaarts, *Quantum Theory*, DOI 10.1007/978-3-319-09561-5_17

Regularities in the properties of the various elements led to the realization, by Dmitri Mendeleyev and others, that they could be placed in a scheme with rows and columns, with here and there gaps, for unknown elements, which indeed were later discovered and had the right properties. This model was purely phenomenologic; there was no way of explaining or deriving its properties from a more basic model. This became possible when quantum mechanics arrived on the scene in the 1920s and was able to describe in detail atoms and their structure and interactions.

This is exactly the present situation with the Standard Model. There is not yet an underlying theoretical model which can explain, for instance, why similar elementary particles can have masses which are orders of magnitude apart, or from which one could derive the value of the 16 or so different parameters in the model. All this is not to deride or minimize the enormous merits of constructing this model, over several decades, out of a chaotically large number of subatomic particles and their various interactions, nor to think lightly of the very sophisticated experiments underpinning it and the unbelievably complicated calculations connecting its properties with experimental data. But nevertheless, when all is said, it is a 20th century Periodic System, with no underlying model in sight yet. And it has been like this for over 40 years.

## 17.2 Theoretical Physics of the 21th Century

### 17.2.1 Introduction

Which parts of (theoretical) physics are at this moment truly evolving with a good chance of leading to really new and interesting insights in the near future? Applied physics, together with computer science, is developing very fast, as the necessary basis for the ongoing revolution in the biological sciences, and as the great stimulus of further important changes—improvements, let us hope—in our daily life. But what about fundamental theoretical physics? As I just argued, particle physics, is stagnant, its very great technical achievements not withstanding.

Actual topics in physics that are of fundamental interest and developing are cosmology (if I take the definition of physics broad enough to include it), and the complex of ideas that I shall denote as "The Einstein-Podolsky-Rosen Paradox and All That", a variation on a well-known historical book title [1]. Then there is a number of speculative theories that I shall briefly discuss under the heading of "Science Fiction".

### 17.2.2 Cosmology

Speculations about the nature of heaven and earth go probably back to the beginning of humanity. In this sense cosmology is the oldest of all exact sciences, even though it only began as a science in a serious manner at the beginning of our era with

Ptolemaeus, was further developed by Arabic astronomers and mathematicians, and became a science in the modern sense during the West European Scientific Revolution in the 15th and 16th centuries.

At present astronomy with cosmology is an undertaking of almost industrial proportions: gigantic optical telescopes, high up in the mountains or in Antarctica or on artificial satellites, and radio telescopes consisting of networks of receivers covering hundred of square miles. The general problem at which it directs its attention is the birth ('Big Bang') and the subsequent development of the cosmos.

Specific problems.

1. *Inflation.* It is now generally believed that the universe expanded right after the Big Bang (13.8 billion years ago), very fast, faster than the velocity of light—which seems to contradict Einstein's theory of relativity, but it does not, because the metric expansion of space is a different notion from the relative motion of two objects in the flat spacetime. It did this over a very short time (from $10^{-36}$ till roughly $10^{-33}$ seconds after the Big Bang). This phenomenon is supposed to have been important for the further development of the universe, such as galaxy formation. All this was suggested for the first time by Guth in [2]. A remainder of these events is a cosmic microwave background which has indeed been observed. Further refined observations are planned. Some cosmologists and in particular philosophers remain however very critical of the concept of inflation. Some critics use the term 'Standard Cosmos'. See for a long, critical and pertinent review Ref. [3]. Recently (May 2014) there have been observations at a radio telescope in Antarctica that cast doubt on the notion of inflation, or at least on the reality of some of its possible consequences. See [4].

2. *Dark energy.* A form of energy which is supposed to permeate all of space. It is connected with inflation. There are three sources of evidence for its existence.

   a. Measurements of distances and the relation of their results with the *redshift*, meaning that because of the continuing expansion of the universe light from far away sources has a larger wavelength when it reaches us.
   b. It is suggested by the outcome of measurements of large scale wave-patterns of mass density in the universe.
   c. It is theoretically needed to explain the global flatness of the universe.

   Dark energy is evenly and very thinly spread in space. It is not sensitive to any of the known forces, which is the reason why it has not yet been observed directly.

3. *Dark matter.* Its existence is inferred from effects it seems to have on other masses. Already in 1932 the Dutch astronomer Jan Oort postulated its existence, after measuring the motions of stars in the Milky Way. Roughly 27 % of the matter in the universe is dark matter. The amount of dark energy is more or less 68 %. This leaves some 5 % for 'ordinary' matter. A staggering fact.

Both dark energy and dark matter remain a great mystery.

General conclusion: even though some of its explanations and statements are very speculative and some appear rather outlandish, on the borderline with science fiction, astronomy—cosmology is nevertheless true science. It has precisely formulated theories and there is a wealth of very precise observations to contradict or confirm these theories.

### 17.2.3 The Einstein-Podolsky-Rosen Paradox—and All That

#### The EPR Paradox

Einstein made important contributions to the origins of quantum mechanics. He nevertheless came to feel deeply unhappy about the form given to it later in the 1920s, in particular in the interpretation of the theory by Bohr. He wrote in 1935, with Boris Podolsky and Nathan Rosen as coauthors, a short paper [5], which had great influence on all later discussions, so much that the subject is now often just called the EPR-paradox. It marked the beginning of a ten year long discussion between Einstein and Bohr, consisting of a succession of proposals for thought experiments by Einstein and their subsequent refutal by Bohr. In this paper Einstein and his coauthors state that quantum mechanics is 'incomplete', by which they mean that not every element of the physical reality of a system has a counterpart in the quantum mechanical theory, which is supposed to describe the system.

The essence of their argument is that an objective reality, which exists independently from observers is denied by quantum mechanics.

Although Einstein recognized the validity of quantum mechanics as a tool to calculate the results of physical processes in a very precise and efficient way, he nevertheless disliked what he called the incompleteness of the theory and above all the lack of causality, a dislike that stayed with him till the end of his life.

It should be noted that for many years the great majority of physicists adhered to the 'Copenhagen Interpretation' and thought of Einstein—with all the great respect they had for his genius—as old-fashioned. Recently, the tide seems to be turning: quite a few prominent theoreticians have said that maybe Einstein, after all, . . .. The jury is still out.

#### Schrödinger's Cat

Shortly after the appearance of the EPR article, and clearly inspired by it, Schrödinger published a paper in which he proposed a rather curious thought experiment [6].

A cat is penned up in a steel box, together with what Schrödinger calls a "diabolical device": a small amount of radioactive matter coupled to a triggering mechanism that, when an atom decays with the emission of a fast electron, kills the cat by the release of poison coming from a little flask broken by the electron. When the cat is put in the box, it is alive. Some time later the state of the system, i.e. its wave function, gives a certain probability that the cat is dead, a probability which increases with time. The observer cannot see what happens inside the box. When he finally opens it he will find either a dead cat or a cat which is still alive. Before that, the only thing he has as description of the situation inside the box is the 'state' of the system: a wave function giving probabilities for the two possibilities. One might say that the cat is partially dead and partially alive. Opening the box gives certainty.

In the Copenhagen interpretation there is nothing problematic about this situation. It does not make sense to speak of the reality inside the box when it is not observed. The only reality that is relevant is the fact that the cat was alive at the beginning of the experiment and the outcome of the observation when the box is finally opened. All this was unacceptable to Schrödinger.

It will be clear that the main theme of Schrödinger's discussion of this thought experiment, even more so than of the EPR paper, is the question "What is the meaningof reality?" This is an old question, discussed by philosophers since the beginning of philosophical discourse. Perusing the literature one gets the impression that most physicists writing on the foundations of quantum theory are not aware of this historical background.

In the past, over the years, a whole spectrum of answers to this question has been given. On one side there is, what may be called 'naive objectivism', the idea that reality is just there, independent of us, with definite properties, even if we do not (yet) know everything about it. It is clear that Einstein and Schrödinger both belong to this school, without saying so explicitly. That is the source of their problems with quantum mechanics. At the other end of the spectrum is 'solipsism', which holds that an objective reality does not exist. All there is are the unconnected individual sensual impressions of separate individuals. Solipsism is in equal measure perfectly logical and totally absurd. In the middle, but not very far from solipsism is the position of Immanuel Kant. He believed that certain fundamental concepts and properties of our mind structure our experience; the world consists of 'Dingen an sich' ('things in itself'), which are unknowable.

One might take a position near Kant, but farther from solipsism, namely, that there are indeed objectively existing objects outside us, and that we can effectively know them. This knowledge is however not complete and never will be. It is the total of the combination of reasoning and observation at a certain moment. In fact that is what science is, a collective process of discovering more and more properties of nature, by theoretical arguments coupled to experimental observations. The ever growing totality of this knowledge represents the reality of our physical world. Whether or not there is a fixed stable basis carrying this may be considered irrelevant. Such an attitude is perhaps a useful background or point of departure for the discussions on what 'reality' means in quantum theory, even though it does not in itself solve the problems connected with it.

A second point that worried Einstein and Schrödinger is *locality*, the fact that quantum mechanics seems to allow instant interaction over great distances, which would be in contradiction with special relativity. But it opens the possibility for *teleportation*, a phenomenon that will briefly be touched upon further on.

*Bell's Theorem*

Already in 1935 Bohr replied to the EPR paper—and implicitly to Schrödinger's paper—in his usual courteous but incisive style, expressing his disagreement with the opinion of Einstein and his coauthors [7]. A very different reaction came much later, in 1964, from the Northern Ireland particle physicist John Bell, with a paper that contained a proposal that might lead to an experimental confirmation or refutation of the EPR ideas. It was the first in Bell's series of important publications on the philosophical and possibly experimental consequences of the EPR [8–10]. One of the main conclusions of the EPR paper was that quantum mechanics in its standard form was incomplete and that there must therefore be 'hidden variables' in the theory. Bell began his work in the conviction that the arguments of Einstein, Podolsky and Rosen were correct. In his paper he tried to think of experiments by which the possibility of hidden variables could be established. For this he designed an ingenious but essentially simple experimental situation with two entangled spin-$\frac{1}{2}$ particles, in which this possibility could be tested. (Note that this means two particles that together form a single quantum state with spin 0. See Sect. 8.4.2. Such a state need not be localized, but can be extended over a long distance in space. Entangled states are also an important tool in studying quantum teleportation. See below.) For such experiments he found correlation inequalities that could be measured and when satisfied would confirm the possibility of hidden variables.

Bell's ideas were generalized in various directions. What is more important is that experimental physicists took up his challenge and succeeded in doing experiments of the kind he had suggested. The best known is that by Alain Aspect and his group [11–13], which showed that Bell's inequalities were violated, excluding in this manner the possibility of hidden variables. Other experimental work by others followed, in the main confirming this result. It remains to this day a fruitful research area of which the end is not yet in sight. See for further discussions on Bell's work and its implications [14–16]. Bell's theorem remains a subject of great theoretical and experimental importance.

*Teleportation*

Teleportation is a notion which comes from science fiction, such as from the TV-series "Star Trek". It means that a person can be transported in the form of radiation of some kind from one location to another, preferably a distant one. Quantum mechanics can do the same although in a more down to earth style. In fact, what is transported are not human beings but characteristics and information. This is done by using typical properties of quantum theory, in particular the possibility of having *entanglement* of quantum states. See our earlier remark. A difficult point is the unavoidable decoherence—the enemy of entanglement—due to the coupling between the quantum system and the environment, which may gradually destroy entanglement and will restore a classical or semiclassical situation.

A pioneer in this field is Zeilinger [17]. His group was in 1997 the first to demonstrate the possibility of teleportation. In 2005 they obtained teleportation of photons through a tunnel under the Danube in Vienna. Recent experiments involved teleportation between two of the Canary islands, a distance of 144 km.

This whole field is very promising for its possible applications, such as *quantum computing* and *quantum cryptography*, ideas which may however need another 20 years or so to become full reality. An excellent review of all this can be found in a book edited by Bouwmeester et al. [18].

From all this it is clear that what I have denoted as "Einstein-Podolsky-Rosen paradox—And All That" represents truly 21st century physics: it combines deep and well-formulated theories with difficult and precise experiments, and is aimed at solving a complex of fundamental problems.

## 17.2.4  Science Fiction

A physical theory has to meet two requirements:

1. It can be written in the form of a consistent mathematical model provided with clear and unambiguous rules for its interpretation.
2. Using these rules it can be tested by experiments—or in the case of astronomy— by observations.

If a theory is confirmed by such experiments—or rather not contradicted, in the spirit of the philosopher Karl Popper—it is a *valid* physical theory, for the time being— again according to Popper. If this is not the case, we have a *false* theory. A model of an aspect of the physical world that satisfies only the first requirement, and moreover is sufficiently imaginative, should be called *science fiction*. Of course a speculative theory deserves an ample period of time before this judgement should be made. But there is a reasonable limit to the length of this period.

### String Theory

String theory is a fundamental approach  to elementary particle physics dating from the late 1960s. Since then it has developed into an all encompassing description of the foundations of the physical world. Its basic idea is that the spacetime elements on which the theory is built are not points, like in ordinary quantum field theory, but tiny one dimensional objects, 'strings', open or closed, living in a spacetime of more than four dimensions. It pretends to be able to bridge the gap between quantum theory and relativity.

A problem is that string theory does not offer realistic suggestions for experiments that could confirm or contradict its predictions, because these would involve very high energies that, even in a distant future, cannot be realized. This has been the case for at least 40 years. The situation is aggravated by the fact that there are now many different models in string theory, so there is no means of selecting one by experiment. The conclusion is inescapable: string theory is not a physical theory; it is science fiction.

As a topic string theory has been and still is very popular among physicists. Some particle physics departments consist only of string theorists. Why? One answer is that if young ambitious theoreticians—of which there are many—can choose between doing hard and tedious calculations in 'ordinary' elementary physics or working on a new theory with a lot of interesting mathematics, a theory that moreover promises to be the 'Theory of Everything', their choice will be an easy one.

Mathematics is not a natural science. The second requirement for a physical theory is in mathematics replaced by the duty to give precise proofs of all statements made. Again, there is a place for conjectures that may take many years before they are proved or disproved. A good example is Fermat's conjecture, which took three centuries to prove, by Andrew Wiles in 1994. It is a good illustration how in mathematics conjectures may stimulate the development of the field.

String theory is not mathematics, but it contains beneath a layer of heuristic formulations some very interesting mathematical ideas. Understanding these is not facilitated by the style of most string theory papers, which in fact are inaccessible to most mathematicians. They are written in a kind of heuristic mathematics particular to the subject; often it is not clear whether a statement is a definition or a theorem. In the last case it is then not clear whether it is proved or not.

But the contributions of string theory to mathematics are real and substantive, due, for instance, to string theorists like Edward Witten and Nathan Seiberg. The first in particular is a brilliant mathematician whose heuristic but very original ideas have provided much work for pure mathematicians to rigorously prove his conjectures. His Fields Medal in 1990 was therefore well-deserved.

In conclusion it can be said that the real contributions of string theory to physics are minimal, but that it has been of great importance for mathematics.

There are many textbooks on string theory, for example, the two-volume treatise by Polchinsky [19]. Well worth reading are two critical books: that of Woit [20] and of Smolin [21].

*Supersymmetry*

Supersymmetry proposes a kind of symmetry between bosons and fermions. It was developed in the 70s, first by Russian physicists whose work was published in Russian language journals and consequently was not noticed in the Western world, and then further developed by Julius Wess and Bruno Zumino. See [22].

It promised, among other things, to make the renormalization of certain quantum field theories easier. Supersymmetry predicts for every known particle a supersymmetric companion, for example a selectron with an electron and a sneutrino with a neutrino. Up till now none of these particles, nor any other effect of supersymmetry, have been detected experimentally. String theory depends heavily on supersymmetry, so the fate of both theories is intimately connected.

A further even more speculative development of supersymmetry is *supergravity*, which is a super form of general relativity.

Supersymmetry and supergravity have led to interesting new mathematical notions: super Lie algebras (See Supp. Sect. 24.10) and supermanifolds (See Sect. 13.9.3.3).

A good introduction to supersymmetry and supergravity is the book by Wess and Bagger [23].

*Many Worlds*

The many worlds interpretation of quantum mechanics was proposed by Hugh Everett in 1957 as an answer to the reality problem posed by Schrödinger's Cat. In Everett's point of view the result of the measuring, i.e. the opening of the box, with the cat either alive or death, is real. The opposite possibility that we do not see is also real. It exists in a parallel world. The same is true with every quantum mechanical measurement. What we find is real, but all the possible other outcomes are equally real. In this manner there is a great ever growing multitude of simultaneous real worlds. All this is, of course, totally unobservable. Everett's theory shares with solipsism (preceding section) the combination of being logical and at the same time absurd. See [24].

*Emergent*

Recently the idea that spacetime and gravity are not fundamental notions but that they are 'emergent' has circulated and drawn attention, both positive and negative. It is not clear from what spacetime is supposed to emerge. Vague references to 'entropy' and 'information' are not helpful in this.

There are quite a few other theories, that all combine great ingenuity with a total lack of a possibility of tests by experiments. They are very well covered in a book by Baggott [25].

## 17.3  Sociological Aspects of Modern Physics

Finally a few words on what might be called the sociology of modern physics. Before the Second World War professors were appointed to do teaching. Most did just that— usually competently—but not much else. A minority did also research. These are the people whose names we still know: Gibbs, Planck, Einstein, Heisenberg, and others. Nowadays the main task of a professor or lecturer, at least unofficially, is research. In the universities teaching has no great prestige anymore, even though this will be rarely admitted. Appointments and promotions are made on the basis of the quality of research papers, more recently on their number. Even more recently the ability tot get research funding has become an important recommendation for advancement. The quality of teaching does not play a role in this respect.

All this has led to great progress, in science in general and in specific areas of theoretical physics in particular. But it has also led to an ever more narrow specialization and to short term thinking.

The position of young scientists is unenviable. There is an enormous output of bright and energetic PhD graduates, who almost all want to stay in academia and hope to obtain a permanent university position after a number of post-doc periods. The capacity of university departments is such that at most 15 % will eventually reach this goal. The others will finally have to take jobs outside the university, at industrial research institutes, in applied computer science, many at banks and other financial institutes. They would have been better advised to do this right after getting their PhD; however their advisors seldom gave them an advice, based on a more realistic view on the market for academic scientists. Trying to get a permanent university position is a rat race, in which only a few will be successful. Personal life is difficult for someone who is, for instance, post-doc in Amsterdam, while his wife holds a similar position in New York, implying that for a couple of years they see each other twice a year. The times that a wife automatically followed her husband to whatever country he took a job in are definitely over.

## 17.4 The Future

There is a definite problem with universities as institutions. Many people there do excellent work. But there is a growing bureaucracy—with too many managers— in a curious manner connected with the belief that a university should be run as a business organization. Top scientists may spend more than half of their time on writing research proposals or on refereeing research proposals of their colleagues. The managers who run public research funds believe in planning of research and in the advantages of working in large groups. Fine for particular parts of experimental physics, for instance for operating a large particle accelerator. But it should be remembered that almost all important discoveries in the past century were the work of separate individuals, or small networks of individuals. There is also a heavy publication pressure, especially on young researchers. Eventually a price will be paid for all this.

## 17.5 Mathematics

There does not seem to be any sort of stagnation in mathematics. It flowers and branches out in all directions. New applications in other areas appear frequently, often of mathematical topics that were considered absolutely pure. Number theory applied to encryption, for example. However, physics has lost its privileged position in this respect. Most of the great mathematicians of the past where deeply interested in physics and made important contributions to it. The younger generation of mathematicians now is in general completely ignorant of modern physics. The situation in which mathematics and physics as sciences were twins will not return.

## References

1. Sellar, W.C., Yeatman, R.J.: 1066 and All That: a memorable history of England, comprising all the parts you can remember, including 103 Good Things, 5 Bad Kings and 2 Genuine Dates. Methuen, London (1930)
2. Guth, A.H.: The inflationary universe: a possible solution to the horizon and flatness problem. Phys. Rev. **D23**, 347–356 (1981) (The 1980 SLAC preprint http://www.slac.stanford.edu/cgi-wrap/getdoc/slac-pub-2576.pdf.)
3. Earman, J., Mosterin, J.: A critical look at inflationary cosmolology. Philos. Sci. **66**, 1–49 (1999)

4. Aschenbach, J.: BICEP2's cosmological conundrum. The Washington Post, Washington (2014). (http://www.washingtonpost.com/blogs/achenblog/wp/2014/05/19/bicep2s-cosmological-conundrum/?wpisrc=nl%5Fpopns. 19 May 2014)

5. Einstein, A., Podolsky, B., Rosen, N.: Can quantum-mechanical description of physical reality be considered complete? Phys. Rev. **47**, 777–780 (1935) (http://journals.aps.org/pr/pdf/10.1103/PhysRev.47.777.)

6. Schrödinger, E.: Die gegenwärtige Situation in der Quantenmechanik. Naturwissenschaften **23**, 807–812 (1935). (Translated: The present situation in quantum mechanichs. Proc. Am. Phil. Soc. **124**, 323–338 (1980) http://hermes.ffn.ub.es/luisnavarro/nuevo_maletin/Schrodinger_1935_cat.pdf or http://www.tuhh.de/rzt/rzt/it/QM/cat.html.)

7. Bohr, N.: Can quantum-mechanical description of physical reality be considered complete? Phys. Rev. **48**, 696–702 (1935) (http://www-f1.ijs.si/~ramsak/teaching/eprbohr.pdf.)

8. Bell, J.S.: On the Einstein-Podolsky-Rosen Paradox. Physics **1**, 195–290 (1964) (http://philoscience.unibe.ch/documents/TexteHS10/bell1964epr.pdf, http://www.drchinese.com/David/Bell_Compact.pdf.)

9. Bell, J.S.: On the problem of hidden variables in quantum mechanics. Rev. Mod. Phys. **38**, 447–452 (1966)

10. Bell, J.S.: Speakable and unspeakable in quantum mechanics. Collected papers on quantum philosophy, 2nd edn. 2004. Cambridge University Press, Cambridge (1987) (Excerpts from this book, which contains the papers mentioned above and which give a good idea of Bell's way of thinking can be found at: http://www.phys.washington.edu/users/vladi/CommonBook/Bell.pdf.)

11. Aspect, A., Grangier, P., Roger, G.: Experimental tests of realistic local theories via Bell's theorem. Phys. Rev. Lett. **47**, 460–463 (1981) (http://www.ino.it/~azavatta/References/PRL47p460.pdf.)

12. Aspect, A., Grangier, P., Roger, G.: Experimental realization of Einstein-Podolsky-Rosen-Bohm Gedankenexperiment: a new violation of Bell's inequalities. Phys. Rev. Lett. **49**, 91–94 (1982)

13. Aspect, A.: Bell's inequality test: more ideal than ever. Nature **398**, 189–190 (1999) (http://www.ece.rice.edu/~kono/ELEC565/Aspect_Nature.pdf.)

14. Clauser, J.F., Shimony, A.: Bell's theorem. Experimental tests and implications. Rep. Progr. Phys. **41**, 1881–1927 (1978)

15. John, F.: Clauser early history of Bell's theorem. Coherence Quantum Opt. **VIII**, 19–43 (2003)

16. Shimony, A.: Bell's theorem. Stanford Encyclopedia of Philosophy (2009) (http://plato.stanford.edu/entries/bell-theorem/.)

17. Zeilinger, A.: Experiment and the foundations of quantum physics. Rev. Mod. Phys. **71**, 288–297 (1999) (http://qudev.ethz.ch/content/courses/phys4/studentspresentations/epr/zeilinger.pdf.)

18. Bouwmeester, D., Ekert, A.K., Zeilinger, A. (eds.): The Physics of Quantum Information: Quantum Cryptography, Quantum Teleportation, Quantum Computation. Springer, New York (2000)

19. Polchinski, J.: String Theory I, II. Cambridge University Press, Cambridge (2005)

20. Woit, P.: Not even wrong: the failure of string theory and the search for unity in physical law. Basic Books (2007) (The main title refers to the opinion that is attributed to Wolfgang Pauli after attending a certain seminar, that he disapproved of. The book is characterized by a strong sense of humor.)

21. Smolin, L.: The Trouble With Physics: The Rise of String Theory, The Fall of Science, and What Comes Next. Mariner Books, Boston (2007) (This book is particular good at describing the sociological consequence of string theory in the high-energy physics community.)

22. Wess, J., Zumino, B.: Supergauge transformations in four dimensions. Nucl. Phys. B **70**, 39–50 (1974)

23. Wess, J., Bagger, J.: Supersymmetry and Supergravity. Princeton University Press, Princeton (1992). Revised edition

24. Everett III, H.: Relative state formulation of quantum mechanics. Rev. Mod. Phys. **29**, 454–462 (1957) (http://jamesowenweatherall.com/SCPPRG/EverettHugh1957PhDThesis_ BarrettComments.pdf. This is a abridged summary of his 1956 Princeton PhD thesis The Theory of the Universal Wavefunction, http://www-tc.pbs.org/wgbh/nova/manyworlds/pdf/ dissertation.pdf.)
25. Baggott, J.: Farewell to Reality: How Modern Physics has Betrayed the Search For Truth. Pegasus (2013) (This book gives also a good review of the development of elementary particle physics as discussed in the preceding chapter.)

# Part II
# Supplementary Material

# Chapter 18
# Topology

## 18.1 Introduction

Topology is a generalization of geometry, in which 'distance' is generalized to a more qualitative idea of 'proximity'. It looks in general at geometrical objects from the point of view that such objects remain essentially the same under deformations which do not involve tearing. For example, an open cube in $\mathbb{R}^3$, i.e. a set

$$\{(x_1, x_2, x_3) \in \mathbb{R}^3 \mid -1 < x_j < +1, \ j = 1, 2, 3\},$$

is, as far as topology is concerned, the same object as an open ball,

$$\{(x_1, x_2, x_3) \in \mathbb{R}^3 \mid \sum_{j=1,2,3} |x_j|^2 < 1\}.$$

Removing the origin $(0, 0, 0)$ in each of these objects changes both into something that is topologically different.

Topology as a subject can be divided into:

a. *General* or *point set topology.* This generalizes notions used in standard calculus such as *limit* and *continuity*. It is an important tool in functional analysis, in particular in Hilbert space theory, the main mathematical ingredient of quantum theory.

b. *Algebraic topology.* This studies the classification of topological spaces, to be defined in the next subsection, in the spirit of the example just given, with special algebraic methods.

This chapter is devoted to point set topology, which is usually not included in the undergraduate physics curriculum. It should be seen, certainly by readers from physics, as an introduction to the other supplementary chapters.

© Springer International Publishing Switzerland 2015
P. Bongaarts, *Quantum Theory*, DOI 10.1007/978-3-319-09561-5_18

Good books for point set topology are, for example, Refs. [1–5]. Very useful is also the first volume of the series of books by Reed and Simon. See Ref. [1] of Supp. Chap. 21.

## 18.2 Basic Definitions

The basic notion in general topology is that of a *topological space*, in which an idea of 'nearness' is defined by a certain system of subsets of a given set.

**(18.2,a) Definition** A topological space is a nonempty set $X$, in which one has specified a *topology*, i.e. a system $T$ of subsets of $X$, with the following properties:

1. The set $X$ itself and the empty set $\emptyset$ belong to $T$.
2. The union of an arbitrary system of sets in $T$ is in $T$.
3. The intersection of a finite system of sets in $T$ is in $T$.

The sets in $T$ are called the *open* sets in $X$. The above definition says that a topological space is a pair $(X, T)$ consisting of a nonempty set $X$ and a topology $T$ in $X$. One usually denotes such a topological space just by $X$, unless more than one topology is used in $X$.

**(18.2,b) Definition** A subset of $X$ is *closed* iff its set-theoretical complement $A^c = \{x \in X \mid x \notin A\}$ is an open set.

**(18.2,c) Problem** A topological space can be equivalently defined with the closed sets as primary objects and open set as derived notion. Give the three axioms for such a definition.

Note that $X$ and the empty set $\emptyset$ are both open and closed. A topological space is called *connected* iff $X$ and $\emptyset$ are the only sets that are both open and closed.

**(18.2,d) Definition** A subset $A$ of a topological space $X$ is a *neighbourhood* of a point $x$ of $X$ iff there is an open set $U$ with $x \in U \subset A$.

The notion of neigbourhood of a point is an additional way of expressing 'nearness' in the sense of topological spaces. It also gives a useful criterium for proving that a set is open.

**(18.2,e) Problem** Prove that a subset $A$ of $X$ is open iff each point in $A$ has a neighbourhood which is contained in $A$.

A topological space $X$ is a *Hausdorff space* iff for every pair of distinct points $x$ and $y$ in $X$ there are neighbourhoods $U$ of $x$ and $V$ of $y$ that are disjunct. This property is an example of a *separation axiom*. A Hausdorff space is also called a $T_2$-space. There exists a sequence of ever stronger separation axioms. The Hausdorff property is so basic that we assume all topological spaces in this book to be Hausdorff.

A topology can usefully be generated by a smaller system of open sets:

**(18.2,f) Definition** A system $B$ of open subsets of $X$ is called a *basis* for the topology $T$ iff each open set can be written as a union of sets from $B$.

The set $X$ itself is open, wich implies that $X = \cup_{U \in \mathcal{B}} U$, i.e. the system $\mathcal{B}$ covers $X$. There is a second equivalent definition of basis.

(18.2,g) **Problem** Show that (18.2,f) is equivalent to: a system $\mathcal{B}$ of open subsets of $X$ containing the empty set is a basis for the topology $\mathcal{T}$ iff for every nonempty open set $U$ and every point $x$ in $U$ there is a set $U_1$ from $\mathcal{B}$ such that $x \in U_1 \subset U$.

Not every system of subsets of a given set $X$ can be a basis for a topology. It can be shown that a system $\mathcal{B}$ of subsets of $X$ is a basis for a topology in $X$ iff every finite intersection of sets in $\mathcal{B}$ can be written as a union of sets in $\mathcal{B}$.

There are two trivial topological spaces, an arbitrary set with as open sets only the set itself and the empty set, and, slightly less trivial, an arbitrary set in which all subsets are defined to be open. The topology of the last example is called the *discrete topology*.

(18.2,h) *Example* The obvious examples of topological spaces are the Euclidean spaces $\mathbb{R}^n$, for $n = 1, 2, \ldots$. As a basis for the standard topology in $\mathbb{R}^n$ we can use the system of open balls $\{B_{x_0,r}\}$, for every real positive $r$ and every point $x_0$ in $X$,

$$B_{x_0,r} = \{x \in \mathbb{R}^n \mid ||x - x_0|| < r\},$$

with the Euclidean length in $\mathbb{R}^n$, given by $||x|| = (\sum_{k=1}^{n} ||x_j||^2)^{1/2}$.

This system remains a basis for the same topology if we restrict $r$ to the positive rational numbers, or even to the sequence $r = 1/k$, $k = 1, 2, \ldots$. More sophisticated examples of topological spaces exist, such as spaces of functions. They play a central role in the mathematical formulation of quantum theory. See Examples (18.4,b) and (18.4,c) in Supp. Sect. 18.4 and the more detailed discussions in Supp. Chap. 21.

There is a useful way to combine the ideas of basis and neighbourhood. For this we need:

(18.2,i) **Definition** A *basis of neighbourhoods of a point* $x$ in $X$ is a collection of neighbourhoods of $x$ with the property that every neighbourhood of $x$ contains a neighbourhood from this system.

A topological space is called *first countable* iff every point has a countable basis of neighbourhoods. If its topology has a countable base it is called *second countable*. It is clear that the second notion implies the first. The set of real numbers $\mathbb{R}^n$ as a topological space described in Example (18.2,h) is first countable, as is obvious from the basis $\mathcal{B}$ in Example (18.2,h), and also second countable as will be shown further on in this section. In fact all the topological spaces in this book are second countable.

(18.2,j) **Problem** Choose for each point $x$ of a topological space $X$ a basis $\mathcal{B}_x$ of *open* neighbourhoods of $x$. Prove that the total system $\mathcal{B} = \cup_{x \in X} \mathcal{B}_x$ is a basis for the topology of $X$.

Such a basis $\mathcal{B}$ is called a *basis of open neighbourhoods of a topology*. The basis for the topology of $\mathbb{R}^n$ given in Example (18.2,h) is clearly of this type.

Each subset $A$ of a topological space has an *interior*, the largest open set that is contained in $A$, or, equivalently, the union of all open sets contained in $A$. This interior may, of course be empty. The set $A$ also has a *closure*, the smallest closed set that contains $A$, or the intersection of all closed sets that contain $A$.

There is also a notion of 'nearness' between sets and single points.

**(18.2,k) Definition** A point $x$ in a topological space $X$ is a *point of accumulation* of a subset $A$ of $X$ iff every neighbourhood of $x$ contains at least one point of $A$ different from $x$.

One can show that the closure of a set can be obtained by adding all its accumulation points, in so far as they are not already in it; in fact a set is closed iff it contains all its accumulation points. In a topological space a set $A$ is said to be *dense* in a set $B$ iff the closure of $A$ contains $B$. A topological space $X$ that contains a countable subset which is dense in $X$ is called *separable*. A first countable topological space is separable if and only if it is second countable. The rational numbers form a countable dense set in $\mathbb{R}^1$; in general the set of $n$-tuples of rational numbers form a dense set in $\mathbb{R}^n$, so $\mathbb{R}^n$ as topological space (according to Example (18.2,h)) is separable and therefore also second countable.

There is an important notion of 'tightness' of sets:

**(18.2,l) Definition** A subset of a topological space is called *compact* iff every covering by open sets has a finite subcover. This means that every system $\{U_\alpha\}_\alpha$ of open sets which covers a compact set $A$, i.e. has the property $A \subset \cup_\alpha U_\alpha$, contains a finite subsystem $\{U_{\alpha_j}\}_{j=1}^n$ which still covers $A$.

The space $X$ itself may be compact; one then calls $X$ a compact topological space. A compact set is closed, due to the Hausdorff property. In $\mathbb{R}^n$ the notion of compact is equivalent to closed and bounded. In any case, it is clear that in general a set consisting of a finite number of points is compact. In fact, in applications, speaking intuitively, compactness is often a first step away from finiteness into infiniteness.

A topological space is called *locally compact* iff every point has a compact neighbourhood. There is a useful standard procedure to compactify a locally compact space, the so-called *one-point compactification*. This goes as follows:

Let $X$ be a Hausdorff locally compact space. Form a new space $\widehat{X}$ by adding a single point to $X$, called 'infinity'. So $\widehat{X} = X \cup \{\infty\}$. A subset $O$ of $\widehat{X}$ will be defined as open if it is either a subset which comes from an open set in $X$ or if it is the union of $\{\infty\}$ with an open set from $X$ with compact complement in $X$.

**(18.2,m) Problem** Verify that this defines a compact topology in $\widehat{X}$.

One says that a continuous function $f$ on $X$ has the limit $\lambda$ 'at infinity' iff there is, for every positive $\epsilon$, a compact set $K \subset X$ such that $|f(x) - \lambda| < \epsilon$ for $x$ not in $K$. Such a function can clearly be extended to a continuous function on $\widehat{X}$ with the value $f(\infty) = \lambda$.

New topological spaces can be constructed from given ones:

1. *Product topology*: Let $X$ and $Y$ be topological spaces., with $X \times Y$ their Cartesian product i.e. the set of all ordered pairs $(x, y)$, for $x$ in $X$ and $y$ in $Y$. The sets of

the form $U \times V$, $U \subset X$ and $V \subset Y$, form a basis for a topology in $X \times Y$, the product topology. Product topological spaces can be constructed in this manner from a finite collection of topological spaces.

2. *Relative topology*: Let $X$ be a topological space and $A$ an arbitrary (nonempty) subset of $X$. $A$ on itself can be made into a topological space, by calling a subset $U$ of $A$ open iff it is the intersection of $A$ with an open set in $X$. This topology is called the relative topology in $A$ induced by $X$.

## 18.3 Limits and Continuity

Limits and continuity occupy a central place in standard calculus. Here they are set in the more general context of topology. For first countable spaces all topological properties can be expressed in terms of sequences.

(18.3,a) **Definition** A sequence of points $\{x_n\}_n$ in a topological space is said to *converge* to a point $x$ iff for every neighbourhood $U$ of $x$ there is an integer $N$ such that $x_n$ is in $U$ for all $n > N$. One writes $\lim_{n \to \infty} x_n = x$; the point $x$ is called the *limit* of the sequence $\{x_n\}_n$. The Hausdorff property ensures that a sequence can have at most one limit.

Continuity of a map $f: X \to Y$ means that 'small' deviations from a point $x$ in $X$ result in 'small' deviations for the image $f(x)$.

(18.3,b) **Definition** Let $X$ and $Y$ be topological spaces. A map (function) $f: X \to Y$ is *continuous at a point $x$* from $X$ iff for every neighbourhood $V$ of the image point $f(x)$ of $x$ in $Y$ there exists a neigbourhood $U$ of $x$ in $X$ such that $f(U) \subset V$. The map $f$ is *continuous on $X$* if and only if it is continuous at every point $x$ of $X$.

There is a simpler but less intuitive equivalent definition for continuity of a map.

(18.3,c) **Problem** Prove that the map $f: X \to Y$ is continuous iff the inverse image $f^{-1}(A)$ of every open set $A$ in $Y$, i.e. the set $\{x \in X \mid f(x) \in A\}$ in $X$, is open in $X$.

Continuous maps can be composed:

(18.3,d) **Problem** Let $X, Y$ and $Z$ be topological spaces, and $f: X \to Y$ and $g: Y \to Z$ be continuous maps. Show that the composed map $g \circ f: X \to Z$ is continuous. Hint: Use the definition of continuity from the preceding problem.

If a continuous map between topological spaces is invertible, then its inverse need not be continuous. If a continuous map has a continuous inverse, it is called a *homeomorphism*. Two spaces that are homeomorphic are topologically the same.

(18.3,e) **Problem** Show that a complex-valued continuous function $f$ on a compact topological space $X$ is bounded, i.e. that there is a (nonnegative) constant $C$ such that $|f(x)| \leq C$, for all $x$ in $X$.

In the context of locally compact spaces one often considers continuous functions $f$ that vanish at infinity, i.e. such that there is for every positive $\epsilon$ a compact subset such that $|f(x)| < \epsilon$, for all $x$ outside this subset. See the remark on functions with a limit at $\infty$ in the preceding section.

## 18.4 Metric Spaces

In this book most but not all of the examples of topological spaces are of a special type; they have a 'distance', even though they do not need to be Euclidean in a strict geometric sense.

(18.4,a) **Definition** Let $X$ be a nonempty set. A *metric* or *distance function* on $X$ is an assignment $d\colon X \times X \to \mathbb{R}^1$ with the following properties:

1. $d(x, y) = d(y, z)$ (symmetry),
2. $d(x, y) \geq 0$, with $d(x, y) = 0 \Leftrightarrow x = y$,
3. $d(x, z) \leq d(x, y) + d(y, z)$, (triangle inequality),

for all $x$, $y$ and $z$ in $X$.

Such a metric can be used to define a basis $\mathcal{B}$ of open neighbourhoods for a topology in $X$. It consists of systems $\mathcal{B}_{x_0} = \{B_{x_0,\varepsilon}\}_k$, for each $x_0$ in $X$, containing the sets $B_{x_0,\varepsilon} := \{x \in X \mid d(x, x_0) < \varepsilon\}$, for all $\varepsilon > 0$. The resulting topology makes $X$ into a *metric (topological) space*, which is obviously a generalization of $\mathbb{R}^n$ as topological space given in Example (18.2,h). It is Hausdorff, and is clearly first countable, but not necessarily separable like $\mathbb{R}^n$. More sophisticated examples of metric topological spaces.

(18.4,b) *Example* Consider the space $C_b(\mathbb{R}^1)$ of bounded complex-valued continuous functions on the real line. Define, for each pair of functions in $C_b(\mathbb{R}^1)$, a distance

$$d(f, g) = \sup_{x \in \mathbb{R}^1} |f(x) - g(x)|.$$

This is indeed a metric; it defines a topology in the space $C_b(\mathbb{R}^1)$.

(18.4,c) *Example* Consider the space $C_c(\mathbb{R}^1)$ of all complex-valued continuous functions with compact support, i.e. which vanish outside a compact subset of $\mathbb{R}^1$. Define for this space the metric as

$$d(f, g) = \int_{-\infty}^{+\infty} |f(x) - g(x)| \, dx.$$

With this the space $C_c(\mathbb{R}^1)$ becomes a metric topological space.

For metric spaces the definition of convergence of a sequence becomes more or less that of standard calculus.

(18.4,d) **Problem** Show that a sequence $\{x_n\}_n$ in a metric space converges to a point $x$ iff $\lim_{n \to \infty} d(x, x_n) = 0$.

(18.4,e) **Definition** A sequence $\{x_n\}_n$ in a metric space is called a *Cauchy sequence* iff $\lim_{n,m \to \infty} d(x_m, x_n) = 0$, i.e. if there exists for every $\varepsilon > 0$ an integer $N$ such that for $m, n > N$ one has $d(x_m, x_n) < \varepsilon$.

A convergent sequence is a Cauchy sequence, but not every Cauchy sequence has a limit.

(18.4,f) **Definition** A metric space is called *complete* if and only if every Cauchy sequence converges.

Every metric space can be completed, i.e. for a given metric space $X$ one can construct a unique complete metric space, the completion $\widehat{X}$, such that $X$ is homeomorphic to, and can therefore be identified with a dense subset of $\widehat{X}$. The prime example of this construction is the way the set of real numbers are generated from the rational numbers.

Here is a brief sketch of the completion procedure: Let $X$ be a topological space with metric $d$. Let $X'$ be the set of all Cauchy sequences in $X$. Define two such sequences $\{x_n\}_n$ and $\{y_n\}_n$ to be equivalent if $d(x_n, y_n)$ converges to 0. Inject $X$ into $X'$ by assigning to a point $x_0$ in $X$ the sequence $\{x_n\}_n$ with $x_n = x_0$, for $n = 1, 2, \ldots$, and extend finally $d$ from $X$, as subset of $X'$, to the full $X'$.

For metric spaces there is a useful stronger form of continuity:

(18.4,g) **Definition** A map $f$ from a metric space $X$ to a second metric space $Y$ is called *uniformly continuous* iff for every pair of points $(x, y$ in $X$ and every $\varepsilon > 0$ there exists a number $\delta > 0$, such that $d(x, y) < \delta$ implies $d(f(x), f(y)) < \varepsilon$.

It can be shown that a uniformly continuous map $f$ from a metric space $X$ to a metric space $Y$ can be uniquely extended to a uniformly continuous map $\widehat{f}$ from the completion $\widehat{X}$ to the completion $\widehat{Y}$.

In the function space $C_b(\mathbb{R}^1)$ of Example (18.4,b) convergence of a sequence $\{f_n\}_n$ to a function $f$ is just uniform convergence in the variable $x$. The set of real numbers $\mathbb{R}^1$ is a complete metric space; a Cauchy sequence $\{f_n\}_n$ therefore converges pointwise. From elementary analysis we know that a uniform limit of a sequence of continuous functions is continuous. Since a uniform limit of bounded functions it is bounded, so the sequence $\{f_n\}_n$ converges to a function $f$ in the sense of convergence in the metric $d$ in $C_b(\mathbb{R}^1)$. So $C_b(\mathbb{R}^1)$ is complete. The function space $C_c(\mathbb{R}^1)$ from Example (18.4,c) is *not* complete; completion gives the space of all integrable functions on $\mathbb{R}^1$, usually denoted as $L^1(\mathbb{R}^1, dx)$. (Strictly speaking the elements of $L^1(\mathbb{R}^1, dx)$ are equivalence classes of functions. See Supp. Chap. 21, Example (21.4,d).) There exists a sequence of such function spaces $L^p(\mathbb{R}^1, dx)$, for all $p = 1, 2, \ldots$, all complete as metric topological spaces, with metric

$$d_p(f, g) = \left( \int_{-\infty}^{+\infty} |f(x) - g(x)|^p dx \right)^{1/p}.$$

In fact $L^p(\mathbb{R}^1 dx)$ is a complete metric space for all real $p$ with $p \geq 1$. The metric of these spaces is called a *norm*. The space $L^2(\mathbb{R}^1, dx)$ is an example of a *Hilbert space*.

(18.4,h) *Remark* In a *metrizable topological space* the topology can be given by a metric, even though it is not explicitly given or used. Knowing that a topological space is metrizable is sometimes very useful.

## 18.5  Topological Spaces with Additional Structure

Topological spaces may have additional structure, or one should say that certain other types of spaces like manifolds (Supp. Chap. 20), Hilbert spaces (Supp. Chap. 21), or Lie groups (Supp. Chap. 24), have an underlying topological structure. In these cases there are additional compatibility requirements between the two structures. Of particular importance for the mathematical formulation of quantum theory are infinite dimensional vector spaces with an underlying topology, which are, of course, more sophisticated than an obvious finite dimensional example like $\mathbb{R}^n$. Such infinite dimensional spaces, in particular Hilbert space, the principle mathematical vehicle for quantum theory will be extensively discussed in Supp. Chap. 21.

## 18.6  Limits and Convergence

It can be shown that for first countable topological spaces continuity at a point can be expressed by using sequences: the map $f: X \to Y$ is continuous at $x$ in $X$ iff $\lim_{n\to\infty} x_n = x$ implies $\lim_{n\to\infty} f(x_n) = f(x)$. This definition is in fact the definition used in elementary calculus where $X$ is usually (a subset of) $\mathbb{R}^n$, which is certainly first countable.

For topological spaces that are not first countable there is a similar property, if a more general definition of convergence is used—we shall discuss it here for the sake of completeness. For this, and also for the definition of *lattice* in Sect. 12.5.3, we need the notion of a *poset*—short for *partially ordered set*.

(18.6,a) **Definition** A non-empty set $X$ is called a *poset* iff there is a relation $\prec$ between some pairs of elements satisfying the following requirements:

1. For every $x$ in $X$ one has $x \prec x$ (*reflexivity*).
2. If $x \prec y$ and $y \prec x$ then $x = y$ (*antisymmetry*).
3. If $x \prec y$ and $y \prec z$ then $x \prec z$ (*transitivity*).

Note that one may also use, in an equivalent manner, the opposite relation $\succ$, with $x \prec y \iff y \succ x$. If the relation $\prec$ or $\succ$ holds for all pairs then $X$ is said to be *totally ordered*.

(18.6,b) **Definition** An element $x$ in a poset $X$ is called a *lower bound* of a set $A \subset X$ iff $x \prec y$, for all $y$ in $A$; it is an *upper bound* of $A$ iff $y \prec x$, for all $Y$ in $A$. A subset $A$ may have a *largest lower bound* (*infinum*) (*infinum*) and a *smallest upperbound* (*supremum*). If they exist they are unique.

These definitions enable us to introduce a *net* as a generalization of the notion of sequence:

(18.6,c) **Definition** A nonempty set is called *directed* iff it is partially ordered such that each pair of elements has an upper bound. Given a nonempty set $X$ and a directed index set $\mathcal{I}$, a *net* in $X$ is then a set $\{x_\alpha\}_{\alpha \in \mathcal{I}}$ in $X$.

The set of natural numbers $N = \{1, 2, \ldots\}$ is an ordered and therefore a fortiori a partially ordered set. A net $\{x_n\}_{n \in N}$ is therefore just a sequence.

(18.6,d) **Definition** A net $\{x_\alpha\}_{\alpha \in \mathcal{I}}$ in a topological space $X$ converges to a point $x$ in $X$ iff for each neighbourhood $U$ of $x$ there is an index $\alpha_0$ such that $x_\alpha \in U$ for all $\alpha$ with $\alpha_0 \prec \alpha$.

This type of convergence is called *net convergence* or *Moore-Smith convergence*. In the case of a net $\{x_n\}_{n \in Z}$ this reduces, of course, to *sequential convergence*.

All this makes it finally possible to formulate a generalization of the statement at the beginning of this section:

(18.6,e) **Theorem** *Let* $X$ *and* $Y$ *be topological spaces and* $f$ *a map from* $X$ *to* $Y$. *The map* $f$ *is continuous at a point* $x$ *in* $X$ *iff* $\lim_{\alpha \to \infty} x_\alpha = x$ *implies* $\lim_{\alpha \to \infty} f(x_\alpha) = f(x)$.

Partially ordered sets and directed sets are not only used in topology but also in many other topics in mathematics. In this book we use partially ordered sets in Sect. 12.5.3, and directed sets in Supp. Sect. 27.6.

In this chapter topological spaces are defined by specifying the open sets, which is the most common method. It is also possible to use for this the notion of convergence of nets, which has advantages in certain situations.

Suppose that we have a notion of convergence of nets in the set $X$ (This notion has to meet a few fairly obvious requirements, that we do not spell out here.). Then a subset $A$ of $X$ is defined to be closed iff all convergent nets in $A$ converge to points in $A$. Open sets in $X$ are the complements of these closed sets.

A very useful book for this chapter is the first volume of the series of books by Reed and Simon (Ref. [1] of Supp. Chap. 21).

# References

1. Jänich, K.: Topology. Springer, Heidelberg (1984) (A good introductory textbook on point set and algebraic topology.)
2. Kelley, J.L.: General Topology. Van Nostrand, New York (1955) (Re-issued by Springer in 1975. A classic. It remains an excellent textbook on point set topology.)
3. Dugundji, J.: Topology. Allyn and Bacon, Boston (1966) (Similar in spirit as Kelley, but more advanced. Out of print.)
4. Pederson, G.K.: Analysis Now. Springer, Heidelberg (1989) (A book on functional analysis with an introductory chapter on general topology that covers most of the material discussed in this chapter.)
5. Buskes, G., van Rooij, A.: Topological Spaces: From Distance to Neighborhood. Springer, Heidelberg (1997) (A recent introductory textbook).

# Chapter 19
# Measure and Integral

## 19.1 Introduction

In standard physics courses integration usually means Riemann integration, while students in mathematics are taught from the beginning the more sophisticated Lebesgue integral. Functional analysis, which for this book means in the first place Hilbert space theory, requires Lebesgue integration, which in turn needs measure theory. This chapter will therefore provide the basic ingredients of the theory of measure and integration as an introduction to Supp. Chap. 21, serving a similar purpose as Supp. Chap. 18.

## 19.2 Measurable Space

Let $X$ be a nonempty set.

**(19.2,a) Definition** A nonempty system $\mathcal{F}$ of subsets of $X$ is called a *σ-algebra* iff it has the following properties:

1. $A$ in $\mathcal{F}$ implies that the complement $A^c = \{x \in X \mid x \notin A\}$ is also in $\mathcal{F}$,
2. a countable union of sets in $\mathcal{F}$ is again in $\mathcal{F}$.
3. $X$ is in $\mathcal{F}$.

**(19.2,b) Problem** Use the assumption that $\mathcal{F}$ is a nonempty system of subsets of $X$ to prove that condition 3 in (19.2,a) follows from 1 and 2.

**(19.2,c) Problem** Show that a σ-algebra $\mathcal{F}$ of subsets of $X$ contains the empty set $\emptyset$ and that the intersection of a countable system of sets in $\mathcal{F}$ is again in $\mathcal{F}$.

A σ-algebra is sometimes called a *σ-field*—that is why it is denoted by $\mathcal{F}$, and the members of $\mathcal{F}$ *measurable sets*.

© Springer International Publishing Switzerland 2015
P. Bongaarts, *Quantum Theory*, DOI 10.1007/978-3-319-09561-5_19

There are two trivial examples of $\sigma$-algebras in any nonempty $X$: the system of subsets consisting of $\emptyset$ and $X$ and the system of all subsets of $X$.

**(19.2,d) Problem** Show that the intersection of an arbitrary collection of $\sigma$-algebras is again a $\sigma$-algebra.

Let $\mathcal{A}$ be an arbitrary nonempty system of subsets of $X$. The result in the above problem allows us to conclude that the intersections of all $\sigma$-algebras $\mathcal{F}$ containing $\mathcal{A}$ is again a $\sigma$-algebra. It is obviously the smallest $\sigma$-algebra containing $\mathcal{A}$ and is called the $\sigma$-algebra generated by $\mathcal{A}$.

**(19.2,e) Definition** A nonempty set $X$ together with a $\sigma$-algebra $\mathcal{F}$ is called a *measurable space*. It may be denoted as a pair $(X, \mathcal{F})$; if in a discussion $\mathcal{F}$ remains fixed, just denoting it as $X$ is sufficient.

**(19.2,f) Example** Let $X$ be a topological space (See Supp. Chap. 18), with a topology $\mathcal{T}$, the system of subsets that are by definition 'open'. According to Problem (19.2,c) we have the $\sigma$-algebra $\mathcal{F}$ generated by $\mathcal{T}$, the intersection of all $\sigma$-algebras containing $\mathcal{T}$. It is called the *Borel $\sigma$-algebra* of the topological space $X$, with the sets of $\mathcal{F}$ called *Borel sets*, or sometimes *Borel measurable sets*.

**(19.2,g) Definition** Let $(X_1, \mathcal{F}_1)$ and $(X_2, \mathcal{F}_2)$ be two measurable spaces. A map $f : X_1 \to X_2$ will be called *measurable* iff the inverse image $f^{-1}(A)$ of each measurable set $A$ in $X_2$ is measurable in $X_1$.

Compare this definition with the definition of continuity of a map between two topological spaces in Supp. Chap. 18.

## 19.3  Measure

To avoid complicated formulations in terms of limits to infinity it is convenient in the context of measure and integration theory to extend the set of real number $\mathbb{R}^1$ with the two additional symbols $-\infty$ and $+\infty$. One assumes the following relations, for all finite $x$:

$$\pm\infty = x + (\pm\infty) = (\pm\infty) + (\pm\infty),$$
$$x\,(\pm\infty) = \pm\infty, \text{ for } x > 0; = \mp\infty, \text{ for } x < 0,$$
$$(\pm\infty)(\pm\infty) = \infty, (\pm\infty)(\mp\infty) = -\infty,$$
$$x/(\pm\infty) = 0, -\infty < x < +\infty.$$

The set $\{-\infty\} \cup \{\mathbb{R}^1\} \cup \{+\infty\}$ is called the set of *extended real numbers*; the nonnegative number together with $+\infty$ are the *extended nonnegative real numbers*, etc.

(19.3,a) **Definition** Let $(X, \mathcal{F})$ be a measurable space. A *measure* on $(X, \mathcal{F})$ is a map $\mu$ from the sets in $\mathcal{F}$ to the extended nonnegative real numbers with the properties:

1. $\mu(\emptyset) = 0$,
2. for a countable system $\{A_n\}_{n=1}^{\infty}$ of disjunct sets in $\mathcal{F}$,

$$\sum_{n=1}^{\infty} \mu(A_n) = \lim_{n \to +\infty} \sum_{j=1}^{n} \mu(A_j) = \mu(\cup_{n=1}^{\infty} A_n).$$

This property of $\mu$ is called $\sigma$-additivity. The triple $(X, \mathcal{F}, \mu)$ is called a *measure space*.

To exclude pathological cases, one assumes that $\mu(A)$ is finite, for at least one set $A \in \mathcal{F}$. If $\mu(X)$ is finite $\mu$ is called a *finite measure*. In that case $\mu$ can be rescaled such that the total measure of $X$ becomes 1, which would make it a *probability measure*. Probability measure spaces are usually denoted as $(\Omega, \mathcal{F}, P)$. They will be discussed in Supp. Chap. 22.

(19.3,b) **Problem** Show that a measure is *monotonic*, i.e. that for $A$ and $B$ in $\mathcal{F}$ with $A \subset B$ one has $\mu(A) \leq \mu(B)$.

(19.3,c) *Example* Let $X$ be a countable set $\{x_n\}_n$, $\mathcal{F}$ the system of all subsets of $X$ and $\mu$ defined as $\mu(x_n) = c_n$, for arbitrarily chosen nonnegative real numbers $c_n$. Then $(X, \mathcal{F}, \mu)$ is a *discrete measure space*. For $\sum_n c_n = 1$ it is a *discrete probability space*.

(19.3,d) *Example* Let $\mathcal{F}$ be the system of Borel measurable sets in $\mathbb{R}^n$. The Euclidean volume in $\mathbb{R}^n$ determines a measure on $\mathcal{F}$, which is called the *Lebesgue measure on* $\mathbb{R}^n$. The explicit construction of this measure on arbirary Borel sets is decsribed in [1, Chap. II].

Let $(X, \mathcal{F}, \mu)$ be a measure space. A set in $\mathcal{F}$ with $\mu(A) = 0$ is called a *null set*. For technical reasons it is desirable that every subset of a null set is a member of $\mathcal{F}$ and is consequently then also a null set. In general this is not the case, but this can be remedied by a standard procedure by which $\mathcal{F}$ is extended to a possibly larger $\sigma$-algebra $\mathcal{F}^*$, with a corresponding extension of $\mu$ on $\mathcal{F}$ to a measure $\mu^*$ on $\mathcal{F}^*$. Note that this extension of $\mathcal{F}$ depends on the measure $\mu$ on $\mathcal{F}$.

The Borel $\sigma$ algebra in $\mathbb{R}^n$ extends in this way to what is called the *Lebesgue-Borel* $\sigma$-algebra; the extended measure is usually again called Lebesgue measure. It can be shown that there are sets in $\mathbb{R}^n$ which are not Lebesgue measurable. However, such sets cannot be constructed in an explicit way. See [1, p.69].

(19.3,e) **Definition** The *support* of a Borel measure is the set consisting of the points for which all open neighbourhoods have positive measure.

(19.3,f) **Problem** Show that the support of a Borel measure is a closed set.

In this book the basic measurable space is almost always $(\mathbb{R}^n, \mathcal{F})$, with $\mathcal{F}$ the $\sigma$-algebra of Borel sets in $\mathbb{R}^n$. On this measurable space different measures are considered, in the first place, the Lebesgue measure. The resulting measure spaces are always assumed to be completed, which means that for the Lebesgue measure $\mathcal{F}^*$ is the $\sigma$-algebra of Lebesgue-Borel sets.

## 19.4 Measurable Functions

Let $(X, \mathcal{F}, \mu)$ be a measure space. A *real measurable function* is a measurable map, in the sense of (19.2,f), from $X$ to the extended real numbers. One can show that $f$ is a measurable function if and only if the set $\{x \in X \mid f(x) < \lambda\}$ is measurable, i.e. is a member of the $\sigma$-algebra $\mathcal{F}$ in $X$, for all extended real $\lambda$. See [1, p.79]. To avoid pathological cases one assumes that a measurable function is $-\infty$ or $+\infty$ only on a set of measure 0. A complex-valued function is measurable iff its real and imaginary parts are measurable. A measurable function on a probability measure space is called a *random variable*. See for this notion Supp. Chap. 22, Definition (22.2.2,b).

*Reminder*: Let $Y$ be a nonempty set. An *equivalence relation* $\sim$ is a relation between elements of $Y$, with the properties, for all $a, b, c$ in $Y$,

1. $a \sim a$, (identity),
2. $a \sim b \Rightarrow b \sim a$, (symmetry),
3. $a \sim b, b \sim c \Rightarrow a \sim c$, (associativity).

An equivalence relation divides a set into *equivalence classes*. The equivalence class $[a]$ is the class which contains the element $a$; $a$ is a *representative* of this class. One has $[a] = [b] \Leftrightarrow a \sim b$. The system of equivalence classes forms a new set: the *quotient space* of $Y$ with respect to $\sim$. It is usually denoted as $Y/\sim$. There is an obvious surjective map $s : Y \to Y/\sim$. The construction of a quotient space with respect to an equivalence relation is very widely used in mathematics.

There is an important equivalence relation between measurable functions on a given measure space.

(19.4,a) **Definition** Two measurable functions $f$ and $g$ on a measure space $(X, \mathcal{F}, \mu)$ are said to be *equal almost everywhere* iff they are equal except on a set of measure 0. For random variables on a probability space one uses the term *equal almost surely*.

In the context of Lebesgue integration theory measurable functions which are equivalent in this sense are considered as equal. Therefore in the remainder of this chapter 'function' will usually mean 'equivalence class of functions' in the sense of (19.4,a), and with respect to the measure space at hand.

A real measurable function $f$ on a measure space $(X, \mathcal{F}, \mu)$ induces a measure $\mu_f$ on $\mathbb{R}^1$, defined by extension of $\mu_f((a, b)) := \mu(f^{-1}((a, b)))$, for all $a$ and $b$

with $-\infty < a < b < +\infty$. If $\mu$ is finite, i.e. if $\mu(X) < \infty$, this induced measure can be completely characterized by the real-valued function $F_f$ given by

$$F_f(x) = \lim_{\varepsilon \downarrow 0} \mu_f((-\infty, x + \varepsilon)) = \mu_f((-\infty, x]).$$

(19.4,b) **Problem** Show that $F_f$ is a monotonically nondecreasing function, or short, nondecreasing function, meaning that $x_1 < x_2$ implies $F_f(x_1) \le F_f(x_2)$, that it is continuous from the right, i.e. with $F_f(x_0) = \lim_{x \downarrow x_0} F_f(x)$, and that $\lim_{x \to -\infty} F_f(x) = 0$ and $\lim_{\to +\infty} F_f(x) = \mu_f(R^1)$.

There is a one-to-one correspondence between the collection of all functions with the properties stated in Problem (19.4,b) and the set of all finite measures on the Borel sets in $\mathbb{R}^1$. A function $F$ with the above properties gives a measure $\mu_F$ on $\mathbb{R}^1$ by extension of $\mu_F((a, b]) = F(b) - F(a)$. If $\lim_{x \to +\infty} F(x) = 1$ the measure is a probability measure. This case will be discussed in more detail in Supp. Chap. 22, where the function $F$ is called a *distribution function*.

Note that the right continuity of $F$ is a matter of convention. The definition $F_f(x) = \lim_{\varepsilon \uparrow x} \mu_f((-\infty, x - \varepsilon) = \mu_f((-\infty, x))$ would have given continuity from the left.

For later use, it may be helpful to remind the reader of some well-known properties of monotonic functions.

1. A monotonic function has finite limits for $x \to \pm\infty$, when it is bounded,
2. it has in every point a limit from the right and from the left; if these limits are equal the function is continuous,
3. it has only a countable number of points of discontinuity, so called 'jumps',
4. it is differentiable almost everywhere, i.e. the points where the function is not differentiable form a set with Lebesgue measure 0,
5. it is Riemann integrable on a bounded interval. See the next subsection.

## 19.5 Integration

### 19.5.1 The Riemann Integral

In standard physics calculus courses the Riemann integral is the integral of choice. Riemann integration is typically applied to continuous real- or complex-valued functions on finite intervals in $R^1$.

The definition of such an integral $\int_a^b f(x)dx$ goes as follows:

1. Choose a finite *partition* of $[a, b]$, i.e. sets of points $\{t_j\}_j$ with the property

$$t_0 = a < t_1 < \cdots < t_n, t_{n+1} = b,$$

with numbers $s_j$, such that $t_j < s_j < t_{j+1}$, for $j = 0, \ldots, n$.

2. Define for every partition the *Riemann sum* $\sum_{j=0}^{n} f(s_j)(t_{j+1} - t_j)$.
3. Show that with an appropriate definition of limit these partial sums converge to a limit, for increasing $n$ and the length of the subintervals $[t_j, t_{j+1}]$ going to 0. This limit is the Riemann integral $\int_a^b f(x)dx$.

The Riemann integral defined in this manner gives an intuitively clear and satisfactory picture of what an integral should be. No measure theory is needed.

The integral above may be called a *proper* Riemann Integral. For functions with singularities or by integration over infinite intervals one defines *improper* Riemann integrals as limits of proper ones, e.g.

$$\int_a^{+\infty} \frac{1}{x^2}\, dx := \lim_{b \to +\infty} \int_a^b \frac{1}{x^2}\, dx.$$

### 19.5.2 The Lebesgue Integral

Courses for mathematics students usually offer the *Lebesgue integral*. This integral has several advantages, the most important being the fact that more general functions can be integrated, on a broader range of spaces than $\mathbb{R}^n$. It also has sharper and more general limit theorems. A function that is Riemann integrable is also Lebesgue integrable. For some of the material in this book the Riemann integral would be sufficient. However Lebesgue integration is a standard ingredient of Hilbert space theory, the main ingredient of the mathematical formalism of quantum theory.

The Riemann integral is defined by approximating a function by step functions on finite partitions of an interval, or more generally of a similar subset of $R^n$, and then taking a limit for increasingly finer partitions. For the Lebesgue integral one uses also step functions, but now defined by partitions of the target space, the extended real line.

Let $f$ be a real-valued measurable function on a measure space $(X, \mathcal{F}, \mu)$. The Lebesgue integral $\int_X f\,d\mu$, if it exists, is defined in the following four steps:

1. The function $f$ can be uniquely written as a sum of a nonnegative function $f_+$ and a negative function $f_-$, according to
   $f_+(x) = f(x)$ for $\{x \in X \mid f(x) \geq 0\}$ and $= 0$ for $\{x \in X \mid f(x) < 0\}$,
   $f_-(x) = f(x)$ for $\{x \in X \mid f(x) < 0\}$ and $= 0$ for $\{x \in X \mid f(x) \geq 0\}$.
2. Choose a finite sequence of real numbers
   $0 = \alpha_0 \leq \alpha_1 < \cdots < \alpha_n = +\infty,$
   Define subsets of $X$ as
   $X_j = f_+^{-1}([\alpha_j, \alpha_{j+1}))$, for $j = 0, \ldots, n-1$, $X_n = f_+^{-1}([\alpha_n, +\infty])$.
   These subsets may be wild, but they are measurable, disjoint, with $\cup_{j=0}^n X_j = X$, so they form a finite partition $S$ of $X$. This partition defines a *step function* or *simple function* $f_+^S(x) = \alpha_j$, for $x$ in $X_j$, $j = 0, \ldots, n$, and a partial sum

$\sum_{j=0}^{n} \alpha_j \mu(X_j)$. The supremum over all such partial sums is by definition the Lebesgue integral $\int_X f_+ \, d\mu$. It is also the limit of these partial sums, in an appropriate sense. Two outcomes are possible: this limit is (a) finite, or (b) infinite, i.e. $+\infty$.

3. One defines similarly $\int_X f_- \, d\mu$ with possible outcomes: (c) finite or (d) infinite, i.e. $-\infty$.

4. The results in 2 and 3 combined give the Lebesgue integral

$$\int_X f \, d\mu = \int_X f_+ \, d\mu + \int_X f_- \, d\mu,$$

with four possibilities:

   (a) + (c): the integral is finite,
   (a) + (d): it is $-\infty$,
   (b) + (c): it is $+\infty$,

in which three cases the function $f$ is said to be *integrable*, and finally

   (b) + (d): the integral is not defined, in which case $f$ is said to be *not integrable*.

For details and proofs, see [1, Chap.V].

### 19.5.3 The Stieltjes Integral

The Stieltjes integral $\int f \, dg$ depends on two functions $f$ and $g$. There are again two versions, the Riemann-Stieltjes and the Lebesgue-Stieltjes integral. The former can, in the case of $X = R^1$, be defined as limit over sums of the form $\sum_s f(x_s)(g(x_{s+1}-g(x_s)$ In Supp. Sect. 21.10.2 and subsequent subsections of Supp. Chap. 21, where we use the Stieltjes integral to formulate the spectral theorem for a single selfadjoint operator, the function $f$ is simply $f(x) = x$, with $g$ a bounded, nonnegative, monotonically nondecreasing function which is moreover continuous from the right. For this the Riemann-Stieltjes integral is sufficient. It exists as a finite real number. The same is true for the formulation of the spectral theorem for a system of commuting selfadjoint operators, with an obvious generalization of the Riemann-Stieltjes integral to $n$ variables. For the further sections, where arbitrary functions $f$ of systems of commuting selfadjoint operators are discussed, implying that $f$ is an arbitrary measurable function, we need the Lebesgue version.

In the Lebesgue-Stieltjes integral we need that the function $g$ to have the same properties as in the simple case above. This means that it defines a measure in $R^1$, not the standard Lebesgue measure, characterized by the prescription on finite intervals $(a, b]$ as $\mu((a, b]) = b - a$, but a different one determined by $\mu_g((a, b]) = g(b) - g(a)$. This new measure may have a different completion on a different extension of the Borel $\sigma$-algebra.

(19.5.3,a) **Problem** Show that in this case a function that is measurable with respect to the Borel $\sigma$-algebra remains measurable with respect to its completion with respect to $\mu_g$.

With this result the Lebesgue-Stieltjes integral $\int_{-\infty}^{+\infty} f\,dg$ becomes an ordinary Lebesgue integral $\int_{-\infty}^{+\infty} f\,d\mu_g$. The case of $X = \mathbb{R}^n$, i.e. with $n$ variables $x_1, \ldots, x_n$, is completely similar.

## 19.6  Absolute Continuity: The Radon-Nikodym Theorem

(19.6,a) **Definition** Let $\mu$ and $\nu$ be two measures on the same measurable space $(X, \mathcal{F})$. Then $\mu$ is said to be *absolutely continuous* with respect to $\nu$ iff $\nu(A) = 0$, for $A$ in $\mathcal{F}$, implies $\mu(A) = 0$, with notation $\mu \ll \nu$.

(19.6,b) *Example* Let $\nu$ be Lebesgue measure $dx$ on $\mathbb{R}^1$ and $\mu$ the measure which assigns to every $\{n\}$, $n = 1, 2, \ldots$, a positive number $c_n$. Then $\mu$ is obviously not absolutely continuous with respect to $\nu$, because $\nu(\{n\}) = 0$, but $\mu(\{n\}) = c_n \neq 0$.

(19.6,c) *Example* Let $\nu$ be again the Lebesgue measure on $\mathbb{R}^1$, and let $\mu$ be given by the integral $\mu(A) = \int_A f(x)\,dx$, for every Borel set $A$ and with $f$ a nonnegative integrable real function on $R^1$. Then $\mu$ is clearly absolutely continuous with respect to the Lebesgue measure.

(19.6,d) **Definition** A Borel measure $\mu$ on $\mathbb{R}^1$ is called *singular with respect to Lebesgue measure* iff for every Borel set $S$ in $\mathbb{R}^1$ with Lebesgue measure zero one has $\mu(S) = 0$.

(19.6,e) **Theorem** (Lebesgue decomposition theorem) *A Borel measure $\mu$ on $\mathbb{R}^1$ can be written in a unique way as a sum of a pure point part $\mu_{pp}$ and a continuous part $\mu_{ct}$; the second can be written uniquely as the sum of an absolutely continuous part $\mu_{act}$ and a singularly continuous part $\mu_{sct}$, all with respect to Lebesgue measure.*

One has an important general theorem:

(19.6,f) **Theorem** (Radon-Nikodym) *Let $\mu$ and $\nu$ be measures on the measurable space $(X, \mathcal{F})$ and let $\mu$ be absolutely continuous with respect to $\nu$. Then there is a nonnegative measurable function $f$ on $X$ such that*

$$\mu(A) = \int_A f(x)\,d\nu,$$

*for all $A$ in $\mathcal{F}$. The function $f$ is called the Radon-Nikodym derivative of $\mu$ with respect to $\nu$ and is sometimes denoted as $\frac{d\mu}{d\nu}$.*

If the measures $\mu$ and $\nu$ are probability measures $f$ is called a *probability density*. This will be discussed in Supp. Chap. 22. The proof of the Radon-Nikodym theorem can be found in [1, Sect. 31].

The use of the word 'absolute continuity' in this context has its origin in real analysis. A real-valued function on an interval $\mathcal{I}$ of $\mathbb{R}^1$ is called absolutely continuous on $\mathcal{I}$ iff for every positive $\varepsilon$ there exists a positive $\delta$ such that for every finite set of disjoint intervals $[x_j, y_j]$, $j = 1, \ldots, n$, in $\mathcal{I}$, the inequality $\sum_{x_j j=1}^n (y_j - x_j) < \delta$ implies $\sum_{j=1}^n |f(y_j) - f(x_j)| < \varepsilon$. For a finite measure $\mu$ on the Borel sets in $\mathbb{R}^1$ the nondecreasing function $F_\mu$ introduced in Supp. Sect. 19.4 is absolutely continuous iff $\mu$ is absolutely continuous with respect to the Lebesgue measure on $\mathbb{R}^1$.

## 19.7 Measure and Topology

A topological space is in a natural way a measurable space. Given a topological space, then there is a smallest $\sigma$-algebra of sets containing the open sets in this space. The sets in this $\sigma$-algebra are called the *Borel sets*. A function measurable with respect to the Borel sets is called *Borel measurable* or a *Borel function*, a measure on the Borel sets is a *Borel measure*. Measure spaces of this kind are called *standard Borel measure spaces*. Almost all measure spaces in this book are in fact such standard measure spaces.

## 19.8 Algebraic Integration Theory

An important mathematical idea underlying our approach to quantum theory is that a 'space', i.e. a set with additional structure, can be characterized by an appropriate (commutative) algebra of functions on this set. This leads to a generalization in which a corresponding noncommutative algebra. suggests a 'space', which does not exist as a point set, but which leads nevertheless, as an intuitive notion, to interesting new structures, with properties analogous to the commutative case. This is the basic idea of Alain Connes' noncommutative geometry. This is discussed in Chap. 12, where this idea is fully developed.

In this picture an integral of a function, with respect to a measure, is the result of applying a linear functional on an algebra of measurable functions. For a noncommutative integral one uses a linear functional on a noncommutative algebra.

As already noted, a measure $\mu$ on $X$ with $\mu(X) = 1$ is called a probability measure. In fact the theory of measure and integration forms the basis of probability theory. In Chap. 12 (Physical Theories as Algebraic Systems) we argue that quantum theory can be seen as a form of *noncommutative probability theory*.

Between the case of commutative algebras (classical theory) and that of fully noncomutative algebras (quantum theory) there is the case of what we call *almost commutative algebras*, i.e. algebras that are commutative up to minus signs. They are defined and their use is explained in Sect. 13.9.3.2. There is a particular type of algebraic integral, the *Berezin integral*. All this is part of the theory of superalgebras and supermanifolds, discussed in Sects. 13.9.3.2 and 13.9.3.3.

A very useful book also for this chapter is the first volume of the series of books by Reed and Simon. See Supp. Chap. 21, [1].

- Most books on probability theory have extensive introductions to the theory of measure and integration. See the references in Supp. Chap. 22.

## Reference

1. Paul, R.: Halmos Measure Theory. Van Nostrand, New York (1950) (Re-issued by Springer in 1974) [Specialists in measure theory may find this clearly written book slightly outdated. Nevertheless, it remains, after more than fifty years, unsurpassed as a general textbook on measure and integration.]

# Chapter 20
# Manifolds

## 20.1 Definition of a Manifold

### 20.1.1 Introduction

The simplest nontrivial example of a differentiable manifold is a smooth two dimensional surface in three dimensional Euclidean space. It can be obtained as the subset of points $(x, y, z)$ of $\mathbb{R}^3$, such that $F(x, y, z) = 0$, for a suitable differentiable function $F$. The function $F(x, y, z) = x^2 + y^2 + z^2 = 1$ gives in this manner the 2-sphere centered in the origin. Solutions of systems of $m - n$ equations $F_j(x^1, \ldots, x^m) = 0$, $j = 1, \ldots, m - n$, give differentiable manifolds of dimension $n < m$ as subsets of $\mathbb{R}^m$. In the development of differential geometry it was at a certain point realized that it is possible and convenient to define a manifold not as a subset of a higher dimensional Euclidian space, but in a more intrinsic manner as an object on its own, an idea first described by Riemann in an explicit manner. See [1]. In this picture an $n$-dimensional differentiable manifold can be intuitively visualized as a space which is smoothly patched together from open pieces of $R^n$.

The definition of manifold will be given in two steps. In Supp. Sect. 20.1.2 a topological manifold is defined; providing such a manifold with a differential structure leads in Supp. Sect. 20.1.3 to a differentiable manifold. Tangent vectors at each point of a manifold give globally defined vector fields, sections of the tangent bundle. This is discussed together with more general vector bundles and their modules of sections in Supp. Sect. 20.2.

The second most important vector bundle on a manifold, the cotangent bundle is treated in Supp. Sect. 20.3, and general tensor fields in Supp. Sect. 20.4. Differential forms, a very important notion in differential geometry, with the exterior derivative as the main tool, appear in Supps. Sects. 20.5 and 20.6 briefly discusses the notion of symmetry for a manifold. In Supp. Sect. 20.7 symplectic manifolds are defined; the central notion in this chapter, because it provides the mathematical basis of classical mechanics. Finally, in Supp. Sect. 20.8, we present an algebraic reformulation of differential geometry, which is in particular relevant for Chap. 12. Note that in all

© Springer International Publishing Switzerland 2015
P. Bongaarts, *Quantum Theory*, DOI 10.1007/978-3-319-09561-5_20

that follows the terms *differentiable, infinitely differentiable, $C^\infty$-, and smooth* will be synonyms.

### 20.1.2 Topological Manifolds

In the precise discussion that follows a manifold will be defined as a *topological space* on which a *differentiable structure* is superimposed.

A *topological space* is a set $X$ provided with a system of *open sets*, satisfying certain requirements, as explained in Supp. Chap. 18. An *n-dimensional topological manifold*, is a topological space which is locally homeomorphic with $\mathbb{R}^n$, i.e. in which every point has an open neighbourhood homeomorphic with an open set in $\mathbb{R}^n$. For technical reasons we include in this definition the requirement that $X$ has the *Hausdorff property* and is *second countable*. (See Supp. Sect. 18.2, after Definitions (18.2,e) and (18.2,i)).

Let $\mathcal{M}$ be topological space. A *chart* on $\mathcal{M}$ is a pair $(U, \phi)$, consisting of an open set $U$ in $\mathcal{M}$ and a homeomorphism $\phi$ from $U$ onto an open subset of $\mathbb{R}^n$. Such a chart defines a system of *local coordinates* $x^1, \ldots, x^n$, as a function from $U$ to $\mathbb{R}$. An *atlas* is a collection of charts $\{U_\alpha, \phi_\alpha\}_\alpha$ such that the sets $U_\alpha$ cover $X$. Let $(U_\alpha, \phi_\alpha)$ and $(U_\beta, \phi_\beta)$ be two charts on $\mathcal{M}$. The map $\phi_{\beta\alpha} = \phi_\beta \circ \phi_\alpha^{-1}$ is a homeomorphism from $\phi_\alpha(U_\alpha \cap U_\beta)$, the image under $\phi_\alpha$ of the overlap of $U_\alpha$ and $U_\beta$, onto the image of the same overlap set under $\phi_\beta$, for all pairs of charts on $\mathcal{M}$. It is called the *transition function*, associated with the two charts. This map $\phi_{\beta\alpha}$ and its inverse $\phi_{\alpha\beta}$, real-valued functions of real variables, are homeomorphisms of open sets of $\mathbb{R}^n$ onto open sets of $\mathbb{R}^n$.

Topological manifolds are discussed in [2].

### 20.1.3 Smooth Manifolds

Let $\mathcal{M}$ be a topological manifold. A *differentiable structure* on $\mathcal{M}$ is an atlas for which all transition functions and their inverses are differentiable.

(20.1.3,a) **Definition** A *differential manifold* (or, from now on *manifold*) is a topological manifold provided with a differentiable structure.

A real-valued function $f$ on $\mathcal{M}$ is differentiable (smooth, etc.) iff for all the charts $(U_\alpha, \phi_\alpha)$ of the atlas which defines the differentiable structure of $\mathcal{M}$ the compositions $f \circ \phi_\alpha^{-1}$ are differentiable functions on the appropriate open sets of $\mathbb{R}^n$, in the ordinary sense. A continuous map from an manifold $\mathcal{M}_1$ into a second manifold $\mathcal{M}_2$ will be called differentiable iff the corresponding maps from open sets in $\mathbb{R}^{n_1}$ to open sets in $\mathbb{R}^{n_2}$ are differentiable. If such a map has a differentiable inverse it is called a *diffeomorphism*.

## 20.2 Tangent Vectors and Vector Fields

### 20.2.1 The Tangent Space at a Point of a Manifold

An $n$-dimensional manifold $\mathcal{M}$ has in every point $p$ a *tangent space* $T_p(\mathcal{M})$. If one thinks of $\mathcal{M}$ as a submanifold of some $\mathbb{R}^m$, with $n < m$, it is intuitively clear what $T_p(\mathcal{M})$ is: a first order approximation of $\mathcal{M}$ at $p$ by an $n$-dimensional linear space spanned by the tangents of all curves in $\mathcal{M}$ passing through $p$. We make this precise in the picture in which a manifold is an intrinsic object, as defined in the preceding section.

Consider *curves* in $\mathcal{M}$, passing through the point $p$, i.e. smooth maps $\gamma$ from a real interval $(-\varepsilon, +\varepsilon)$, for some $\varepsilon > 0$, into $\mathcal{M}$, such that $\gamma(0) = p$. On a coordinate neighbourhood $U$ of $p$, with local coordinates $x^1, \ldots, x^n$, a curve $\gamma$ is given by $n$ smooth functions $x^1(t), \ldots, x^n(t)$. One defines an equivalence relation between such curves: two curves $\gamma_1$ and $\gamma_2$ are equivalent if and only if in some system of local coordinates one has, for $s = 1, \ldots, n$,

$$\left( \frac{d}{dt} x_1^s(t) \right)_{t=0} = \left( \frac{d}{dt} x_2^s(t) \right)_{t=0}.$$

(20.2.1,a) **Problem** Check that this definition is independent of the choice of coordinates.

Strictly speaking, curves through $p$ cannot be added; however, the equivalence classes can. Choose again local coordinates and represent two curves $\gamma_1$ and $\gamma_2$ by functions $x_1^s(\tau)$ and $x_2^s(\tau)$. Define the curve "$\gamma_1 + \gamma_2$" by the equivalence class of the function $(x_1 + x_2)^s(\tau) = x_1^s(\tau) + x_2^s(\tau)$, for $s = 1, \ldots, n$. The definition of $\gamma_1 + \gamma_2$ depends on the choice of the local coordinates, the equivalence class $[\gamma_1 + \gamma_2]$ is however intrinsic.

(20.2.1,b) **Problem** Verify this.

One similarly defines scalar multiplication on the equivalence classes. Using this one defines the tangent space $T_p(\mathcal{M})$ as the space of these equivalence classes. An equivalence class $[\gamma]$ represents the tangent of a curve $\gamma$ at $p$. $T_p(\mathcal{M})$ is clearly a vector space of dimension $n$.

### 20.2.2 Vector Fields

A *vector field* on $M$ is a map which assigns in a smooth manner to each point $p$ of $M$ a tangent vector from $T_p(\mathcal{M})$. Smooth means here that for a local coordinate neighbourhood $(U, \phi)$ a vector field $X$ is given by $n$ smooth local functions $X^s(x^1, \ldots, x^n)$. This map acts on a locally defined function $f$ as a first order linear differential operator as

$$X(f) = \sum_{s=1}^{n} X^s \frac{\partial(f \circ \phi^{-1})}{\partial x^s}.$$

(20.2.2,a) **Problem** Show that this definition is independent of the choice of local coordinates and defines a linear map

$$X : C^\infty(\mathcal{M}) \mapsto C^\infty(\mathcal{M}).$$

(20.2.2,b) **Problem** Show also that it is a *derivation* of the algebra $C^\infty(\mathcal{M})$, meaning that it satisfies *Leibniz property*

$$X(fg) = X(f)\,g + f\,X(g), \quad \forall f, g \in C^\infty(\mathcal{M}).$$

A nontrivial theorem states that each derivation, i.e. each linear map from $C^\infty(\mathcal{M})$ into itself which has this property comes from a unique smooth vector field. This important fact allows us to see and treat vector fields as first order linear differential operators, or, in a completely equivalent manner, as derivations of the algebra of smooth functions on the manifold, a simple, purely algebraic characterization that has many advantages. From now on the space of vector fields will be denoted as $V(\mathcal{M})$.

Vector fields can be added and multiplied by real numbers. They can also be multiplied by functions, $X \mapsto fX$, or $(fX)(g) = f(X(g))$, for all $f$ and $g$ from $C^\infty(\mathcal{M})$. This means that the space of vector fields is not only a real linear space but also a *module*—a notion that will be discussed in Supp. Sect. 20.2.5—over the algebra $C^\infty(\mathcal{M})$. (The notion of a module over an algebra or ring $\mathcal{A}$ is a generalization of that of a vector space over a field like $\mathbb{R}$ or $\mathbb{C}$, with elements of $\mathcal{A}$ playing the role of scalars, as will be discussed in Supp. Sect. 20.2.5)

(20.2.2,c) **Problem** The composition two derivations $X$ and $Y$, $X \circ Y$, written simply as $XY$, gives a linear map but *not* a derivation. The commutator $[X, Y] = XY - YX$ is however a derivation. Show this.

With respect to this commutator the linear space of vector fields is an infinite dimensional *Lie algebra*. See for the notion of 'Lie algebra' Supp. Sect. 24.5.

### 20.2.3 The Tangent Bundle

The set of all tangent vectors for all points of the $n$-dimensional manifold $\mathcal{M}$ can be given in a natural way the structure of a $2n$-dimensional manifold. This manifold is called the *tangent bundle* of $\mathcal{M}$, and is denoted as $T(\mathcal{M})$. A chart $(U, \phi)$ on $\mathcal{M}$, with $\phi$ a system of local coordinates $x^1, \ldots, x^n$, gives a chart $(\widehat{U}, \hat{\phi})$ on $T(\mathcal{M})$, with $\widehat{U}$ the set of all tangent vectors at points of $U$, and $\hat{\phi}$ a system of coordinates $x^1, \ldots, x^n, y^1, \ldots, y^n$. The additional coordinates $y^1, \ldots, y^n$ are defined as follows:

A tangent vector $T_p$, at a point $p$ of $U$, can be represented by a curve through $p$, which means in terms of the coordinates $x^1, \ldots, x^n$ by $n$ functions $x^1(t), \ldots, x^n(t)$. The value of the coordinate $y^j$ at $p$ is then $y^j(p) = \left(\frac{d}{dt}x^j(t)\right)_{t=0}$. A vector field $Y$, a map $\mathcal{M} \mapsto T(\mathcal{M})$, assigns a tangent vector to each point $p$, and is in these coordinates given by $n$ (smooth) functions $Y^j(x^1, \ldots, x^n)$.

### 20.2.4  A Few General Remarks on Vector Bundles

The manifold $T(\mathcal{M})$ of tangent vectors is called the tangent bundle because it is a particular example of a more general notion, that of a *vector bundle* over a manifold, of which we shall see other examples in the next subsections.

A *vector bundle* over a manifold $\mathcal{M}$ of dimension $m$ is a smooth assignment of copies of the same $n$-dimensional vector space $V$ to all points of a given manifold. In this manner one gets a new manifold of dimension $m + n$. This manifold is called the *bundle space*; the underlying manifold the *base manifold*. It is locally a product space; if it is globally a product then the bundle is called *trivial*. Vector bundles over $\mathbb{R}^n$ are trivial. There is an obvious projection map $\pi$ from the bundle space to the base space. The vector space attached to a point $p$, the inverse image $\pi^{-1}(p)$, is called the *fibre* at $p$. A smooth choice of a vector from each fibre is called a *section*. For a trivial bundle sections are just smooth functions $s$ from the base manifold $\mathcal{M}$ to the vector space $V$. A vector field is a section of the tangent bundle.

The notion of a vector bundle is a specific example of a general concept, that of a *fibre bundle*, in which the fibres are copies of some general space, for example a Lie group. We do not need more general fibre bundles in this book.

### 20.2.5  Modules

In a linear space vectors can be multiplied by real or complex numbers. Using instead multiplication by elements of a general (commutative) algebra $\mathcal{A}$ leads to the notion of a $\mathcal{A}$-*module*. This is very useful when working with vector bundles in differential geometry, because the space of sections of a vector bundle over a manifold $\mathcal{M}$ is a $C^\infty(\mathcal{M})$-module.

A vector space generated by a system of finitely many vectors has a (finite) basis. Every such basis has the same number of vectors, the dimension of the vector space. Finitely generated modules with this property are called *free*. The module of sections of a trivial bundle is free, so the vector fields on $\mathbb{R}^n$ form a free module. For an arbitrary manifold $\mathcal{M}$ the modules of sections of vector bundles are not necessarily free, but are in any case finitely generated. Apart from this and related properties there is very little difference between the linear algebra of vector spaces and of modules; all constructions and definitions are very similar. For instance, the dual $V^*$ of an $\mathcal{A}$-module $V$ is defined as the space of $\mathcal{A}$-linear maps $V \to \mathcal{A}$. One defines tensor products of modules in the same way as tensor products of vector spaces. (See for vector space tensor products Supp. Chap. 22.)

The $C^\infty(U)$-module of the vector fields on a local coordinate neighbourhood $U$ is obviously free. Using the coordinates $\{x^j\}_j$ one defines a local basis $\{\frac{\partial}{\partial x^j}\}_j$ for this module as the derivations $\frac{\partial}{\partial x^j} f$, for all $f$ in $C^\infty(U)$.

## 20.3 The Cotangent Bundle

For each point $p$ of an $n$-dimensional manifold $\mathcal{M}$ the tangent space $T_p(\mathcal{M})$ was in the preceding chapter defined as an $n$-dimensional vector space 'attached' to $\mathcal{M}$ in $p$. The dual of $T_p(\mathcal{M})$, the $n$-dimensional vector space of linear maps from $T_p(\mathcal{M})$ into $\mathbb{R}$, is denoted as $T_p^*(\mathcal{M})$, and is called the *cotangent space* at the point $p$. By joining in a smooth manner all the cotangent spaces one gets the *cotangent bundle* $T^*(\mathcal{M})$. The sections of $T^*(\mathcal{M})$ might be called *co-vector fields*; the standard name is 1-*forms*, for reasons that will soon become clear. The space of sections of $T^*(\mathcal{M})$ is denoted as $\Omega_1(\mathcal{M})$.

One alternatively defines a 1-form, without explicit reference to the cotangent vectors at separate points, as a map from $V(\mathcal{M})$ to $C^\infty(\mathcal{M})$, linear in the sense of $C^\infty(\mathcal{M})$-modules, meaning that it is a linear map $\alpha : V(\mathcal{M}) \to C^\infty(\mathcal{M})$ in the ordinary sense, with the additional property $\alpha(fX) = f\alpha(X)$, for all $X$ in $V(\mathcal{M})$ and $f$ in $C^\infty(\mathcal{M})$, or in other words, it is the *dual* of $V(\mathcal{M})$ in the sense of $C^\infty(\mathcal{M})$-modules

(20.3.1,a) **Problem** Show that the space of 1-forms is indeed a $C^\infty(\mathcal{M})$-module, i.e. that multiplication of a 1-form $\alpha$ by a function $f$, $\alpha \mapsto f\alpha$, or $(f\alpha)(X) = f(\alpha(X))$, gives again a 1-form.

## 20.4 General Tensor Fields

In every point $p$ of $\mathcal{M}$ one has a tangent space $T_p(\mathcal{M})$ and a cotangent space $T_p^*(\mathcal{M})$. From these one can construct arbitrary tensor products. See Supp. Chap. 23 for the notion of tensor products. Joining the tensor product spaces of the same type for all points $p$ smoothly together gives a new manifold. Its sections are called *general tensor fields*. They will be called *contravariant* when built from tensor products of tangent spaces, *covariant* when built from cotangent spaces, and *mixed* when built from both.

Two important special examples: symmetric covariant two tensors for spacetime metrics in general relativity and an antisymmetric two tensor in the mathematical formulation of classical mechanics. The first example appears briefly in Chap. 15; the second one is introduced in the next section and forms the basis for Supp. Sect. 20.6, and consequently for Chap. 2.

Tensor fields are extensively used in mathematical physics, both quantum and classical, usually in terms of a system of *local coordinates* $\{x^1, \ldots, x^n\}$, defined on an open set $U \subset \mathcal{M}$, in fact a chart in the sense of Sect. 20.1.2. All differential geometric objects can be expressed in such coordinates. A typical mixed tensor could be a section from, for example, the vector bundle $T_p(\mathcal{M}) \otimes T_p(\mathcal{M}) \otimes T_p^*(\mathcal{M}) \otimes T_p^*(\mathcal{M})$. Only manifolds diffeomorphic to $\mathbb{R}^n$ can be covered by a single set of *global coordinates*. Most physics books, especially books on general relativity, ignore this. When performing explicit calculations, local coordinates are often indispensable, even in purely mathematical work. For this physics has come up with some useful conventions. The indices for covariant tensor fields are written with upper indices, contravariant tensors with lower indices. The mixed tensor field above will in this manner be written as $A^{j_1 j_2}{}_{k_1 k_2}(x^1, \ldots, x^n)$. In a summation over the same index, appearing simultaneously as a covariant and a contravariant index, one follows the *Einstein summation convention* according to which the sum symbol is omitted, i.e. the formula $(AB)_k = \sum_j A_{jk} B^j$ is written as $(AB)_k = A_{jk} B^j$.

## 20.5  Differential Forms

### 20.5.1  Introduction

A further possibility in the construction, as was noted in the preceding section, is to specialize to symmetric and antisymmetric tensor products. The antisymmetric case is in the context here the most important one, with vector bundles that can be denoted as $\wedge^k T^*(\mathcal{M})$ or $(\otimes_a^k T^*(\mathcal{M})$ in the notation of Supp. Sect. 23.3). The sections of these bundles are called $k$-forms; they form $C^\infty(\mathcal{M})$-modules denoted as $\Omega_k(\mathcal{M})$, with $\Omega_0(\mathcal{M}) = C^\infty(\mathcal{M})$ and $\Omega_k(\mathcal{M}) = \{0\}$, for $k > n$, because of antisymmetry. We consider the direct sum

$$\Omega(\mathcal{M}) = \Omega_0(\mathcal{M}) \oplus \cdots \oplus \Omega_n(\mathcal{M}),$$

which is an (associative) algebra, with the product of two forms $\alpha$ and $\beta$ denoted as $\alpha \wedge \beta$. It is 'supercommutative' or 'graded commutative', i.e. with

$$\alpha_j \wedge \beta_k = (-1)^{jk} \beta_k \wedge \alpha_j, \quad \forall \alpha_j \in \Omega_j(\mathcal{M}), \quad \beta_k \in \Omega_k(\mathcal{M}).$$

See for the general context of grading and graded commutativity Sect. 13.9.3.2. For a local coordinate system $\{x^1, \ldots, x^n\}$ one has a local basis of vector fields $\{X_1 = \frac{\partial}{\partial x^1}, \ldots, X_n = \frac{\partial}{\partial x^n}\}$, together with a dual basis of 1-forms $\{dx^1, \ldots, dx^n\}$, connected by the formula $dx^j(X_k) = \delta_k^j$. A general $p$-form can be locally written as

$$\alpha_p = \frac{1}{p!} \sum_{j_1,\ldots,j_p=1}^{n} (\alpha_p)_{j_1,\ldots,j_p} \, dx^{j_1} \wedge \cdots \wedge dx^{j_p}$$

$$= \sum_{1 \le j_1 < \ldots < j_p \le n} (\alpha_p)_{j_1,\ldots,j_p} \, dx^{j_1} \wedge \cdots \wedge dx^{j_p}.$$

Note that the components $(\alpha_p)_{j_1,\ldots,j_p}$ of $\alpha_p$ are functions, i.e. elements of the function algebra $C^\infty(U)$, but can also be viewed as 0-forms.

### 20.5.2 The Exterior Derivative

(20.5.2,a) **Theorem** *There is a linear map* $d : \Omega(\mathcal{M}) \mapsto \Omega(\mathcal{M})$, *the exterior derivative, defined by the following three properties* :

(1) $d^2 = 0$ *(nilpotence)*.
(2) $d(\alpha_p \wedge \beta_q) = d\alpha_p \wedge \beta_q + (-1)^{pq} \alpha_p \wedge d\beta_q$, *for all $p$-forms $\alpha_p$ and $q$-forms $\beta_q$ (graded or super Leibniz property)*.
(3) $d\alpha_0(X) = X(\alpha_0)$, $\forall X \in V(\mathcal{M})$ *(action of $d$ on the 0-forms)*.

(20.5.2,b) **Problem** Show that the action of $d$ on $\Omega_0(\mathcal{M})$ in local coordinates is given by $(d\alpha_0)_j = \frac{\partial}{\partial x^j}\alpha_0$, for $j = 1, \ldots, n$. In elementary calculus this is just the gradient of a function.

It is not hard to prove that the properties (1), (2) and (3) define indeed a unique linear operator, and to derive the form of the action of it on general $p$-forms, both in terms of local coordinates and in an intrinsic form. All this can be found in the literature. Here are a few results:

• The formula for $d$ carrying a 1-form into a 2-form:

$$d\alpha_1(X, Y) = X(\alpha_1(Y)) - Y(\alpha_1(X) - \alpha_1([X, Y]), \quad \forall X, Y \in V(\mathcal{M}),$$

and in components with respect to local coordinates

$$(\alpha_1)_j \;\mapsto\; (d\alpha_1)_{j_1 j_2} = \frac{\partial}{\partial x^{j_1}}(\alpha_1)_{j_2} \, dx^{j_1} \wedge dx^{j_2}.$$

• The formula for $d$ carrying a 2-form into a 3-form:

$$(d\alpha_2)(X, Y, Z) = X(\alpha_2(Y, Z)) + Y(\alpha_2(Z, X)) + Z(\alpha_2(X, Y)$$
$$+ \alpha_2([X, Y], Z) + \alpha_2([X, Z], Y)$$
$$+ \alpha_2([Y, Z], X), \quad \forall X, Y, Z \in V(\mathcal{M}),$$

and in terms of local coordinates

$$(\alpha_2)_{jk} \mapsto (d\alpha_2)_{jkl} = \frac{\partial}{\partial x^j}(\alpha_2)_{kl} + \frac{\partial}{\partial x^k}(\alpha_2)_{lj} + \frac{\partial}{\partial x^l}(\alpha_2)_{jk}.$$

- One can show that there is a general property (4) of $d$, already visible in the above results, which could be added to our list of basic properties, though it is not independent from the other three, namely,

(4) $d : \Omega_p \rightarrow \Omega_{p+1}$ ( $d$ is a graded linear map of degree $+1$).

(20.5.2,a) *Remark* For $\mathcal{M}$ a three dimensional real inner product space the numbers of independent components of the forms are, respectively, 1, 3, 3, 1; the forms are called functions, vectors, vectors, functions. So in $\mathbb{R}^3$ the sequence

$$\Omega_0(\mathcal{M}) \xrightarrow{d} \Omega_1(\mathcal{M}) \xrightarrow{d} \Omega_2(\mathcal{M}) \xrightarrow{d} \Omega_3(\mathcal{M})$$

translates into

$$functions \xrightarrow{\nabla} vectors \xrightarrow{curl} vectors \xrightarrow{divergence} functions,$$

in which, strictly speaking, the second class of vectors are pseudo-vectors and the last functions pseudo-scalars because of their behaviour with respect to spatial reflections.

(20.5.2,b) **Definition** A $k$-form $\omega_k$ is called *closed* iff $d\omega_k = 0$; it is called *exact* iff there exists a $(k-1)$-form $\sigma_{k-1}$ such that $\omega_k = d\sigma_{k-1}$. An exact form is closed; the converse does not need to be true. The quotient spaces of the spaces of closed forms over that of the exact forms give important information on the topology of the underlying manifold. They are called *de Rham cohomology spaces* see below.

(20.5.2,c) *Remark* The system of differential forms together with the exterior derivative is an example of a large class of more general systems consisting of a sequence of spaces with an nilpotent linear map satisfying the Leibniz relation. They are studied in *cohomology theory*, a broad and important mathematical field, which lies however outside the scope of this book. In this general context the system of differential forms on a manifold is called its *de Rham cohomology*. There is also *homology theory*, with the arrows of a map $\partial$ running backwards.

## 20.6 Symmetries of a Manifold

The natural morphisms between manifolds are smooth maps

$$\phi : \mathcal{M}_1 \rightarrow \mathcal{M}_2, \quad p \mapsto \phi(p).$$

If such a map has a smooth inverse it is called a *diffeomorphism*. Of particular interest are one parameter groups $\phi_t$ of diffeomorphisms from a manifold onto itself.

Under certain conditions of regularity such a group generates a *flow* on the manifold. The group $\phi_t$ is generated by a vector field $X$ which defines an ordinary first order differential equation, with the flow as solution. This carries all mathematical objects on the manifold along by means of induced one parameter groups of linear maps. They can be written as $e^{t\mathcal{L}_X}$, with as generators the *Lie derivatives* $\mathcal{L}_X$, acting on the algebra of smooth functions $C^\infty(\mathcal{M})$ as $\mathcal{L}_X f = X(f)$, on the module of vector fields $V(\mathcal{M})$ as $\mathcal{L}_X Y = [X, Y]$, and on the modules of 1-forms $\Omega_1(\mathcal{M})$ as $(\mathcal{L}_X \alpha_1)(Y) = X(\alpha_1(Y)) + \alpha_1([Y, X])$, etc. For the formula for general $p$-forms, and for proofs, see [3, p.70].

There are many good books on differential geometry, i.e. the theory of manifolds. Examples are [3], [4] (very thorough) and [5].

## 20.7  Symplectic Manifolds

### 20.7.1  Introduction

This subsection discusses the mathematical background of classical mechanics. The central notion in this is that of a *symplectic manifold*, in physics the phase space of a classical mechanical system.

(20.7.1,a) **Definition** A 2-form $\omega$ on the $C^\infty$-manifold $\mathcal{M}$ is called a *symplectic form* if it is closed and nondegenerate. This means that $d\omega = 0$ and moreover that $\omega(X, Y) = 0$, for all vector fields $Y$, implies $X = 0$. One can show that the requirement of nondegeneracy forces $\mathcal{M}$ to have even dimension. A *symplectic manifold* is a manifold provided with a symplectic form.

(20.7.1,b) **Problem** Show that by looking at the properties of $\omega$ at each point $p$ of $\mathcal{M}$, i.e. as a nondegenerate antisymmetric form $\omega_p$ on the tangent space $T_p(\mathcal{M})$, one can derive that the dimension of $\mathcal{M}$ is even.

(20.7.1,c) *Remark* Dropping the requirement that $\omega$ is nondegenerate gives a weaker notion, that of a *Poisson manifold*.

A symplectic manifold has a very useful canonical $2n$-form. It is simply defined as

$$\theta_\omega = \frac{1}{n!} (-1)^{\frac{1}{2}n(n-1)} \underbrace{\omega \wedge \cdots \wedge \omega}_{n\ times}.$$

The $2n$-form $\theta_\omega$ is a *volume form*, i.e. it is everywhere nonzero. This implies that $\mathcal{M}$ is orientable. It plays an important role in classical statistical mechanics, in particular in Liouville's theorem.

The natural morphisms between symplectic manifolds are *symplectic transformations*, i.e. smooth maps that preserve the symplectic forms. If such maps have smooth inverses they are called *symplectomorphisms*. For a given symplectic manifold they form a group. Of particular interest are one parameter groups $\phi_t$. With sufficient smoothness they can be generated by a vector field $X$, which is by definition a *symplectic vector field*. These vector fields can also be defined by the formula $\mathcal{L}_X\omega = 0$, using the concept of Lie derivative, introduced in the preceding section.

In applications to mechanics, the main purpose of this chapter, the following notion is central. For every $h$ in $C^\infty(\mathcal{M})$ there is a vector field $X_h$ such that

$$\omega(X_h, Y) = Y(h), \quad \forall Y \in V(\mathcal{M}).$$

$X_h$ is called the *Hamiltonian vector field* associated with $h$, the *Hamiltonian* or *Hamiltonian function* for $X_h$. A Hamiltonian vector field is symplectic. Both notions coincide on sufficiently small neigbourhoods of points. They may differ globally; therefore a symplectic vector field is sometimes called *locally Hamiltonian*. In our applications the manifolds that we consider are supposed to be simply connected; in that case both notions coincide.

### 20.7.2 The Poisson Bracket

Let $\mathcal{M}$ be a symplectic manifold with symplectic form $\omega$. One defines a bilinear map

$$\{\cdot, \cdot\} : C^\infty(\mathcal{M}) \times C^\infty(\mathcal{M}) \to C^\infty(M)$$

by the formula

$$\{f, g\} = \omega(X_f, X_g), \quad \forall f, g \in C^\infty(\mathcal{M}).$$

This is the *Poisson bracket* associated with the symplectic manifold $\mathcal{M}$. It has the following properties:

(1) $X_{\{f,g\}} = [X_f, X_g]$.
(2) $\{f, gh\} = \{f, g\}h + \{f, h\}g$.
(3) $\{f, g\} = -\{g, f\}$ (*antisymmetry*).
(4) $\{f, \{g, h\}\} + \{g, \{h, f\}\} + \{h, \{f, g\}\} = 0$ (*Jacobi identity*).
(5) $\{f, g\} = 0, \forall g \in C^\infty(\mathcal{M})$ implies that $f$ is a constant function.

for all $f, g$ and $h$ in $C^\infty(M)$. Properties (2)–(4) mean that the Poisson bracket makes the associative algebra of functions $C^\infty(\mathcal{M})$ into a *Lie algebra*. See for this notion Supp. Chap. 24 (*Lie groups and Lie algebras*). Property (1) then states that the map $f \mapsto X_f$ is a Lie algebra homomorphism from $C^\infty(\mathcal{M})$ into $V(\mathcal{M})$, relating the Poisson bracket with the commutator of vector fields.

**(20.7.2,a) Problem** Prove the properties (1)–(5).

### 20.7.3 Darboux Coordinates

Explicit calculations in differential geometry are often facilitated by the use of local coordinates, as we have seen. For symplectic manifolds there exist special coordinates in which the symplectic form becomes particularly simple. This is stated by the *Darboux theorem*, which we give here without proof.

**(20.7.3,a) Theorem** *A 2n-dimensional symplectic manifold possesses a system of coordinate neighbourhoods, each with coordinates $\{p_j\}_j$, $\{q^k\}_k$ such that the symplectic form $\omega$ is locally represented by the expression $\omega = \sum_{j=1}^{n} dp_j \wedge dq^j$.*

Note that this theorem implies that all symplectic manifolds with the same dimension are locally the same.

**(20.7.3,b) Problem** Show that with respect to a system of Darboux coordinates a Hamiltonian vector field $X_h$ has components

$$(X_h)_j = \frac{\partial h}{\partial p_j}, \qquad (X_h)_k = -\frac{\partial h}{\partial q^k}, \quad j, k = 1, \ldots, n.$$

**(20.7.3,c) Problem** Show that the Poisson bracket of two functions $f$ and $g$ takes the local form

$$\{f, g\} = \sum_{j=1}^{n} \left( \frac{\partial f}{\partial p_j} \frac{\partial g}{\partial q^j} - \frac{\partial g}{\partial p_j} \frac{\partial f}{\partial q^j} \right).$$

**(20.7.3,d) Problem** Show that the canonical volume form becomes

$$\theta_\omega = dp_1 \wedge \cdots \wedge dp_n \wedge dq^1 \wedge \cdots \wedge dq^n$$

Note that this gives, with an appropriate choice of orientation, the integration measure $dp_1 \ldots d_n \, dq^1 \ldots dq^n$.

For useful material on symplectic manifolds, see [6].

## 20.8 An Algebraic Reformulation of Differential Geometry

A purely algebraic form of differential geometry was proposed in 1962 by Jean-Louis Koszul. See [7]. The main idea in this approach is that all geometrical objects on a manifold $\mathcal{M}$ can be defined algebraically starting from the algebra $C^\infty(\mathcal{M})$ of smooth functions on $\mathcal{M}$. One introduces in the first place vector fields as derivations of $C^\infty(\mathcal{M})$, a purely algebraic definition, made possible by a nontrivial theorem which states that every derivation of $C^\infty(\mathcal{M})$ is a vector field on $\mathcal{M}$. From this one develops, in the linear-algebraic language of $C^\infty(\mathcal{M})$-modules, other notions such as cotangent vector fields (1-forms), defined as elements of the dual of the module of vector fields, tensor fields, both contravariant and covariant, by taking tensor products, everything in the sense of $C^\infty(\mathcal{M})$-modules.

A symplectic manifold is defined by specifying a symplectic 2-form $\omega$, a nondegenerate covariant antisymmetric 2-tensor field; a Riemannian manifold by a Riemannian metric, a strictly positive definite symmetric covariant 2-tensor field. A symplectic manifold has a Poisson bracket, making $C^\infty(\mathcal{M})$ into a *Poisson-Lie algebra* or, for short, a *Poisson algebra*, an algebra with two multiplications, an associative one: $(f, g) \mapsto f g$, and a nonassociative one: $(f, g) \mapsto [f, g]$, which makes it into a Lie algebra.

The characterization of a symplectic manifold by its Poisson algebra is central to the comparison of classical and quantum mechanics as algebraic physical theories, as is explained in Chap. 12.

All this is strengthened by a remarkable theorem, unfortunately in an unpublished preprint, stating that two manifolds $\mathcal{M}_1$ and $\mathcal{M}_2$ are diffeomorphic if and only if the corresponding algebras of functions $C^\infty(\mathcal{M}_1)$ and $C^\infty(\mathcal{M}_2)$ are isomorphic, and that moreover a manifold $\mathcal{M}$ can be reconstucted from its function algebra $C^\infty(\mathcal{M})$. See [8].

A natural further development of this general idea is the possibility to build a similar formalism on suitable *noncommutative* algebras and defining in this way *noncommutative differential geometry*, in which there is, strictly speaking, no underlying manifold as a point set, but in which an intuitive notion of a 'noncommutative manifold' gives interesting suggestions for an algebraic formulation of various geometric objects living on this 'manifold'. This idea has found its fullest expression in the work on noncommutative geometry of Alain Connes. See [9].

# References

1. Riemann, B.: Über die Hypothesen, welchen der Geometrie zu Grunde liegen. Habilitationsschrift, 1854. Abhandlungen der Königlichen Gesellschaft der Wissenschaften zu Göttingen, 13. http://www.maths.tcd.ie/pub/HistMath/People/Riemann/Geom/Geom.pdf [Trans.: On the Hypotheses which lie at the Bases of Geometry. http://www.maths.tcd.ie/pub/HistMath/People/Riemann/Geom/WKCGeom.html. Still fascinating reading.]
2. Lee, J.M.: Introduction to Topological Manifolds, 2nd edn. Springer, New York (2010)
3. Lee, J.M.: Introduction to Smooth Manifolds. Springer, New York (2002) [The two volumes of [7] provide a solid, comprehensive, compactly written text on all aspects of differential geometry. The references [8] and [2] are more accessible, and cover in particular all the subjects discussed in this chapter.]
4. Kobayashi, S., Nomizu, K.: Foundation of Differential Geometry I, II. Original Edition. Wiley 1963, 1969. Paperback Edition Wiley-Blackwell (2009)
5. Warner, F.W.: Foundation of Differentiable Manifolds and Lie Groups. Original edition: Scott, Foresman and Company 1971. Second printing, Springer (1983) [Less forbidding than [4]]. Covers a wide area in a fairly compact but clear way.
6. da Silva, A.C.: Lectures on Symplectic Geometry. Second corrected printing. Springer 2001. Revised edition 2006. http://citeseerx.ist.psu.edu/viewdoc/download?doi=10.1.1.146.4370&rep=rep1&type=pdf [A broad review of symplectic geometry.]

7. Koszul, J.L.: Lectures on Fibre Bundles and Differential Geometry. Tata Institute of Fundamental Research. Bombay. http://www.math.tifr.res.in/~publ/ln/tifr20.pdf (1960) [A fundamental paper, not very well-known.]
8. Thomas, E.G.F.: Characterization of a manifold by the $*$-algebra of its $C^\infty$ functions. Preprint Mathematical Institute, University of Groningen (Unpublished)
9. Connes, A.: Noncommutative Geometry. Academic Press, San Diego (1994)

# Chapter 21
# Functional Analysis: Hilbert Space

## 21.1 Introduction

In the early development of quantum mechanics in two different forms, as 'matrix mechanics' by Heisenberg and as 'wave mechanics' by Schrödinger, it was soon realized, in particular by mathematicians, that a unified formulation of the theory could be given in terms of Hilbert space concepts. Such a framework was developed in the early thirties by John von Neumann in a series of papers and in his book 'Grundlagen der Quantenmechanik', later translated into English as 'The Mathematical Foundations of Quantum Mechanics', [1], contributing very much to a better conceptual understanding of quantum theory. The mathematics needed for this will be reviewed in this chapter.

After a short introduction to infinite dimensional vector spaces in Supp. Sect. 21.2, Supp. Sect. 21.3 treats inner product spaces. Hilbert space is introduced in Supp. Sect. 21.4, with direct sums of Hilbert spaces in Supp. Sect. 21.5. General properties of linear operators in Hilbert space are discussed in Supp. Sect. 21.6. Next, unitary operators in Supp. Sect. 21.7 and projection operators in Supp. Sect. 21.8. Supp. Section 21.9 begins the discussion of unbounded operators, with as topics the closure of operators, Hermitian symmetry and in particular selfadjointness, essential for quantum theory. In Supp. Sect. 21.10 the centre piece of von Neumann's Hilbert space formalism for quantum mechanics, the spectral theorem for general unbounded selfadjoint operators and—with a slight generalization—for commuting systems of such operators is discussed. Related to this and only second in importance is the theorem of Stone and von Neumann about continuous one parameter groups of unitary operators, to be found in Supp. Sect. 21.11. Both theorems, the general spectral theorem and the Stone–von Neumann theorem, natural but far-reaching generalizations to infinite dimensional spaces of well-known properties of finite dimensional linear algebra and matrix theory, will be carefully and rigorously formulated. Their proofs, for which references will be given, will be made plausible, by first looking at the finite dimensional situation and then rewriting all the formulas in a manner which suggests infinite dimensional analogues and generalizations.

© Springer International Publishing Switzerland 2015
P. Bongaarts, *Quantum Theory*, DOI 10.1007/978-3-319-09561-5_21

Readers from both mathematics and physics are assumed to be familiar with standard calculus and linear algebra. Calculus courses include notions such as continuity, limit and convergence of sequences. For Hilbert space theory and functional analysis concepts from the more general context of *topology* are needed, which are usually not included in the undergraduate physics curriculum. The same is true for the theory of measure and integration. For this reason this supplementary chapter is supplemented by Supp. Chap. 18 (Topology) and Supp. Chap. 19 (Measure and integration).

A good standard textbook on Hilbert space theory is [4]. A very thorough treatment of the general theory of linear operators can be found in the three-volume treatise by Dunford and Schwartz [6].

Also useful for special topics is [8].

## 21.2  Infinite Dimensional Vector Spaces: Norms

Vector spaces with infinite dimension are indispensable as basis of the mathematical formulation of quantum theory. The study of what essentially is infinite dimensional linear algebra is called *functional analysis*, because it started originally with studying functionals, linear functions on vector spaces. To operate in a sensible manner in such spaces, to be able to use infinite sums, for example, one needs notions as limits and convergence, i.e. general topological notions of the sort discussed in Supp. Chap. 18.

(21.2,a) **Definition** A *topological vector space* is a vector space which has a topology such that the two vector space operations, addition $(x, y) \mapsto x + y$, and scalar multiplication $(\lambda, x) \mapsto \lambda x$, are continuous, simultaneously in the two variables, and with respect to this topology.

(21.2,b) **Problem** Show that a linear map $T$ from a topological vector space $V$ to a second topological vector space $W$ is continuous if it is continuous in the origin of $V$. Hint: Use the fact that in a topological vector space translation of an open set gives an open set.

The function spaces given in Supp. Chap. 18, as Examples (18.4,b) and (18.4,c) are topological vector spaces. Their topology is given by a *metric*, which depends, because of translation invariance, on a single argument, the difference of two vectors. Instead of a distance one has a length or *norm* of a vector, written as $||f|| = d(0, f)$. This norm has an additional property expressing compatibility with scalar multiplication. Adding this to the requirements for a metric, stated in Definition (18.4,a), the properties of a norm in a *normed vector space* $V$ become:

1. $||x|| \geq 0$, with $||x|| = 0 \Leftrightarrow x = 0$,
2. $||\lambda x|| = |\lambda| \, ||x||$,
3. $||x + y|| \leq ||x|| + ||y||$ (triangle inequality),

for all $x$, $y$ in $V$, and all $\lambda$ in $\mathbb{C}$ or $\mathbb{R}$.

(21.2,c) **Problem** Show that the triangle inequality for a norm follows from that of a metric, as given in Definition (18.4,a).

The space $C_c(\mathbb{R}^1)$ in Example (18.4,c) is complete. Complete normed vector spaces are called *Banach spaces*. The complete function space $L^2(\mathbb{R}^1, dx)$ has a norm which comes from an inner product according to $\|f\| = (f, f)^{1/2}$ and is called a *Hilbert space*, as will be discussed further on. There is a class of topological vector spaces more general than Hilbert and Banach spaces: the *locally convex topological vector spaces*, or just *locally convex spaces*. Their topology is not given by a single norm, but by an infinite system of *seminorms*, a notion in which the requirement 1 in the definition of norm is weakened to just $\|x\| \geq 0$. Such spaces are quite useful and appear, explicitly or implicitly, in many places in mathematical physics. In this book a particular type of such spaces, Fréchet and LF spaces play a role in Chap. 12 and Supp. Chap. 27.

For each vector space $V$ one defines a *dual space*, denoted as $V'$, and consisting of all linear functionals on $V$, i.e. complex (or real) linear maps from $V$ to $\mathbb{C}$ (or $\mathbb{R}$). $V'$ is in an obvious manner again a vector space.

(21.2,d) **Problem** Show that the dual of a finite dimensional vector space has the same (finite) dimension as the given space.

For a topological vector space $V$ one distinguishes two duals, the *algebraic dual* $V'$, and the *topological dual*, usually denoted as $V^*$, consisting of all *continuous* linear functionals on $V$. One has, obviously, $V^* \subset V'$. For finite dimension $V'$ and $V^*$ coincide. As we are dealing mainly with infinite dimensional vector spaces, in most but not all cases with Hilbert spaces, we will only be interested in topological duals. For a normed vector space a continuous linear functional $F$ has a norm, defined, not surprisingly, as

$$\|F\| = \sup_{\|\psi\|=1} |F(\psi)|.$$

This makes $V^*$ into a normed vector space. The topological dual of a Banach space is again a Banach space. The dual $V^*$ of $V$ has again a dual, the second dual $(V^*)^* = V^{**}$.

(21.2,e) **Problem** Show that the vectors of $V$ can be identified with a linear subspace of $V^{**}$.

If $V$ can be identified with all of $V^{**}$, then $V$ is called *reflexive*. Finite dimensional vector spaces are reflexive, which is trivial; Hilbert spaces–to be defined shortly–are reflexive (with respect to topological duals), which is nontrivial.

## 21.3 Inner Product Spaces

We start with the definition of what may be seen as a complex version of an Euclidean space:

(21.3,a) **Definition** A vector space $V$ over the complex numbers is called a complex or Hermitian *inner product space* iff it is provided with a Hermitian inner product, i.e. with a sesquilinear map

$$(\cdot, \cdot) : V \times V \to \mathbb{C},$$

sesquilinear meaning that it is conjugate linear in first and linear in the second variable—the standard convention in physics, Hermitian symmetric, i.e. with

$$\overline{(\psi_1, \psi_2)} = (\psi_2, \psi_1),$$

for all $\psi_1$ and $\psi_2$ in $V$, the bar meaning complex conjugation, and positive definite, i.e.

$$(\psi, \psi) \geq 0,$$

for all $\psi$ in $V$, with moreover $(\psi, \psi) = 0$ implying $\psi = 0$.

In an inner product space the topology is determined by a *norm*, which is obtained from the inner product as

$$||\psi|| = (\psi, \psi)^{1/2}.$$

We prove an important inequality:

(21.3,b) **Theorem** (Inequality of Schwarz) *Let $\psi_1$ and $\psi_2$ be two vectors in a complex inner product space $V$. Then*

$$|(\psi_1, \psi_2)| \leq (\psi_1, \psi_1)^{1/2} (\psi_2, \psi_2)^{1/2}.$$

*This is an equality if and only if one vector is a scalar multiple of the other.*

*Proof* The first part of the statement holds trivially for $\psi_1 = 0$ or $\psi_2 = 0$; the second part is meaningless in this case. We therefore assume $\psi_1$ and $\psi_2$ to be nonzero. Consider the inner product

$$\left( \psi_2 - \frac{(\psi_1, \psi_2)}{(\psi_1, \psi_1)} \psi_1, \psi_2 - \frac{(\psi_1, \psi_2)}{(\psi_1, \psi_1)} \psi_1 \right) = (\psi_2, \psi_2) - \frac{|(\psi_1, \psi_2)|^2}{(\psi_1, \psi_1)} \geq 0.$$

Taking square roots this formula gives

$$|(\psi_1, \psi_2)| \leq ||\psi_1|| \, ||\psi_2||,$$

which is Schwarz's inequality. This proves the first part of the theorem. If the above expression is zero one gets

$$|(\psi_1, \psi_2)| = ||\psi_1|| \, ||\psi_2||.$$

For any pair $\psi_1$ and $\psi_2$ with $\psi_1 = \lambda \psi_2$, $\lambda \neq 0$, this equality holds. On the other hand if the above expression is zero, then the first part gives

$$\psi_2 = \frac{(\psi_1, \psi_2)}{(\psi_1, \psi_1)}\psi_1,$$

which was to be proved.

The Schwarz inequality is sometimes called the *Cauchy-Schwarz inequality*, or even *Cauchy-Schwarz-Bunyakovsky* inequality.

(21.3,c) **Problem** Show that the expression $||\psi|| = (\psi, \psi)^{1/2}$ satisfies indeed the requirements for a norm, as given in the preceding section. Show in particular that the triangle inequality holds. Hint: Use for this the Schwarz inequality.

(21.3,d) **Problem** Show that the inner product is simultaneously continuous in both variables, i.e. as a function from the product space $V \times V$ to the complex numbers. Hint: Use the notion of product topology defined in Supp. Sect. 18.2, Schwarz's inequality and some $\varepsilon, \delta$ footwork from standard analysis.

(21.3,e) *Remark* Inner product spaces which do not have the positivity property appear in the description of relativistic quantum field theory for the Maxwell field. This will be discussed very briefly in Chap. 16. Such indefinite inner product spaces are required to have a *nondegenerate inner product*, i.e. an inner product $(\cdot, \cdot)$ with: if $(\psi_1, \psi_2) = 0$, for all $\psi_2$ in $V$, then $\psi_1 = 0$.

(21.3,f) **Problem** Suppose $V$ has a degenerate inner product. Show that there is a subspace $V_0$ of $V$ such that the inner product on $V$ descends to a nondegenerate one on the quotient space $V/V_0$.

(21.3,g) **Problem** Use the Schwarz inequality to prove that the inner product of a positive definite inner product space is nondegenerate.

In an indefinite inner product space $V$ the inner product $(\cdot, \cdot)$ does not define a topology because it does not give a norm, so one usually specifies an additional topology on $V$ such that $(\cdot, \cdot)$ is simultaneously continuous in the two variables.

## 21.4 Hilbert Space: Elementary Properties

An additional requirement leads from an inner product space to the in many ways simplest and, certainly in this book, the most important example of an infinite dimensional topological vector space, namely *Hilbert space*. Here is the definition:

(21.4,a) **Definition** A Hilbert space is a *complete* inner product space.

An inner product space that is not complete is called a *pre-Hilbert space*; it can be completed by a standard procedure to a unique Hilbert space, along the lines of the construction of the completion of a metric space, sketched in Supp. Sect. 18.4.

There are real and complex Hilbert spaces, however the Hilbert spaces in quantum theory are always complex, so only these are discussed in this book. A Hilbert space is called *separable* iff the topology associated with the norm is separable. This means that a Hilbert space has a countable dense set. See Supp. Sect. 18.2 for the notion of dense. By a well-known orthogonalization procedure a countable *orthonormal basis* can be constructed from such a set. This is a sequence of vectors $\phi_1, \phi_2, \ldots$, with $(\phi_j, \phi_k) = \delta_{jk}$, for $j, k = 1, 2, \ldots$, such that an arbitrary vector $\psi$ can be expanded as $\psi = \sum_{j=1}^{\infty} c_j \phi_j$, with uniquely determined complex coefficients $c_j$ and the infinite sum convergent in the sense of the inner product norm.

(21.4,b) **Problem** Show that the coefficient $c_k$ of a vector $\psi$ with respect to the orthonormal base $\{\phi_n\}_{n=1}^{\infty}$ is equal to $(\phi_k, \psi)$. Hint: Use the continuity of the inner product, proved in Problem (D.3,d).

The Hilbert spaces used in quantum theory are separable, and usually but not always infinite dimensional. In this book separability—but not infinite dimensionality—will therefore be added to the definition of Hilbert space. All (separable) infinite dimensional Hilbert spaces are isomorphic, so abstractly speaking there is only one such Hilbert space. There are of course many different concrete realizations of the same 'abstract' Hilbert space. The Hilbert spaces in quantum theory, in particular in quantum mechanics, are usually but not always *function spaces*.

The following examples of Hilbert spaces are basic:

(21.4,c) *Example* The space $l^2$ of all infinite sequences $\{a_j\}_{j=1}^{\infty}$ of complex numbers, such that the series $\sum_{j=1}^{\infty} |a_j|^2$ is absolutely convergent, with inner product

$$(a, b) = \sum_{j=1}^{\infty} \bar{a}_j b_j,$$

for two such sequences $a$ and $b$. One shows fairly easily that this space is complete with respect to the norm $\|a\| = (a, a)^{1/2} = (\sum_{j=1}^{\infty} |a_j|^2)^{1/2}$.

(21.4,d) *Example* The space $L^2(\mathbb{R}^1, dx)$ of all complex-valued measurable functions $\psi(x)$ of the real variable $x$, square integrable in the sense of Lebesgue, and with inner product

$$(\psi, \phi) = \int_{-\infty}^{+\infty} \overline{\psi(x)} \phi(x) dx.$$

Strictly speaking, $L^2(\mathbb{R}^1, dx)$ consists of *equivalence classes* of square integrable functions that are equal, almost everywhere. This point will be glossed over in most of the remainder of this book. The proof that $L^2(\mathbb{R}^1, dx)$, is complete with respect to the norm $\|\psi\| = (\psi, \psi)^{1/2}$ is not completely elementary.

## 21.5  Direct Sums of Hilbert Spaces

One can construct new Hilbert spaces from given ones, namely *direct sums* and *tensor products*. Here we discuss the construction of a direct sum of two and more Hilbert spaces. Tensor products are also very important in physics; they are a bit more difficult as mathematical concepts and will be dealt with separately in Supp. Chap. 23.

Let $\mathcal{H}_1$ and $\mathcal{H}_2$ be two given Hilbert spaces, with inner product $(\cdot, \cdot)_1$ and $(\cdot, \cdot)_2$. The direct sum $\mathcal{H} = \mathcal{H}_1 \oplus \mathcal{H}_2$ is in the first place the direct sum of $\mathcal{H}_1$ and $\mathcal{H}_2$ in the usual vector space sense, i.e. it is the Cartesian product $\mathcal{H}_1 \times \mathcal{H}_2$, consisting of all pairs $\{\cdot, \cdot\}$ of elements from $\mathcal{H}_1$ and $\mathcal{H}_2$, with the vector space structure given by

$$\{\psi_1, \psi_2\} + \{\phi_1, \phi_2\} = \{\psi_1 + \phi_1, \psi_2 + \phi_2\},$$

and

$$\lambda\{\psi_1, \psi_2\} = \{\lambda\psi_1, \lambda\psi_2\},$$

for all pairs $\{\psi_1, \psi_2\}$ and $\{\phi_1, \phi_2\}$ in $\mathcal{H}_1 \times \mathcal{H}_2$ and all complex numbers $\lambda$. This vector space direct sum is a Hilbert space with respect to the inner product

$$(\{\psi_1, \psi_2\}, \{\phi_1, \phi_2\}) = (\psi_1, \phi_1)_1 + (\psi_2, \phi_2)_2.$$

Note that, conversely, a given Hilbert space $\mathcal{H}$, or for that matter, a general vector space $V$ is the direct sum $V = V_1 \oplus V_2$ of linear subspaces $V_1$ and $V_2$, iff every vector $\psi$ in $V$ can be *uniquely* written as a sum $\psi = \psi_1 + \psi_2$, with $\psi_1$ in $V_1$ and $\psi_2$ in $V_2$.

(21.5,a) **Problem** Show that $V = V_1 \oplus V_2$ implies $V_1 \cap V_2 = \{0\}$.

The generalization to a direct sum of a finite or even a countably infinite number of Hilbert spaces is rather obvious. In the latter case the 'algebraic direct sum' $\oplus_{j=1}^{\infty} \mathcal{H}_j$, consisting of all infinite sequences $\{\psi_1, \psi_2, \ldots\}$, with only a finite number of nonzero vectors $\psi$, is not complete and has to be completed to a Hilbert space direct sum. There is also a useful notion of *direct integral* of Hilbert spaces. This will not be discussed here.

For a pair of linear operators, $A_1$ in $\mathcal{H}_1$ and $A_2$ in $\mathcal{H}_2$, one has the *direct sum operator* $A_1 \oplus A_2$ defined by $(A_1 \oplus A_2)\{\psi_1, \psi_2\} = \{A_1\psi_1, A_2\psi_2\}$.

## 21.6  Operators in Hilbert Space: General Properties

### 21.6.1  Operators and Their Domains

Linear maps from a Hilbert space $\mathcal{H}$ into itself are called *linear operators* or just *operators*. It turns out that most of the operators that come up naturally in quantum mechanics cannot be defined on all vectors in $\mathcal{H}$, but only on a dense linear subspace of $\mathcal{H}$. This makes working with Hilbert space operators complicated, with many subtle technical points that have to be taken care of.

We consider therefore in general linear operators that are defined on a dense linear subspace of $\mathcal{H}$, which may sometimes be $\mathcal{H}$ itself; in fact that is what the word 'operator' will from now on mean in this text, unless the contrary is explicitly stated. An operator has—strictly speaking—to be specified by a *pair* $(A, \mathcal{D})$, with $\mathcal{D}$ a dense linear subspace, and $A$ a linear map from $\mathcal{D}$ into $\mathcal{H}$. $\mathcal{D}$ is called the *domain* of the operator. Two operators will be considered equal if their domains coincide and if they are the same as maps on this common domain. Simple algebraic manipulations with operators may be complicated and nontrivial. The product $AB$ of two operators $A$ and $B$, for instance, makes sense only if Im $(B)$, the image or range of $B$, is a subset of $\mathcal{D}_A$; the sum $A + B$ of two operators is defined if the two domains $\mathcal{D}_A$ and $\mathcal{D}_B$ coincide, or have at least a sufficiently large intersection $\mathcal{D}_A \cap \mathcal{D}_B$.

The operators that are important in quantum theory are either *bounded*, unitary operators (in Supp. Sect. 21.7), and projections (in Supp. Sect. 21.8), or *unbounded* (in Supp. Sect. 21.9), in particular unbounded selfadjoint operators (in Supp. Sect. 21.10).

### 21.6.2 Bounded Operators

We start with operators, for which such domain problems do not occur, or are inessential and can be removed in a simple and unique way.

(21.6.2,a) **Definition** An operator $A$ is called *bounded* iff there exists a positive constant $C$ such that $||A\psi|| \leq C\,||\psi||$, for all $\psi$ in its domain $\mathcal{D}_A$.

(21.6.2,b) **Problem** Show that an operator is bounded iff it is continuous as a map $\mathcal{D}_A \to \mathcal{H}$. Hint: Note first that $A$ is a linear map from the linear space $\mathcal{D}_A$ with as topology the relative topology induced on $\mathcal{D}_A$ by the topology of $\mathcal{H}$, to the full space $\mathcal{H}$. See Supp. Sect. 18.2 for the notion of relative topology. Note also that normed vector spaces are first countable, see Supp. Sect. 18.3, and that therefore continuity and limits can be expressed in terms of convergence of sequences.

An operator $A'$ with domain $\mathcal{D}_{A'}$ is called an *extension* of an operator $A$ with domain $\mathcal{D}_A$ if $\mathcal{D}_A$ is a subset of $\mathcal{D}_{A'}$ and $A'\psi = A\psi$, for all vectors $\psi$ in $\mathcal{D}_A$. This situation is denoted as $A \subseteq A'$.

Everything that sofar has been said about linear operators in $\mathcal{H}$ remains valid, with obvious adaptations in terminology, for linear maps from one Hilbert space $\mathcal{H}_1$ into a second Hilbert space $\mathcal{H}_2$, including *linear functionals* on $\mathcal{H}$, i.e. linear maps from $\mathcal{H}$ to $\mathbb{C}^1$.

(21.6.2,c) **Problem** Show that a bounded operator on a dense domain has a unique extension to a bounded operator defined on all of $\mathcal{H}$.

It is clear that if two bounded operators $A_1$ and $A_2$ with domains $\mathcal{D}_{A_1}$ and $\mathcal{D}_{A_2}$ are the same on $\mathcal{D}_{A_1} \cap \mathcal{D}_{A_2}$, and this intersection is dense in $\mathcal{H}$, then their extensions $A_1'$ and $A_2'$ to all of $\mathcal{H}$ will be the same. This means that for a bounded operator $A$ the actual (dense) domain $\mathcal{D}_A$ is irrelevant; $A$ can be uniquely extended to the full Hilbert space $\mathcal{H}$. Speaking of a bounded operator we shall therefore always assume that it has already been extended to a bounded operator on all of $\mathcal{H}$.

(21.6.2,d) **Problem** A bounded operator $A$ is completely determined by the inner products $(\psi_1, A\psi_2)$ for all vectors $\psi_1$ and $\psi_2$ in $\mathcal{H}$. Show that $A$ is in fact determined by the inner products $(\psi, A\psi)$ for all $\psi$ in $\mathcal{H}$. Hint: Express by 'polarization' the general inner product $(\psi_1, A\psi_2)$ in the 'diagonal' inner products $(\psi_1 \pm \psi_2, A(\psi_1 \pm \psi_2))$ and $(\psi_1 \pm i\psi_2, A(\psi_1 \pm i\psi_2))$.

(21.6.2,e) **Problem** Show that for a vector $\psi$, given by the coefficients $c_n$ with respect to an orthonormal basis $\{\phi_n\}_n$, $A\psi$ has the coefficients $c_n' = \sum_{k=1}^{\infty} A_{nk} c_k$, with $A_{nk} = (\phi_n, A\phi_k)$ the *matrix elements* of $A$ with respect to $\{\phi_n\}_n$.

If the quotient $||A\psi||/||\psi||$ is bounded by a positive constant $C$, than it is bounded by all constants $C' > C$. This enables us to define

$$||A|| =: \inf_{\psi \neq 0} \frac{||A\psi||}{||\psi||} = \inf_{||\psi||=1} ||A\psi||,$$

the *operator norm* of $A$ which has, as the notation and the name suggests, indeed all the properties of a norm, as can be checked easily. The vector space of bounded operators in $\mathcal{H}$ is usually denoted as $\mathcal{B}(\mathcal{H})$; with respect to the operator norm it is a *complete normed space*, i.e. $\mathcal{B}(\mathcal{H})$ is a *Banach space*, see Supp. Sect. 21.2. Since the product of two bounded operators is again bounded with $||A||\,||B|| \leq ||AB||$, it is also a *Banach algebra*, a notion that will be used in Supp. Chap. 27.

The operator norm is used to define convergence for sequences of bounded operators. One says that a sequence $\{A_n\}_n$ of such operators *converges in norm*, or *uniformly*, to a bounded operator $A$ iff $\lim_{n \to \infty} ||A - A_n|| = 0$. Norm convergence is too strong for many purposes. A weaker notion of convergence, called—strangely enough—strong operator convergence, is often more useful: A sequence $\{A_n\}_n$ of bounded operators *converges strongly* to a bounded operator $A$ iff $\lim_{n \to \infty} A_n\psi = A\psi$, i.e. iff $\lim_{n \to \infty} ||(A - A_n)\psi|| = 0$, for all $\psi$ in $\mathcal{H}$. We shall usually denote such a limit as s- lim. There is also *weak operator convergence*, meaning that $\lim_{n \to \infty} (\psi_1, (A - A_n)\psi_2) = 0$, for all $\psi_1$ and $\psi_2$ in $\mathcal{H}$.

### 21.6.3 Hermitian Adjoints: Selfadjoint Operators

Two bounded operators $A$ and $B$ in a Hilbert space $\mathcal{H}$ are called each others *Hermitian adjoint* iff, for each pair $\psi_1$ and $\psi_2$ in $\mathcal{H}$, one has

$$(A\psi_1, \psi_2) = (\psi_1, B\psi_2).$$

This relation between $A$ and $B$ is written as $A = B^*$, or $B = A^*$. The operator $A^*$ is called the *Hermitian adjoint* of $A$. (In this we follow general usage although it would be linguistically more correct to write 'Hermitean', as this notion is named after the nineteenth century French mathematician Charles Hermite.) Each bounded operator $A$ has a unique Hermitian adjoint $A^*$. To prove this we need the *Riesz representation theorem*:

(21.6.3,a) **Theorem** A continuous linear functional $F$ on $\mathcal{H}$ can be written as $F(\psi) = (\varphi_F, \psi)$, for all $\psi$ in $\mathcal{H}$ and with $\varphi_F$ a fixed vector in $\mathcal{H}$, uniquely determined by $F$.

*Proof*

1. We first prove the uniqueness. Suppose that for a given continuous linear functional $F$ there are vectors $\varphi_F$ and $\varphi'_F$ such that $F(\psi) = (\varphi_F, \psi)$ and $F(\psi) = (\varphi'_F, \psi)$, so $F(\psi) = (\varphi_F - \varphi'_F, \psi) = 0$, for all $\psi$ in $\mathcal{H}$. Since the inner product in $\mathcal{H}$ is nondegenerate, $\varphi_F - \varphi'_F = 0$, i.e. $\varphi_F = \varphi'_F$.
2. For the zero functional we have immediately $F(\psi) = (0, \psi)$, for every $\psi$. Suppose we have a continuous linear functional $F$ which is not the zero functional. The continuity of $F$ implies that Ker $F$ is a closed subspace of $\mathcal{H}$. It is a Hilbert space in its own right. Choose an orthonormal basis in $\mathcal{H}$ containing a subsequence $\{\psi_k\}_k$ which is an orthonormal basis for Ker $F$. $F$, as a map from $\mathcal{H}$ to $\mathbb{C}^1$, can be written as a direct sum of a map from Ker $F$ to $\{0\}$, and a map from $(\text{Ker } F)^{\perp}$, the orthogonal complement of Ker $F$, to $\mathbb{C}^1$. The last map is injective, so $(\text{Ker } F)^{\perp}$ is one dimensional and is therefore spanned by a single basis vector, say $\psi_{k_0}$. An arbitrary $\psi$ in $\mathcal{H}$ can be expanded as $\psi = \sum_{k \neq k_0} c_k \psi_k + c_{k_0} \psi_{k_0}$, with $F(\psi) = c_{k_0} F(\psi_{k_0})$. Define $\varphi_F = \overline{F(\psi_{k_0})} \psi_{k_0}$. Then

$$(\varphi_F, \psi) = (\overline{F(\psi_{k_0})} \psi_{k_0}, \psi)$$
$$= (\overline{F(\psi_{k_0})} \psi_{k_0}, c_{k_0} \psi_{k_0}) = c_{k_0} F(\psi_{k_0}) = F(\psi).$$

This proves the theorem.

We use this theorem to construct the Hermitian adjoint $A^*$ of a given $A$. Consider the expression $(\psi_1, A\psi_2)$. For fixed $\psi_1$ it is a continuous linear functional in the variable $\psi_2$, so it can be written as $F(\psi_2) = (\psi'_1, \psi_2)$, for a unique vector $\psi'_1$. This means that we assign to a given $\psi_1$ a $\psi'_1$, which by definition is $A^* \psi_1$. It can easily be checked that this assignment is linear and defines an operator with the desired properties.

(21.6,3,b) **Problem** Prove the following properties of Hermitian adjoints: $(A+B)^* = A^* + B^*$, $(A^*)^* = A$, $(\lambda A)^* = \overline{\lambda} A^*$, $(AB)^* = B^* A^*$ and $||A^*|| = ||A||$.

A bounded operator $A$ is called *selfadjoint* iff $A^* = A$. The term *Hermitian operator* is also used, especially in physics books. Its mathematical meaning there is usually not precise, so we shall avoid it. See the remark at the beginning of Supp. Sect. 21.9.2.

(21.6.3,c) **Problem** Show that for a bounded operator $A$ both $A^*A$ and $AA^*$ are selfadjoint.

There is a more general class of operators, which share important properties with selfadjoint operators: a bounded operator $A$ is called *normal* iff $A^*A = AA^*$. Unitary operators, to be discussed in the next section, are normal.

(21.6.3,d) **Problem** Show that the conditions $(A\psi_1, \psi_2) = (\psi_1, B\psi_2)$, for all $\psi_1, \psi_2 \in \mathcal{H}$, and $(A\psi, \psi) = (\psi, B\psi)$, for all $\psi \in \mathcal{H}$, are equivalent.

## 21.7 Unitary Operators

Each class of mathematical objects has natural isomorphisms: linear isomorphisms for vector spaces, homeomorphisms for topological spaces, continuous linear isomorphisms for topological vector spaces. For inner products one has linear isomorphisms that preserve inner products, which for Hilbert spaces are called *unitary transformations* in the case of isomorphisms and *unitary operators* for automorphisms, i.e. maps from a space onto itself, in our context the most important situation. We have the following obvious geometrical definition:

(21.7,a) **Definition** An invertible bounded operator $U$ on a Hilbert space $\mathcal{H}$ is unitary iff $(U\psi_1, U\psi_2) = (\psi_1, \psi_2)$, for all $\psi_1$ and $\psi_2$ in $\mathcal{H}$.

(D.21,b) **Problem** Show that it is sufficient to require $(U\psi, U\psi) = (\psi, \psi)$, for all $\psi$ in $\mathcal{H}$. Use this to show that $\|U\| = 1$.

An equivalent very practical algebraic definition employs Hermitian adjoints:

(21.7,c) **Problem** Show that $U$ is unitary iff $U^*U = 1$ and $UU^* = 1$. Explain why one needs both formulas, unless $\mathcal{H}$ is finite dimensional. Hint: Use basic linear algebra.

(21.7,d) **Problem** Choose an orthonormal basis $\phi_1, \phi_2, \ldots$ in $\mathcal{H}$. Define the operator $U$, a so-called *shift operator*, by $U\phi_j = \phi_{j+1}$, for $j = 1, 2 \ldots$ This operator clearly has the property $U^*U = 1$. Show that it is nevertheless not unitary.

## 21.8 Projection Operators

### 21.8.1 Definition and Simple Properties

Orthogonal projections on linear subspaces are important tools in the geometry of finite dimensional Euclidean spaces; this remains true for Hilbert space. In particular, projection operators can be used as building blocks to construct general, i.e. not necessarily bounded selfadjoint operators, as will be discussed in subsequent sections of this chapter. There is a simple algebraic definition of projection operators:

**(21.8.1,a) Definition** An *orthogonal projection* in a Hilbert space $\mathcal{H}$ is a selfadjoint idempotent operator, i.e. an operator $P$ with $P^* = P$ and $P^2 = P$.

In this book nonorthogonal projections, i.e. idempotent nonselfadjoint operators, will not be used, so from now on 'projection' will always mean 'orthogonal projection'.

**(21.8.1,b)** *Example* A simple but important example of a projection in the function space $\mathcal{H} = L^2(\mathbb{R}^1, dx)$ is the operator $P_\alpha$ defined as $(P_\alpha \psi)(x) = \psi(x)$ for $x \geq \alpha$, and $(P_\alpha \psi)(x) = 0$ for $x < \alpha$.

**(21.8.1,c) Problem** Show that a projection $P$ is a bounded operator which has norm 1, unless $P$ is the zero projection. Hint: Calculate $||P\psi||^2$ and use the Schwarz inequality.

Note that $P = 0$ and $P = 1$ are called the *trivial projections*.

A projection operator $P$ projects on a linear subspace $\mathcal{M}_P$, the image of $P$. The expression $P^c = 1 - P$ is also a projection and is called the *complementary projection* or *complement* of $P$. One checks easily that $\mathcal{M}_P = \text{Im } P = \text{Ker } P^c$. Problem (D.8.1,c) implies that $P^c$ is continuous. Because in Problem (A.3,c) the word 'closed' can be substituted for 'open', $(P^c)^{-1}(0) = \text{Ker } P^c = \text{Im } P$ is closed, so $\mathcal{M}_P$ is a closed linear subspace of $\mathcal{H}$.

Define the *orthogonal complement* of a closed subspace $\mathcal{M}$ as

$$\mathcal{M}^\perp = \{\psi \in \mathcal{H} \mid (\psi, \psi_1) = 0, \forall \psi_1 \in \mathcal{M}\}.$$

**(21.8.1,d) Problem** Show that $\mathcal{M}^\perp$ is a closed subspace of $\mathcal{H}$, with

1. $\mathcal{M}^\perp \cap \mathcal{M} = \{0\}$,
2. $(\mathcal{M}^\perp)^\perp = \mathcal{M}$,
3. $\mathcal{H}$ is the direct sum $\mathcal{M} \oplus \mathcal{M}^\perp$ (See Supp. Sect. 21.5 for 'direct sum'),
4. If $P$ projects on $\mathcal{M}$, then $P^c$ projects on $\mathcal{M}^\perp$.

Hint: Use the characterization of closed sets in terms of accumulation points given in Supp. Sect. 18.2. Choosing orthonormal bases in $\mathcal{M}$ and $\mathcal{M}^\perp$ is useful for proving 2 and 3.

A projection $P$ projects on a closed subspace $\mathcal{M}_P$. Conversely, given a closed subspace $\mathcal{M}$ there is a unique projection $P_\mathcal{M}$ on $\mathcal{M}$. The result is a one-to-one correspondence between projections in $\mathcal{H}$ and closed (linear) subspaces of $\mathcal{H}$.

**(21.8.1,e) Problem** Show that the operator $P_{\{\psi_j\}_j}$ which projects on the closed subspace generated by a (possibly infinite) sequence $(\psi_1, \psi_2, \dots)$ of linearly independent, mutually orthogonal vectors is given by the formula

$$P_{\{\psi_j\}_j}\psi = \sum_k \frac{(\psi_k, \psi)}{(\psi_k, \psi_k)}\psi_k,$$

which in the case of the one dimensional subspace generated by a single vector $\psi_0$ with unit norm reduces to $P_{\psi_0}\psi = (\psi_0, \psi)\psi_0$.

**(21.8.1,f) Problem**

1. Let $\{P_n\}_n$ be a sequence of projections which converges to a bounded operator $P$ in the sense of the strong operator limit. Show that $P$ is a projection.
2. Suppose that in addition $\{P_n\}_n$ is a sequence of commuting projections. Show that $P$ then commutes with all $P_n$.

## 21.8.2 Systems of Commuting Projections

In this section we consider systems $\mathcal{P} = \{P_\alpha\}_\alpha$ of *commuting* projections. The first example that comes to mind is the system consisting of the projections $P_\alpha$ from Example (21.8.1,b), for all real values of $\alpha$, with $P_{-\infty} = 0$ and $P_{+\infty} = 1$.

Any system of projections in Hilbert space, in particular the system of *all* projections, has a *partial ordering*. We restrict the discussion of this ordering to commuting systems.

*Reminder:* A set $X$ is said to be partially ordered if there is a binary relation $x \prec y$ between elements of $X$ with the following properties:

1. $x \prec x$ (identity property),
2. $x \prec y$, $y \prec x \Rightarrow x = y$ (reflexivity),
3. $x \prec y$, $y \prec z \Rightarrow x \prec z$ (transitivity),

for all $x$, $y$ and $z$ in $X$ for which these statements are meaningful.

If for each pair $x$ and $y$ in $X$ one has either $x \prec y$ or $y \prec x$ then $X$ is called *totally ordered*.

**(21.8.2,a) Definition** A system $\mathcal{P} = \{P_\alpha\}_\alpha$ of commuting projection operators in $\mathcal{H}$ has a partial ordering defined by

$$P_\alpha \prec P_\beta \iff P_\alpha P_\beta = P_\alpha.$$

**(21.8.2,b) Problem** Show that $\prec$ indeed defines a partial ordering in $\mathcal{P}$ and that the following four statements are equivalent:

1. $P_\alpha \prec P_\beta$,
2. $\mathcal{M}_\alpha \subset \mathcal{M}_\beta$,
3. $\|P_\alpha\psi\| \leq \|P_\beta\psi\|$, for all $\psi$ in $\mathcal{H}$,
4. $(\psi, P_\alpha\psi) \leq (\psi, P_\beta\psi)$, for all $\psi$ in $\mathcal{H}$.

Hint: Use for 1, 2, 3, 4 simple properties of norm and inner product. Use in 3 $\Rightarrow$ 1 the complementary projection $P_\beta^c = 1 - P_\beta$.

A given system of commuting projections $\mathcal{P}$ may be extended in four different ways, namely, by taking

1. complements: $P \mapsto P^c = 1 - P$,
2. products: $(P_1, P_2) \mapsto P_1 P_2$,
3. sums of two projections, if their product is 0,
   as $(P_1, P_2) \mapsto P_1 + P_2$,
4. limits of strongly convergent sequences,
   again projections according to Problem (21.8.1,f).

From Problem (21.8.2,b) we see that the partial ordering of projections corresponds geometrically to the partial ordering of closed subspaces by inclusion. The constructions of new projections from given ones have also obvious translations in the language of closed subspaces.

(21.8.2,c) **Problem** Prove the following relations:

1. $\mathcal{M}_{P^c} = \mathcal{M}_P^\perp$,
2. $\mathcal{M}_{P_\alpha P_\beta} = \mathcal{M}_{P_\alpha} \cap \mathcal{M}_{P_\beta}$,
3. $\mathcal{M}_{P_\alpha + P_\beta} = \mathcal{M}_{P_\alpha} \oplus \mathcal{M}_{P_\beta}$, for $P_\alpha P_\beta = 0$.

Hint: Use the general relation $P\psi = \psi \Leftrightarrow \psi \in \mathcal{M}_P$.

If a system of commuting projections $\mathcal{P}$ cannot be extended in this way, i.e. if it is invariant under the operations 1, 2, 3 and 4, we may call it *full*. An extremely important notion in the application of systems of commuting projections is the following:

(21.8.2,d) **Definition** A system $\mathcal{P}$ of commuting projections will be called *maximal* if and only if every projection that commutes with all projections in $\mathcal{P}$ belongs to $\mathcal{P}$.

The system of all projections $P_\alpha$ mentioned in Example (D.8.1,b) is not full, but can be extended to a full system, by using the above operations 1, 2, 3, 4. It is then maximal. The rather trivial example of the system consisting of the projections 0 and 1 is full but not maximal.

# 21.9 Unbounded Operators

## 21.9.1 Closed Operators

Physical observables are in quantum theory represented by selfadjoint operators. Most of these are unbounded. In this and the subsequent sections 'operator' will always mean 'linear operator defined on a dense linear subspace'. When necessary an operator $A$ will be denoted as $(A, \mathcal{D}_A)$, with $\mathcal{D}_A$ the domain of definition of $A$. See the remarks in Supp. Sect. 21.6.1. The notion of unbounded operator is too general to be useful. We need to specialize by adding a property, weaker than boundedness, the notion of *closed operator*.

There is a mathematically simple and elegant but not very intuitive definition. Remember that in general a *graph* of a map $f$ from a set $X$ to a second set $Y$ is the subset of the Cartesian product $X \times Y$ consisting of the pairs $(x, f(x))$, for all $x$ in $X$. For a linear operator $A$ in $\mathcal{H}$, with domain $\mathcal{D}_A$, this means the subset of $\mathcal{H} \times \mathcal{H}$ consisting of the pairs $(\psi, A\psi)$, for all $\psi$ in $\mathcal{D}_A$.

**(21.9.1,a) Definition** An operator $A$ on a domain $\mathcal{D}_A$ is *closed* iff its graph is a closed set in the Cartesian product $\mathcal{H} \times \mathcal{H}$.

A second equivalent definition is sometimes more practical. Note that if a sequence $\{\psi_n\}_n$ is convergent, the corresponding sequence $\{A\psi_n\}_n$ may not converge, because of the lack of continuity of $A$.

**(21.9.1,b) Problem** Show that an operator $A$ is closed iff the following holds: for every sequence $\{\psi_n\}_n$ in $\mathcal{D}_A$, which is convergent and for which the corresponding sequence of images $\{A\psi_n\}_n$ is also convergent, the limit $\psi$ of $\{\psi_n\}_n$ is in $\mathcal{D}_A$ with the limit of $\{A\psi_n\}_n$ equal to $A\psi$.

If an operator $A$, with domain $\mathcal{D}_A$, is not closed, it may be possible to extend it to a closed operator $\overline{A}$, with domain $\mathcal{D}_{\overline{A}}$. There is an obvious definition: an operator $A$ is *closable* iff the closure of its graph in $\mathcal{H} \times \mathcal{H}$ is again the graph of an operator. From this definition follows that such an extension, if it exists, is unique.

**(21.9.1,c) Problem** Give an alternative definition of 'closable' in the spirit of the answer in Problem (21.9.1,b).

In this book we shall have not much use for closable operators as such; closable operators will usually be assumed to have been extended to its closure. We have no use for operators that are not closable.

## 21.9.2 Symmetric and Selfadjoint Operators

In standard physics textbooks observables in quantum mechanics are represented by what is called 'Hermitian operators', which are supposed to have complete orthonormal bases of eigenvectors, with real eigenvalues, interpreted as the possible results of measurements. This looks like an obvious generalization of what is known of finite dimensional hermitian matrices to the case of operators in infinite dimensional Hilbert space. It is an intuitively appealing picture but the actual mathematical situation in infinite dimensional space is much more complicated, even more so because most observables in quantum theory are represented by unbounded operators—however even bounded operators may have continuous spectrum in addition to discrete eigen-values. The in this context physically obvious notion of 'spectrum' will be given a precise mathematical meaning further on.

In this section we discuss the complications appearing in the notions of Hermitian adjoint and selfadjoint operator in the case of unbounded operators.

Let $A$ be an operator with dense domain $\mathcal{D}_A$. Define the linear subspace

$$\mathcal{D} := \{\psi \in \mathcal{H} \mid \exists \, \psi' \text{ such that } (\psi', \varphi) = (\psi, A\varphi), \forall \varphi \in \mathcal{D}_A\}.$$

The assignment $\psi \mapsto \psi'$ defines an operator $A^*$ on the domain $\mathcal{D}_{A^*} = \mathcal{D}$. It can be shown that this $\mathcal{D}_{A^*}$ is dense in $\mathcal{H}$ if and only if $A$ is closed or closable. See for this Ref. [2], p. 252.

(21.9.2,a) **Problem** Show that the Hermitian adjoint $A^*$ of $A$ is a closed operator. Show moreover that $(A^*)^* = \overline{A}$, with $\overline{A}$ the closure of $A$ and that for two operators $A$ and $B$ with $A \subseteq B$ one has $B^* \subseteq A^*$. Hint: Use the second definition of a closed operator, formulated in Problem (21.9.1,b), to prove that a Hermitian adjoint is closed. See for the meaning of the relation $\subseteq$ for operators Supp. Sect. 21.6.2.

For unbounded operators there is a notion that looks like selfadjointness but is in fact weaker:

(21.9.2,b) **Definition** An operator $A$ with domain $\mathcal{D}_A$ is called *symmetric* iff one has, for all $\psi$ and $\varphi$ in $\mathcal{D}_A$,

$$(\psi, A\varphi) = (A\psi, \varphi).$$

A compact formulation of this definition is that $A$ is symmetric iff $A \subseteq A^*$.

(21.9.2,c) **Problem** Show the equivalence of Definition (21.9.2,b) and this compact formula.

The spectral theorem, the centre piece of the mathematical formulation of quantum theory, is not valid for symmetric operators. We therefore need the more refined notion of selfadjointness, more appropriate for the case of unbounded operators.

(21.9.2,d) **Definition** An operator $A$ on a domain $\mathcal{D}_A$ is *selfadjoint* iff the situation with a vector $\psi$ in $\mathcal{H}$ with $(\psi, A\varphi) = (\psi', \varphi)$, for some vector $\psi'$ in $\mathcal{H}$ and all $\varphi$ in $\mathcal{D}_A$, always implies $\psi \in \mathcal{D}_A$ and $\psi' = A\psi$.

There is again a compact formulation, deceptively simple, namely $A = A^*$. Note that this includes the requirement $\mathcal{D}_A = \mathcal{D}_{A^*}$. Note also that from this form of the definition it follows immediately that selfadjointness implies symmetry, but not vice versa.

(21.9.2,e) **Problem** Show the equivalence of Definition (21.9.2,d) with the compact formulation $A = A^*$.

For a bounded operator the definitions of symmetry and selfadjointness coincide, in which case one may fall back on the simple notion of Hermiticity as used in physics textbooks.

There are various criteria for the selfadjointness of a symmetric operator. A simple one is the following: The closed symmetric operator $A$ is selfadjoint iff Ker $(A^* \pm i) = \{0\}$. See for a proof of this [2], p. 257.

An operator may be symmetric but not selfadjoint, for instance when it is not closed. If its closure is selfadjoint it is called *essentially selfadjoint*. We have a less trivial situation when after closure it is still not selfadjoint. In that case there are two possibilities:

a. The operator may have further symmetric extensions but does *not* possess a *selfadjoint* extension. Such an operator is useless for quantum theory.
b. The operator has a selfadjoint extension. In that case there are in fact infinitely many different ones. The choice which of these has to be used will depend on the physical context. This is discussed for instance in Sect. 5.4.

(21.9.2,f) *Example* Define in $\mathcal{H} = L^2(\mathbb{R}^1, dx)$ the multiplication operator $M$ as $(M\psi)(x) = x\,\psi(x)$ and the differentiation operator $D$ as $(D\psi)(x) = -i\frac{d}{dx}\psi(x)$. Both operators are essentially selfadjoint on the domain $\mathcal{D} = \mathcal{S}(\mathbb{R}^1)$, the space of Schwartz functions, i.e. the infinitely differentiable complex-valued functions on $\mathbb{R}^1$, which go to 0, for $x \to \pm\infty$, faster than any inverse polynomial in $x$. The closure of $M$ is selfadjoint on all square integrable functions $\psi$, such that $\int_{-\infty}^{+\infty} x^2 |\psi(x)|^2 dx < \infty$; the closure of $D$ on all functions $\psi$ which are almost everywhere differentiable, i.e. except on a set of measure 0, such that $\int_{-\infty}^{+\infty} |\frac{d}{dx}\psi(x)|^2 dx < \infty$. In quantum mechanics the operator $M$ describes the position of a one dimensional particle, the operator $D$, after insertion of Planck's constant $\hbar$ for dimensional reasons, its linear momentum. See Sect. 4.1. Similar statements hold, of course, for the three dimensional case.

(21.9.2,g) *Example* Consider in $\mathcal{H} = L^2([a, b], dx)$, with $-\infty < a < b < +\infty$, the differentiation operator $D$ as in (D.9.2,f), with as domain $C^\infty([a, b])$, the infinitely differentiable functions $\psi$ on the interval $[a, b]$, with $\psi(a) = \psi(b) = 0$. This operator is symmetric but not yet closed. After closure it can be extended to a selfadjoint operator in infinitely many different ways. This example describes a one dimensional quantum mechanical particle in a box. The different selfadjoint extensions of $D$ correspond with different boundary conditions at the walls of the box. This is discussed in Sect. 5.4.

(21.9.2,h) *Example* Let $\mathcal{H} = L^2([0, +\infty), dx)$ and $D$ the same operator as before, now defined on the infinitely differentiable functions $\psi$ on $[0, +\infty)$, with $\psi(0) = 0$. This operator has no selfadjoint extensions and cannot be used to represent a quantum mechanical observable.

In a certain very generalized sense there exists for a selfadjoint operator a complete orthonormal basis of eigenvectors with real eigenvalues. This is stated by the *spectral theorem*, a theorem which is valid for selfadjoint but not for symmetric operators. It is of central importance for quantum theory and will be discussed in the next section.

## 21.10 The Spectral Theorem for Selfadjoint Operators

### 21.10.1 Introduction

The spectral theorem for selfadjoint operators plays a central role in Hilbert space theory; as such it is the basis for the physical interpretation of the mathematical framework of quantum theory in terms of states and observables as presented in this book. Intimately connected with this theorem is a cluster of mathematical notions such as *eigenvalue* and *spectrum, spectral resolution, projection-valued measure, multiplicity* and *diagonalization*, all having important physical interpretations.

In understanding the spectral theorem, it is helpful to approach it—together with the notions that are associated with it—as a generalization of an elementary theorem from finite dimensional linear algebra to the technically more sophisticated context of infinite dimensional Hilbert space. This is the strategy that will be adopted in this section. In Supp. Sect. 21.10.2 the general theory of the spectrum of operators in Hilbert space is reviewed—first briefly for the finite dimensional case, just in terms of eigenvalues and eigenvectors of hermitian matrices—and is then brought in a more general mathematical form suitable to infinite dimension. In Supp. Sect. 21.10.3 illustrative examples of selfadjoint operators in infinite dimensional space are given, which have continuous spectrum, illustrating the difference between finite and infinite dimension.

In Supp. Sect. 21.10.4 the reader is reminded of the spectral theorem in finite dimension, which is just the statement that a selfadjoint $A$ can be uniquely written as a sum $A = \sum_j \alpha_j P_j$, with an increasing finite sequence of real eigenvalues $\alpha_j$ and projections $P_j$ on the corresponding eigenspaces. In Supp. Sect. 21.10.5 this expression is transformed by a succession of simple steps into a Riemann-Stieltjes integral $\int \alpha \, dE_\alpha$ over a 'spectral resolution' $\{E_\alpha\}_{\alpha \in \mathbb{R}^1}$, a result which is completely equivalent to the discrete sum formula, although a bit far-fetched in this finite dimensional context. Using this idea of a vector-valued Riemann-Stieltjes integral over a one parameter family of projections, the main theorem of this section, the general spectral theorem for a selfadjoint operator in a Hilbert space of arbitrary finite or infinite dimension is formulated and made plausible in Supp. Sect. 21.10.6. This theorem is extended to the case of a system of a finite number of commuting selfadjoint operators in Supp. Sect. 21.10.7.

In the remaining subsections various ideas related to the spectral theorem are discussed. In Supp. Sect. 21.10.8 we briefly treat projection-valued measures, a mathematically elegant generalization of spectral resolutions, in Supp. Sect. 21.10.9 general functions of a selfadjoint operator or of a commuting system of such operators, in Supp. Sect. 21.10.10 multiplicity properties, with the introduction of the notion of a complete system of commuting selfadjoint operators, important for its application to the labelling of quantum states in physics, and finally in Supp. Sect. 21.10.11 diagonalization of selfadjoint operators.

Further details, as well as proofs not given here, can be found in [2, 3], the main references for this section.

A good introduction to spectral theory in Hilbert space can be found in [7].

## *21.10.2 The Spectrum of an Operator*

Consider first a finite dimensional Hilbert space $\mathcal{H}$, which can, of course, be identified with $\mathbb{C}^n$ for some $n$, after a choice of an orthonormal basis, meaning that we are discussing elementary linear algebra. A number $\lambda$ is called an *eigenvalue* of the linear operator $A$ if there is a nonzero vector $\psi$ with $A\psi = \lambda\psi$. Such a $\psi$ is called an *eigenvector*.

(21.10.2,a) **Problem** Show that for a selfadjoint operator the eigenvalues are real and that eigenvectors for different eigenvalues are mutually orthogonal.

In infinite dimension the situation is much more complicated. The definition of eigenvalue and eigenvector remains valid. However, many important operators in quantum theory do not have eigenvalues or eigenvectors in the strict sense; they still have generalized eigenvalues and eigenvectors and exhibit 'spectral properties'. A review of the general notion of spectrum may therefore be useful.

Before this we may briefly look at how the notion of spectrum—very important in quantum mechanics—is treated in most physics textbooks. Two types are mentioned; the intuitive meaning of both is clear:

1. *Discrete spectrum.* By this one means the collection of isolated eigenvalues associated with normalizable eigenfunctions. This is in fact the discrete spectrum of the rigorous description below, not what is called there the point spectrum.
2. *Continuous spectrum.* In physics one needs the absolutely continuous spectrum defined below; there is no use for the singular continuous spectrum.

Let us now give the rigorous classification for the various kind of spectra for operators. An important notion in this context is that of a *resolvent*.

a. The earlier definition of eigenvector and eigenvalue is valid for an arbitrary linear operator in an arbitrary vector space.
   Let $\mathcal{H}$ be a (separable, complex) Hilbert space.
b. Suppose $A$ to be a densely defined operator in $\mathcal{H}$.

(21.10.2,b) **Definition** The *resolvent set* $\rho(A)$ of $A$ consists of all complex numbers $z$ such that the operator $A - z$ has a bounded inverse. This operator is called the *resolvent* and is denoted as

$$R(A - z) = (A - z)^{-1}.$$

It is an operator-valued analytic function on $\rho(A)$. One can show that $\rho(A)$ is an open set.

(21.10.2,c) **Definition** The *spectrum* $\sigma(A)$ of $A$ is the complement of $\sigma(A)$ in the complex plane. The set $\rho(A)$ is obviously closed.

c. Let $A$ be selfadjoint operator in $\mathcal{H}$.

The spectrum $\sigma(A)$ is a subset of the real line.

The different types of spectrum of $A$ can be classified by looking at the properties of the map $A - \lambda$.

1. The *point spectrum* $\sigma_{pt}(A)$: The map $A - \lambda$ as a map from $\mathcal{D}_{A-\lambda} = \mathcal{D}_A$ into $\mathcal{H}$ is not injective.

   (21.10.d) **Problem** Show that $\sigma_{pt}(A)$ consists of all the eigenvalues of $A$.

   The eigenvalues that have finite multiplicity and are isolated points of $\sigma(A)$ form the *discrete spectrum* $\sigma_{disc}(A)$, so $\sigma_{disc}(A) \subset \sigma_{pt}(A)$.
   Finite multiplicity of an eigenvalue means that the corresponding eigenspace is finite dimensional. A point $\lambda_0$ is an isolated point of the spectrum iff there exists an $\epsilon > 0$ such that the interval $(\lambda_0 - \epsilon, \lambda_0 + \epsilon)$ does not contain other points of the spectrum.

2. The *continuous spectrum*: $\sigma_{ct}(A)$. The map $A - \lambda$ is injective; its image is dense in $\mathcal{H}$.
   The continuous spectrum is the disjunct union of the *absolutely continuous spectrum* $\sigma_{ac}(A)$ and *singular continuous spectrum* $\sigma_{sc}(A)$. The idea of this splitting is that it is an operator version of the Lebesgue decomposition of a Borel measure on $\mathbb{R}^1$, discussed in Supp. Sect. 19.6.

3. The *residual spectrum* $\sigma_{res}(A)$. The map $A - \lambda$ is injective; its image is not dense in $\mathcal{H}$.

The residual spectrum is of no concern to us, as it is absent in a selfadjoint operator. As stated above, the only types of spectrum that are acceptable for physical theories are the discrete and the absolutely continuous spectrum. It is in general rather difficult to prove that a given Hamiltonian operator has no singular continuous spectrum. In view of this we shall in much of our text mean by 'continuous spectrum' just 'absolutely continuous spectrum', unless stated otherwise.

It will be obvious that the spectrum of a unitary operator is a closed subset of the unit circle in the complex plane. Most of the unitary operators in this book belong to continuous one parameter groups; for these the spectrum can be easily found from their selfadjoint generator.

### 21.10.3 Examples

In this section a few examples are given that illustrate the points discussed in the preceding sections.

(21.10.3,a) *Example* Consider in $\mathcal{H} = L^2(\mathbb{R}^1, dx)$ the multiplication operator $M$ from Example (21.9.2,f). The eigenvalue–eigenfunction equation for this operator is $x\psi(x) = \alpha\psi(x)$. For a particular $\alpha = \alpha_0$ this gives $\psi(x) = 0$, for $x \neq \alpha_0$, and leaves $\psi(x)$ undetermined for $x = \alpha_0$. With functions in $\mathcal{H} = L^2(\mathbb{R}^1, dx)$ we really mean equivalence classes of functions differing only on sets of Lebesgue measure zero; this $\psi$ is equivalent to the function which is identically 0, a function not permitted as eigenfunction. The conclusion is clear: the selfadjoint operator $M$ has no eigenvalues.

Nevertheless, the spectral theorem applies to $M$, as will be shown; $M$ has 'generalized' eigenvalues, that in fact fill the real line. This is reflected in an heuristic approach due to P.A.M. Dirac, one of the pioneers in the formulation of quantum theory. To save the idea of eigenfunctions and eigenvectors in the above example he introduced a 'singular' function $\delta$, now known as the *Dirac $\delta$-function*, which is zero at $x \neq 0$ and infinite at $x = 0$. Integrated over the real line it gives 1; its integral with an arbitrary smooth real-valued function $f$ is assumed to give $\int_{-\infty}^{+\infty} f(x)\delta(x)dx = f(0)$. There are translated delta functions $\delta_a$ centered at arbitrary points $a$, obtained as $\delta_a(x) = \delta(x - a)$, which are supposed to give integrals $\int_{-\infty}^{+\infty} f(x)\delta_a(x) = f(a)$. Mathematicians, in the first place the French mathematician Laurent Schwartz, made this idea into a new piece of rigorous mathematics, the theory of *generalized functions* or *distributions*. This is reviewed in Supp. Chap. 26. Dirac's delta function calculus has shown itself to be a powerful heuristic tool for explicit manipulations in quantum mechanics. However, in the mathematically precise approach followed in this book it has only a restricted place.

(21.10.3,b) *Example* Consider the operator $D$ from (21.9.2,f). As a differentiation operator it has the eigenfunctions $\psi_k(x) = e^{ikx}$, for each real eigenvalue $k$.

The functions $\psi_k$ are not square integrable and correspond therefore not with Hilbert space vectors. As a Hilbert space operator $D$ has no eigenvalues and eigenvectors. Just as in the case of the Dirac $\delta$-functions for $M$, the 'improper' eigenvectors of $D$ are useful and are extensively used as *plane waves* for heuristic calculations in quantum mechanics, but again have no place in a mathematically rigorous formulation of quantum theory.

(21.10.3,c) *Example* Let $\mathcal{H}$ be an infinite dimensional Hilbert space, with an orthonormal basis $(\phi_1, \phi_2, \dots)$. Define the selfadjoint operator $A$ by $A\phi_n = \frac{1}{n}\phi_n$. It is clear that $A$ has the $\phi_n$ as eigenvectors, with eigenvalues $\alpha_n = \frac{1}{n}$, for $n = 1, 2, \dots$.

(21.10.3,d) **Problem** Show that the number $\lambda = 0$ is not an eigenvalue of $A$ but belongs nevertheless to its spectrum.

The fact that has to be proved in this example is not surprising. We know already that the spectrum of an operator is a closed set. The point $\lambda = 0$ is the accumulation point of the sequence of eigenvalues $\{\alpha_n = \frac{1}{n}\}_n$. Adding this point makes the sequence into a closed set. See Supp. Sect. 18.2 for the notion of accumulation point.

Example (21.10.3,c) can be modified slightly by enlarging the Hilbert space by adding a vector $\phi_\infty$ to the basis $(\phi_1, \phi_2, \ldots)$, with $A$ acting as $A\phi_\infty = 0$. This does not change the spectrum. The number 0 is again in the spectrum; however 0 is now an eigenvalue.

## 21.10.4 The Spectral Theorem: Finite Dimension

As was already observed, finite dimensional Hilbert space theory is nothing but elementary complex linear algebra. Choosing a fixed orthonormal basis, vectors become finite sequences of coordinates; operators finite matrices, with in particular selfadjoint operators becoming Hermitian matrices $\{A_{jk}\}_{jk}$, i.e. matrices with the matrix element $A_{jk}$ equal to the complex conjugate $\overline{A}_{kj}$ of the matrix element $A_{kj}$.

A central result in this context is that such a Hermitian $n \times n$ matrix can be brought in *diagonal form* by means of a transformation with a suitable unitary matrix $\{U_{jp}\}_{jp}$, i.e. matrices with $(U^{-1})_{jp} = (U^*)^{-1}_{jp} = \overline{U}_{pj}$, according to

$$\sum_{p,q=1}^{n} U_{jp} A_{pq} U_{qk}^{-1} = \delta_{jk}\alpha_j,$$

with the real numbers $\alpha_j$ the *characteristic values* or *eigenvalues* of the Hermitian matrix $\{A_{jk}\}_{jk}$.

Going back to the more intrinsic Hilbert space formulation one says that a selfadjoint operator $A$ in an $n$-dimensional Hilbert space $\mathcal{H}$, has an orthonormal basis of eigenvectors with real eigenvalues, i.e. a system of vectors $\phi_1, \ldots, \phi_n$, with $(\phi_j, \phi_k) = \delta_{jk}$, for $j, k = 1, \ldots, n$, and a unique expansion $\psi = \sum_{j=1}^{n} c_j\phi_j$ for each $\psi$ in $\mathcal{H}$, together with real numbers $\alpha_1, \ldots, \alpha_n$ such that $A\phi_j = \alpha_j\phi_j$, for $j = 1, \ldots, n$. If the eigenvalues $\alpha_j$ are degenerate, this orthonormal basis is not uniquely determined; even in the nondegenerate case there is still the freedom of multiplying each $\phi_j$ by a phase factor. To eliminate this nonuniqeness from the formulation it is better to consider eigenvalues which are all different, $\alpha_1 < \alpha_2 < \cdots < \alpha_p$, with $p \leq n$, and use instead of eigenvectors the eigenspaces $\mathcal{M}_j$, for each eigenvalue $\alpha_j$. Let $P_j$ be the orthogonal projections on the $\mathcal{M}_j$. One has $P_j P_k = P_k P_j = 0$ and $\sum_{j=1}^{p} P_j = 1$. This statement about eigenvalues and eigenvectors, or rather eigenspaces, can be formulated as a theorem.

(21.10.4,a) **Theorem** (Finite dimensional spectral theorem) *For a selfadjoint operator $A$ in a finite dimensional Hilbert space $\mathcal{H}$ there exists a unique set of real numbers $\alpha_1 < \alpha_2 < \cdots < \alpha_p$, forming the spectrum $\sigma(A)$ of the operator $A$, together with* a unique system of mutually orthogonal projection operators $P_1, P_2, \ldots, P_p$ with $P_j P_k = P_k P_j = 0$, for $j, k = 1, \ldots, p$, and $\sum_{j=1}^{p} P_j = 1$, such that

$$A = \sum_{j=1}^{p} \alpha_j P_j.$$

## 21.10.5 Towards the Infinite Dimensional Spectral Theorem

When $\mathcal{H}$ is infinite dimensional, the spectral theorem has in some cases the same form as in the finite dimensional situation, apart from the fact that the sum now runs to $\infty$. An example is the operator $D^2 + M^2$, given as

$$((D^2 + M^2)\psi)(x) = (-\frac{d^2}{dx^2} + x^2)\psi(x),$$

which is, up to physical constants, the operator representing the energy of the one dimensional quantum mechanical harmonic oscillator, as discussed in Chap. 6 of the main text. It has discrete nondegenerate eigenvalues $\alpha_j = 2j + 1$, for $j = 0, 1, \ldots$, with a corresponding orthonormal basis of eigenvectors $\{\phi_j\}_{j=0}^{\infty}$.

Such a case is just one possibility, fairly exceptional. Consider as a typical example of a different type again the operator $D = -i\frac{d}{dx}$ from Example (21.9.2,f). It is symmetric on $\mathcal{S}(\mathbb{R}^1)$ and can be extended to a selfadjoint operator on an appropriate domain. As a differential operator it has eigenfunctions $\phi_k(x) = e^{ikx}$ with eigenvalues $k$, for each real number $k$, as was already noted. An interesting and suggestive fact is that an arbitrary $\psi$ in $\mathcal{H} = L^2(\mathbb{R}^1, dx)$ can be expanded, by using the inverse Fourier transformation formula, discussed in Sect. 4.2, as

$$\psi(x) = \int_{-\infty}^{+\infty} \hat{\psi}(k)e^{+ikx}dk = \int_{-\infty}^{+\infty} c_k \phi_k dk,$$

with the 'expansion coefficients' $c_k$ equal to $\hat{\psi}(k)$. The differential operator $D$ has eigenfunctions, real eigenvalues and a corresponding eigenfunction expansion. One is tempted to write a continuous version of the operator formula for the spectral theorem in finite dimensional space, something like

$$D = \int_{-\infty}^{+\infty} k P_k dk, \quad \text{or, maybe,} \quad D = \int_{-\infty}^{+\infty} k \, dP_k.$$

However, both these formulas are mathematically meaningless; the eigenfunctions $\phi_k$ are not square integrable, i.e. are *not* elements of $\mathcal{H}$, so the projections $P_k$ do not exist. To find a rigorous form for this heuristic idea we first reformulate the spectral theorem for the finite dimensional case, given in the preceding section.

Let $A$ be again a selfadjoint operator in a finite dimensional Hilbert space $\mathcal{H}$, with $\alpha_1 < \cdots < \alpha_p$ its eigenvalues, and $P_1, \ldots, P_p$ the corresponding projections on the eigenspaces $\{\mathcal{M}_j\}_j$. Define new projection operators $E_j = \sum_{s=1}^{j} P_s$, clearly again a system of commuting projection operators, now on the direct sum spaces $\mathcal{N}_j = \oplus_{s=1}^{j} \mathcal{M}_s$. The operators $E_j$ form a sequence of projections which is monotone, nondecreasing, i.e. with $E_j \prec E_{j+1}$, in the sense of the order relation for projections defined in Supp. Sect. 21.8.2, together with the corresponding subspaces $\mathcal{N}_j$, ordered by inclusion according to $\mathcal{N}_j \subset \mathcal{N}_{j+1}$.

If we introduce for notational convenience $E_0 = 0$, then $P_j$ can be recovered from the $E_j$ as $P_j = E_j - E_{j-1}$, for $j = 1, \ldots p$, so the formula for the spectral theorem for the operator $A$ becomes

$$A = \sum_{j=1}^{p} \alpha_j P_j = \sum_{j=1}^{p} \alpha_j (E_j - E_{j-1}).$$

One next converts in a rather trivial way the finite sequence $\{E_j\}_{j=1}^{p}$ into an infinite system of projections $E_\alpha$, depending on a continuous real parameter $\alpha$. Define, for $j = 1, \ldots, p-1$,

$$E_\alpha = E_j, \text{ for } \alpha \in [\alpha_j, \alpha_{j+1}),$$

together with

$$E_\alpha = 0, \text{ for } \alpha < \alpha_1, \quad E_\alpha = 1, \text{ for } \alpha \geq \alpha_p.$$

The result is an infinite system of commuting projections, monotone nondecreasing in the parameter $\alpha$. It has $\lim_{\alpha \to -\infty} E_\alpha = 0$ and $\lim_{\alpha \to +\infty} E_\alpha = 1$ and is (by definition) continuous from the right in $\alpha$. The new system $\{E_\alpha\}_{\alpha \in R^1}$ is completely equivalent to the original sequence $\{E_j\}_{j=1}^{p}$.

The next step, in what looks like a way to bring a simple theorem into a complicated form, is to introduce the notion of a *Stieltjes integral*. See for this Supp. Sect. 19.5.3. The Stieltjes integral of a function $f$ with respect to a second function $g$ is denoted by the expression $\int f(x) dg(x)$.

For an arbitrary vector $\psi$ in $\mathcal{H}$ the expression $(\psi, E_\alpha \psi)$ is such a function $g$, in fact a step function which takes on only a finite number of different values, is equal to 0 for $\alpha < \alpha_1$ and equal to the positive number $||\psi||^2 = (\psi, \psi)$ for $\alpha \geq \alpha_p$. One gets therefore immediately

$$\int_{-\infty}^{+\infty} \alpha \, d(\psi, E_\alpha \psi) = \sum_{j=1}^{p} \alpha_j (\psi, (E_j - E_{j-1})\psi)$$

$$= \sum_{j=1}^{p} \alpha_j (\psi, P_j \psi) = (\psi, A\psi).$$

This can be seen as the definition of the operator relation

$$A = \int_{-\infty}^{+\infty} \alpha \, dE_\alpha,$$

which, if one wishes, can be understood as a way of formulating a collection of relations in terms of real numbers

$$(\psi, A\psi) = \int_{-\infty}^{+\infty} \alpha \, d(\psi, E_\alpha \psi),$$

for all vectors $\psi$ in $\mathcal{H}$. With this we have obtained a formulation of the spectral theorem for a selfadjoint operator in a finite dimensional Hilbert space, which is completely equivalent to the original one and which will carry over to the infinite dimensional case in the next subsection, because it contains nothing that is not meaningful in an infinite dimensional Hilbert space.

## 21.10.6  The General Spectral Theorem

Collecting the data from the preceding subsection, we obtain the following definition valid in a Hilbert space $\mathcal{H}$ of arbitrary finite or infinite dimension.

(21.10.6,a) **Definition** A *one parameter spectral resolution*, or short, *spectral resolution*, in $\mathcal{H}$ is a system $\{E_\alpha\}_{\alpha \in R^1}$ of commuting projections in $\mathcal{H}$ with the following properties:

1. $E_{\alpha_1} \prec E_{\alpha_2}$, for $\alpha_1 \le \alpha_2$,
2. $\lim_{\alpha \to -\infty} E_\alpha = 0$ and $\lim_{\alpha \to +\infty} E_\alpha = 1$,
3. $E_\alpha$ is (strongly) continuous from the right as a function in $\alpha$.

These conditions are equivalent to:

1. $(\psi, E_{\alpha_1} \psi) \le (\psi, E_{\alpha_1} \psi)$, for $\alpha_1 \le \alpha_2$,
2. $\lim_{\alpha \to -\infty} (\psi, E_\alpha \psi) = 0$ and $\lim_{\alpha \to +\infty} (\psi, E_\alpha \psi) = (\psi, \psi)$,
3. $(\psi, E_\alpha \psi)$ is continuous from the right as a function of $\alpha$,

for all vectors $\psi$ in the Hilbert space $\mathcal{H}$. For each $\psi$ the expression $(\psi, E_\alpha \psi)$ is a real-valued monotone, nondecreasing function of $\alpha$, so $\lim_{\alpha \uparrow \alpha_0} (\psi, E_\alpha \psi)$ and $\lim_{\alpha \downarrow \alpha_0} (\psi, E_\alpha \psi)$ exist, for every $\alpha_0$ and all $\psi$. See Supp. Sect. 19.4. This implies that $\lim_{\alpha \uparrow \alpha_0} E_\alpha$ and $\lim_{\alpha \downarrow \alpha_0} E_\alpha$ exist, in the sense of strong operator convergence, as projections $E_{\alpha_0^-}$ and $E_{\alpha_0^+}$. Requirement 3 of (21.10.6,a) means the choice of $E_{\alpha_0} = E_{\alpha_0^+} = \lim_{\alpha \downarrow \alpha_0} E_\alpha$, which is a matter of convention. Note that for unit vectors $\psi$ such monotonic functions have the properties of distribution functions in probability theory—see Supps. Sects. 22.2 and 22.3. That is in fact the role they play in the operator context of quantum theory.

Remember from Supp. Sect. 21.10.4 that the formula for the spectral theorem for a finite dimensional selfadjoint operator $A$ was $A = \sum_{j=1}^{p} \alpha_j P_j$, with $P_j$ the eigenspaces for the eigenvalues $\alpha_j$. By transforming the sequence of eigenprojections $P_j$ into a nondecreasing sequence of projections $E_j = \sum_{s=1}^{j} P_s$, this formula became a sum over differences $\alpha_j (E_j - E_{j-1})$, which by extending the definition of the $E_j$ in a rather trivial manner to all real $\alpha$ could be written as the operator-valued Riemann-Stieltjes integral $\int_{-\infty}^{+\infty} \alpha \, dE_\alpha$. The system $\{E_\alpha\}_\alpha$ defined in this way is a simple finite dimensional example of a spectral resolution in the sense of the above definition, with the requirements on the limits $\alpha \to \pm\infty$ trivially satisfied, because $E_\alpha$ is identically 0 for $\alpha$ smaller as the lowest eigenvalue and identically 1 for $\alpha$ larger than the highest eigenvalue.

All this is a preparation to state the main theorem of this chapter:

(21.10.6,b) **Theorem** (Spectral theorem for a selfadjoint operator) *Let $\mathcal{H}$ be a Hilbert space of arbitrary finite or infinite dimension, and let $A$ be a selfadjoint operator in $\mathcal{H}$, with domain $\mathcal{D}_A$. Then there is a unique spectral resolution $\{E_\alpha\}_{\alpha \in \mathbb{R}^1}$, in the sense of Definition (21.10.6,a), such that*

$$A = \int_{-\infty}^{+\infty} \alpha \, dE_\alpha,$$

or, for all $\psi$ in $\mathcal{D}_A$,

$$(\psi, A\psi) = \int_{-\infty}^{+\infty} \alpha \, d(\psi, E_\alpha \psi).$$

Conversely, a spectral resolution $\{E_\alpha\}_\alpha$ defines a selfadjoint operator on the domain consisting of all vectors $\psi$ such that the Riemann-Stieltjes integral $\int_{-\infty}^{+\infty} \alpha^2 d(\psi, E_\alpha \psi)$ is finite.

The proof of this theorem can be found in [2], Chaps. VII and VIII.3.

The various types of spectrum of the operator $A$ can be read off from the system $\{E_\alpha\}_\alpha$. One has the following cases: the operator valued function $E_\alpha$ is

1. *constant* in $\alpha_0$, i.e. there exists a positive $\varepsilon$ such that $E_{\alpha_0+\varepsilon} = E_{\alpha_0-\varepsilon}$. Such points $\alpha_0$ clearly form an open set, which is the resolvent set $\rho(A)$, consisting of the points which do not belong to the spectrum $\sigma(A)$ of $A$.

2. *discontinuously increasing* in $\alpha_0$, i.e. one has $E_{\alpha_0} \neq \lim_{\alpha\uparrow\alpha_0} E_\alpha$. Such point $\alpha_0$ is an eigenvalue of $A$, with $E_{\alpha_0} - \lim_{\alpha\uparrow\alpha_0} E_\alpha$, the corresponding eigenprojection. The closure of the sets of these points is what we have called in Supp. Sect. 21.10.2 the discrete spectrum $\sigma_d(A)$ of $A$.

3. *continuously increasing* in $\alpha_0$, meaning that $E_{\alpha_0} = \lim_{\alpha\uparrow\alpha_0} E_\alpha$, together with $E_{\alpha_0+\varepsilon} \neq E_{\alpha-\varepsilon}$, for all positive $\varepsilon$. The closure of this set of points is $\sigma_c(A)$, the continuous spectrum of $A$. As noted in Supp. Sect. 21.10.2 we may assume that there is no singularly continuous spectrum. This means that for appropriate vectors $\psi$ the expression $F_\psi(\alpha) = d(\psi, E_\alpha\psi)$ can be written as $\rho_\psi(\alpha)d\alpha$, with $\rho_\psi$ a Radon-Nikodym derivative defined in Supp. Sect. 19.6, or, for a unit vector $\psi$, in probabilistic language, a probability density.

## 21.10.7 The Spectral Theorem for a System of Commuting Operators

It is again helpful to look first at the finite dimensional case. Suppose one has in a $n$-dimensional Hilbert space $\mathcal{H}$ two commuting selfadjoint operators, $A$ and $B$, which can be written by the spectral theorem as

$$A = \sum_{j=1}^{s} \alpha_j P_j, \quad B = \sum_{l=1}^{t} \beta_l Q_l,$$

with $P_j$ and $Q_l$ the projections on the eigenspaces associated with the eigenvalues $\alpha_j$ of $A$ and $\beta_l$ of $B$. Define the operators $R_{j,l} = P_j Q_l$, for $j = 1, \ldots, s$ and $l = 1, \ldots, t$.

(21.10.7,a) **Problem** Show that the $R_{j,l}$ are projections on the common eigenspaces of the operators $A$ and $B$, with the properties

$$R_{j,l} R_{j',l'} = \delta_{jj'}\delta_{ll'} R_{j,l},$$

$$\sum_{j=1}^{s}\sum_{l=1}^{t} R_{j,l} = 1.$$

Show that the spectral theorems for $A$ and $B$ can be written as

$$A = \sum_{j=1}^{s}\sum_{l=1}^{t} \alpha_j R_{j,l}, \quad B = \sum_{j=1}^{s}\sum_{l=1}^{t} \beta_l R_{j,l}.$$

One obtains from this two index system of mutually orthogonal projections $\{R_{j,l}\}_{(j,l)\in\mathbb{N}^2}$ a nondecreasing two-index system $\{E_{j,l}\}_{(j,l)\in\mathbb{N}^2}$ and finally a nondecreasing two parameter system $\{E_{\alpha,\beta}\}_{(\alpha,\beta)\in\mathbb{R}^2}$, in the same manner as in Supp. Sect. 21.10.5 the system $\{E_j\}_j$ was obtained from $\{P_j\}_j$. It shall be called a *two parameter spectral resolution*, and has the properties, for all real values of $\alpha$, $\alpha_1$, $\alpha_2$ and $\beta$, $\beta_1$, $\beta_2$,

1. $E_{\alpha_1,\beta_1} \prec E_{\alpha_2,\beta_2}$, for $\alpha_1 \le \alpha_2$ and $\beta_1 \le \beta_2$.
2. $E_{\alpha,\beta}$ goes (strongly) to 0, separately, for $\alpha$ and $\beta$, going to $-\infty$, and to 1 for $\alpha$ and $\beta$ going to $+\infty$.
3. $E_{\alpha,\beta} = (\lim_{\beta' \to +\infty} E_{\alpha,\beta'})(\lim_{\alpha' \to +\infty} E_{\alpha',\beta})$.
4. $E_{\alpha,\beta}$ is (strongly) continuous from the right, separately in the two variables $\alpha$ and $\beta$.

It allows us to write the two spectral theorems, for $A$ and $B$, in the form of Riemann-Stieltjes integrals:

$$A = \int_{-\infty}^{+\infty} \int_{-\infty}^{+\infty} \alpha \, d E_{\alpha,\beta} = \int_{-\infty}^{+\infty} \alpha \, d F_\alpha,$$

$$B = \int_{-\infty}^{+\infty} \int_{-\infty}^{+\infty} \beta \, d E_{\alpha,\beta} = \int_{-\infty}^{+\infty} \beta \, d G_\beta,$$

with $\{F_\alpha\}_\alpha$ and $\{G_\beta\}_\beta$ the one parameter spectral resolutions for $A$ and $B$, and $E_{\alpha,\beta} = F_\alpha G_\beta$, $d E_{\alpha,\beta} = d F_\alpha d G_\beta$.

All this leads immediately to the spectral theorem for two commuting selfadjoint operators $A$ and $B$ in *infinite dimension*. The two parameter spectral resolutions $\{E_{\alpha,\beta}\}_{\alpha,\beta}$, satisfy 1–4.

(21.10.7,b) **Problem** Show that from a two parameter spectral resolution $\{E_{\alpha,\beta}\}_{\alpha,\beta}$ one can extract two commuting one parameter spectral resolutions $\{F_\alpha\}_\alpha$ and $\{G_\beta\}_\beta$, such that $E_{\alpha,\beta} = F_\alpha G_\beta$ for all $\alpha$ and $\beta$.

The spectral theorem then asserts that every pair of commuting selfadjoint operators define a unique two parameter spectral resolution, which can be used to obtain vector-valued Riemann-Stieltjes integrals as in the above formulas, but now for the infinite dimensional situation. Conversely, a given two parameter resolution defines two commuting selfadjoint operators. The domains of such operators $A$ and $B$ are given by

$$\psi \in \mathcal{D}_A \Leftrightarrow \int_{-\infty}^{+\infty} \int_{-\infty}^{+\infty} |\alpha|^2 d E_{\alpha,\beta} = \int_{-\infty}^{+\infty} |\alpha|^2 d F_\alpha < \infty,$$

$$\psi \in \mathcal{D}_B \iff \int\limits_{-\infty}^{+\infty}\int\limits_{-\infty}^{+\infty} |\beta|^2 dE_{\alpha,\beta} = \int\limits_{-\infty}^{+\infty} |\beta|^2 dG_\alpha < \infty.$$

The proof of all this follows easily from that of the spectral theorem for a single selfadjoint operator. The generalization to a system of $n$ commuting operators $A_1, \ldots, A_n$, is left to the reader.

### 21.10.8 Projection-Valued Measures

Behind the notions of spectral resolution for various values of $n$ there is an interesting more general mathematical idea, exhibiting a connection with measure theory:

(21.10.8,a) **Definition** Let $(X, \mathcal{F})$ be a measurable space, for which we take in this section $\mathbb{R}^n$ with its Borel $\sigma$-algebra. A *projection-valued measure* is a system $\{E(U)\}_{U \in \mathcal{F}}$ of (orthogonal) projections in a Hilbert space $\mathcal{H}$ such that

1. $E(U^c) = E(U)^\perp$, for all $U$ in $\mathcal{F}$,
2. If $\{U_n\}_n$ is a countable family of pairwise disjunct sets in $\mathcal{F}$, then $E(\cup^n U_n) = \sum_n E(U_n) = (\lim_{n\to\infty} \sum_{j=1}^n E(U_j))$, with pairwise orthogonal projections $E(U_j)$,
3. $E(U_1 \cap U_2) = E(U_1)E(U_2)$.

(21.10.8,b) **Problem** Show that 1 and 2 imply

$$E(\emptyset) = 0, \quad E(X) = 1, \quad U_1 \subset U_2 \implies E(U_1) \prec E(U_2).$$

See Supp. Sect. 21.8.2 for the properties of systems of commuting projections.

The requirements 1–2 and further properties are very similar to those for a finite real-valued measure given in Supp. Sect. 19.3; a projection-valued measure can indeed be seen as a structured collection of finite real-valued measures $\mu_\psi$, with $\mu_\psi(U) = (\psi, E(U)\psi)$, for every $\psi \in \mathcal{H}$; a probability measure when $\psi$ is a unit vector. An $n$-parameter spectral resolution leads to a projection-valued measure; however the converse is not true, as will be made clear below.

The notion of *support* of a real-valued measure, defined in Supp. Sect. 19.3, can be generalized in an obvious manner to projection-valued measures. The support of a projection-valued measure $E$ is the closure of the union of the supports of the real-valued measures $\mu_\psi$, for all $\psi$ in $\mathcal{H}$, with $\mu_\psi(U) = (\psi, E(U)\psi)$, for all sets $U$ in $\mathcal{F}$. It can be shown that the support of the spectral resolution $\{E_\alpha\}_\alpha$ of a selfadjoint operator $A$ is $\sigma(A)$, the spectrum of $A$; the support of the two parameter spectral resolution $\{E_{\alpha,\beta}\}_{\alpha,\beta}$ of two commuting operators $A$ and $B$ is the joint spectrum $\sigma(A, B) = \sigma(A) \times \sigma(B)$ of $A$ and $B$, and so on. A projection-valued measure needs not be an $n$-parameter spectral resolution, because the latter has as its support a

Cartesian product of $n$ sets in $\mathbb{R}^1$, while the support of the former can be an arbitrary Borel set in $\mathbb{R}^n$.

Note finally that a projection-valued measure is a full set of commuting projections in the sense of Supp. Sect. 21.8.2.

### 21.10.9  Functions of Selfadjoint Operators

For a selfadjoint operator $A$ polynomial expressions in $A$ are again selfadjoint, with in general smaller but still dense domains of definition. Theorem (21.11.2,c), a companion to the Stone–von Neumann theorem, both to be discussed in Supp. Sect. 21.11.2, will give us a definition of the exponential function $e^{-t\frac{i}{\hbar}A}$, based on the spectral resolution of $A$, which allows us in fact to define a very general class of functions of $A$:

(21.10.9,a) **Theorem** *Let $f$ be a real-valued Borel measurable function on $\mathbb{R}^1$. Then the operator expression*

$$f(A) = \int\limits_{-\infty}^{+\infty} f(\alpha)dE_\alpha$$

gives a selfadjoint operator with domain

$$\mathcal{D}_{f(A)} = \{\psi \in \mathcal{H} \mid \int\limits_{-\infty}^{+\infty} |f(\alpha)|^2 d(\psi, E_\alpha\psi) < \infty\},$$

with $\{E_\alpha\}_\alpha$ the spectral resolution of $A$. The measurable functions $f$ should be finite, except on a set of measure 0.

(21.10.9,b) *Example* For the function $f(x) = 1/x$ the operator $f(A)$ is not defined if $A$ has a purely discrete spectrum $0, 1, \ldots$; it is defined for $A$ with spectrum $1, 2, \ldots$

Note that in this context 'function' stands for 'equivalence class of measurable functions', with respect to the equivalence relation $f \sim g$ if and only if $f(x) = g(x)$ for all $x$ in $U \in \mathcal{F}$, such that $\int_U f(\alpha)dE_\alpha = 0$, meaning $\int_U f(\alpha)d(\psi, E_\alpha\psi) = 0$ for all $\psi$ in $\mathcal{H}$.

For a proof of this theorem and for a further discussion of its validity, see [2], Chap. VIII. Complex-valued measurable functions $f$ give *normal* operators $f(A)$. See Supp. Sect. 21.6.3 for the definition of a normal operator. All such normal operators $f(A)$ together form the commutative algebra of operators generated by $A$. It is a Banach algebra for bounded functions, see Supp. Sect. 21.6.2, in fact a so-called $C^*$-algebra, important in the algebraic approach to quantum theory, discussed in Chap. 12.

An alternative way of characterizing the algebra of operators $f(A)$ is by using the spectral projections of $A$; they are precisely the operators with spectral resolutions that can be made from the projections in the projection-valued measure generated by the spectral resolution of $A$.

(21.10.9,c) **Problem** Let $\{E_\alpha\}_\alpha$ be the spectral resolution of a selfadjoint operator $A$ in a Hilbert space $\mathcal{H}$. A particular projection $E_{\alpha_0}$ from this resolution can be written as a function of $A$, i.e. as $E_{\alpha_0} = f(A)$. Which function?

The idea of a function of an operator can easily be generalized to functions on $n$ commuting selfadjoint operators $A_1, \ldots, A_n$, by using the $n$-parameter spectral resolution $E_{\alpha_1,\ldots,\alpha_n}$, writing, for a measurable function $f$ on $R^n$,

$$f(A_1, \ldots, A_n) = \int_{-\infty}^{+\infty} \cdots \int_{-\infty}^{+\infty} f(\alpha_1, \ldots, \alpha_n) dE_{\alpha_1,\ldots,\alpha_n}.$$

Note that this only makes sense for functions of *commuting* selfadjoint operators. For arbitrary systems of bounded operators one may consider the algebra generated by these operators, by taking first, in the absence of domain problems, all polynomial expressions and then taking limits, for instance in operator norm. So one has algebras generated by arbitrary systems of not necessarily commuting *bounded* operator.

The study of operator-valued functions of one or more *commuting* selfadjoint operators is sometimes called *functional calculus* of operators. Functional calculus for selfadjoint operators in the standard sense is only meaningful for *commuting operators*. However, in physics one has *quantization*, a procedure in which one assigns in a systematic although non unique manner, operators to general expressions in noncommuting operators. One might call this *quantum functional calculus*. It is discussed in Chap. 13.

## 21.10.10 Multiplicity

We start again in finite dimension, so let $\mathcal{H}$ be an $n$-dimensional Hilbert space and $A$ be a selfadjoint operator in $\mathcal{H}$.

(21.10.10,a) **Definition** The operator $A$ will be called *multiplicity free* iff its spectrum is nondegenerate, i.e. iff all its eigenspaces are one dimensional.

It is immediately obvious that $A$ is multiplicity free iff it has an orthonormal basis of eigenvectors which is unique up to phase factors.

There is an equivalent characterization for $A$ to be multiplicity free. For this we define:

(21.10.10,b) **Definition** A vector $\psi$ in $\mathcal{H}$ is called *cyclic* with respect to $A$ iff, for all polynomials $p$, the vectors $p(A)\psi$ are dense in $\mathcal{H}$.

(21.10.10,c) **Theorem** *A selfadjoint operator $A$ in $\mathcal{H}$ is multiplicity free if and only if $\mathcal{H}$ contains a cyclic vector with respect to $A$.*

*Proof*
($\Rightarrow$): Let $A$ be multiplicity free, implying that $A$ has eigenvalues $\alpha_1 < \cdots < \alpha_n$, with an associated orthonormal basis of eigenvectors $\phi_1, \ldots, \phi_n$. Define the vector $\psi_0$ as $\psi_0 = \sum_{j=1}^{n} \phi_j$. It is enough to show that for every $\phi_j$ there is a polynomial $p_j$ such that $p_j(A)\psi_0 = \phi_j$. Choose $j = j_0$, and take

$$p_{j_0}(x) = C_{j_0} (x - \alpha_1) \cdots (x - \alpha_{j-1})(x - \alpha_{j+1}) \cdots (x - \alpha_n),$$

with

$$C_{j_0} = (\alpha_{j_0} - \alpha_1)^{-1} \cdots (\alpha_{j_0} - \alpha_{j_0-1})^{-1}(\alpha_{j_0} - \alpha_{j_0+1})^{-1} \cdots (\alpha_{j_0} - \alpha_n)^{-1}.$$

This gives $p_{j_0}(A)\psi_0 = p_{j_0}(A)\phi_{j_0} = \phi_{j_0}$. This argument is valid for any $j_0$, which proves that $\psi_0$ is cyclic.

($\Leftarrow$): $A$ has real eigenvalues $\alpha_1, \ldots, \alpha_n$, some of which may be equal, and an associated orthonormal basis $\phi_1, \ldots, \phi_n$, not necessarily unique up to phase factors. Suppose that there is a cyclic vector $\psi_0$. It can be expanded as $\psi_0 = \sum_{j=1}^{n} c_j \phi_j$. Assume that two eigenvalues are equal, say $\alpha_1 = \alpha_2$. For an arbitrary polynomial $p$ the operator $p(A)$ acts on $\psi_0$ as $p(A)\psi_0 = \sum_{j=1}^{n} c'_j \phi_j$, with $c'_j = p(\alpha_j)c_j$. The assumption $\alpha_1 = \alpha_2$ implies $c'_1 = c'_2$. This means that for every $p$ the vector $p(A)\psi_0$ has the same coefficient with respect to $\phi_1$ and $\phi_2$, so, for example, a vector like $\phi_1 - \phi_2$ cannot be reached. This would mean that $\psi_0$ is not cyclic, contradicting the assumption $\alpha_1 = \alpha_2$. This proves that all eigenvalues $\alpha_j$ are different, i.e. that $A$ is multiplicity free.

In preparation for the discussion of the infinite dimensional case we reformulate Definition (21.10.10,b):

(21.10.10,d) **Definition** (*alternative to* (D.10.10,b)) A vector $\psi$ in $\mathcal{H}$ is called *cyclic* with respect to $A$ iff, for all real measurable functions $f$ in one variable, the vectors $f(A)\psi$ form a dense linear subspace of $\mathcal{H}$.

Note that again 'function' stands for 'equivalence class of functions' in the manner discussed in the preceding section. In finite dimension there is for every measurable function a polynomial in the same equivalence class; moreover, as can be shown by arguments similar to the ones in the proof of Theorem (21.10.10,c), it is enough to require that all the $p(A)\psi_0$ or $f(A)\psi_0$ span a dense subspace of $H$. So in finite dimension (21.10.10,b) and (21.10.10,d) are equivalent.

For a selfadjoint operator in an infinite dimensional Hilbert space, in particular when the operator has continuous spectrum and no eigenvalues or eigenvectors, it is not immediately clear what multiplicity free should mean. The operator $M$ : $\psi(x) \mapsto x\psi(x)$ is a simple example of a selfadjoint operator that one would, for various reasons, like to call multiplicity free, which it is not however, in any case not according to Definition (21.10.10,a), which is obviously too narrow in infinite dimension.

There is a simple solution to this problem, which consists of using the property stated in Theorem (21.10.10.c) as a general definition:

(21.10.10,e) **Definition** A selfadjoint operator $A$ in an arbitrary finite or infinite dimensional Hilbert space $\mathcal{H}$ is multiplicity free iff $\mathcal{H}$ contains a vector that is cyclic with respect to $A$, in the sense of (21.10.10,d).

(21.10.10,f) **Problem** The multiplication operator $M$ in acting in $L^2(\mathbb{R}^1, dx)$ as $\psi(x) \mapsto x\psi(x)$ is multiplicity free, in the sense of Definition (D.10.10,e), while $M_1$ acting in $L^2(\mathbb{R}^3, d\mathbf{x})$ as $\psi(x_1, x_2, x_3) \mapsto x_1\psi(x_1, x_2, x_3)$ is not. Prove this by giving in $L^2(\mathbb{R}^1, dx)$ an example of a cyclic vector with respect to $M$, and show that no such vector exists for $M_1$ in $L^2(\mathbb{R}^3, d\mathbf{x})$.

It is not difficult to generalize all this to systems of $n$ commuting selfadjoint operators. Consider, for instance, two commuting selfadjoint operators $A$ and $B$ in the finite dimensional Hilbert space $\mathcal{H}$. Multiplicity free means here that all the common eigenspaces for pairs of eigenvalues $(\alpha_j, \beta_k)$ are one dimensional, or equivalently, that there is an orthonormal basis $\{\psi_{\alpha_j, \beta_k}\}_{\alpha_j, \beta_k}$ of common eigenvectors of $A$ and $B$, uniquely determined up to phase factors. A generalization of Theorem (21.10.10,c) says that the system $(A, B)$ is multiplicity free iff there is a vector $\psi$ in $\mathcal{H}$, which is cyclic with respect to $A$ and $B$, i.e. such that the subspace of $\mathcal{H}$ spanned by vectors $f(A, B)\psi$, for all real-valued measurable functions on $\mathbb{R}^2$, is dense in $\mathcal{H}$. In an infinite dimensional Hilbert space this theorem becomes a definition for the system $(A, B)$ to be multiplicity free.

The following concept is of great practical importance in the application to quantum theory, because it makes it possible, at least in the case of discrete spectrum, to give quantum states, i.e. unit vectors, an unambiguous labelling in terms of common eigenvalues. See Sect. 3.4.

(21.10.10,h) **Definition** A system of commuting selfadjoint operators is called *complete* iff it is multiplicity free.

Such a complete system $(A_1, \ldots, A_n)$ is a *maximally commuting system*, which means that every selfadjoint operator that commutes with all $A_j$ is a function $f(A_1, \ldots, A_n)$.

(21.10.10,i) **Problem** Consider the operators $M_j$, for $j = 1, 2, 3$, acting in $\mathcal{H} = L^2(\mathbb{R}^3, d\mathbf{x})$ as $M_j : \psi(x_1, x_2, x_3) \mapsto x_j\psi(x_1, x_2, x_3)$. Show that the pair $(M_1, M_2)$ is not a complete commuting system, but becomes one after adding the third operator $M_3$.

### 21.10.11 *Diagonalization of Selfadjoint Operators*

In Supp. Sect. 21.10.4 we started the discussion of the spectral theorem in finite dimension by looking at the *diagonalization* of hermitian matrices.

In the more intrinsic language of vector spaces this means that for a given self-adjoint operator $A$ in a $n$-dimensional Hilbert space $\mathcal{H}$, one can find a unitary map $U$ from $\mathcal{H}$ to a second Hilbert space $\widehat{\mathcal{H}} = \mathbb{C}^n$, the space of finite sequences $\{c_j\}_j$, such that the transform $\widehat{A} = UAU^{-1}$ is diagonal, i.e. is a multiplication operator, acting as $(\widehat{A}c)_j = \alpha_j c_j$, for $j = 1, \ldots, n$, with the $\alpha_j$ the real eigenvalues of $A$. This 'diagonal representation' of $A$ is unique iff the spectrum is nondegenerate, i.e. if $A$ is multiplicity free, in the terminology of the preceding section.

There is a generalization to infinite dimension. Two examples with continuous spectrum:

(21.10.11,a) *Example* The multiplication operator $M$ introduced in (21.9.2,f), acting in $L^2(\mathbb{R}^1, dx)$ as $(M\psi)(x) = x\psi(x)$, is the simplest example of an operator in a 'continuous' diagonal representation.

We have also an obvious example of an operator which is not given in diagonal form, but can easily be 'diagonalized' by a unitary transformation:

(21.10.11,b) *Example* The differentiation operator $D$ defined in (21.9.2,f) acts in $\mathcal{H} = L^2(\mathbb{R}^1, dx)$ as $(D\psi)(x) = -i\frac{d}{dx}\psi(x)$. The Fourier transformation given in Sect. 4.2 as

$$\hat{\psi}(k) = \frac{1}{\sqrt{2\pi}} \int\limits_{-\infty}^{+\infty} \psi(x)\, e^{-ikx} dx,$$

is a unitary map $U$ from $\mathcal{H}$ to a second Hilbert space $\widehat{\mathcal{H}} = L^2(\mathbb{R}^1, dk)$ which results in $(\widehat{D}\hat{\psi})(k) = (UDU^{-1}\hat{\psi})(k) = k\hat{\psi}(k)$, i.e. the transformed operator $\widehat{D}$ is the diagonal representation of $D$.

Just as for the finite dimensional case the spectral theorem for a selfadjoint operator is equivalent to the possibility of diagonalizing the operator. There is a theorem which says roughly that for a selfadjoint $A$ in $\mathcal{H}$ there is a unitary transformation $U$ to a second Hilbert space $\widehat{\mathcal{H}}$, of the general form $\widehat{\mathcal{H}} = L^2(\sigma(A), d\mu)$, such that $(UAU^{-1}\phi)(\alpha) = \alpha\phi(\alpha)$.

There is also a more specific general theorem on diagonalization of *multiplicity free* selfadjoint operators, using the fact that such operators have a cyclic vector, as was explained in the preceding section. It gives in fact another criterium for a selfadjoint operator to be multiplicity free.

(21.10.11,c) **Theorem** *Let $A$ be a multiplicity free selfadjoint operator in a Hilbert space $\mathcal{H}$, with cyclic vector $\psi_0$. Then $A$ can be brought in diagonal form by a unitary* map $U$ from $\mathcal{H}$ onto $\widehat{\mathcal{H}} = L^2(\sigma(A), d\mu_{\psi_0})$, such that the transform $\widehat{A} = UAU^{-1}$ acts as $(\widehat{A}\widehat{\phi})(\alpha) = \alpha\phi(\alpha)$. This representation is unique up to unitary maps.

Note that this theorem is quite general; it covers selfadjoint operators in arbitrary dimension, with discrete or continuous spectrum. The Fourier transformation in (21.10.11,a) acting on the operator $M$ is an example with continuous spectrum; the operator $D^2 + M^2$ would give an example of discrete spectrum, with $\widehat{\mathcal{H}} = l^2$, the sequence space of Example (21.4,c).

Operators that are not multiplicity free can also be brought in diagonal form, but this procedure is in this case nonunique and in general more complicated. See for this [2], Chaps. VII and VIII.3.

It is not difficult to obtain generalizations to systems of commuting selfadjoint operators. For example, Theorem (21.10.11,c) becomes for a multiplicity free pair of selfadjoint operators $A$ and $B$:

(21.10.11,d) **Theorem** Consider a multiplicity free pair of commuting selfadjoint operators $A$ and $B$ in $\mathcal{H}$, with cyclic vector $\psi_0$. Then $A$ and $B$ can be brought simultaneously in diagonal form by a unitary map $U$ from $\mathcal{H}$ onto the Hilbert space

$$\widehat{\mathcal{H}} = L^2(\sigma(A) \times \sigma(B), d\mu_{\psi_0}),$$

and with $d\mu_{\psi_0}$ the product measure, such that the transform $\widehat{A}$ of $A$ acts as

$$(\widehat{A}\widehat{\psi})(\alpha, \beta) = (UAU^{-1}\phi)(\alpha, \beta) = \alpha\phi(\alpha, \beta)$$

and the transform $\widehat{B}$ of $B$ as

$$(\widehat{B}\widehat{\psi})(\alpha, \beta) = (UBU^{-1}\phi)(\alpha, \beta) = \beta\phi(\alpha, \beta).$$

(21.10.11,e) **Problem** Give similar formulas for the three commuting operators $(D_j\psi)(\mathbf{x}) = -i\frac{\partial}{\partial x_j}\psi(\mathbf{x})$ in $\mathcal{H} = L^2(\mathbb{R}^3, d\mathbf{x})$.

## 21.11 Systems of Unitary Operators

### 21.11.1 One Parameter Groups of Unitary Operators

Consider a continuous complex-valued function $f : \mathbb{R}^1 \to \mathbb{C}$ with the property

$$f(t_1 + t_2) = f(t_1)f(t_2),$$

for all $t_1$ and $t_2$ in $\mathbb{R}^1$. One can show, using elementary analysis, that continuity here implies differentiability and that $f(t)$ has the form $f(t) = e^{tb}$, for some complex number $b$. If the image of $f$ is contained in the unit circle, $b$ has the form $i\beta$ with $\beta$

real. Conversely, each complex number $b$ gives an exponential function $f(t) = e^{tb}$, which can be defined, for instance as the sum of the absolutely convergent power series

$$e^{tb} = \sum_{n=0}^{\infty} \frac{1}{n!} (tb)^n.$$

This idea can be generalized, from complex-valued functions to operator-valued functions in Hilbert space. It can indeed be shown that a map $F$ from $\mathbb{R}^1$ into $\mathcal{B}(\mathcal{H})$, the space of bounded operators in $\mathcal{H}$, which is continuous in $t$, with respect to the *operator norm topology* in $\mathcal{B}(\mathcal{H})$, and which satisfies

$$F(t_1 + t_2) = F(t_1)F(t_2),$$

for all $t_1$ and $t_2$ in $\mathbb{R}^1$, can be written as $F(t) = e^{tB}$, for some bounded operator $B$, with the operator-valued exponential function $e^{tB}$ defined by the series

$$e^{tB} = \sum_{n=0}^{\infty} \frac{1}{n!} (tB)^n,$$

absolutely convergent in the operator norm topology. Moreover $F(t)^* = e^{tB^*}$. Every bounded operator $B$ gives a function $e^{tB}$ defined as a convergent power series. The proof of this is an easy generalization in which arguments using absolute values of complex numbers are replaced by similar arguments using operator norms. However, as was already noted at the end of Supp. Sect. 21.6.2, norm convergence is too strong for most physical applications; moreover the selfadjoint operators that have to be exponentiated, such as the Hamiltonian operator, which generate the time-evolution in quantum systems, are usually unbounded, which means that the terms in the exponential series do not make sense. We need the weaker 'strong' operator convergence. The following special notion is of great importance in quantum theory:

(21.11.1,a) **Definition** A system $\{U(t)\}_{t \in \mathbb{R}^1}$ of unitary operators in $\mathcal{H}$ is called a *strongly continuous one parameter group of unitary operators* or short a *one parameter group of unitary operators* in $\mathcal{H}$ iff

1. $U(t_1 + t_2) = U(t_1)U(t_2)$, for all real numbers $t_1$ and $t_2$,
2. $U(t)$ is continuous in $t$, in the strong operator topology, meaning that $U(t)\psi$ is continuous in $t$, for every $\psi$ in $\mathcal{H}$.

Note that property 1 in this definition implies that all the $U(t)$ *commute*.

(21.11.1,b) **Problem** Show that one has $U(0) = 1$, and also $U(t)^{-1} = U(-t)$, for all $t \in \mathbb{R}^1$.

(21.11.1,c) **Problem** Show that continuity of $U(t)\psi$ in $t = 0$ implies continuity for all real values of $t$.

## 21.11.2 The Stone–von Neumann Theorem

Definition (21.11.1,a) leads to a theorem that is in importance for quantum theory only second to the spectral theorem:

(21.11.2,a) **Theorem** *Let* $\{U(t)\}_{t \in \mathbb{R}^1}$ *be a strongly continuous one parameter group of unitary operators. The vectors* $\psi$ *in* $\mathcal{H}$, *for which the limit*

$$\lim_{t \to 0} -i \, \frac{U(t) - 1}{t} \psi$$

*exists, form a dense linear subspace* $\mathcal{D}$ *of* $\mathcal{H}$. This limit defines an operator $A$, which is selfadjoint on $\mathcal{D}$. The operator $A$ is called the *infinitesimal generator* of the one parameter group.

The result of this theorem can be written as

$$\left( \frac{d}{dt} U(t) \right)_{t=0} = i A,$$

with an obvious meaning of this derivative in terms of a limit of vectors in $\mathcal{H}$, in the domain of $A$.

(21.11.2,b) **Problem** Show that a one parameter group of unitary operators $U(t)$ can in this spirit be seen as a solution of the operator-valued differential equation $\frac{d}{dt} U(t) = i A U(t)$.

In quantum theory this equation gives the Schrödinger equation, as is discussed in Sects. 3.5 and 4.8.

Theorem (21.11.2,a) is completed by the statement that a one parameter group of unitary operators can in an appropriate sense be written as an exponential function $U(t) = e^{itA}$, a statement which is based on the spectral theorem for selfadjoint operators, in particular on the functional calculus discussed in Supp. Sect. 21.10.9.

(21.11.2,c) **Theorem** *Let* $A$ *be a selfadjoint operator, with spectral resolution* $\{E_\alpha\}_{\alpha \in \mathbb{R}^1}$. *Then the operators defined by*

$$U(t) = \int_{-\infty}^{+\infty} e^{it\alpha} \, dE_\alpha,$$

or, more explicitly,

$$(\psi, U(t)\psi) = \int_{-\infty}^{+\infty} e^{it\alpha} \, d(\psi, E_\alpha \psi),$$

for all $\psi$ in $\mathcal{H}$, form a one parameter group of unitary operators with $A$ as infinitesimal generator.

The proof of both theorems, nontrivial exercises in standard Hilbert space theory, can be found in [2], pp. 266–268.

Theorem (21.11.2,c) gives a precise meaning to the notion of an exponential of an unbounded selfadjoint operator. For this the exponential series used for a bounded operator need not make sense, as will be clear from the following example.

(21.11.2,d) *Example* Consider in $\mathcal{H} = L^2(\mathbb{R}^1, dx)$ the translation operators $U(a)$, defined as $(U(a)\psi)(x) = \psi(x - a)$. These operators form a *strongly continuous one parameter* group of unitary operators, with as infinitesimal generator minus the differential operator $D = -i\frac{d}{dx}$, in agreement with the Taylor expansion of a function $\psi(x)$, so we can write $U(a)\psi = e^{-iaD}\psi$. The operator $D$ and all its powers $D^n$ are well-defined on the dense linear subspace of infinitely differentiable functions with compact support in $\mathbb{R}^1$, so all the partial sums $S_n = 1 - iaD + \frac{1}{2}(-iaD)^2 + \cdots + \frac{1}{n!}(-iaD)^n$ are well defined on this domain. Nevertheless, the action of the successive $S_n$ will in the limit of $n \to \infty$ clearly not result in a translation of such functions, because the support of each $S_n\psi$ will remain contained in the support of $\psi$.

## 21.11.3 Two Parameter Systems

The most import application of one parameter groups of unitary operators in quantum theory is to describe the time development of *autonomous systems*, systems for which the dynamics is time translation invariant. For nonautonomous systems, e.g. systems evolving under the influence of a time dependent force, one needs a more general system of unitary operators:

(21.11.3,a) **Definition** A system $\{U(t_2, t_1)\}_{t_1, t_2 \in \mathbb{R}^1}$ of unitary operators in $\mathcal{H}$ is called a *strongly continuous two paramater evolution system of unitary operators*, or short, a *two parameter unitary evolution system* iff

1. $U(t_3, t_2)U(t_2, t_1) = U(t_3, t_1)$, for all $t_1$, $t_2$ and $t_3$ in $\mathbb{R}^1$,
2. $U(t_2, t_1)$ is continuous in the strong operator topology, which means that $U(t_1, t_2)\psi$ is simultaneously continuous in $t_1$ and $t_2$ for every $\psi$ in $\mathcal{H}$.

Note that in this case the unitary operators need *not* commute. The system $\{U(t_2, t_1)\}_{t_1, t_2}$ is not a group, but something more general; it is a simple example of what is called a *groupoid*. In the case of translation invariance, i.e. if $U(t_2, t_1) = U(t_2 + a, t_1 + a)$, for all real $a$, we are back in the situation of a one parameter group, with $U'(t_2 - t_1) = U(t_2, t_1)$ or $U'(t) = U(t, 0)$.

(21.11.3,b) **Problem** Show that $U(t, t) = 1$ and that $U(t_2, t_1)^{-1} = U(t_1, t_2)$, for all $t$, $t_1$ and $t_2$ in $\mathbb{R}^1$.

It is obvious that $U(t_2, t_1)$ cannot be written in the form of an exponential function. Under reasonable mathematical assumptions it can nevertheless be shown that there is an analogue of the Stone–von Neumann theorem (21.11,d), with a time dependent generator $A(t)$. This also leads to an operator-valued differential equation for $U(t_2, t_1)$, in fact, because of noncommutativity, to two such equations, namely

$$\frac{\partial}{\partial t_2} U(t_2, t_1) = i A(t_2) U(t_2, t_1),$$

and

$$\frac{\partial}{\partial t_1} U(t_2, t_1) = i U(t_2, t_1) A(t_1).$$

### 21.11.4 General Groups of Unitary Operators

In quantum theory more general groups of unitary operators are used, in particular *unitary repesentations* of *Lie groups*, in discussions of symmetry principles. See Sect. 3.6, and for a general discussion of the mathematical properties of groups and their representations Supp. Chap. 24.

A typical example of such a Lie group is $O(3)$, the three dimensional orthogonal group, consisting of rotations in three dimensional real Euclidean space, the representations of which play a role in quantum mechanical models such as that of the hydrogen atom, discussed in Chap. 7.

## References

1. von Neumann, J.: Mathematical Foundations of Quantum Mechanics. Princeton University Press, Princeton (1996) ((Translated from the original 1932 German Springer edition) A great classic, written by the creator of the mathematical formalism of quantum mechanics. The part on the theory of measurement may have been overtaken by more recent developments, but the general discussion of quantum theory, in particular of selfadjoint operators and the spectral theorem remains instructive and fresh. https://app.box.com/s/fsp81filfebei98umd6u)
2. Reed, M., Simon, B.: Methods of Mathematical Physics. I: Functional Analysis. Academic Press, New York (1972) (This book is because of its readability our main reference for this chapter. A textbook on functional analysis, with the theory of operators in Hilbert space as its central topic, especially written for applications in mathematical physics. It is the first of a series of four books by the same authors. The other volumes are more specialized, but contain nevertheless useful material, in particular II and IV.)
3. Schmüdgen, K.: Unbounded Operators in Hilbert Space. Springer, Hardcover (2012) (The best recent book on the theory of unbounded operators.)
4. Akhiezer, N.I., Glazman, I.M.: Theory of Linear Operators in Hilbert Space Translated from the Russian Dover (1993). A classical standard reference on Hilbert space theory. The first five chapters of this book can be found at http://www.stat.wisc.edu/~wahba/stat860/pdf2/ag1.35.pdf
5. Prugovecki, E.: Quantum Mechanics in Hilbert Space, 2nd edn. Dover Publications, New York (1981) (Even though its general approach may be somewhat dated, it contains a wealth of precise

and clearly formulated additional material on the mathematical aspects of quantum theory. The modest price of the Dover edition makes it especially worthwhile.)

6. Dunford, N., Schwartz, J.T.: Linear Operators, vol. I–III. Wiley-Interscience, New York (1957, 1963, 1971) (Reissued as paperbacks in 1988, at outrageous prices. A formidable set, a monument of scholarship. Maybe not suitable for quick browsing, but it remains, after 40 years, an indispensable tool for serious research in linear operator theory.)

7. Helmberg, G.: Introduction to Spectral Theory in Hilbert Space North-Holland 1969. Reissued by Dover (2008) A leisurely and clear introduction

8. Gustafson, S.J., Michael, I.: Sigal Mathematical Concepts of Quantum Mechanics Springer, Berlin (2006) (The level of mathematical rigour in this book is variable; however its merit is that it discusses a number of interesting advanced topics)

# Chapter 22
# Probability Theory

## 22.1 Introduction

### 22.1.1 History

Probability theory is a set of ideas and procedures which allows us to act rationally in situations where we have incomplete information. It arose, as a definite mathematical discipline, in the seventeenth century, its main applications being gambling and insurance, but it had to wait until the first half of the twentieth century before its general principles were put together in a rigorous mathematical formalism. This was provided in the thirties by the Russian mathematician Andrei Kolmogorov; his 'axiomatic' formulation in terms of the theory of measure and integration and, more generally, functional analysis, has become the generally accepted standard form of probability theory.

Good books on probability theory are [1–4]. See also the classic book by Kolmogorov [5].

### 22.1.2 Probability Theory in Physics

Probability theory has played an important role in physics since the second half of the nineteenth century, when it was realized that matter is composed of a very large number of separate particles, atoms and molecules, and that therefore the properties of systems of gases and liquids could to a large extent be understood from averaging statistically over the behaviour of the individual particles. In this connection one should mention the names of Ludwig Boltzmann, James Clerk Maxwell, also known for his fundamental theory of electromagnetism, and above all Josiah Willard Gibbs, the first American theoretical physicist of international fame, who laid the foundations of what is now known as *statistical mechanics*. He was working before the mathematical formalization of probability theory by Kolmogorov; some of the terminology he introduced for basic probabilistic notions has only survived in physics.

© Springer International Publishing Switzerland 2015
P. Bongaarts, *Quantum Theory*, DOI 10.1007/978-3-319-09561-5_22

A good example is the notion of *ensemble*; a mathematician reading a modern physics textbook will need some time to realize that this is just a probability measure on the phase space, i.e. the space of all positions and momenta of the particles of a classical system.

A further use of probability theory in physics came with quantum mechanics. This is in fact the reason that this chapter is included in this book. Quantum theory implies a fundamental lack of determinism in physical systems. Philosophers of science do not agree on what this precisely means, but this does not lead to problems at the level of applications to standard physical situations. In any case the use of probability theory in quantum mechanics is much more intriguing and fundamental than in the rest of physics. In fact, as a mathematical model quantum theory can be regarded as a nontrivial 'noncommutative' generalization of classical probability theory. This aspect is studied in what is called 'quantum probability theory', a lively area of specialized research which is outside the scope of this book, even though our algebraic approach to classical and quantum statistical physics, explained in particular in Chap. 12, is close to it.

## 22.2  Kolmogorov's Formulation of Probability Theory

### 22.2.1  Discrete Probability Theory

Before setting up the general formalism it is useful to consider briefly the simple case in which one assigns probabilities to a finite number of $n$ elementary events, for example the six sides of a dice. Let such possible outcomes of an experiment, be denoted by the integers $1, \ldots, n$. The probability that the $j$th event turns up in an experiment should be equal to a nonnegative real number $\rho_j$, with of course $\sum_{j=1}^n \rho_j = 1$. For a given function $f_j$ of the outcomes, one defines the average value of the 'random variable' $f$ as $\sum_{j=1}^n f_j \rho_j$. The standard interpretation of such probabilities and averages is based on the idea of repeating the same experiment many times, but that does not concern us here. One may ask for the probability that any one of a subset of the basic events $1, \ldots, n$ turns up. The answer is of course the sum of the numbers $\rho_j$ with the $j$ belonging to the subset. It is therefore sensible to call all the subsets of $1, \ldots, n$ *events*. This is a good stepping stone for going to 'continuous' situations, say that of throwing a dart arrow instead of a dice. The probability of hitting a given point inside a circle is not a useful concept, as it is clearly 0. To have a nonzero probability one needs as events sets with a nonzero volume. This leads to probabilities as *measures* of sets, in the technical sense of measure theory, as discussed in Supp. Chap. 19; averages become integrals with respect to this measure. This is precisely the starting point of Kolomogorov's 'axiomatic' formulation of probability theory.

## 22.2.2 The General Framework

Following the ideas of Kolmogorov, we consider the following set up:

(22.2.2,a) **Definition** A *probability theory* is a system $(\Omega, \mathcal{F}, P)$, a probability measure space, as defined in Supp. Sect. 19.3. The subsets in the $\sigma$-algebra $\mathcal{F}$, measurable sets in standard measure theory, are called here *events*, with the set $\Omega$ itself the *certain event* and the empty set the *impossible event*. The real number $P(A)$, nonnegative and smaller or equal to 1, is the *probability* for the event $A$, satisfying the two requirements of Definition (22.3,a) in Supp. Sect. 22.3, with the additional condition $P(\Omega) = 1$.

(22.2.2,b) **Definition** A (real-valued) *random variable* or *stochastic variable*, with respect to this probability theory is a real-valued measurable function on $\Omega$, with the understanding that 'function' in this context means 'equivalence class of functions' with respect to the relation which considers measurable functions equivalent when they are identical 'almost everywhere'—or 'almost surely' in probabilistic language, i.e. except on sets of measure 0. The average or *expectation* of a random variable is defined as $E(f) = \int_\Omega f(\omega) P(d\omega)$ or $\int_\Omega f(\omega) dP$. This expectation exists if the integral is finite. More generally, $E(f^p)$, when it exists, is called the *p*th *moment* of $f$. In the applications in the main text of this book the expectation $E(f)$—or 'expectation value' as it is called in physics—will often be denoted by $\overline{f}$.

(22.2.2,c) **Definition** The expression $E\left((f - E(f))^2\right) = E(f^2) - (E(f))^2$ is called the *variance* of $f$; its square root is usually denoted as $\sigma$ or $\sigma(f)$—in physical applications as $\Delta f$—and called the *standard deviation*.

(22.2.2,d) **Definition** A stochastic variable $f$ has a *distribution function*, or *cumulative distribution function*. We denote it here as $F_f$. For each real $x$ the value $F_f(x)$ is defined as the probability that the random variable takes a value smaller or equal to $x$, i.e.

$$F_f(x) := P(f^{-1}((-\infty, x]) = P(\{\omega \in \Omega \mid f(\omega) \leq x\}).$$

One often considers *systems* $\{f_\alpha\}_\alpha$ of random variables. Such systems can be arbitrary; a system of random variables indexed by a continuous parameter $t$, usually time, is called a *stochastic process*. Here we restrict the discussion to finite systems $f_1, \ldots, f_n$. Such a system has a *joint* or *simultaneous distribution function*. It is a real valued function on $\mathbb{R}^n$ defined simply as

$$F(x_1, \ldots, x_n) := P\left(\{\omega \in \Omega \mid f_1(\omega) \leq x_1, \ldots, f_n(\omega) \leq x_n\}\right).$$

(22.2.2,e) **Definition** The expression $E\left((f - E(f))(g - E(g))\right)$ is denoted as $\mathrm{Cov}\,(f, g)$ and called the *covariance*, of $f$ and $g$. It has a normalized form $\mathrm{Cov}\,(f, g)/\sigma(f)\sigma(g)$, called the *coefficient of correlation*; it takes values between $-1$ and $+1$. Two random variables $f$ and $g$ are said to be *uncorrelated* if $\mathrm{Cov}(f, g) = 0$.

## 22.3 Mathematical Properties of Distribution Functions

### 22.3.1 Introductory Remark

It is worthwhile to discuss the mathematical properties of distribution functions in somewhat more detail. We consider first the case of distribution functions of a single variable, and then the general case of $n$ variables.

### 22.3.2 Distribution Functions of a Single Variable

It was already indicated in Supp. Sect. 19.4, that a distribution function associated with a random variable $f$, as defined in the preceding section, is a real-valued, monotone nondecreasing function $F$ on the real line, continuous from the right, with moreover $\lim_{x \to -\infty} F(x) = 0$ and $\lim_{x \to +\infty} F(x) = 1$. One proves easily that the collection of such functions $F$ is in one-to-one correspondence with the collection of probability measures on the real line, with as $\sigma$-algebra of subsets the system of *Borel* sets, the smallest $\sigma$-algebra containing all open sets in $\mathbb{R}^1$, as was defined in Supp. Sect. 19.2. On the one hand one defines, for a given function $F$, a probability measure $P_F$ by extension of the formula $P_F((x_1, x_2]) = F(x_2) - F(x_1)$, for all half-open intervals $(x_1, x_2]$, on the other hand one obtains from a probability measure $\mu$ on $\mathbb{R}$ a function $F_\mu$ as $F_\mu(x) = \mu((-\infty, x])$. These formulas mean that each $F$ is the distribution function for the random variable $x$ with respect to a unique probability theory $(X, \mathcal{B}, P_F)$ with $\Omega = \mathbb{R}$, $\mathcal{B}$ the Borel $\sigma$-algebra of $\mathbb{R}$ and $P_F$ a unique probability measure. Because of this it is reasonable to use the term *distribution function* for arbitrary functions in $\mathcal{F}$. Note that for a distribution function in this sense monotonicity implies the existence, for each $x$ in $\mathbb{R}$, of $F_-(x) := \sup_{x' < x} F(x')$ and $F_+(x) := \inf_{x' > x} F(x')$, this without using the right continuity. One has of course $F_-(x) \leq F_+(x)$, for all $x$ in $\mathbb{R}$. It should also be remarked that it is clear from this that continuity from the *right* in this set up is a matter of convention.

Consider the union of all open intervals $(x_1, x_2)$ on which a given distribution function is constant. Its complement is a nonempty closed set in $R$, the *support* of $F$, denoted as Supp $(F)$. It is also the support of the probability measure in $\mathbb{R}$ associated with $F$. Note that the support of $F$ can equivalently be defined as $\{x \in \mathbb{R} \mid F(x-\varepsilon) < F(x+\varepsilon), \forall \varepsilon > 0\}$. Points $x$ for which $F_-(x) < F_+(x)$ obviously belong to the support of $F$. They are called the *discrete points* of Supp $(F)$. If the support of $F$ has only discrete points, i.e. if the function $F$ is a step function, we are back in discrete probability theory. For points $x$ with $F_-(x) = F_+(x)$ the function $F$ is continuous in $x$. If $F$ is not only continuous but *absolutely continuous*, a stronger requirement, then $F$ is differentiable, except in a set of points with Lebesgue measure 0; the derivative is an integrable function $\rho$, nonnegative, with integral $\int_{-\infty}^{+\infty} \rho(x)dx = 1$; it is called a *probability density*. In physics one usually considers probabilities to be discrete or to be given by probability densities. Not much is lost there by this simplification.

### 22.3.3 The Normal Distribution

The most widely used probability distribution in physics, as in almost all other applications of probability theory, is the *normal* or *Gaussian distribution*. In the one dimensional case it is given by the density

$$\rho(x) = \frac{1}{\sqrt{2\pi\sigma^2}} e^{-\frac{(x-\bar{x})^2}{2\sigma^2}},$$

with mean $\bar{x}$ and standard deviation $\sigma$. For mean $\bar{x} = 0$ and standard deviation $\sigma = 1$, i.e. with

$$\rho(x) = \frac{1}{\sqrt{2\pi}} e^{-\frac{x^2}{2}},$$

it is called the *standard normal distribution*. See, for example, Sect. 4.5, Problem (4.5).

### 22.3.4 Distribution Functions in n Variables

For a system $f_1, \ldots, f_n$ of random variables the *n-variable distribution function* is a real-valued function $F_n$ on $\mathbb{R}^n$, monotone nondecreasing and continuous from the right in each variable $x_j$, with moreover

$$\lim_{x_j \to -\infty} F(x_1, \ldots, x_j, \ldots, x_n) = 0$$

for every $j$, and again for each $j$

$$\lim_{x_j \to +\infty} F(x_1, \ldots, x_n) = F_{n-1}(x_1, \ldots, x_{j-1}, x_{j+1}, \ldots, x_n),$$

leading to the *marginal distribution function* in the $n - 1$ variables $x_1, \ldots, x_{j-1}, x_{j+1}, \ldots, x_n$. The marginal distribution functions of a single random variable, e.g. $x_1$, can be obtained by taking the limit $x_2, \ldots, x_n \to +\infty$. In this manner all the information on the $f_j$ as separate random variables is contained in the $n$-variable function; the converse is of course not true. Similar to the single variable case there is a one-to-one correspondence between $n$-variable distribution functions $F_n$ and probability measures on $\mathbb{R}^n$. Each $F_n$ can then be regarded as the $n$-variable distribution function associated with the $n$ random variables $x_1, \ldots, x_n$ with respect to a probability theory $(\Omega, \mathcal{B}, P_f)$, with $\Omega = \mathbb{R}^n$, $\mathcal{B}$ the Borel $\sigma$-algebra of $\mathbb{R}^n$ and $P_{F_n}$ a unique probability measure. One can again define the notion of support of a distribution function and correspondingly of the associated probability measure on $\mathbb{R}^n$, with similar formulas. This is left to the reader.

(22.3.4,a) **Problem** Give the formula for normal distribution in $n$ variables.

### 22.3.5 *Stochastic Variables as a Basic Notion*

The general definition of a stochastic (or random) variable is a measurable function on a probability measure space. In working with stochastic variables this measure space is often not mentioned, let alone explicitly described; one just assumes that it exists and is present somewhere in the background. What is given is the joint distribution function of the variables. As long as one has a fixed set of stochastic variables in mind, this is all one needs: the distribution function contains all information. It is not hard to show, for instance for a single stochastic variable $f$, using the results from the preceding subsections, that the expectation of $f$, defined in terms of the underlying probability space $(\Omega, \mathcal{B}, P)$ as $E(f) = \int_\Omega f(\omega) P(d\omega)$, can be written as a Lebesgue-Stieltjes integral over $F$ as $\int_{-\infty}^{+\infty} x \, dF(x)$. See for the Stieltjes integral Supp. Sect. 19.5.3. In the case of absolute continuity, with a probability density $\rho$, this becomes the ordinary Lebesgue integral $\int_{-\infty}^{+\infty} x\rho(x)dx$. One has similarly, for a system $f_1, \ldots, f_n$ of stochastic variables with $n$-variable distribution function, the expression $E(f_j) = \int_{-\infty}^{+\infty} \ldots \int_{-\infty}^{+\infty} x_j dF(x_1, \ldots, x_n)$, for $j = 1, \ldots, n$, which is of course equal to the one dimensional integral $\int_{-\infty}^{+\infty} x_j dF_j(x_j)$, with $F_j$ the marginal distribution function for $f_j$. The covariance of two variables $f_j$ and $f_k$, another important quantity, defined as an integral over $\Omega$ with respect to $P$, becomes a Lebesgue-Stieltjes integral over $\mathbb{R}^n$, which can be reduced immediately to a two dimensional integral. All this means that for a fixed system of $n$ stochastic variables one can, instead of the original probability space $(\Omega, \mathcal{F}, P)$ which is supposed to be in the background, effectively use an explicit probability theory in $\mathbb{R}^n$, determined by a joint distribution function $F(x_1, \ldots, x_n)$ in the manner described in the preceding sections. It finally means that for practical purposes a system of $n$ random variables can be *defined* as an $n$-variable distribution function $F$ on $\mathbb{R}^n$.

## 22.4 Stochastic Processes

A *stochastic process* is a collection of random variables $\{X_t\}_{t \in \mathcal{I}}$ with $\mathcal{I}$ an index set, for a *continuous stochastic process* usually the time $\mathbb{R}^1$ or $\mathbb{R}^1_+$, for a *discrete process* $\mathbb{N}$ or $\mathbb{N}_+$. A stochastic process can, as any system of random variables be described by a set of simultaneous distribution functions as discussed in the preceding sections for the continuous case. A *Markov process* is a stochastic process which has a *state space* and a system of transition probabilities, in the continuous case denoted by functions $P(x_2, t_2; x_1, t_1)$ which satisfy the *Chapman-Kolmogorov equation*

$$\int_{x_2=-\infty}^{+\infty} P(x_3, t_3; x_2, t_2) P(x_2, t_2; x_1, t_1) \, dx_2 = P(x_3, t_3; x_1, t_1),$$

for $t_1 < t_2 < t_3$. The most basic example of a discrete Markov process is the *random walk* and of its continuous version the *Wiener process* for Brownian motion, both discussed in Supp. Sect. 13.10.4, in the context of the relation between Feynman's path integral in quantum mechanics and classical diffusion theory.

There is an abundance of excellent books on mathematical probability theory.

# References

1. Chung, K.L.: A Course in Probability Theory, Second revised edition. Academic Press, San Diego (2000) (A comprehensive modern textbook.)
2. Klenke, A.: Probability Theory. Springer, London (2006) (Another comprehensive modern textbook.)
3. Billingsley, P.: Probability and Measure, 3rd edn. Wiley-Interscience, New York (1995) (Combines probability and measure theory.)
4. Rosenthal, J.S.: First Look at Rigorous Probability Theory, 2nd edn. World Scientific, Singapore (2006) (A short course which stresses the mathematical structure of probability theory.)
5. Kolmogorov, A.N.: Foundation of the Theory of Probability, Second English edition. Chelsea, New York (1956). http://statweb.stanford.edu/~cgates/PERSI/Courses/Phil166-266/Kolmogorov-Foundations.pdf. A short book of great historical interest. Original German edition by Springer, 1933

# Chapter 23
# Tensor Products

New Hilbert spaces can be constructed from given ones in several ways. *Direct sums* were used in Sect. 9.2; its mathematical properties given in Supp. Sect. 21.5. This chapter will treat the mathematics of *tensor products*, extensively used in Sect. 9.2. We start with the general context: tensor products of vector spaces. Useful for this chapter are Refs. [1–3].

## 23.1 Tensor Products of Vector Spaces

At an elementary level, in finite dimension, vectors can be represented by finite sequences of numbers, coordinates with respect to a basis. A tensor product of two such spaces, $V_1$ and $V_2$, with vectors as coordinate sequences $\{\xi_j\}_j$ and $\{\eta_k\}_k$, is then the vector space of double sequences $\{\chi_{jk} = \xi_j \eta_k\}_{jk}$.

There is a more intrinsic definition. Let $V_1$ and $V_2$ be arbitrary not-necessarily finite dimensional vector spaces. Consider the Cartesian product $V_1 \times V_2$ as a point set and construct the *free vector space* generated by this set, i.e. the vector space consisting of all formal finite linear combinations of points $(x_1, x_2)$ in $V_1 \times V_2$. Form the quotient space $W$ over the linear subspace defined by the equivalence relations

$$(x_1 + y_1, x_2) \sim (x_1, x_2) + (y_1, x_2),$$
$$(x_1, x_2 + y_2) \sim (x_1, x_2) + (x_1, y_2),$$
$$\lambda(x_1, x_2) \sim (\lambda x_1, x_2),$$
$$\lambda(x_1, x_2) \sim (x_1, \lambda x_2),$$

for all $x_1, y_1$ in $V_1, x_2, y_2$ in $V_2$ and $\lambda$ in $\kappa = \mathbb{C}$ or $\mathbb{R}$, but always $\mathbb{C}$ in the present text. This quotient space is the *tensor product* of $V_1$ and $V_2$ and is denoted as $V_1 \otimes V_2$. It has as special elements equivalence classes $[(x_1, x_2)]$, called *pure tensors* and denoted as $x_1 \otimes x_2$. The canonical projection on the quotient space defines a bilinear map

© Springer International Publishing Switzerland 2015
P. Bongaarts, *Quantum Theory*, DOI 10.1007/978-3-319-09561-5_23

$$m : V_1 \times V_2 \to V_1 \otimes V_2, \quad (x_1, x_2) \mapsto x_1 \otimes x_2.$$

One can show that this map is universal in the sense that for any other vector space $W'$ and bilinear map $m' : V_1 \times V_2 \to W'$ there is a unique linear homomorphism $u : V_1 \otimes V_2 \to W'$ such that $m' = u \circ m$. This property is used in the modern definition of tensor product of vector spaces:

(23.1,a) **Definition** The *tensor product* of two vector spaces $V_1$ and $V_2$ is a vector space $W$, denoted as $V_1 \otimes V_2$, together with a bilinear map

$$m : V_1 \times V_2 \to V_1 \otimes V_2, \quad (x_1, x_2) \mapsto m((x_1, x_2)) := x_1 \otimes x_2,$$

such that for any other linear space $W'$ and bilinear map $m' : V_1 \times V_2 \to W'$ there is a linear homomorphism $u : W \to W'$ with $m' = u \circ m$.

The construction as a quotient space shows that such a tensor product exists; the definition implies that it is unique up to isomorphisms.

There is an obvious generalization of this definition to the notion of a tensor product $V_1 \otimes \cdots \otimes V_n$ of $n$ vector spaces $V_1, \cdots, V_n$, with an $n$-linear map

$$m : V_1 \times \cdots \times V_n \to V_1 \otimes \cdots \otimes V_n.$$

Note that the tensor product is associative, e.g.

$$(V_1 \otimes V_2) \otimes V_3 \simeq V_1 \otimes (V_2 \otimes V_3) \simeq V_1 \otimes V_2 \otimes V_3.$$

Linear transformations $T_j$ in the factors $V_j$ induce a linear transformation in the tensor product space $V_1 \otimes \cdots \otimes V_n$ in two different ways:

1. As a *product operator*

$$\otimes_{j=1}^{n} T_j = T_1 \otimes \cdots \otimes T_n,$$

acting, by linear extension of

$$x_1 \otimes \cdots \otimes x_n \mapsto T_1 x_1 \otimes \cdots \otimes T_n x_n,$$

2. As a *sum operator*, also by linear extension,

$$\oplus_{j=1}^{n} T_j = (T_1 \otimes 1_{V_2} \otimes \cdots \otimes 1_{V_n}) + \cdots + (1_{V_1} \otimes 1_{V_2} \otimes \ldots \otimes T_n)$$

of

$$x_1 \otimes \cdots \otimes x_n \mapsto (T_1 x_1 \otimes x_2 \otimes \cdots \otimes x_n) + \cdots + (x_1 \otimes x_2 \otimes \cdots \otimes T_n x_n).$$

## 23.2  Tensor Products of Hilbert Spaces

Let $\mathcal{H}_1, \ldots, \mathcal{H}_n$ be Hilbert spaces. Form the tensor product

$$\otimes_{j=1}^n \mathcal{H}_j = \mathcal{H}_1 \otimes \cdots \otimes \mathcal{H}_n$$

in the sense of the preceding section. It is a pre-Hilbert space with respect to the inner product $(\cdot, \cdot)$, determined by the inner products $(\cdot, \cdot)_{\mathcal{H}_j}$ in the $\mathcal{H}_j$, by linear extension of

$$(\psi_1 \otimes \cdots \otimes \psi_n, \phi_1 \otimes \cdots \otimes \phi_n) = (\psi_1, \phi_1)_{\mathcal{H}_1} \ldots (\psi_n, \phi_n)_{\mathcal{H}_n}.$$

(23.2,a) **Problem** Verify for $n = 2$ that this defines a proper, i.e. positive definite inner product.

This inner product space can be completed in the standard way to a Hilbert space. Such a completed tensor product is sometimes denoted as $\mathcal{H}_1 \hat{\otimes} \cdots \hat{\otimes} \mathcal{H}_n$. In this text, in order not to overload the notation, the symbol $\otimes$ will for Hilbert spaces mean the completed tensor product, unless explicitly stated otherwise.

(23.2,b) *Example* Let $\mathcal{H} = L^2(\mathbb{R}^1, dx)$. The uncompleted tensor product $\mathcal{H} \otimes \mathcal{H}$ consists of all functions that are finite sums $\sum_j f_j g_j$ with the $f_j$ and $g_j$ in $L^2(\mathbb{R}_1, dx)$; its completion is $L^2(\mathbb{R}^2, dx_1 dx_2)$.

## 23.3  Symmetric and Antisymmetric Tensor Products

For the physical applications in this book we consider tensor products of a vector space $V$ with itself, 'tensor powers' of $V$, i.e.

$$\otimes^n V = \underbrace{V \otimes \cdots \otimes V}_{n \text{ times}}.$$

Of particular importance in this respect are *symmetric* and *antisymmetric* tensor products.

(23.3,a) **Definition** The $n$-fold *symmetric* tensor product of a vector space $V$ with itself, or the $n$th symmetric tensor power of $V$, is a vector space denoted as $\otimes_s^n V$ together with a (totally) symmetric $n$-linear map $m_s : \times^n V \to \otimes_s^n V$ which is universal in the sense that for any other vector space $W_s'$ and symmetric $n$-linear map $m_s' : \times^n V \to W'$ there exists a unique linear homomorphism $u_s : \otimes_s^n V \to W_s'$ with $m_s' = u_s \circ m_s$.

Similarly:

**(23.3,b) Definition** The $n$-fold *antisymmetric* or *alternating* tensor product of a vector space $V$ with itself, or the $n$th antisymmetric tensor power of $V$, is a vector space denoted as $\otimes_a^n V$, together with a (totally) antisymmetric $n$-linear map $m_a : \times^n V \to \otimes_a^n V$, which is universal in the sense that for any other vector space $W_a'$ and antisymmetric $n$-linear map $m_a' : \times^n V \to W_a'$ there exists a unique linear homomorphism $u_a : \otimes_a^n V \to W_a'$ with $m_a' = u_a \circ m_a$.

Existence is proved by construction as before. In the construction of the quotient space additional equivalence relations are added: for the symmetric case

$$(x_1, \ldots, x_n) \sim (x_{\sigma(1)}, \ldots, x_{\sigma(j)}),$$

and for the antisymmetric case

$$(x_1, \ldots, x_n) \sim \text{sign}(\sigma)(x_{\sigma(1)}, \ldots, x_{\sigma(n)}),$$

for all $x_j$ in $\mathcal{H}$ and all permutations $\sigma$, with $\text{sign}(\sigma)$ the sign of the permutation $\sigma$. The definitions of $\otimes_s^n V$ and $\otimes_a^n V$ imply again uniqueness up to isomorphisms.

There is no generally accepted notation for the symmetric tensor product of two vectors $x_1$ and $x_2$; for the antisymmetric case the symbol for exterior or Grassmann multiplication $\wedge$ is often used. It is convenenient to use here the symbols $\otimes_s$ and $\otimes_a$. This enables us to write formulas in an economic way simultaneously for both cases, like for instance,

$$m_{s/a}((x_1, \ldots, x_n)) = x_1 \otimes_{s/a} \cdots \otimes_{s/a} x_n.$$

Note that we have

$$x_1 \otimes_s \cdots \otimes_s x_n = x_{\sigma(1)} \otimes_s \cdots \otimes_s x_{\sigma(n)},$$

$$x_1 \otimes_a \cdots \otimes_a x_n = \text{sign}(\sigma)(x_{\sigma(1)} \otimes_a \cdots \otimes_a x_{\sigma(n)}),$$

for all $x_j$ and all permutations $\sigma$.

The symmetric and antisymmetric tensor products can be realized as subspaces of the ordinary tensor product. Define in $\otimes^n V$ the linear maps $P_s$ and $P_a$ by linear extension of

$$P_s(x_1 \otimes \cdots \otimes x_n) = \frac{1}{n!} \sum_\sigma (x_{\sigma(1)} \otimes \cdots \otimes x_{\sigma(n)}),$$

$$P_a(x_1 \otimes \cdots \otimes x_n) = \frac{1}{n!} \sum_\sigma \text{sign}(\sigma)(x_{\sigma(1)} \otimes \cdots \otimes x_{\sigma(n)}).$$

**(23.3,c) Problem** Check that the maps $P_{s/a}$ are projections.

The unique linear isomorphism from $\otimes^n_{s/a} V$ to $P_{s/a}(\otimes^n V) \subset \otimes^n V$ is given by

$$x_1 \otimes_{s/a} \cdots \otimes_{s/a} x_n \;\to\; P_{s/a}(x_1 \otimes \cdots \otimes x_n).$$

This means, of course, that the (symmetric or antiysymmetric) multiplication in the subspace $P_{s/a}(x_1 \otimes \cdots \otimes x_n)$ of $\otimes^n V$ defined in this manner is different from the original multiplication in $\otimes^n V$.

For $\mathcal{H}$ a Hilbert space with inner product $(\cdot, \cdot)_{\mathcal{H}}$ the tensor product space $\otimes^n_{s/a} \mathcal{H}$ is a pre-Hilbert space with respect to the inner product $(\cdot, \cdot)$, defined by linear extension of

$$(\psi_1 \otimes_s \cdots \otimes_s \psi_n, \phi_1 \otimes_s \cdots \otimes_s \phi_n) = \sum_\sigma (\psi_1, \phi_{\sigma(1)})_{\mathcal{H}} \cdots (\psi_n, \phi_{\sigma(n)})_{\mathcal{H}},$$

for the symmetric, and

$$(\psi_1 \otimes_a \cdots \otimes_a \psi_n, \phi_1 \otimes_a \cdots \otimes_a \phi_n) = \sum_\sigma \operatorname{sign}(\sigma)(\psi_1, \phi_{\sigma(1)})_{\mathcal{H}} \cdots (\psi_n, \phi_{\sigma(n)})_{\mathcal{H}}$$

$$= \operatorname{Det}(\psi_i, \psi_j)$$

for the antisymmetric case. The pre-Hilbert space $\otimes^n_{s/a} \mathcal{H}$ can be completed to a Hilbert space, also denoted as $\otimes^n_{s/a} \mathcal{H}$, in the usual way.

**(23.3,d) Problem** Show that for an orthogonal sequence of vectors $\{\psi_j\}^n_{j=1}$ in $\mathcal{H}$ the norm of the tensor product vector $\psi_1 \otimes_{s/a} \cdots \otimes_{s/a} \psi_n$ is given by

$$\|\psi_1 \otimes_{s/a} \cdots \otimes_{s/a} \psi_n\| = \|\psi_1\|_{\mathcal{H}} \cdots \|\psi_n\|_{\mathcal{H}}.$$

## 23.4 Tensor Algebras

Consider for a given vector space $V$ the $n$th tensor product powers $\otimes^n V$, for $n = 1, 2, \ldots$, to which we add for convenience a one dimensional vector space $\otimes^0 V$, containing a preferred vector $w_0$. Define

$$T(V) := \oplus^\infty_{n=0}(\otimes^n V),$$

the *general tensor algebra over* $V$, an algebra because it has an obvious multiplication, given by linear extension of

$$(x_1 \otimes \cdots \otimes x_m)(y_1 \otimes \cdots \otimes y_n) = x_1 \otimes \cdots \otimes x_m \otimes y_1 \otimes \cdots \otimes y_n,$$

$$\omega_0 (x_1 \otimes \cdots \otimes x_n) = (x_1 \otimes \cdots \otimes x_n) \omega_0 = x_1 \otimes \cdots \otimes x_n.$$

$T(V)$ is an associative algebra with unit element $\omega_0$. This type of direct sum is in the context of this book of particular importance in the case of symmetric and antisymmetric tensor products. Define therefore

$$T_{s/a}(V) := \oplus_{n=0}^{\infty}(\otimes_{s/a}^n V),$$

the *symmetric*, respectively *antisymmetric tensor algebras* over $V$, with again the inclusion of a one dimensional vector space $\otimes_{s/a}^0 V$ with a preferred vector $\omega_0$. Note that $\otimes_{s/a}^1 V$ can, of course, be identified with $V$. $T_s(V)$ is a commutative algebra; $T_a(V)$ is graded-commutative, which means that

$$a_m \in T_a^m(V) = \otimes_a^m V, \qquad b_n \in T_a^n(V) = \otimes_a^n V$$

implies $a_m b_n = (-1)^{mn} b_n a_m$. $T_a(V)$ is in the literature often denoted as $\Lambda(V)$ and called the *exterior* or *Grassmann algebra* over $V$. The algebra properties of $T_{s/a}(V)$ will not be used in this book.

For a Hilbert space $\mathcal{H}$ the symmetric and antisymmetric tensor algebras in the sense of vector spaces are completed to Hilbert space tensor algebras, to be denoted as $T_{s/a}(\mathcal{H})$. The $\omega_0$ is assumed to be a unit vector.

The spaces $T_{s/a}(\mathcal{H})$ appear in quantum theory as state spaces for many particle systems; $T_s(\mathcal{H})$ is then the many particle state space of a *bosonic* system, $T_a(\mathcal{H})$ that of a *fermionic* one. All this is fully discussed in Chap. 9 of the main text. In the remainder of this chapter it will be convenient to use occasionally the terms 'bosonic' and 'fermionic' for the symmetric, respectively the antisymmetric case.

## 23.5 Raising and Lowering Operators

### 23.5.1 The Symmetric Case

Let $\mathcal{H}$ be a Hilbert space, and $T_s(\mathcal{H})$ the symmetric tensor algebra over $\mathcal{H}$. Define, for each $\psi$ in $\mathcal{H} = \otimes_s^1 \mathcal{H} \subset T_s(\mathcal{H})$, a pair of linear operators in $T_s(\mathcal{H})$, a *raising operator* $a^*(\psi)$ by linear extension of

$$a^*(\psi)(\psi_1 \otimes_s \cdots \otimes_s \psi_n) = \psi \otimes_s \psi_1 \otimes_s \cdots \otimes_s \psi_n,$$

for $n = 1, \ldots$, with $a^*(\psi)\omega_0 = \psi$, and a *lowering operator* $a(\psi)$ by

$$a(\psi)(\psi_1 \otimes_s \cdots \otimes_s \psi_n) = \sum_{j=1}^{n} (\psi, \psi_j)(\psi_1 \otimes_s \cdots \otimes_s \psi_{j-1} \otimes_s \psi_{j+1} \otimes_s \cdots \otimes_s \psi_n),$$

for $n = 2, \ldots$, with

$$a(\psi)\psi_1 = (\psi, \psi_1)\omega_0, \qquad a(\psi)\omega_0 = 0,$$

for all $\psi_j$ in $\mathcal{H}$. It is clear that $a^*(\psi)$ and $a(\psi)$ depend linearly, respectively conjugate linearly, on $\psi$ in $\mathcal{H}$. Each operator $a^*(\psi)$ and $a(\psi)$ has as common invariant dense domain of definition the subspace $T_s(\mathcal{H})_{\text{fin}}$ of $T_s(\mathcal{H})$ consisting of vectors that have only a finite number of non-zero components from each $\otimes_s^n \mathcal{H}$.

(23.5.1,a) **Problem** The commutator of two operators $A$ and $B$ is defined, as usual, as $[A, B] = AB - BA$.

1. Prove $[a^*(\psi), a^*(\phi)] = [a(\psi), a(\phi)] = 0$, $[a(\psi), a^*(\phi)] = (\psi, \phi)$.

2. Prove that $a^*(\psi)$ and $a(\psi)$ are indeed each others hermitian conjugates, i.e. that, for all $\psi$ in $\mathcal{H}$, one has, with respect to the inner product in $T_s(\mathcal{H})$,

$$(a^*(\psi)\Psi_1, \Psi_2) = (\Psi_1, a(\psi)\Psi_2), \quad \forall \psi \in \mathcal{H}, \quad \forall \Psi_1, \Psi_2 \in T_s(\mathcal{H})_{\text{fin}}.$$

3. Use the commutation relations for $a(\cdot)$ and $a^*(\cdot)$ to prove that the norm of the vector $\Psi_n = (a^*(\psi))^n \omega_0$ is equal to $\sqrt{n!} \, ||\psi||^n$, for $n = 1, 2, \ldots$ . Conclude from the result that the operator $a^*(\psi)$, and therefore also $a(\psi)$, is an unbounded operator, for all $\psi \neq 0$.

The space $T_s(\mathcal{H})$ is always infinite dimensional, even when $\mathcal{H}$ has finite dimension. Think of the Hilbert space of the harmonic oscillator discussed in Chap. 6. There the infinite-dimensional Hilbert space of the system has a 1-dimensional $\mathcal{H}$, spanned by the energy eigenvector $\psi_1$, and is generated by $\omega_0$, the ground state $\psi_0$, by a single pair of operators $a^*$ and $a$.

It is useful to choose an orthonormal basis $\{\psi_j\}_j$ in the 'one particle space' $\mathcal{H}$. The raising and lowering operators are then denoted as $a_j^* = a^*(\psi_j)$ and $a_j = a(\psi_j)$. The commutation relations become

$$[a_j^*, a_k^*] = [a_j, a_k] = 0, \quad [a_j^*, a_k] = \delta_{jk},$$

With the operators $a_j^*$ one defines vectors of the form

$$a_{j_1}^* \ldots a_{j_n}^* \omega_0.$$

Because all the $a_j^*$ commute, changing the order of the operators does not change the vector. Note that the same operator may occur more than once. A unique way of writing these vectors is, for positive integers $n_{j_s}$,

$$(a_{j_1}^*)^{n_{j_1}} \ldots (a_{j_p}^*)^{n_{j_p}} \omega_0, \qquad j_1 < \cdots < j_p.$$

In the context of Dirac's so-called bra-ket formalism, an ingenious, partially heuristic, but very practical way of writing the basic concepts of quantum theory, there is a simple and efficient notation for these vectors, namely, after normalization, as

$$(n_{j_1}!)^{-1/2} \ldots (n_{j_p}!)^{-1/2} (a_{j_1}^*)^{n_{j_1}} \ldots (a_{j_p}^*)^{n_{j_p}} \omega_0 = \mid n_{j_1} \ldots n_{j_p} >,$$

with $\omega_0$ denoted as $\mid 0 >$. Dirac's bra-ket formalism is explained in Supp. Chap. 27.

(23.5.1,b) **Problem**
1. Prove the commutation relation $[a_j, (a_k^*)^n] = n(a_k^*)^{n-1}\delta_{jk}$ and use this relation to show that

$$\|(a_{j_1}^*)^{n_{j_1}} \ldots (a_{j_p}^*)^{n_{j_p}} \omega_0\| = (n_{j_1}!)^{1/2} \ldots (n_{j_p}!)^{1/2}.$$

2. Prove also that the system of normalized vectors $\mid n_{j_1} \ldots n_{j_p} >$, with $\mid 0 >$ added, is an orthonormal basis for $T_s(\mathcal{H})$.

3. Show that the $\mid n_{j_1} \ldots n_{j_p} >$, with $n = \sum_{s=1}^{p} n_{j_s}$ span $\otimes_s^n \mathcal{H} \subset T_s(\mathcal{H})$.

The vectors $\mid n_{j_1} \ldots n_{j_p} >$ can be characterized by infinite sequences of arbitrary non-negative integers $\{n_1, n_2, \ldots\}$, with only a finite number different from zero. In physics these $n_j$ are called *occupation numbers*.

(23.5.1,c) **Problem** The operator $N = \sum_{n=1}^{\infty} a_n^* a_n$ in $T_s(\mathcal{H})$ is called the *total number operator*. Prove the commutation relation $[N, a_j^*] = a_j^*$ and use this to prove that

$$N\mid n_{j_1} \ldots n_{j_p} >= (n_{j_1} + \cdots + n_{j_p})\mid n_{j_1} \ldots n_{j_p} >$$

Verify that this means that $N$ is a selfadjoint operator with discrete eigenvalues $n$ and eigenspaces $\otimes_s^n \mathcal{H}$, for $n = 0, 1, \ldots$.

## 23.5.2 The Antisymmetric Case

The antisymmetric case is similar to the symmetric one. Define, for each $\psi$ in $\mathcal{H}$, a pair of linear operators in $T_a(\mathcal{H}) = \oplus_{n=0}^{\infty}(\otimes_a^n \mathcal{H})$, a *raising operator* $a^*(\psi)$ by linear extension of

$$a^*(\psi)(\psi_1 \otimes_a \cdots \otimes_a \psi_n) = \psi \otimes_a \psi \otimes_a \psi_1 \otimes_a \cdots \otimes_a \psi_n,$$

for $n = 1, \ldots$, with $a^*(\psi)\omega_0 = \psi$, and a *lowering operator* $a(\psi)$ as

$$a(\psi)(\psi_1 \otimes_a \cdots \otimes_a \psi_n)$$

$$= \sum_{j=1}^{n} (\psi, \psi_j)(-1)^j (\psi_1 \otimes_a \cdots \otimes_a \psi_{j-1} \otimes_a \psi_{j+1} \otimes_a \cdots \otimes_a \psi_n).$$

The operators $a^*(\psi)$ and $a(\psi)$ depend again linearly, respectively anti-linearly on $\psi$ in $\mathcal{H}$, with in first instance $T_a(\mathcal{H})_{\text{fin}}$ as a common invariant dense domain of definition.

### (23.5.2,a) Problem
1. The expression $[a^*(\psi), a(\phi)]_+ = a^*(\psi)a(\phi) + a(\phi)a^*(\psi)$ is called the *anticommutator* of $a^*(\psi)$ and $a(\phi)$. Prove, for all $\psi, \phi$ in $\mathcal{H}$,

$$[a^*(\psi), a^*(\phi)]_+ = [a(\psi), a(\phi)]_+ = 0, \quad [a^*(\psi), a(\phi)]_+ = (\psi, \phi).$$

2. Prove that with respect to the inner product in $T_a(\mathcal{H})$ one has

$$(a^*(\psi)\Psi_1, \Psi_2) = (\Psi_1, a(\psi)\Psi_2), \quad \forall \psi \in \mathcal{H}, \quad \forall \Psi_1, \Psi_2 \in T_a(\mathcal{H})_{\text{fin}}.$$

Choosing an orthonormal basis $\{\psi_j\}_j$ in $\mathcal{H}$ is again useful. The corresponding raising and lowering operators are denoted as $a_j^* = a^*(\psi_j)$ and $a_j = a(\psi_j)$; the anticommutation relations become

$$[a_j^*, a_k^*]_\pm = [a_j, a_k]_\pm = 0, \quad [a_j^*, a_k]_\pm = \delta_{jk}.$$

The operators $a_j^*$ are used to define an orthogonal basis in $T_s(\mathcal{H})$, in a similar but simpler way than in the symmetric case because the relation $[a_j^*, a_j^*]_+ = 0$ implies $(a_j^*)^2 = 0$, so each operator $a_j^*$ can occur only once. We have therefore a system consisting of the vectors $\omega_0$ and all $| n_{j_1} \ldots n_{j_p} \rangle = a_{j_1}^* \ldots a_{j_p}^* \omega_0$, with $j_1 < \cdots < j_p$. Note that the sequence of occupation numbers now consists of the numbers 0 and 1. Note also that the operators $a_j^*$ and $a_k$ are *bounded* and can therefore be extended from $T_a(\mathcal{H})_{\text{fin}}$ to the full tensor algebra $T_a(\mathcal{H})$.

### (23.5.2,b) Problem
Give the result of the action of $a_j^*$ on the vectors $a_{j_1}^* \ldots a_{j_n}^* \omega_0$ and on $\omega_0$ and show that the result implies that $a_j^*$, and therefore also $a_j$, is bounded with operator norm $\|a_j^*\| = \|a_j\| = 1$.

Another difference with the symmetric case is that $T_a(\mathcal{H})$ is finite dimensional if $\mathcal{H}$ is finite dimensional.

(23.5.2,c) **Problem** What is the dimension of $T_a(\mathcal{H})$ if $\mathcal{H}$ has finite dimension $n$?

# References

1. Greub, W.: Multilinear Algebra. Springer, Heidelberg (1967) [Second edition 1978, A good introduction to multilinear algebra. Symmetric tensor algebras are discussed, but not the anti-symmetric ones.]
2. Bourbaki, N.: Elements of Mathematics. Algebra I. Springer, Berlin (1989) [Translated from the original 1970 French edition, Treats both symmetric and antisymmetric tensor algebras. Written in a clear, precise but somewhat severe style.]
3. Gallier, J.: Tensor Algebras, Symmetric Algebras and Exterior Algebras Lecture Notes. University of Pennsylvania, Pennsylvania (2011) [An excellent very readable set of course notes. Available at: http://www.cis.upenn.edu/~cis610/diffgeom7.pdf. Raising and lowering operators are notions from physics; it is probably therefore that they are not mentioned in any of the above texts.]

# Chapter 24
# Lie Groups and Lie Algebras

## 24.1 Introduction

The notion of group—'concrete' as a group of transformations of a set or system of differential equations—goes back to the eigthteenth century. In modern language one may think of a nonempty set $X$ with $\mathcal{G}_X$ the set of all invertible maps of $X$ onto itself. $\mathcal{G}_X$ is a group with the composition $g_1 \circ g_2$ as multiplication and the identity transformation as unit element. It can be shown that an arbitrary group $\mathcal{G}$ is isomorphic to a group of transformations of some set $X$.

The notion of group was developed further in the nineteenth and twentieth century, and became more 'abstract', meaning a set with a certain multiplication operation between elements satisfying certain requirements.

This chapter discusses groups used in this book to describe symmetries in non-relativistic and covariance properties in relativistic quantum theory. Most of these groups are *Lie groups*, i.e. groups that have an underlying manifold structure, implying that each has a *Lie algebra*, the tangent space at its origin, on which an additional bracket structure is defined. In physical applications these Lie algebras are in general more useful than the Lie groups themselves because they lead to linear relations.

There is a wide choice of books on the topics treated in this chapter, the theory of groups, in particular Lie groups and Lie algebras and their representations and their applications in physics. See, for example, Refs. [2–12]. A rather special question is discussed in [1].

## 24.2 Groups: Generalities

### 24.2.1 Definition and Basic Properties

(24.2.1,a) **Definition** A nonempty set $\mathcal{G}$ is called a *group* iff it is provided with a *multiplication*, i.e. a map

© Springer International Publishing Switzerland 2015
P. Bongaarts, *Quantum Theory*, DOI 10.1007/978-3-319-09561-5_24

$$\mathcal{G} \times \mathcal{G} \to \mathcal{G}, \quad (g_1, g_2) \mapsto g_1 g_2,$$

which has the following properties:

1. $g_1(g_2 g_3) = (g_1 g_2) g_3, \quad \forall g_1, g_2, g_3 \in \mathcal{G}$ *(associativity)*.
2. $\exists e \in G$, with $ge = eg = g, \quad \forall g \in \mathcal{G}$ *(existence of a unit)*.
3. $\exists g' \in \mathcal{G}$ with $g'g = gg' = e, \quad \forall g \in \mathcal{G}$ *(existence of an inverse)*.

The element $g'$ is, not surprisingly, denoted as $g^{-1}$.

**(24.2.1,b) Problem** Show that the unit element $e$ and the inverse $g^{-1}$ of $g$ are unique.

A group $\mathcal{G}$ is called *abelian* or *commutative* if $g_1 g_2 = g_2 g_1$, for all $g_1, g_2$ in $G$. An abelian group is often written additively, with the multiplication $g_1 g_2$ as $g_1 + g_2$, with $g^{-1}$ as $-g$ and $e$ as 0.

**(24.2.1,c) Definition** A map $\phi$ from a group $\mathcal{G}_1$ into a second group $\mathcal{G}_2$ is called a *(group) homomorphism* if it maps the identity element of $\mathcal{G}_1$ onto the identity element of $\mathcal{G}_2$ and respects the multiplication, i.e. has $\phi(gg') = \phi(g)\phi(g')$, for all $g$ and $g'$ in $\mathcal{G}_1$. The kernel Ker $(\mathcal{G})$ of such a map is defined as $\phi^{-1}(e) = \{g \in \mathcal{G}_1 \mid \phi(g) = e\}$.

**(24.2.1,d) Problem** Show that this definition implies that $\phi(g^{-1}) = (\phi(g))^{-1}$, for all $g$ in $\mathcal{G}_1$.

A homomorphism that is both injective and surjective is called an *isomorphism*. An *automorphism* of a group $\mathcal{G}$ is an injective and surjective homomorphism of $\mathcal{G}$ onto itself. The set of automorphisms of a group $\mathcal{G}$ is again a group; it is usually denoted as $\mathrm{Aut}(\mathcal{G})$.

## 24.2.2 Constructing New Groups from Given Groups

There are various ways in which new groups can be constructed from known groups.

## 24.2.3 Subgroups

**(24.2.2,a) Definition** A subset $\mathcal{H}$ of $\mathcal{G}$ is a *subgroup* of $\mathcal{G}$ if it is closed under the multiplication $(g_1, g_2) \mapsto g_1 g_2$ and contains with each $g$ its inverse $g^{-1}$.

**(24.2.2,b) Problem** Show that $\mathcal{H}$ contains the unit element $e$, implying that $\mathcal{H}$ itself is indeed a group.

**(24.2.2,c) Definition** A subgroup $\mathcal{H}$ of $\mathcal{G}$ is called a *normal subgroup* or *invariant subgroup* if $ghg^{-1}$ is in $\mathcal{H}$ for all $h$ in $\mathcal{H}$ and all $g$ in $\mathcal{G}$.

**(24.2.2,d) Problem** Let $\phi : \mathcal{G}_1 \to \mathcal{G}_2$ be a homomorphism of groups. Prove that the inverse image $\phi^{-1}(\mathcal{G}_2)$ is a normal subgroup of $\mathcal{G}_1$.

In an abelian group every subgroup is clearly a normal subgroup. In general non-abelian groups normal subgroups are scarce; many groups have only the unit element $\{e\}$ and the group itself as normal subgroups and have no nontrivial normal subgroups. Such groups are called *simple*. A group that has no *abelian* normal subgroups is called *semisimple*.

### 24.2.4 Quotient Groups

Let $\mathcal{G}$ be a group and $H$ a normal subgroup of $\mathcal{G}$. Define an equivalence relation in $\mathcal{G}$: $g_1 \sim g_2$ if and only if there is an element $h$ in $\mathcal{H}$ such that $g_1 = hg_2$. The space of equivalence classes $[g]$ is denoted as $\mathcal{G}/\mathcal{H}$.

(24.2.2,e) **Problem** Show that $\mathcal{G}/\mathcal{H}$ is a group with respect to the obvious multiplication $[g_1][g_2] = [g_1g_2]$, for all $g_1$ and $g_2$ in $\mathcal{G}$, and with $[e]$ as unit element.

The group $\mathcal{G}/\mathcal{H}$ is called a *quotient group*. Note that if the subgroup $\mathcal{H}$ is not invariant, $\mathcal{G}/\mathcal{H}$ is still defined as a space of equivalence classes, but is not a group.

### 24.2.5 Direct Product Groups

Consider two groups $\mathcal{G}_1$ and $\mathcal{G}_2$. The (Cartesian) product $\mathcal{G}_1 \times \mathcal{G}_2$, i.e. the set of all pairs of elements $(g_1, g_2)$, $g_1 \in \mathcal{G}_1$ and $g_2 \in \mathcal{G}_2$, has a natural group multiplication

$$(g_1, g_2)(g_1', g_2') = (g_1 g_1', g_2 g_2'),$$

for all $g_1$, $g_1'$ in $\mathcal{G}_1$ and all $g_2$, $g_2'$ in $\mathcal{G}_2$, with $e = (e_1, e_2)$ as unit element. $\mathcal{G}_1 \times \mathcal{G}_2$, with this group structure, is called a *(direct) product group*.

(24.2.2,f) **Problem** Check that $\mathcal{G}_1 \times \mathcal{G}_2$ is indeed a group, i.e. satisfies the group properties.

### 24.2.6 Semidirect Product Groups

This is a useful generalization of the notion of direct product group. Consider again two groups $\mathcal{G}_1$ and $\mathcal{G}_2$, now together with an homomorphism $\phi$ from $\mathcal{G}_2$ into $\mathrm{Aut}(\mathcal{G}_1)$, the group of all automorphisms of $\mathcal{G}_1$. The product space $\mathcal{G}_1 \times \mathcal{G}_2$ can be given a $\phi$-dependent group structure, in general different from the direct product group structure, by defining the multiplication

$$(g_1, g_2)(g_1', g_2') = (g_1(\phi(g_2)g_1'), g_2 g_2'),$$

for all $g_1$, $g_1'$ in $\mathcal{G}_1$ and $g_2$, $g_2'$ in $\mathcal{G}_2$. The unit element is again $(e_1, e_2)$. This group is called the *semi-direct product* of $\mathcal{G}_1$ and $\mathcal{G}_2$. It depends on the choice of $\phi$, and is usually denoted as $\mathcal{G}_1 \times_\phi \mathcal{G}_2$. If one chooses $\phi$ as the assignment of the identity automorphism of $\mathcal{G}_1$ to all elements of $\mathcal{G}_2$, the corresponding semi-direct product is just the ordinary product $\mathcal{G}_1 \times \mathcal{G}_2$.

(24.2.2,g) **Problem** Check that the inverse $(g_1, g_2)^{-1}$ of an element $(g_1, g_2)$ in $\mathcal{G}_1 \times_\phi \mathcal{G}_2$ is given by the formula

$$(g_1, g_2)^{-1} = (\phi(g_2^{-1})g_1^{-1}, g_2^{-1}),$$

The elements of the form $(g_1, e_2)$, respectively $(e_1, g_2)$, form a subgroup of $\mathcal{G}_1 \times_\phi \mathcal{G}_2$ isomorphic to $\mathcal{G}_1$, respectively $\mathcal{G}_2$. The first subgroup is normal; the quotient of $\mathcal{G}_1 \times_\phi \mathcal{G}_2$ over $\mathcal{G}_1$ is isomorphic to $\mathcal{G}_2$. One has $(g_1, g_2) = (g_1, e_2)(e_1, g_2)$, so an arbitrary element of $\mathcal{G}_1 \times_\phi \mathcal{G}_2$ can be written as a (unique) product of elements from both subgroups.

The semi-direct product multiplication rule is not very transparent at first sight. Its meaning can be better understood from the example of the group of *inhomogeneous* or *affine linear transformations* of a vectorspace. Let $x$ be an element of $\mathbb{R}^n$, with coordinates $x_1, \ldots, x_n$, $a$ also an element of $\mathbb{R}^n$, with coordinates $a_1, \ldots, a_n$, and $T = \{T_{jk}\}$ an $n \times n$ matrix from $GL(n, \mathbb{R})$. An inhomogeneous linear transformation of $\mathbb{R}^n$ onto itself is defined as

$$x'_j = \sum_{k=1}^{n} T_{jk}x_k + a_j.$$

Such transformations are clearly invertible and form a group. It is a semi-direct product $\mathcal{G}_1 \times_\phi \mathcal{G}_2$, with $\mathcal{G}_1 = \mathbb{R}^n$, $\mathcal{G}_2 = GL(n, \mathbb{R})$, and the homomorphism $\phi$ : $\mathcal{G}_2 \to \mathcal{G}_1$ given by $\phi(T)x = Tx$, i.e. by $(Tx)_j = \sum_{k=1}^{n} T_{jk}x_k$. Writing such an inhomogeneous linear transformation as a pair $(a, T)$, with $a$ in $\mathbb{R}^n$ and $T$ in $GL(n, \mathbb{R})$, we have as multiplication rule for two such transformations

$$(a, T)(a', T') = (a + Ta', TT').$$

The identity transformation is $(0, 1)$ and the inverse transformation is written as

$$(a, T)^{-1} = (-T^{-1}a, T^{-1}).$$

## 24.3 Examples of Groups

### 24.3.1 Finite Groups

- *Symmetric groups* $S_n$, consisting of all permutations of a (finite) set of $n$ objects. They are the only finite group in this book; all other groups are (infinite) Lie groups, as such provided with Lie algebras. The group $S_2$ is abelian; the other symmetric groups non-abelian. $S_n$ has $n!$ elements. A subgroup of $S_n$ is $\mathbb{Z}_n$, the group of *cyclic permutations*. We can represent an $n$-tuple $\{a_1, \ldots, a_n\}$ of objects by equidistant points on the unit circle. Cyclic permutations of this set are obtained by turning the system of points clockwise over a distance $k(2\pi/n)$, for $k = 1, \ldots, n$. It is obvious that $\mathbb{Z}_n$ has $n$ elements and is abelian, for each $n$.

The symmetric groups play a role in the description of systems of identical particles in Chap. 9.

There are many other, much more complicated finite groups. In physics, but not in this book, one uses for example the *crystallographic groups* for a systematic description of the properties of crystals.

### 24.3.2 Lie Groups

The underlying set of a Lie group is a finite dimensional smooth manifold. See for the properties of smooth (or differentiable) manifolds Supp. Chap. 20. The multiplication $(g_1, g_2) \mapsto g_1 g_2$ and the taking of inverses $g \to g^{-1}$ are smooth maps. The group multiplication in a Lie group $\mathcal{G}$ induces a bracket on the tangent space at the identity element, a real $n$-dimensional vector space, making it into its Lie algebra $\mathcal{L}_\mathcal{G}$. A Lie algebra $\mathcal{G}$ as a general mathematical object is defined as a vector space provided with a *Lie bracket*, i.e. a bilinear map $\mathcal{G} \times \mathcal{G} \to \mathcal{G}$, satisfying certain requirements. An $n$-dimensional Lie group defines an $n$-dimensional Lie algebra. Each Lie algebra can be "exponentiated" to a Lie group. However, two different Lie groups may have the same Lie algebra. This was already noted in Sect. 7.4.2. This will be further discussed in Supp. Sect. 24.5, together with the general properties of Lie algebras in Supp. Sect. 24.4.1 and with more details on the relations between Lie groups and Lie algebras in Supp. Sect. 24.5. Lie groups form the majority of the groups used in this book.

All Lie groups employed in this book are real; complex Lie groups, with an underlying complex manifold, will not be discussed.

- *Translation groups in dimension n.* The translation group in three dimensional space plays an important role in quantum mechanics—and of course also in the simpler one dimensional case. Symmetry with respect to this group leads to linear momentum as conserved quantity. Note that the one parameter group of time evolution operators $\{U(t)\}_{t \in \mathbb{R}}$ in an autonomous quantum system defines translation in time. This is connected with conservation of energy. We also have the four dimensional spacetime translation group in relativistic quantum theory, connected with the conservation of energy-momentum.

(24.3.2,a) **Problem** Show that the $n$-dimensional translation group can be written as a group of $(n + 1) \times (n + 1)$ matrices.

### 24.3.3 Matrix Lie Groups

The example of Problem (23.3.2,a) is somewhat artificial. Most of the important Lie groups in physics are in a natural manner matrix groups. All such groups are closed subgroups of $GL(n, \mathbb{C})$ or $GL(n, \mathbb{R})$, the general linear groups in $n$ complex or real dimensions, consisting of all invertible real or complex $n \times n$ matrices. Their Lie algebras are denoted as $gl(n, \mathbb{C})$ or $gl(n, \mathbb{R})$. Matrix groups can of course be seen as groups of linear transformations in a finite dimensional vector space in which a basis has been chosen. The following examples are used in this book:

- *Rotation group* $O(3)$ *in dimension* 3: The group of $3 \times 3$ real orthogonal matrices. A matrix from this group has determinant $\pm 1$. We shall mainly use $SO(3)$, the special orthogonal group, in which all matrices have determinant 1, excluding space reflection. Our main application of this group is in Chap. 7, where it is used to simplify the solution of the eigenvalue–eigenfunction problem for the hydrogen atom. Symmetry with respect to $SO(3)$ is related to the conservation of angular momentum.
  The group $SO(3)$ has $SU(2)$ as *universal covering group*—a notion that will be explained further on. $SU(2)$ consists of all $2 \times 2$ unitary matrices with determinant 1. It is needed in Chap. 8 for the discussion of *spin*, an intrinsic angular momentum, typical for quantum theory.
- *Lorentz group* $\mathcal{L}$ or rather $O(1, 3)$: It consists of all real $4 \times 4$ matrices which leave an indefinite bilinear form in $\mathbb{R}^4$ invariant. This group is used in Chaps. 15 and 16 where relativistic quantum physics is discussed. The group $\mathcal{L}$ has a connected subgroup, denoted as $\mathcal{L}_+^\uparrow$, which has as its universal covering group $SL(2, \mathbb{C})$, the group of all complex $2 \times 2$ matrices with determinant $+1$. Combining $\mathcal{L}$ with the group of four dimensional spacetime translations gives $\mathcal{P}$, the *inhomogeneous Lorentz group* or *Poincaré group*.

## 24.4  Lie Algebras

### 24.4.1 Lie Algebras as 'Abstract' Objects: Generalities

(24.4.1,a) **Definition** A (real or complex) vector space $\mathcal{L}$ is called a (real or complex) *Lie algebra* iff it is provided with a *Lie bracket*, i.e. a bilinear map

$$\mathcal{L} \times \mathcal{L} \to \mathcal{L}, \quad (u, v) \mapsto [u, v],$$

which has the following two properties, for all $u$, $v$ and $w$ in $\mathcal{L}$:

1. It is *antisymmetric*, i.e. it has $[u, v] = -[v, u]$,
2. It satisfies the *Jacobi identity*, i.e.

$$[u, [v, w]] + [v, [w, u]] + [w, [u, v]] = 0.$$

If $[u, v] = 0$ for all $u, v$ in $\mathcal{L}$, satisfying in this trivial way the requirements for a Lie bracket, $\mathcal{L}$ is called an *abelian* Lie algebra.

In this chapter all the examples will be finite dimensional. Infinite dimensional Lie algebras appear in Supp. Chap. 20 as Poisson algebras of functions on a symplectic manifold, and as such play a role in classical mechanics. They are also mentioned in Supp. Chap. 27.

Let $\mathcal{L}$ be finite dimensional Lie algebra. Choose a basis $(e_1, \ldots, e_p)$ in $\mathcal{L}$. The Lie bracket of two basis elements, $e_j$ and $e_k$, can be written as $[e_j, e_k] = \sum_{l=1}^{p} c_{jk}^l e_l$. The numbers $c_{jk}^l$, called *structure constants*, determine the Lie algebra $\mathcal{L}$. They are of course basis dependent.

**(24.4.1,b) Problem** Let $\mathcal{L}$ be an $n$-dimensional Lie algebra, with structure constants $c_{jk}^k$ with respect to a given basis $(e_1, \ldots, e_n)$. Show that the number of these constants is $\frac{1}{2}n(n-1)$. Prove that the necessary and sufficient conditions for a system of numbers $c_{jk}^l$ to be the structure constants of an $n$-dimensional Lie algebra are

$$c_{ji}^k = -c_{ij}^k, \quad \sum_{l=1}^{n}(c_{ij}^l c_{lk}^m + c_{ki}^l c_{lj}^m + c_{jk}^l c_{li}^m) = 0.$$

**(24.4.1,c) Definition** A linear subspace $\mathcal{K}$ of the Lie algebra $\mathcal{L}$ is called a *Lie subalgebra* iff it is closed under the operation of taking the bracket of two elements. $K$ is called an *invariant Lie subalgebra* or an *ideal* iff moreover $[u, v]$ is in $K$, for all $u$ in $K$ and all $v$ in $\mathcal{L}$. A Lie algebra is called *simple* iff it has no nontrivial ideals, *semisimple* iff it has no nontrivial abelian ideals.

**(24.4.1,d) Definition** A map $\phi$ from a Lie algebra $\mathcal{L}_1$ into a second Lie algebra $\mathcal{L}_2$ is called a *(Lie algebra) homomorphism* iff it is linear and connects the brackets, i.e. satisfies $\phi([u, v]_1) = [\phi(u), \phi(v)]_2$, for all $u$ and $v$ in $\mathcal{L}_1$.

There are obvious definitions of the notions of *isomorphism* and *automorphism*. The automorphisms of a Lie algebra $\mathcal{L}$ form a group denoted as $\mathrm{Aut}(\mathcal{L})$.

### 24.4.2 Constructing New Lie Algebras from Given Ones

There are Lie algebra analogues for the group constructions discussed in Supp. Sect. 24.2.2:

### 24.4.3 Quotient Lie Algebras

An ideal in a Lie algebra $\mathcal{L}$ is a linear subspace $\mathcal{K}$ of $\mathcal{L}$ such that for every $u \in \mathcal{L}$ and every $v \in \mathcal{K}$ the commutator $[u, v]$ is in $\mathcal{K}$. Usually ideal means nontrivial ideal, i.e. not the zero ideal $\{0\}$ or $\mathcal{L}$ itself. An ideal is the Lie algebra analogue of a normal subgroup. Given an ideal $\mathcal{K}$ in $\mathcal{L}$ one defines $\mathcal{L}/\mathcal{K}$ as the usual quotient of two vector spaces.

(24.4.2,a) **Problem** Show that $\mathcal{L}/\mathcal{K}$ is a Lie algebra when provided with the obvious Lie bracket in terms of equivalence classes.

### 24.4.4  Direct Sum Lie Algebras

Given two Lie algebras $\mathcal{L}_1$ and $\mathcal{L}_2$. Define their vector space *direct sum* $\mathcal{L}_1 \oplus \mathcal{L}_2$.

(24.4.2,b) **Problem** Show that this vector space direct sum is made into a Lie algebra by defining the Lie bracket, for all $u_1$, $v_1$ in $\mathcal{L}_1$ and $u_2$, $v_2$ in $\mathcal{L}_2$, as

$$[(u_1, v_1), (u_2, v_2)] := ([u_1, u_1], [v_1, v_2]).$$

### 24.4.5  Semidirect Sum of Two Lie Algebras

For the construction of a semi direct sum of two Lie algebras we need the notion of a derivation of a Lie algebra. This is a linear map $D : \mathcal{L} \to \mathcal{L}$ such that

$$D[a, b] = [Da, b] + [a, Db], \ \forall a, b \in \mathcal{L}.$$

A derivation of an associative algebra is defined in Supp. Sect. 27.2.

(24.4.2,c) **Problem** Show that Der $(\mathcal{L})$, the linear space of derivations of $\mathcal{L}$, is itself a Lie algebra, as a subspace of $gl(\mathcal{L})$, the Lie algebra of linear maps from $\mathcal{L}$ into $\mathcal{L}$, with respect to the standard commutator of such maps.

Given two Lie algebras $\mathcal{L}_1$ and $\mathcal{L}_2$, together with a linear map $\phi$ from $\mathcal{L}_2$ into Der $(\mathcal{L}_1)$, which preserves the Lie brackets, one introduces a Lie algebra structure in the vector direct sum $\mathcal{L}_1 \oplus \mathcal{L}_2$ by defining the bracket

$$[(u_1, v_1), (u_2, v_2)] := ([u_1, u_2] + \phi(v_1)u_2 - \phi(v_2)u_1, [v_1, v_2]).$$

(24.4.2,d) **Problem** Show that this indeed defines a Lie algebra bracket.

The semidirect sum of two Lie algebras $\mathcal{L}_1$ and $\mathcal{L}_2$, for a given $\phi$ is denoted as $\mathcal{L}_1 \oplus_\phi \mathcal{L}_2$.

### 24.4.6  Lie Algebras: Concrete Examples

An obvious example is the space of all linear operators in a vector space with the commutator $[S, T] = ST - TS$ as Lie bracket. It may be denoted as $gl(V)$. If $V$ has finite dimension $n$, one can choose a basis in $V$. This leads to a matrix Lie algebra, denoted as $gl(n, R)$ for real $V$ and $gl(n, C)$ for complex $V$. This notation is of course suggestive of the connection of Lie algebras with Lie groups, to be discussed further on. An example of an infinite dimensional Lie algebra is the space of vector fields on a smooth manifold, with as Lie bracket the bracket of vector fields. See for this Supp. Chap. 20. Another infinite dimensional Lie algebra is the space of smooth functions on a symplectic manifold, with as Lie bracket the Poisson bracket, defined for the use of classical mechanics in Chap. 2, and also in Chap. 12.

Most of the Lie algebras needed in our presentation of quantum theory are *matrix Lie algebras*, as such obvious Lie subalgebras of either $GL(n, \mathbb{R})$ or $GL(n, \mathbb{C})$ with its matrix commutator as Lie bracket. We have the following representative list:

- $gl(n, \mathbb{R})$: all real $n \times n$ matrices; dimension $n^2$.
- $sl(n, \mathbb{R})$: the real $n \times n$ matrices with trace 0; dimension $n^2 - 1$.
- $sl(n, \mathbb{C})$: the complex $n \times n$ matrices with trace 0; dimension as a *real* Lie algebra $2(n^2 - 1)$.
- $o(n) = so(n)$: the real antisymmetric $n \times n$ matrices; dimension $\frac{1}{2}n(n - 1)$.
- $u(n)$: the complex antihermitian $n \times n$ matrices; real dimension $n^2$.
- $su(n)$: the complex antihermitian $n \times n$ matrices with trace 0; real dimension $n^2 - 1$.

The notation for these Lie algebras suggest, of course, the relation with the Lie groups listed in Supp. Sect. 24.3. The next section is devoted to this relation.


## 24.5  Relation Between Lie Groups and Lie Algebras

Within the framework of manifold theory there is a general formulation of the relation between Lie groups and Lie algebras. It is elegant and natural, but mathematically fairly sophisticated. The fact that all finite dimensional Lie groups in this book are matrix groups—or can be written as matrix groups, as in the case of translation groups—allows a simpler, more pedestrian approach, which is, however, completely rigorous.

Let $\mathcal{G}$ be an $n \times n$ matrix Lie group, and let $t \mapsto g(t)$ be a smooth curve in $\mathcal{G}$ with $g(0) = 1_{\mathcal{G}}$. Smooth means in this context that all matrix elements of $g(t)$ are smooth functions of $t$. Such a curve has a derivative in $t = 0$, $\left(\frac{d}{dt}g(t)\right)_{t=0} = A$, with $A$ an $n \times n$ matrix. Two such curves are considered to be equivalent iff their derivatives in $t = 0$ are the same. We call the collection of these matrices $\mathcal{L}_{\mathcal{G}}$. One can prove that an equivalence class with a matrix $A$ as derivative contains as curve the exponential matrix-valued function $g_A(t) = e^{tA}$, defined as a convergent power series.

For $A$ and $B$ in $\mathcal{L}_{\mathcal{G}}$ and two real numbers $\lambda$ and $\mu$ the exponential function $g_{\lambda A + \mu B}(t) = e^{t(\lambda A + \mu B)}$ is well-defined, so represents an element of $\mathcal{L}_{\mathcal{G}}$. This means that $\mathcal{L}_{\mathcal{G}}$ is a real vector space. Consider, for $A$ and $B$ in $\mathcal{L}_{\mathcal{G}}$, $g_A(t) = e^{tA}$ and $g_B(t) = e^{tB}$ as curves in $\mathcal{G}$. The product expression $e^{tA}e^{tB}e^{-tA}$ is also a curve in $\mathcal{G}$. Its derivative in $t = 0$ is $AB - BA = [A, B]$, which implies that this matrix commutator is an element of $\mathcal{L_G}$, which proves that $\mathcal{L_G}$ is a Lie algebra, the Lie algebra of the Lie group $\mathcal{G}$. Note how this argument reflects the way the tangent space at a point of a manifold was defined in Supp. Sect. 20.2.1.

Going in the other direction, we start from a real $n \times n$-matrix Lie algebra $\mathcal{L}$. (Note that 'real' means here that $\mathcal{L}$ is a real vector space, which does not exclude the possibility that matrices in $\mathcal{L}$ have complex matrix elements. Think, for example, of $su(2)$). For every element $A$ in $\mathcal{L}$, the exponential matrix function $g(t) = e^{tA}$ is well-defined as a convergent power series. It is clear that the set of finite products of all such exponential matrix functions form a group; it is the simply connected group $\mathcal{G}$ generated by $\mathcal{L}$. It can be shown that for every connected Lie group a finite number of $N$ of such exponentials is sufficient. Compact groups, such as $SU(2)$ have $N = 1$; the Lorentz group has $N = 2$. In other cases the exponential map $\mathcal{L_G} \rightarrow \mathcal{G}, A \mapsto e^A$ covers only a neighbourhood of the unit element. For a discussion of this covering problem, see [1].

The general relation between Lie algebras and Lie groups is stated in what is usually called *Lie's Third Theorem*:

*Every real finite dimensional Lie algebra is the Lie algebra of a Lie group with the same dimension.*

Lie's Third Theorem does *not* tell us that there is a one-to-one correspondence between Lie algebras and Lie groups. One has the following sharpening of the theorem:

*There is a one-to-one correspondence between real finite dimensional Lie algebras and simply connected real Lie groups.*

The proofs, which are highly nontrivial, can be found in most of the standard textbooks on Lie groups and Lie algebras mentioned in the references, although they are, in most cases, not formulated in a very transparent and clearcut manner.

The notion of a connected topological space was defined in Supp. Sect. 18.2; such a space is *simply connected* iff every closed curve in it can be continuously contracted to a point. A connected Lie group $\mathcal{G}$ has a *universal covering group*. This is a simply connected Lie group for which there is a (smooth) surjective homomorphism, the covering map, onto $\mathcal{G}$. Universal means here that it is the unique Lie group with this property. It is unique up to diffeomorphisms. The group $\mathcal{G}$ has the same Lie algebra as its covering group.

An elementary example of this situation is the group $SO(2)$, the rotation group of two dimensional real space. A second example is $U(1)$, the rotation group of the unit circle in one dimensional complex space. Both are as a manifold just the circle. They are obviously not simply connected; their covering group is the one dimensional translation group, the real line as manifold. Both have the same trivial one dimensional Lie algebra. Of more interest is the case of the group $O(3, \mathbb{R})$ of rotations in three dimensional space, or rather its connected component $SO(3, \mathbb{R})$. It has the same Lie algebra as $SU(2)$, as was observed in Sect. 7.4.2, and is its universal covering group. Another example is the Lorentz group $O(1, 3)$ or its connected versions, with $SL(2, \mathbb{C})$ as its universal covering group, groups that play a role in Chaps. 16 and 17.

## 24.6 Representations of Lie Groups

### 24.6.1 General Remarks

'Abstract' groups or algebras may be represented, i.e. realized as 'concrete' groups or algebras of linear transformations, with preservation of the multiplicative structure—and in the case of algebras the additive structure.

The interesting point of representations is that a single group or algebra has in general many 'inequivalent', i.e. essentially different representations. This is very important in physical applications, in particular for the understanding of symmetries.

In the next subsection we start with the notion of representations for groups. After that representations of Lie algebras. The representations of associative algebras are discussed in Supp. Chap. 27 and applied in Chap. 12.

### 24.6.2 Representations of Groups

(24.6.1,a) **Definition** Let $\mathcal{G}$ be a group and $V$ a real or complex vector space. A *(linear) representation* $\Pi$ *of* $\mathcal{G}$ *in* $V$ is a homomorphism $\Phi$ of $\mathcal{G}$ into the group $GL(V)$ of invertible linear transformations of $V$ onto itself.

More explicitly: a representation $\Pi$ of $\mathcal{G}$ in $V$ assigns to every element $g$ of $\mathcal{G}$ an invertible linear map $\Pi(g)$ of $V$ onto itself such that $\Pi(e_G) = 1_V$ and $\Pi(g_1 g_2) = \Pi(g_1)\Pi(g_2)$, for all $g_1$ and $g_2$ in $\mathcal{G}$. Note that this implies that $\Pi(g^{-1}) = (\Pi(g))^{-1}$, for every $g$ in $\mathcal{G}$. In a finite dimensional vector space a choice of a basis means that a representation $\Pi$ becomes the assignment of a matrix to every element of the group: it becomes a *matrix representation*.

(24.6.2,b) **Definition** A representation of a group $\mathcal{G}$ in a complex Hilbert space $\mathcal{H}$ is called *unitary* if the representing operators $\Pi(g)$ are unitary, for all $g$ in $\mathcal{G}$.

A representation of a group $\mathcal{G}$ is a homomorphism $\Pi$ of $\mathcal{G}$ onto a group $\Pi(\mathcal{G})$ of linear transformations. For two injective representations $\Pi_1$ and $\Pi_2$ of $\mathcal{G}$ these groups of linear transformations are clearly isomorphic. They may however be different *as representations*. What it means for representations to be the same or different is expressed by the following definition:

(24.6.2,c) **Definition** Two representations $\Pi_1$ and $\Pi_2$ of a group $\mathcal{G}$, in vector spaces $V_1$, respectively $V_2$, are called *equivalent* iff there is an invertible linear map $S$ from $V_1$ onto $V_2$ such that $\Pi_2(g) = S\Pi_1(g)S^{-1}$, for all $g$ in $\mathcal{G}$. For unitary representations the linear map $S$ is required to be unitary.

Two representations of a group in vector spaces of different dimensions are obviously inequivalent. However, representations in vector spaces with the same dimensions may also be inequivalent.

### 24.6.3 Reducible and Irreducible Representations

(24.6.3,a) **Definition** A representation $\Pi$ of a group $\mathcal{G}$ in $V$ is called *reducible* iff there is a nontrivial linear subspace of $V$, i.e. not $V$ itself and not $\{0\}$, which is invariant under $\Pi(g)$ for all $g$ in $\mathcal{G}$. The representation $\Pi$ is called *irreducible* if $V$ has no such subspace.

There is a stronger notion than reducibility:

(24.6.3,b) **Definition** A representation $\Pi$ of $\mathcal{G}$ in $V$ is called *completely reducible* iff each invariant subspace has an invariant complement, i.e. if $W$ is an invariant subspace, then there exists a second invariant subspace $W'$ such that $V$ can be written as the direct sum $V = W \oplus W'$.

Complete reducibility of a representation $\Pi$ in a finite dimensional vector space $V$ implies that $V$ can be written as a direct sum representation

$$\Pi = \Pi_1 \oplus \cdots \oplus \Pi_p, \quad V = V_1 \oplus \cdots \oplus V_p,$$

with the $\Pi_j$ irreducible representations of $G$ in the invariant subspaces $V_j$. This reduction is unique up to equivalence.

Definition (24.6.3,b) is valid for infinite dimensional vector spaces, in particular for Hilbert spaces, with a decomposition as a (countably) infinite direct sum.

### 24.6.4 The Central Problem of Group Representation Theory

A central problem in the representation theory of groups is to find *all* irreducible representations of a given group, or class of groups, up to equivalence. The case of finite dimensional representations is the simplest. For many important groups all finite dimensional representations are completely reducible. In that case the irreducible representations are the building blocks of representation theory and the problem of finding all representations reduces to finding all *irreducible* representations. For large classes of groups this problem has been completely solved. The problem is more difficult for infinite dimensional representations; the notions of equivalence and irreducibility remain crucial in this case and much is known.

An important problem in the application of group theory to physics is that of finding explicitly the reduction of a given reducible representation into its irreducible components. A standard example of this problem is the reduction of a tensor product of two irreducible reprentations, which is in general reducible.

### 24.6.5 Constructing New Representations from Given Ones

Let $\Pi_1$ and $\Pi_2$ be representations of the group $\mathcal{G}$ in vector spaces $V_1$, respectively $V_2$. There are two important ways of constructing a new representation from $\Pi_1$ and $\Pi_2$.

1. *Direct sum*: Consider the vector space $V_1 \oplus V_2$, the direct sum of $V_1$ and $V_2$. Define for each $g$ in $\mathcal{G}$ the linear operator $(\Pi_1 \oplus \Pi_2)(g)$ in $V_1 \oplus V_2$ as

$$((\Pi_1 \oplus \Pi_2)(g))(\psi_1, \psi_2) := ((\Pi_1(g))\psi_1, (\Pi_2(g))\psi_2),$$

for all $\psi_1$ in $V_1$ and all $\psi_2$ in $V_2$. This defines the direct sum representation $\Pi_1 \oplus \Pi_2$.

2. *Tensor product*: Consider the vector space $V_1 \otimes V_2$, the tensor product of $V_1$ and $V_2$. Define for each $g$ in $\mathcal{G}$ the linear operator $(\Pi_1 \otimes \Pi_2)(g)$ in $V_1 \otimes V_2$ by linear extension of

$$((\Pi_1 \otimes \Pi_2)(g))(\psi_1 \otimes \psi_2) := (\Pi_1(g))\psi_1 \otimes (\Pi_2(g))\psi_2,$$

for all $\psi_1$ in $V_1$ and all $\psi_2$ in $V_2$. This defines the tensor product representation $\Pi_1 \otimes \Pi_2$.

There are obvious generalizations to the notions of direct sum and tensor product representations of a finite number of given representations $\Pi_1, \ldots, \Pi_n$. There is also a technically more subtle generalization of the notion of direct sum to certain infinite sets of representations: the *direct integral*.

## 24.7 Representations of Lie Algebras

**(24.7,a) Definition** Let $\mathcal{L}$ be a Lie algebra and $V$ a real or complex vector space. A representation $\pi$ of $\mathcal{L}$ in $V$ is a Lie algebra homomorphism of $\mathcal{L}$ into the Lie algebra of linear operators in $V$, with the operator commutator as Lie bracket.

This can be stated in a more explicit fashion. A representation $\pi$ of $\mathcal{L}$ in $V$ assigns to every element $u$ of $\mathcal{L}$ a linear map $\pi(u)$ of $V$ into itself such that $\pi([u, v]) = [\pi(u), \pi(v)]$, for all $u$ and $v$ in $\mathcal{L}$. The bracket on the left is the Lie bracket in $\mathcal{L}$, that on the right the commutator for operators in $V$.

**(24.7,b) Definition** A representation $\pi$ of the Lie algebra $\mathcal{L}$ in the Hilbert space $\mathcal{H}$ is called *unitary* iff, for each $u$ in $\mathcal{L}$, the operator $\pi(u)$ is *antihermitian*, i.e. iff $(\pi(u)\psi_1, \psi_2) = -(\psi_1, \pi(u)\psi_2)$, for all $\psi_1$ and $\psi_2$ in $\mathcal{H}$.

In physics the operators $\pi(u)$ are often connected with important physical quantities such as energy, momentum or angular momentum, which, from general principles, are supposed to be represented by selfadjoint and not anti-selfadjoint (anti-hermitian) operators. For this reason one uses either $i\pi(u)$, $i\pi(v)$, or a complexified version $\mathcal{L}_C$ of the real Lie algebra $\mathcal{L}$, a formal, but rather trivial procedure which does add very little to the above discussion.

Let $V$ be finite dimensional. With a choice of basis in $V$ the operators $\pi(u)$ become matrices. This leads to the notion of *matrix representation* of a Lie algebra. All Lie algebra representations in this book are matrix representations.

We leave aside here the additional technicalities connected with unbounded operators, required by this definition in case the Hilbert space $\mathcal{H}$ is infinite dimensional. It should also be remarked that the term 'unitary' looks a bit strange at this point. It becomes more meaningful after reading the discussion of the relation between Lie groups and Lie algebras in Supp. Sect. 24.5.

A representation $\pi$ of a finite dimensional Lie algebra $\mathcal{L}$ is completely determined by the representatives $\pi(u_1), \ldots, \pi(u_p)$ of a basis $u_1, \ldots, u_p$ in $\mathcal{L}$, because of the fact that the homomorphism $\pi$ is a *linear* map. For this reason applying the representation theory of Lie algebras is much simpler than applying that of Lie groups.

Let $\pi_1$ and $\pi_2$ be representations of the Lie algebra $\mathcal{L}$ in $V_1$, respectively $V_2$. The definition of the direct sum representation is the same as in the group case, namely as

$$((\pi_1 \oplus \pi_2)(u))(\psi_1, \psi_2) := ((\pi_1(u))\psi_1, (\pi_2(u))\psi_2),$$

for all $u$ in $\mathcal{L}$ and all $\psi_1$, $\psi_2$ in $V_1 \oplus V_2$. The definition of the tensor product representation is slightly different. It is by linear extension of

$$((\pi_1 \otimes \pi_2)(u))(\psi_1 \otimes \psi_2) := (\pi_1(u))\psi_1 \otimes \psi_2 + \psi_1 \otimes (\pi_2(u))\psi_2,$$

for all $\psi_1$ in $V_1$ and $\psi_2$ in $V_2$.

The definitions of *irreducibility*, *reducibility* and *complete reducibility* are the same as in the case of group representations.

## 24.8 The Representations of $su(2)$ and $so(3, \mathbb{R})$

### 24.8.1 Generalities

The Lie algebra $so(3, \mathbb{R})$, isomorphic with $su(2)$, is compact, meaning that the corresponding Lie group is compact. (There is an equivalent algebraic definition of compactness of a Lie algebra in terms of an algebraic property of the algebra itself. We do not give it here). As such its representations are completely reducible, meaning that any representation can be written as a direct sum of irreducible representations. The irreducible representations are all finite dimensional and equivalent to unitary representations.

The Lie groups in this book are *real* Lie groups, so their Lie algebras are *real* Lie algebras. This is obvious for the rotation group $SO(3, \mathbb{R})$, a three dimensional real Lie group, with the three dimensional real Lie algebra $so(3, \mathbb{R})$. The Lie algebra $su(2)$, although consisting of complex matrices, is also a real Lie algebra.

The full group $SU(2)$ can be covered by the one parameter groups $U(t) = e^{tA}$, with $A$ elements from $su(2)$. By expanding such exponential functions up to first order in the parameter $t$ one finds immediately that the $A$ are antihermitian. As the generators of such groups correspond quite generally in physics to observables one prefers to write $U(t) = e^{itA}$, with the $A$ hermitian (or selfadjoint). This means that one uses in a rather trivial manner the complexification of $su(2)$, which was hinted at in the preceding section. The same is of course true for the isomorphic Lie algebra $so(3, \mathbb{R})$.

The construction of the irreducible representations of $su(2)$ (isomorphic to $so(3, \mathbb{R})$) is purely finite dimensional linear algebra. It is reminiscent of a procedure used for the solution of the eigenvalue–eigenfunction problem for the harmonic oscillator in Sect. 6.4.

We start by repeating the basic commutation relations for the three generators $J_j$ of $su(2)$, provided with the imaginary $i$ and Planck's constant, for application to physics as angular momentum operators, in particular for the hydrogen atom in Chap. 7 and for spin in Chap. 8,

$$[J_1, J_2] = i\hbar J_3, \quad [J_2, J_3] = i\hbar J_1, \quad [J_3, J_1] = i\hbar J_2,$$

together with for $\mathbf{J}^2 = J_1^2 + J_2^2 + J_3^2$

$$[\mathbf{J}^2, J_j] = 0, \quad j = 1, 2, 3.$$

## 24.8.2 Ladder Operators

We introduce *ladder operators*

$$J_+ = J_1 + iJ_2, \quad J_- = J_1 - iJ_2.$$

(24.8.2,a) **Problem** Prove the following relations

$$[J_+, J_-] = 2\hbar J_3, \quad [J_3, J_\pm] = \pm\hbar J_\pm, \quad [\mathbf{J}^2, J_\pm] = 0.$$

Let $V$ be the carrier space of an irreducible representation of $su(2)$; a finite dimensional Hilbert space. It is spanned by joint eigenvectors of $\mathbf{J}^2$ and $J_3$. The operator $\mathbf{J}^2$ commutes with all operators in $V$, so it must be a scalar multiple of the identity operator; all non-zero vectors in $V$ are eigenvectors of $\mathbf{J}^2$ for the same eigenvalue. We denote it as $\hbar^2\lambda$. The value of $\lambda$ will be determined further on. In order not to overburden the notation, we use the same symbols for elements of the Lie algebra $su(2)$ and its representing operators. As an operator in the finite dimensional representation space, $L_3$ has a lowest eigenvalue. Let $\phi_a$ be a joint eigenvector, for this lowest eigenvalue $\hbar a$ of $L_3$, i.e. with

$$\mathbf{J}^2\phi_a = \hbar^2\lambda\phi_a, \quad J_3\phi_a = \hbar a\phi_a.$$

**(24.8.2,b) Problem** Use the commutation relation of $J_+$ with $J_3$ to prove that if the vector $J_+\phi_a$ is not equal to zero, than it is an eigenvector of $J_3$ with eigenvalue $\hbar(a+1)$. More generally, show that if for a certain $j$ the vector $(J_+)^j\phi_a$ is unequal to zero, it is an eigenvector of $J_3$ for eigenvalue $\hbar m = \hbar(a+j)$.

This procedure ends with a number $n$ such that

$$(J_+)^{n-1}\phi_a \neq 0, \quad (J_+)^n\phi_a = 0,$$

resulting in a 'ladder' of $n$ joint eigenvectors of $J_3$ and $\mathbf{J}^2$, with eigenvalues $\hbar a, \hbar(a+1), \ldots, \hbar(a+n-2), \hbar(a+n-1)$ of $J_3$.

In a similar way repeated application of $J_-$ on $(J_+)^{n-1}\phi_a$ gives a finite sequence of eigenvectors $(J_-)^j(J_+)^{n-1}\phi_a$, with eigenvalues $\hbar(a+n-j)$, with at the end $(J_-)^{n-1}(J_+)^{n-1}\phi_a$ with eigenvalue $\hbar a$. It is clear that this sequence consists of the same vectors as above, except for scalar factors. Because $J_3$ is selfadjoint they form an orthogonal system. It is invariant under both $J_1$ and $J_2$, so its span is an $n$-dimensional linear subspace of $V$, carrying a representation of $su(2)$, which, due to the irreducibility of the representation, is equal to $V$.

**(24.8.2,c) Problem** Prove that $\mathbf{J}^2$ can be written as

$$\mathbf{J}^2 = J_\mp J_\pm + J_3^2 \pm \hbar J_3.$$

Prove that this result applied to the eigenvector $(J_+)^{n-1}\phi_a$ at the top of the ladder, gives

$$\mathbf{J}^2(J_+)^{n-1}\phi_a = \hbar^2(a+n)(a+n-1)(J_+)^{n-1}\phi_a,$$

which, when combined with

$$\mathbf{J}^2(J_+)^{n-1}\phi_a = \hbar^2\lambda(J_+)^{n-1}\phi_a,$$

leads to $\lambda = (a+n)(a+n-1)$. Show also that a similar argument gives $\mathbf{J}^2\phi_a = \hbar^2 a(a-1)\phi_a$, and when combined with $\mathbf{J}^2\phi_a = \hbar^2\lambda\phi_a$, leads to $\lambda = a(a-1)$.

## 24.8.3 Eigenvalues and Eigenvectors of $L^2$ and $L_3$

Our next task is to find the values of $\lambda$ and $a$ for a given $n$. The two relations for $\lambda$ and $a$ give $(2a+n-1)n = 0$, which is equivalent to $n = 0$ or $a = -\frac{1}{2}(n-1)$. The first is unacceptable, because $n \geq 1$. This leaves us with $a = -\frac{1}{2}(n-1)$. Defining $n = 2j+1$ we obtain that $\lambda = j(j+1)$, $a = -j$; the eigenvalues $\hbar m$ of $J_3$ form a sequence

$$-\hbar j, -\hbar(j+1), \ldots, \hbar(j-1), \hbar j.$$

The vectors $(J_+)^j \phi_a$ are not normalized. Assuming that $\phi_a$ is a unit vector, we can calculate the normalized eigenvector, in physics textbooks usually denoted as $|j, m>$, with the Dirac bra-ket notation, on the use of which more is said in Supp. Chap. 26.

**(24.8.3,a) Problem** Prove, with a judicious choice of phase,

$$|j, -j> = \phi_a, \quad |j, -j+1> = \frac{1}{\hbar\sqrt{2j}} J_+ \phi_a,$$

$$|j, -j+2> = \frac{1}{2\hbar^2\sqrt{j(2j-1)}} (J_+)^2 \phi_a.$$

### 24.8.4 Matrix Elements

Explicit expressions for the matrix elements of the operators $J_1$, $J_2$ and $J_3$ can be easily calculated.

**(24.8.4,a) Problem** Prove, for $j = \frac{1}{2}$,

$\langle \frac{1}{2}, \frac{1}{2}|J_1|\frac{1}{2}, -\frac{1}{2}\rangle = \langle \frac{1}{2}, -\frac{1}{2}|J_1|\frac{1}{2}, \frac{1}{2}\rangle = \frac{1}{2}\hbar,$

$\langle \frac{1}{2}, \frac{1}{2}|J_1|\frac{1}{2}, \frac{1}{2}\rangle = \langle \frac{1}{2}, -\frac{1}{2}|J_1|\frac{1}{2}, -\frac{1}{2}\rangle = 0,$

$\langle \frac{1}{2}, \frac{1}{2}|J_2|\frac{1}{2}, -\frac{1}{2}\rangle = -\langle \frac{1}{2}, -\frac{1}{2}|J_2|\frac{1}{2}, \frac{1}{2}\rangle = -\frac{i}{2}\hbar,$

$\langle \frac{1}{2}, \frac{1}{2}|J_2|\frac{1}{2}, \frac{1}{2}\rangle = \langle \frac{1}{2}, -\frac{1}{2}|J_2|\frac{1}{2}, -\frac{1}{2}\rangle = 0,$

$\langle \frac{1}{2}, \frac{1}{2}|J_3|\frac{1}{2}, \frac{1}{2}\rangle = -\langle \frac{1}{2}, -\frac{1}{2}|J_3|\frac{1}{2}, -\frac{1}{2}\rangle = \frac{1}{2}\hbar,$

$\langle \frac{1}{2}, \frac{1}{2}|J_3|\frac{1}{2}, -\frac{1}{2}\rangle = \langle \frac{1}{2}, -\frac{1}{2}|J_3|\frac{1}{2}, \frac{1}{2}\rangle = 0,$

or

$$J_1 = \frac{1}{2}\hbar \begin{pmatrix} 0 & 1 \\ 1 & 0 \end{pmatrix}, \quad J_2 = \frac{1}{2}\hbar \begin{pmatrix} 0 & -i \\ i & 0 \end{pmatrix}, \quad J_3 = \frac{1}{2}\hbar \begin{pmatrix} 1 & 0 \\ 0 & -1 \end{pmatrix}$$

Note that the three matrices appearing in these expressions are usually denoted as $\sigma_j$ and called the *Pauli matrices*.

**(24.8.4,b) Problem** Determine in a similar manner the matrices representing $J_1$, $J_2$ and $J_3$ for the case $j = 1$. Compare these results with the formulas for the spin operators in Chap. 8 and the infinitesimal rotations in Chap. 7.

### 24.8.5 Summary

In this manner all the irreducible representations of $su(2) \equiv so(3, \mathbb{R})$ can be explicitly determined. The result of this section can be resumed as follows:

*The irreducible representations of $su(2) \equiv so(3, \mathbb{R})$ are characterized by the eigenvalue $j$ for the Casimir operator $\mathbf{J}^2$; they have dimension $2j + 1$ and have eigenvectors and eigenvalues for the commuting operators $\mathbf{J}^2$ and $J_3$ given by*

$$\mathbf{J}^2 |j, m> = \hbar^2 j(j + 1)|j, m>, \quad J_3|j, m> = \hbar m|j, m>,$$

*for $j = 0, \frac{1}{2}, 1, 1\frac{1}{2}, \ldots$ and $m = -j, -j + 1, \ldots, j - 1, j$.*

Finally it should be observed that all representations can be exponentiated to representations of the group $SU(2)$ (See Chap. 8: Spin); the representations of $SO(3, \mathbb{R})$ are obtained for $j = 1, 2, \ldots$ (See Chap. 7: The Hydrogen Atom).

## 24.9 Reducing Tensor Product Representations of $su(2)$

Suppose we have two irreducible representations $\pi_1$ and $\pi_2$ of $su(2)$, characterized by the numbers $j_1$ and $j_2$, and acting in representation spaces $V_{j_1}$ and $V_{j_2}$, of dimension $2j_1 + 1$ and $2j_2 + 1$. One may construct a tensor product representation $\pi_1 \otimes \pi_2$ in the $(2j_1 + 1)(2j_2 + 1)$-dimensional space $V_{j_1} \otimes V_{j_2}$. An element $a$ in $su(2)$ is represented in this space by

$$\pi(a)(\psi_1 \otimes \psi_2) = \pi_1(a)\psi_1 \otimes \psi_2 + \psi_1 \otimes \pi_2(a)\psi_2.$$

The representation $\pi$ will in general be reducible, in fact completely reducible, and is therefore a direct sum of irreducible representations. It can be shown that the representation space can be written as

$$V_{j_1} \otimes V_{j_2} = V_{|j_1-j_2|} \oplus V_{|j_1-j_2|+1} \oplus \cdots \oplus V_{j_1+j_2},$$

meaning that in this sum only the representations with

$$j = |j_1 - j_2|, |j_1 - j_2| + 1, \ldots, j_1 + j_2$$

occur and each of these just once. For a proof of this, see e.g. [2, p. 450]. It is clear that this sum has $2j_2 + 1$ terms, for $j_2 < j_1$.

Before analyzing this situation we observe that in this context it is convenient to use Dirac's bra-ket notation: a vector $\psi$ is written as a *ket* $| \psi >$, an inner product of two vectors $\psi_1$ and $\psi_2$ as $\langle \psi_1 | \psi_2 \rangle$. For more background on this see Supp. Chap. 26 in which Dirac's bra-ket formalism is explained in detail.

To study the reduction of $\pi_1 \otimes \pi_2$ we consider two orthonormal bases in the representation space $V_{j_1} \otimes V_{j_2}$. The first is the obvious one, consisting of the tensor product vectors

$$|j_1, m_1; j_2, m_2> = |j_1, m_1> \otimes |j_2, m_2>,$$

for fixed $j_1$ and $j_2$, and $m_1$ and $m_2$ running from $-j_1$ to $+j_1$ and from $-j_2$ to $+j_2$. The second basis exhibits the direct sum properties of the representation space; it consists of orthonormal bases in each of the direct summand. They are eigenvectors of the operators

$$(\mathbf{J}^{(1)})^2, (\mathbf{J}^{(2)})^2, \mathbf{J}^2, \quad (J^{(1)})_3, (J^{(2)})_3, J_3.$$

(24.9,a) **Problem** Show that these six operators commute.

The vectors of this second basis are denoted as $|j_1, j_2; j, m\rangle$, with $j_1$ and $j_2$ given and with $j, m$ the values mentioned above. The correspondence between the two bases is expressed by the general formula

$$|j_1, j_2; j, m\rangle = \sum_{m_1+m_2=m} |j_1, m_1; j_2, m_2\rangle \langle j_1, m_1; j_2, m_2 \,|\, j_1, j_2; j, m\rangle.$$

The matrix elements

$$\langle j_1, m_1; j_2, m_2 \,|\, j_1, j_2; j, m\rangle$$

are called *Clebsch-Gordan coefficients*. Finding these in explicit form is what reduction of the tensor product representation in practice means. Note that for given basis vectors of the first system, phase factors in the choice of basis vectors of the second system are arbitrary. It can be shown that this can be used to make all coefficients real and positive.

There is a recursive procedure for obtaining these coefficients, involving the use of the ladder operators from Supp. Sect. 24.8.2. The details of this are in general fairly complicated and can be found in most quantum mechanics textbooks and in some of the books on Lie groups and Lie algebras, mentioned in the references. Here we shall just describe a few simple examples in a qualitative manner.

1. A trivial example: $j_1 = j_2 = 0$. The representation space is the tensor product of the one dimensional trivial representation of $su(2)$ with itself, and as such again the same one dimensional representation space. There is a single trivial coefficient, which is—or can be chosen to be 1.
2. A slightly less trivial example: $j_1 = \frac{1}{2}, j_2 = 0$. The representation space is the tensor product of the two dimensional $j_1 = \frac{1}{2}$ representation with the one dimensional trivial representation, as such again the $j = \frac{1}{2}$ representation. One can show that, after a suitable choice of phases, the Clebsch-Gordan coefficients form the two dimensional unit matrix.
3. An example of physical interest—a system of two spinning electrons : $j_1 = j_2 = \frac{1}{2}$, or in the usual notation $s_1 = s_2 = \frac{1}{2}$. The representation space is four dimensional, the tensor product of two times the $s = \frac{1}{2}$ representation. Reduction gives a direct sum of the one dimensional trivial representation and the three

dimensional representation that generates the basic representation of the group $SO(3)$.

4. A second example of general physical interest—a particle with both orbital and intrinsic angular momentum : $j_1$ an integer $\geq 1$ and $j_2 = \frac{1}{2}$, in the usual notation $l \geq 1$ integer and $s = \frac{1}{2}$. The representation is the tensor product of a $(2l + 1)$-dimensional orbital angular momentum representation and the two dimensional basic spin representation. At the same time it is the direct sum of two general angular momentum representations, with $j$ is $l - \frac{1}{2}$ and $l + \frac{1}{2}$.

For the explicit form of the coefficients in example 3 and a good background of the general situation, in the style of the present presentation, see [3, Sect. 6.3].

Summarizing one may say that all this has important applications in physics: the hydrogen atom with its orbital angular momentum $\mathbf{L}$, with $l = 0, 1, 2, \ldots$, as discussed in Chap. 8, then combined with an intrinsic angular momentum, its spin $\mathbf{S}$, with $s = \frac{1}{2}$ (example 4), also discussed in Chap. 8; the situation of two electrons in an atom, like helium, spins $\mathbf{S}^{(1)}$ and $\mathbf{S}^{(2)}$, with $s_1 = s_2 = \frac{1}{2}$, is discussed in Sect. 9.4.

## 24.10  Super Lie Algebras

There is an interesting generalization of the notion of Lie algebra: *super Lie algebra*. It originated and is used in super symmetry (See Sect. 17.2.4). For this we need *super algebra* with in particular the notion of $\mathbb{Z}_2$-*grading* (See Sect. 13.9.3.2).

(24.10,a) **Definition** A (real or complex) $\mathbb{Z}_2$-graded vector space $\mathcal{L}$ is called a (real or complex) *super Lie algebra* iff it is provided with a *super Lie bracket*, i.e. a bilinear map

$$\mathcal{L} \times \mathcal{L} \to \mathcal{L}, \qquad (u, v) \mapsto [u, v],$$

which has the following three properties, for all homogeneous elements $u$, $v$ and $w$ in $\mathcal{L}$:

1. $([u, v]) = (u) + (v)$,
2. it is *super antisymmetric*, i.e. it has $[u, v] = -(-1)^{(u)(v)}[v, u]$,
3. it satisfies the *super Jacobi identity*, i.e.

$$(-1)^{(u)(w)}[u, [v, w]] + (-1)^{(v)(u)}[v, [w, u]] + (-1)^{(w)(v)}[w, [u, v]] = 0,$$

with $(x)$ the $\mathbb{Z}_2$-degree of an element $x$.

Good references to super Lie algebras are [2, Vol. III] and [4].

There are many excellent books on group theory in general, on Lie groups, and Lie algebras, or both. Here is a selection.

# References

1. Dokovic, D.Z., Hofmann, K.H.: The surjectivity question for the exponential function of real lie groups. J. Lie Theory **7**, 171–199 (1997)
2. Cornwell, J.F.: Group Theory in Physics I, II, III. Academic Press, Waltham 1984, 1986, 1989 (The first volume covers basic group theory, Lie groups, representations and applications to nonrelativistic physics, the second volume Lie algebras, the connection between Lie algebras and Lie groups, and the third volume discusses supersymmetry and infinite dimensional Lie algebras.)
3. Jones, H.F.: Groups, Representations and Physics. Institute of Physics Publishing (1990) (Not aimed at completeness, clear in its presentation, and therefore very useful for physical applications.)
4. Corwin, I., Ne'eman, Y., Sternberg, S.: Graded lie algebras in mathematics and physics (Bose-Fermi symmetry). Rev. Mod. Phys. **47**, 573–603 (1975)
5. Baker, A.: Matrix Groups: An Introduction to Lie Group Theory. Springer, New York (2003) (This is a particularly useful text because the groups used in this book are mostly matrix groups.)
6. Varadarajan, V.S.: Lie Groups, Lie Algebras, and Their Representations, 1st edn. Prentice-Hall, New Jersey (1974) (Springer 1984, 2009)
7. Humphreys, J.E.: Introduction to Lie Algebras and Representation Theory. Springer, New York (1972) (A standard textbook on the subject.)
8. Hall, B.C.: Lie Groups, Lie Algebras, and Representations: An Elementary Introduction. Springer, New York (2003)
9. Warner, F.W.: Foundations of Differential Manifolds and Lie Groups Scott, Foreman 1971, Springer, New York (2010)
10. Gilmore, R.: Lie Groups, Lie Algebras and Some of Their Applications. Original edition. Wiley, New York (1974), Dover (2006)
11. Barut, A.O., Raczka, R.: Theory of Group Representations and Applications. World Scientific, 2nd revised edition (1986)
12. Heine, V.: Group Theory in Quantum Mechanics: An Introduction to Its Present Usage. Pergamon Press, New Jersey 1960. Dover 1993, reprinted in 2007 (A modern version of Weyl's classical book on quantum theory and group theory, with almost the same title, mentioned in the list of references of Chap. 3.)

# Chapter 25
# Generalized Functions

## 25.1 Introduction

For his formulation of quantum mechanics Dirac introduced in the late 1920s so-called $\delta$-functions, objects that could be manipulated as functions of real variables, could be differentiated and integrated, without however being functions in a strict sense. Others, such as the mathematician Sobolev in 1936 [1] and the applied mathematicians Temple in 1955 [2] and Lighthill in 1958 [3] later came up with similar ideas, the last two inspired by Dirac's work. However, it is the version of Dirac that has eventually become widely used in quantum physics. Dirac's $\delta$-function, together with its derivatives, is an ingenious heuristic idea; it was developed later into a rigorous mathematical theory by Laurent Schwartz, whose work was the beginning of a completely new and important area of functional analysis of which only a small part is needed here. Schwartz used the term 'distributions'. Because we employ in this book this term in the context of probability theory, we shall stick as much as possible to 'generalized functions'. Dirac's approach was heuristic, but very effective. See for this his book on quantum mechanics [4], the Third Edition, in particular pp 58–61. Schwartz's point of view is explained in his book [5]. For modern books on the theory and application of generalized functions, see the books by Richards [6], by Blanchard and Brüning [7] and by Treves [8].

Dirac $\delta$-functions and its variants are universally used in physics textbooks as an elegant heuristic language and as a convenient tool in calculations. In most cases there is no objection to this as long as one is aware of the limitations. Nevertheless, its use in quantum field theory conceals serious mathematical problems, problems which one has learnt to live with, but which remain to this day unsolved. See for a brief discussion in Chap. 15. In this book $\delta$-functions will be mainly used for the discussion of the many particle Fock space formalism in Chap. 9 and for a discussion of relativistic quantum field theory in Chap. 16.

© Springer International Publishing Switzerland 2015
P. Bongaarts, *Quantum Theory*, DOI 10.1007/978-3-319-09561-5_25

## 25.2 The Dirac $\delta$-function: Heuristics

In the simplest situation Dirac considers a real-valued function $\delta$ on $\mathbb{R}^1$ with the properties $\delta(x) = 0$ for $x \neq 0$ and $\delta(0) = +\infty$, something which in classical analysis—in particular in the context of measure theory—would be just equivalent to a function which is identically zero. Dirac however integrates it to give

$$\int_{-\infty}^{+\infty} \delta(x)dx = 1.$$

Integration with an 'ordinary' function $f$ gives

$$\int_{-\infty}^{+\infty} f(x)\delta(x)dx = f(0),$$

and more generally

$$\int_{-\infty}^{+\infty} f(x)\delta(x - x_0)dx = f(x_0).$$

Dirac goes further: the $\delta$-function can be differentiated to give a derivative $\delta^{(1)}(x-x_0)$ for which standard partial differentiation suggests the property

$$\int_{-\infty}^{+\infty} f(x)\delta^{(1)}(x - x_0)dx = -f^{(1)}(x_0),$$

and for higher derivatives $\delta^{(n)}(x - x_0)$,

$$\int_{-\infty}^{+\infty} f(x)\delta^{(n)}(x - x_0)dx = (-1)^n f^{(n)}(x_0).$$

Note that a distribution does not have a derivative *in a point*, as in the case of a differentiable function; it is an overall object. One can multiply $\delta$-functions, but not in the same point. The product $\delta(x)\delta(y)$, for instance, is a generalized function in two variables $x$ and $y$ and can be given a rigorous meaning, but the expression $\delta(x)^2$ as a generalized function in a single variable does not make sense. In the rigorous formulation in the next section in which a generalized function is a linear functional on a space of test functions, the expression $\delta(x)\delta(y)$ corresponds with a tensor product of two such functionals. For $\delta(x)^2$ this is meaningless.

All the definitions and prescriptions just given can be generalized immediately to $n$ variables $\mathbf{r} = (x_1, \ldots, x_n)$: for a δ-function $\delta(\mathbf{r})$ one has

$$\int\limits_{-\infty}^{+\infty} \delta(\mathbf{r}) d\mathbf{r} = 1,$$

and

$$\int\limits_{-\infty}^{+\infty} f(\mathbf{r}) \delta(\mathbf{r} - \mathbf{r}_0) d\mathbf{r} = f(\mathbf{r}_0),$$

and for (partial) derivatives, for instance for $\frac{\partial}{\partial x_j}$,

$$\int\limits_{-\infty}^{+\infty} f(\mathbf{r}) \frac{\partial}{\partial x_j} \delta(\mathbf{r} - \mathbf{r}_0) \, d\mathbf{r} = -\frac{\partial}{\partial x_j} f(\mathbf{r}_0).$$

The heuristic δ-function formalism is of great practical value in all sorts of calculations in physics, as is clear from the fact that it is almost universally used in the physics literature.

It is worth saying a few words about the approach to generalized functions of Lighthill, whose work in this particular field has been overshadowed, in physics by Dirac's approach, and in mathematics by that of Schwartz. See Lighthill's book [3]. It is mathematically less sophisticated than Schwartz's work but is nevertheless rigorous and moreover quite appealing intuitively.

The basic idea of Lighthill's approach is that the object that corresponds with the Dirac δ-function—concentrated, for the sake of simplicity around the point $x = 0$—is a sequence of smooth functions $\{F_n\}_n$, with increasingly sharp peaks around 0, such that

$$\int\limits_{-\infty}^{+\infty} F_n(x) dx = 1, \quad \text{for } n = 1, 2, \ldots$$

leading to

$$\delta(f) = \lim_{n \to \infty} \int\limits_{-\infty}^{+\infty} F_n(x) f(x) dx = f(0),$$

for every 'test function' $f$. In this way this procedure gives Dirac's result, but in a completely rigorous manner. There is an equivalence relation between sequences $\{F_n\}_n$ that give the same limit for $n \to \infty$ for the integral $\int_{-\infty}^{+\infty} F_n(x) f(x) dx$, for all $f$. An obvious candidate for a representative of the equivalence class is the sequence of Gaussian distribution functions, centered around $x = 0$,

$$F_n(x) = \left(\frac{n}{\pi}\right)^{1/2} e^{-nx^2}$$

with $n = (2\sigma^2)^{-1}$ the standard deviation, a measure for the width of the function $F_n$, which goes to 0 for $n$ going to infinity. Many other examples can be given.

The next step is the definition of the derivative. This is simply given by

$$\delta^{(1)}(f) = \lim_{n \to \infty} \int_{-\infty}^{+\infty} F_n^{(1)}(x) f(x) dx = -f^{(1)}(0),$$

a result obtained by partial integration *before the limit $n \to \infty$ is taken*. And so on.

(25.2,a) **Problem** The Dirac $\delta$-function is even, i.e. $\delta(-x) = \delta(x)$. Use the Lighthill approach to show that the first derivative of the $\delta$-function is odd, i.e. that $\delta^{(1)}(-x) = -\delta^{(1)}(x)$.

## 25.3  The Rigorous Formalism

Generalized functions are not functions in the strict sense. The $\delta$-function $\delta(x)$ is a measure, a probability measure on the real line, concentrated on the point $x = 0$, and as such usually denoted as $\delta_0$, with $\delta_a$ for $\delta(x-a)$. However, the derivatives $\delta^{(n)}(x)$ are *not* measures. Generalized functions, as introduced heuristically by Dirac and defined rigorously by Schwartz are objects that are dual with respect to functions. They are linear functionals on spaces of suitable smooth functions, called *test functions*. We shall restrict the discussion here to functions on $\mathbb{R}^1$; the generalization to $\mathbb{R}^n$ is fairly obvious and is left to the reader.

Define $\mathcal{D}(\mathbb{R}^1)$ as the vector space of smooth, i.e. infinitely differentiable, functions on $\mathbb{R}^1$, with compact support, which means that they vanish outside a compact set, in the case of $\mathbb{R}^1$ outside a bounded closed interval. It is not difficult to find such functions. A well-known example is

$$f(x) = a e^{-\frac{1}{(b^2-x^2)}}, \quad \text{for } |x| \le |b|, \qquad f(x) = 0, \quad \text{for } |x| > |b|,$$

with $a$ and $b$ non-zero real constants. It is obvious that this $f$ is smooth for all values of $x$ with $|x| \ne |b|$, but a precise analysis of the limit behaviour of $f$ at the points $x = \pm b$ is needed to become convinced that the formula defines a function that is smooth everywhere. This analysis, a standard exercise in advanced calculus, can be found in the literature.

The function space $\mathcal{D}(\mathbb{R}^1)$ is a vector space which can be made into a *topological vector space*. Remember that in Supp. Sect. 21.2, a class of topological vector spaces was mentioned, more general than Hilbert or Banach spaces, *locally convex spaces*, in which the topology and therefore the notion of continuity is defined not by a norm but by an infinite system of *seminorms*. A seminorm $p$ has the same properties as a norm, see Supp. Sect. 21.2, except that $p(f) = 0$ does not imply $f = 0$. The space $\mathcal{D}(\mathbb{R}^1)$ can be given a locally convex topology by specifying a system of seminorms. However, it turns out that the system of seminorms that one needs here is not countable, so we follow Schwartz in producing the relevant topology in a different manner.

In Supp. Sect. 18.6 we discussed an alternative method of introducing a topology on a space, namely by giving a definition of convergence of sequences. This we do here for $\mathcal{D}(\mathbb{R}^1)$:

**(25.3,a) Definition** A sequence $\{f_n\}_{n \in N}$ converges to a function $f$ iff the supports of all $f_n$ are contained in a closed bounded interval, independent of $n$, and iff the $f_n$ together with all their derivatives separately converge uniformly on this interval. As is explained in Supp. Sect. 18.6, this defines a topology, indeed a locally convex vector space topology on $\mathcal{D}(\mathbb{R}^1)$, as may be checked.

**(25.3,b) Definition** A *distribution* (in the sense of generalized function theory) is a linear functional on $\mathcal{D}(\mathbb{R}^1)$, which is continuous in this topology.

The $\delta$-function becomes in this set up the linear functional $F_\delta(f) = f(0)$, the derivatives of $\delta(x)$ the functionals $F_{\delta^{(n)}}(f) = (-1)^n f^{(n)}(0)$. A distribution has a well-defined functional as Fourier transform. It is given by the formula $\widehat{F}(f) = F(\hat{f})$, with $\hat{f}$ the Fourier of $f$. However, the Fourier transform of a function with compact support does not need to have compact support, so $\widehat{F}$ is not necessarily a distribution.

Luckily there are other types of generalized functions obtained by choosing other test function spaces. Define, for instance, $\mathcal{S}(\mathbb{R}^1)$, the space of *Schwartz functions*— an unfortunate but generally accepted name, as the space of all smooth functions that, together with all their derivatives, go to 0 for $x \to \pm\infty$ faster than any inverse polynomial. The space $\mathcal{S}(\mathbb{R}^1)$ is again a locally convex topological vector space, with a topology which in this case is easy to formulate and is given by a double sequence of seminorms

$$p_{m,n}(f) = \sup_{x \in R^1} |x^m f^{(n)}(x)|,$$

for $n, m = 1, 2, \ldots$ and with $f^{(n)}$ the $n^{\text{th}}$ derivative of $f$. With these seminorms $\mathcal{S}(\mathbb{R}^1)$ is a Fréchet space; $\mathcal{D}(\mathbb{R}^1)$ is only a LF-space. See for these notions Supp. Sect. 27.6. The continuous linear functionals on $\mathcal{S}(\mathbb{R}^1)$ are called *tempered distributions*. The tempered distributions have an advantage: the Fourier transform of a Schwartz function is again a Schwartz function, and therefore the Fourier transform of a tempered distribution is a tempered distribution. One has $\mathcal{D}(\mathbb{R}^1) \subset \mathcal{S}(\mathbb{R}^1)$; consequently a tempered distribution is a distribution; the converse is not true. However, the $\delta$-function and its derivatives are not only distributions, but also tempered distributions.

Tempered distributions are widely used in mathematical physics, in particular in the axiomatic formulation of relativistic quantum field theory (See Chap. 16).

The notion 'distribution' is indeed a generalization of the notion of 'function'. A function $F$ on $\mathbb{R}^1$ that is locally integrable, i.e. is integrable on every finite closed interval, defines a distribution $\tilde{F}$ by integration with a test function $f$ from $\mathcal{D}(\mathbb{R}^1)$ according to

$$\tilde{F}(f) = \int\limits_{-\infty}^{+\infty} F(x) f(x) dx,$$

so the space of locally integrable functions can be identified with a subspace of the space of distributions.

For general mathematical information on topological vector spaces, generalized functions and test function spaces, see the book by Treves [8].

## 25.4  Problems

In this section we collect a few properties of the $\delta$-function in the guise of problems. They can be solved by using the heuristic formulation, which can be justified by keeping an eye on the rigorous formalism.

(25.4,a) **Problem** The Heaviside function $\theta(x)$ is defined as $\theta(x) = 0$, for $x \leq 0$, and $= 1$ for $x > 0$. Show that the derivative of $\theta(x)$ is $\delta(x)$, in the sense of derivatives as defined above.

(25.4,b) **Problem** Show that, with the conventions of Sect. 25.4.2, the Fourier transform of the generalized function $\delta_a(x) = \delta(x - a)$ is equal to the function $\hat{\delta}_a(k) = \frac{1}{\sqrt{2\pi}} e^{-ika}$.

(25.4,c) **Problem** Prove $\delta(ax) = \frac{1}{|a|}\delta(x)$, for $a \neq 0$.

## References

1. Sobolev, S.: Méthode nouvelle à résoudre le problème de Cauchy pour les équations linéaires hyperbolyques normales Rec. Math. [Math. Sbornik] N.S. **1**, 39–72 (1936)
2. Temple, G.: Proc. Roy. Soc. A **228**, 175–190 (1955)
3. Lighthill, M.J.: Introduction to Fourier analysis and generalised functions. Cambridge University Press, Cambridge (1958) (Lighthill was one of the pioneers of modern aerodynamics and fluid dynamics. He is less well-known for his approach to generalized functions, at least among physicists. For him it was in the first place a tool to be used in handling formal expressions in Fourier analysis, as the title of his book indicates.)
4. Dirac, P.A.M.: The principles of quantum mechanics, 4th edn. Oxford University Press, Oxford (1982) (The heuristics of the Dirac $\delta$-function and its derivatives are explained and extensively used in all standard physics books on quantum mechanics. None of these improve on the explanation given in Dirac's classical book. The first edition is from 1930. The present fourth edition is still in print.)

5. Schwartz, L.: Théorie des distributions. vol. I, II Hermann, 1950–1951, New edition (1966) (In this 'magnum opus' Schwartz developed his mathematical theory of generalized functions - 'distributions' as he called them - for which he received the Fields Medal in 1950. His boek seems never to have been translated in English. Fortunately, since its appearance several excellent books in English have appeared. Three examples are [6, 7, 8].)

6. Richards, J.I., Youn, H.K.: The theory of distributions: a nontechnical introduction. Cambridge University Press, Cambridge (1995) (A clear introduction to the mathematically rigorous theory.)

7. Blanchard, P., Brüning, E.: Mathematical methods in physics: distributions, Hilbert space operators, and variational methods. Birkhuser, Boston (2002) (The first part of this book gives a rigorous, more comprehensive, slightly more technical but equally clear exposition of the theory.)

8. Treves, F.: Topological vector spaces, distributions and kernels. Academic Press, New York (1967) (A standard textbook on the topics mentioned in the title.)

# Chapter 26
# Dirac's Bra-Ket Formalism

## 26.1 Introduction

This chapter is devoted to an elegant formalism for operations with vectors and operators in the Hilbert space of quantum mechanics, which is universally used in physics textbooks. It is called the *bra-ket formalism* and is due to Dirac. It appeared first in a 1949 paper [1], and after that in the later editions of his famous book on quantum mechanics [2]. The bra-ket formalism is mathematically rigorous in the case of a finite dimensional Hilbert space—a rare occurrence in quantum theory—but is heuristic in the standard infinite dimensional case. We shall explain it first for the finite dimensional case, in which we can use ordinary linear algebra and do not need functional analysis, and then make a few remarks on its heuristic character in the case of infinite dimensional Hilbert space, as it is used in the physics literature.

## 26.2 A Vector Space $V$ and Its Dual $V^*$

Let $V$ be a complex vector space of (finite) dimension $n$, and $V^*$ its dual, of course also of dimension $n$. Elements of $V$ will be denoted by $\psi$, $\phi$, and elements of $V^*$ by $f$, $g$, etc. A linear operator $A$ in $V$ induces a linear operator $A^*$ in $V^*$, $f \mapsto A^* f$, according to $(A^* f)(\cdot) = f(A \cdot)$. Assume that there is an antilinear isomorphism $j : V \to V^*$. This gives $V$ the structure of an inner product space, with the inner product in $V$ defined by

$$(\psi_1, \psi_2) = [j(\psi_1)](\psi_2), \quad \forall \psi_1, \psi_2 \in V,$$

with $V$ being a Hilbert space if this inner product is positive definite. In fact, such an anti-isomorphism between $V$ and $V^*$ is the defining characteristic for a vector space $V$ to be a Hilbert space.

© Springer International Publishing Switzerland 2015
P. Bongaarts, *Quantum Theory*, DOI 10.1007/978-3-319-09561-5_26

## 26.3 Dirac's Use of This Triple $(V, V^*, J)$

Dirac introduced a special notation for the elements of this scheme. Vectors $\psi$ in $V$ are called *kets* and denoted as $|\psi\rangle$, elements of $V^*$ are *bras*, which could be denoted as $\langle f|$, but which Dirac writes, by using the anti-isomorphism $j$, as $\langle\psi|$, which means in fact $\langle j(\psi)|$. In this notation the inner product $(\psi_1, \psi_2)$ becomes $\langle\psi_1|\psi_2\rangle$, i.e. the functional $j(\psi_1)$ applied to the vector $\psi_2$. The formula $\langle\psi_1|A|\psi_2\rangle$, for an operator $A$ in $V$, clearly means the functional $j(\psi_1)$ in $V^*$ applied to the vector $A\psi_2$, but is, of course, equal to $(\psi_1, A\psi_2)$, which is also equal to $(A^*\psi_1, \psi_2)$, with $A^*$ the hermitian adjoint of $A$.

There is what in physics books is sometimes called an 'exterior product' of a ket $|\phi_1\rangle$ and a bra $\langle\phi_2|$, and written as $|\phi_1\rangle\langle\phi_2|$. Mathematically it is a tensor product of two vectors, one from $V$, the other from $V^*$. Such a product can be identified with an operator in $V$, acting on a ket $|\psi\rangle$ as

$$|\phi_1\rangle\langle\phi_2|\psi\rangle = (\phi_2, \psi)\phi_1.$$

(26.3,a) **Problem** Let $\phi$ be a unit vector. Show that $|\phi\rangle\langle\phi|$ is the projection operator on $\phi$.

(26.3,b) **Problem** Let $(\phi_1, \ldots, \phi_n)$ be an orthonormal basis in $V$. Prove that the expression $\sum_{j=1}^n |\phi_j\rangle\langle\phi_j|$ is the unit operator.

This expression is called a 'resolution of the identity'.

The matrix elements of an operator $A$ with respect to an orthonormal basis $(\phi_1, \ldots, \phi_n)$ are $A_{jk} = (\phi_j, A\phi_k)$ or in Dirac notation $\langle\phi_j|A|\phi_k\rangle$; the transformation formula to a different orthonormal basis $(\psi_1, \ldots, \psi_n)$ is obtained by inserting the resolution of the identity associated with original basis

$$\langle\psi_s|A|\psi_t\rangle = \sum_{j,k=1}^n \langle\psi_s|\phi_j\rangle\langle\phi_j|A|\phi_k\rangle\langle\phi_k|\psi_t\rangle.$$

(26.3,c) **Problem** Write this formula in the standard mathematical manner. Prove that the transformation matrix element $\langle\psi_s|\phi_j\rangle$ and $\langle\phi_k|\psi_t\rangle$ are the elements of a unitary matrix.

## 26.4 The Dirac Formalism in Infinite Dimensional Space

Much of the foregoing remains valid for an infinite dimensional Hilbert space, as long as one is careful with the usual domain problems of unbounded operators. However, orthonormal bases and in particular the formula for resolutions of the identity have only a rigorous meaning for proper vectors.

Improper eigenvectors are widely used in physics textbooks. One obvious elementary example is—in one dimension—the eigenfunctions of the momentum operator $P : \psi(x) \mapsto \frac{\hbar}{i}\frac{d}{dx}\psi(x)$ in Sect. 4.1, the smooth but not square integrable functions $\phi_p(x) = \frac{1}{\sqrt{2\pi\hbar}}e^{\frac{i}{\hbar}px}$. Note that the Fourier $\hat{\psi}(p)$ in Sect. 4.1 can be written in Dirac's language as $\langle p|\psi\rangle$.

Another example, even more heuristic, is the system of eigenfunctions of the position operator $Q : \psi(x) \mapsto x\psi(x)$, not functions, but generalized functions, Dirac $\delta$-functions, another of Dirac's ingeneous, elegant heuristic ideas, in fact strongly tied up with his bra-ket formalism. One has 'eigenfunctions' $\varphi_{x_0}(x) = \delta(x - x_0)$, for the 'eigenvalues' $x_0$. See for an extensive discussion on generalized functions Supp. Chap. 25.

(26.4,a) **Problem** Prove the formula

$$\langle x_1|x_2\rangle = \delta(x_1 - x_2).$$

In the above cases the resolutions of the identity are written as

$$\int\limits_{-\infty}^{+\infty} dp\,|p\rangle\langle p|, \qquad \int\limits_{-\infty}^{+\infty} dx\,|x\rangle\langle x|.$$

Note that Dirac's notation often has typographical advantages. The eigenfunctions of the Hamiltonian of the hydrogen atom, for example, are denoted in Chap. 7 as $\psi_{n,l,m}$ and would look as kets as $|n, l, m\rangle$. This advantage becomes even more apparent when to these three quantum numbers a fourth one, the spin $s$, is added, as is done in Chap. 8. See [3] for the pitfalls inherent in the use of the bra-ket formalism in quantum theory.

# References

1. Dirac, P.A.M.: A new notation for quantum mechanics. Math. Proc. Cambridge Philos. Soc. **35**, 416–418 (1939)
2. Dirac, P.A.M.: The Principles of Quantum Mechanics Original Edition 1930. 4th edn. Cambridge University Press, Cambridge (1982) (The bra-ket formalism did not appear in the first edition but in a later one.)
3. Gieres, F.: Mathematical surprises and Dirac's formalism in quantum mechanics Rep. Prog. Phys. **63**, 1893–1931 (2000) ArXive preprint quant-ph/9907069, July 22, 1999 (This paper shows how lack of rigor in using Dirac's bra-ket formalism can lead to problems.)

# Chapter 27
# Algebras, States, Representations

## 27.1 Introduction

In Chap. 12 we present a formalism for the description of general classical and quantum systems within a single algebraic framework. This chapter provides the necessary mathematical material. Statements will be precise. No proofs will be given, but there will be ample references to suitable literature.

## 27.2 Algebras: Generalities

In this chapter 'algebra' means in first instance 'abstract algebra', i.e. a set of unspecified elements having certain relations, and in second instance 'concrete algebra', i.e. function algebra or operator algebra.

An abstract algebra is a real or complex vector space $\mathcal{A}$ which has, in addition to a multiplication by scalars $(\lambda, a) \mapsto \lambda a$, also a multiplication between pairs of elements, i.e. an assignment $(a, b) \mapsto ab$, with the properties

1. $(ab)c = a(bc)$ (associativity),
2. $a(b + c) = ab + ac$ (distributivity),

for all $a$, $b$ and $c$ in $\mathcal{A}$. Unless otherwise stated, our algebras are assumed to have a unit element $1_{\mathcal{A}}$, usually just denoted as 1, with the property $a1 = 1a$, for all $a$ in $\mathcal{A}$. An element $a$ of $\mathcal{A}$ is called invertible if there exists an element $b$ such that $ab = ba = 1$.

(27.2,a) **Problem** Show that if $a$ is invertible, then this $b$ is unique.

Such an element $b$ is called *the* inverse of $a$, and is denoted, not very surprisingly, as $a^{-1}$.

In this book we also use algebras that are not associative and do not have an identity element, namely Lie algebras, discussed in Supp. Chap. 24, and used in Chap. 7 (symmetry of the hydrogen atom), Chap. 8 (spin), Chap. 16 (relativistic quantum

© Springer International Publishing Switzerland 2015
P. Bongaarts, *Quantum Theory*, DOI 10.1007/978-3-319-09561-5_27

mechanics) and Chap. 17 (relativistic quantum field theory). However, in this chapter as well as in Chap. 12 'algebra' will mean 'associative algebra with unit element', unless explicitly stated otherwise.

A linear map $\phi$ from an algebra $\mathcal{A}$ to an algebra $\mathcal{B}$ is called an *algebra homomorphism* iff $\phi(ab) = \phi(a)\phi(b)$, for all $a$ and $b$ in $\mathcal{A}$. There are obvious definitions for *algebra isomorphism* and algebra automorphism.

A *derivation* of an (associative) algebra $\mathcal{A}$ is a linear map $D$ in $\mathcal{A}$ such that

$$D(ab) = (Da)b + a(Db), \quad \forall a, b \in \mathcal{A}.$$

A derivation is called *inner* iff it has the form $Da = ca - ac$, for some element $c$ in $\mathcal{A}$. Otherwise it is called *outer*. The derivations of $\mathcal{A}$, denoted as Der $(\mathcal{A})$, form a Lie algebra (Lie algebras are extensively discussed in Supp. Sect. 24.4).

An algebra $\mathcal{A}$ is called *abelian* or *commutative* iff $ab = ba$, for all $a$ and $b$ from $\mathcal{A}$. Note that an abelian algebra has no nontrivial $(D = 0)$ inner derivations. In the complex case an algebra is called *involutive* or a *-algebra* if there is an antilinear map, *hermitian conjugation* or *involution*, $a \mapsto a^*$ with $(a^*)^* = a$, $(\lambda a)^* = \overline{\lambda} a^*$, and $(ab)^* = b^* a^*$, for all $a$ and $b$ in $\mathcal{A}$ and $\lambda$ in $\mathbb{C}$. An element $a$ from $\mathcal{A}$ is called *selfadjoint* if $a^* = a$, an element $u$ is *unitary* if $u^* u = u^* u = 1$, and element $p$ with $p^* = p$ and $p^2 = p$ a *projection*.

*Complex versus real algebras.* An algebra over the real numbers can be complexified to a complex algebra. This goes as follows. Let $\mathcal{A}$ be a real algebra. Consider the real tensor product $\mathbb{C} \otimes_{\mathbb{R}} \mathcal{A}$, first as a real vector space, in which the complex numbers $\mathbb{C}$ are seen as a two dimensional real vector space with the multiplication by the imaginary $i$ a real linear map. Make this real space into a complex vector space $\mathcal{A}^{\mathbb{C}}$ by defining multiplication by $i$ as $i(\lambda \otimes_{\mathbb{R}} a) := i\lambda \otimes_{\mathbb{R}} a$. Define an algebra multiplication in $\mathcal{A}^{\mathbb{C}}$ by linear extension of $(\lambda \otimes_{\mathbb{R}} a)(\mu \otimes_{\mathbb{R}} b) = \lambda\mu \otimes_{\mathbb{R}} ab$. This makes $\mathcal{A}^{\mathbb{C}}$ into a complex algebra, the *complexification* of $\mathcal{A}$, which may be written as $\mathcal{H}^{\mathbb{C}} = \mathcal{A} \oplus i \mathcal{A}$. Its unit element is, of course, $1 \otimes_{\mathbb{R}} 1_{\mathcal{A}}$. Not every complex algebra $\mathcal{A}$ is the complexification of a real algebra. The necessary and sufficient condition for this is the existence of an antilinear map $\mathcal{A} \to \mathcal{A}$, $a \mapsto \overline{a}$ with $\overline{(\overline{a})} = a$ and $\overline{ab} = \overline{a}\overline{b}$. Such a map is called a *conjugation*.

**(27.2,b) Problem** Show that a conjugation is one-to-one.

Classical physics is, in principle, formulated in terms of real numbers; in the formalism of quantum theory complex numbers are essential. In the classical case the real algebras can be complexified, with no addition of information, as is clear from the above discussion. In setting up a general algebraic formulation for both we therefore use complex algebras; from now on all algebras in this chapter will be complex.

**(27.2,c) Problem** Show that the conjugation of an algebra $\mathcal{A}$ is an involution if and only if $\mathcal{A}$ is commutative.

(27.2,d) **Problem** Let $\mathcal{A}$ be a complex $*$-algebra. The subset of its selfadjoint elements is a real vector space. Show that this subset is a (real) algebra if and only if $\mathcal{A}$ is commutative.

Obvious concrete examples of $*$-algebras are $M(n, \mathbb{C})$, the algebra of all $n \times n$ complex matrices, in finite dimensions, and in general $B(\mathcal{H})$ with $\mathcal{H}$ a Hilbert space.

The observables of a classical mechanical system are the smooth real-valued functions on the phase spacephase space, a symplectic manifold (See for this notion Supp. Sect. 20.7). They form an infinite dimensional commutative algebra. The observables in quantum theory are selfadjoint operators in an in general infinite dimensional Hilbert space. We need therefore in both cases *infinite dimensional $*$-algebras*, which should have a topology in order to be practically useful.

*Conclusion*: The algebras $\mathcal{A}$ in this chapter and in Chap. 12 will be *topological $*$-algebras*.

The landscape of the theory of general topological algebras is a vast arid desert, with the *locally convex topological algebras* as an inhabited region, containing the connected metropoles of *$C^*$-algebras* and *von Neumann algebras* and the smaller city of *Fréchet algebras* or *Fréchet-like algebras*. We shall consider a sequence of topological algebras, starting with an obvious case and leading to the more sophisticated examples:

- *normed $*$-algebras*
- *Banach $*$-algebras*
- *$C^*$-algebras*
- *von Neumann algebras*
- *smooth $*$-algebras, i.e. Fréchet and LF $*$-algebras.*

There is one type of infinite dimensional algebra, important for our purpose, but usually not discussed in the context of topological algebras:

- *Smooth Poisson $*$-algebras.*

In the next sections we shall review normed $*$-algebras, then Banach $*$-algebras, next $C^*$-algebras, von Neumann algebras, then briefly Fréchet and LF algebras, and finally Fréchet and LF Poisson $*$-algebras.

## 27.3 Normed $*$-Algebras: Banach $*$-Algebras

An algebra $\mathcal{A}$ is called a *normed algebra* if its underlying vector space is a normed vector space (Normed vector spaces are defined in Supp. Sect. 21.2). If $\mathcal{A}$ has an involution it is a *normed $*$-algebra*. The norm has to satisfy additional compatibility conditions for the multiplication and for the unit element, namely

1. $||ab|| \leq ||a|| \, ||b||$, for all $a, b$ in $\mathcal{A}$,
2. $||1_{\mathcal{A}}|| = 1$.
   For a $*$-algebra one adds

3. $||a^*|| = ||a||$, for all $a$ in $\mathcal{A}$.

Every normed algebra can be completed by the standard procedure explained in Supp. Sect. 18.4, applied to normed vector spaces. A complete normed algebra is called a *Banach algebra*, if it possesses an involution it is a *Banach *-algebra* (For completeness see Supp. Sect. 18.4, Definition (18.4,f)).

We call an element $a$ of a *-algebra *positive* iff it can be written as $b^*b$ for some element $b$. There are however other definitions of positivity for general *-algebras. These coincide with this definition for the most important examples of such algebras, namely $C^*$-algebras and von Neumann algebras.

## 27.4  $C^*$-Algebras

From a mathematical point of view a natural first choice for $\mathcal{A}$ in the quantum case is an *abstract complex $C^*$-algebra*, a special type of Banach *-algebra obtained by requiring a seemingly minor additional property on the norm. It is one of the most intensely studied objects in modern functional analysis, a very powerful mathematical object, of which by now very much is known. Together with an abstract $C^*$-algebra one considers its representations as operator algebras in a Hilbert space.

(27.4,a) **Definition** A Banach *-algebra is called a $C^*$-*algebra* iff its norm satisfies $||a^*a|| = ||a||^2$, for all $a$ from $\mathcal{A}$.

It is obvious that this $C^*$-identity is equivalent to $||a^*a|| = ||a^*|| \, ||a||$, for all $a$.

(27.4,b) **Problem** Show that in a $C^*$-algebra $\mathcal{A}$ one has $||1_\mathcal{A}|| = 1$.

In the theory of $C^*$-algebras there are two basic theorems:

1. *The Gelfand-Naimark theorem for the commutative case*: A commutative $C^*$-algebra is *-isometrically isomorphic to the algebra of all complex-valued continuous functions on a compact topological space, provided with the supremum norm. This means that there is a one-to-one correspondence between commutative $C^*$-algebras and compact topological spaces, up to homeomorphisms. For a proof see [1, p. 60], and [2, p. 270].
   The simplest example is $C([0, 1])$, the algebra of all complex-valued continuous functions on the interval $[0, 1]$. One obtains a slight generalization of this theorem if one drops the requirement that a $C^*$-algebra should have a unit element. A commutative $C^*$-algebra, not having a unit element, is *-isometrically isomorphic to the algebra of all complex-valued continuous functions on a locally compact topological space, that vanish at infinity. The simplest example for this case is $C_0(R)$, the algebra of all complex-valued continuous functions $f$ with $\lim_{x \to \pm\infty} |f(x)| = 0$. (For the notions of locally compact and vanishing at infinity, see Supp. Sects. 18.2 and 18.3).
2. *The Gelfand-Naimark theorem for the noncommutative case*: A noncommutative $C^*$-algebra is *-isometrically isomorphic to a norm-closed subalgebra of the algebra of bounded operators in a Hilbert space. See [3, p. 94] and [1, p. 109].

This means that noncommutative $C^*$-algebras can be studied as operator algebras. See [4] for the original Gelfand and Naimark paper in which Theorems (1) and (2) were first stated and proved.

Note that unlike in the commutative case the concrete realization that according to Theorem (2) exists, is far from unique. Looking at operator algebra realizations, i.e. *representations*, is therefore of great additional interest, as in the case of groups where one studies groups together with their representations as linear transformations in suitable vector spaces. Two representations $\pi_1$ and $\pi_2$ of the same abstract $C^*$-algebra $\mathcal{A}$, in Hilbert spaces $\mathcal{H}_1$ and $\mathcal{H}$, are called *unitarily equivalent* iff there exists a unitary map $U : \mathcal{H} \to \mathcal{H}$ such that $\pi_2(a) = U\pi_1(a)U^{-1}$, for all $a$ in $\mathcal{A}$, the definition for unitary equivalence in general. In this case the isomorphism between the two operator $C^*$-algebras $\pi_1(\mathcal{A})$ and $\pi_2(\mathcal{A})$ is called *spatial*. Two (abstract) $C^*$-algebras $\mathcal{A}_1$ and $\mathcal{A}_2$ can be just isomorphic, or in a more special case spatially isomorphic. Note that here in the context of $C^*$-algebra homomorphism, isomorphism, etc., means *-homomorphism, etc.

A few more properties of $C^*$-algebras:

- An isomorphism between $C^*$-algebras $\mathcal{A}_1$ and $\mathcal{A}_2$ is automatically a *-isomorphism
- A *-homomorphism between two $C^*$-algebras is norm decreasing (See for the proof [3, p. 40]). This implies that it is continuous; a *-isomorphism between $C^*$-algebras is obviously an isometry.

  (27.4,c) **Problem** Derive from this that if a *-algebra has a $C^*$-norm then this norm is unique.

- A *-homomorphism of $C^*$-algebras is *positive*, i.e. it maps positive elements into positive elements (See for the proof [5, p. 42]).

References for $C^*$-algebras: the book by Dixmier [6], and more recently the books by Murphy [3] and Arveson [7]. Also useful are volume one of the two-volume book on operator algebras and quantum statistical mechanics by Bratteli and Robinson [5], and general books on operator theory, such as Conway [8], Blackadar [1] and the first volumes of the two series of three books by Kadison and Ringrose [2] and by Takesaki [9].

## 27.5  Von Neumann Algebras

We come next to more general topological algebras, based on so-called locally convex vector spaces, in which the topologies are not defined by a single norm but by systems of *seminorms*. The most important class, in particular for physical interpretation, is formed by *von Neumann algebras*, usually defined as algebras of operators in a Hilbert space, although a definition as abstract algebras is possible. They can also be generated as operator algebras by constructing completions in a suitable topology of represented $C^*$-algebras.

Von Neumann algebras, called originally 'rings of operators', were first system-atically studied in the thirties, in a series of papers by von Neumann and his young co-worker Murray. See for the first paper of this series [10]. Following Dixmier's 1957 book [11], these algebras are now usually called von Neumann algebras.

There is a fairly technical definition of a von Neumann algebra as an abstract algebra, but we will stick to the operator algebra definition. There are actually two such definitions, a purely algebraic and a topological one; their equivalence is one of the basic theorems of the theory.

Let $S$ be a nonempty subset of $B(\mathcal{H})$, the bounded operators in a Hilbert space $\mathcal{H}$. Define the *commutant* of $S$ as

$$S' := \{A \in B(\mathcal{H}) \mid AB = BA, \quad \forall B \in B(\mathcal{H})\}.$$

**(27.5,a) Problem** Show that $S'$ is an algebra, and if $S$ is selfadjoint, i.e. with $A \in S \Rightarrow A^* \in S$, then $S'$ is a $*$-algebra.

**(27.5,b) Problem** Suppose $S \subset T$, for two nonempty subsets $S$ and $T$ of $B(\mathcal{H})$. Show that this implies $T' \subset S'$.

We next define the *double commutant* of $S$ as $S'' = (S')'$. For $S$ nonempty and selfadjoint the double commutant is obviously a $*$-algebra.

**(27.5,c) Problem** Show that for a nonempty subset $S$ of $B(\mathcal{H})$ one has $S \subset S''$.

The special case of this property, in which one takes for $S$ a $*$-algebra $\mathcal{A}$, imme-diately leads to the first definition of a von Neumann algebra:

**(27.5,d) Definition** (algebraic) (1): a $*$-algebra $\mathcal{A} \subset B(\mathcal{H})$ is a *von Neumann algebra* iff $\mathcal{A}'' = \mathcal{A}$.

The linear space of bounded operators in $\mathcal{H}$ has several topologies. The most obvious one is the one given by the operator norm, which is, however, too strong and therefore not very useful in quantum theory. There are weaker topologies, in the first place one which is called, rather bizarre, *strong operator topology*. This is the topology 'par excellence' for quantum theory, as we have seen in Chap. 3. It is a locally convex topology given by a system of seminorms $\{p_\psi\}_{\psi \in \mathcal{H}}$ defined as

$$p_\psi(A) = ||A\psi||, \quad \forall \psi \in \mathcal{H}.$$

A second topology is the still weaker *weak operator topology*, defined by the system of seminorms $\{p_{\psi_1,\psi_2}\}_{\psi_1,\psi_2 \in \mathcal{H}}$ given by the formula

$$p_{\psi_1,\psi_2}(A) = |(\psi_1, A\psi_2)| \quad \forall \psi_1, \psi_2 \in \mathcal{H}.$$

The weak and strong topologies are different but coincide nevertheless on vari-ous subsets of $B(\mathcal{H})$; they can both be used to give, in the same way, equivalent topological definitions of a von Neumann algebra. Von Neumann algebras were—and still are—sometimes called $W^*$-algebras.

There is a third topology, the $\sigma$-weak topology, that will briefly appear in the definition of *normal states*. It is determined by the semi-norms

$$a \mapsto \left| \sum_{i=1}^{\infty} (a\,\xi_i\,,\,\eta_i) \right| ,$$

where $\{\xi_i\}$ and $\{\eta_i\}$ are two sequences in $H$ such that $\sum_{i=1}^{\infty} ||\xi_i|| < \infty$ and $\sum_{i=1}^{\infty} ||\eta_i|| < \infty$. One has the sequence of relations between the topologies:

$\sigma$-*weak* $\subset$ *weak* $\subset$ *strong* $\subset$ *norm*.

There are more topologies on $B(\mathcal{H})$: *ultrastrong*, *strong** and $\sigma$-*strong*, etc., but fortunately we do not need these.

The following important topological property can be proved: For a nonempty selfadjoint subset of $B(\mathcal{H})$ its commutant is weakly closed, in fact a weakly closed *-subalgebra of $B(\mathcal{H})$, and so is its bicommutant. This leads in a natural way to the second (topological) definition of a von Neumann algebra:

(27.5,e) **Definition** (topological) (2): a von Neumann algebra is a weakly closed *-subalgebra of $B(\mathcal{H})$.

(27.5,f) **Problem** Show that a von Neumann algebra is also a $C^*$-algebra. The converse is of course not true.

This means that a commutative von Neumann algebra is also a commutative $C^*$-algebra, and can, according to the Gelfand-Naimark theorem, as such be regarded as the algebra of continuous function on a compact topological space. The topology of this space is what is called 'extremely disconnected', and is a rather exotic type of topology, for which we have no need in this book. Indeed the $C^*$-algebra properties of a von Neumann algebra are secondary features.

Next we have the basic theorem of von Neumann algebra theory:

(27.5,g) **Theorem** *Definitions* (1) *and* (2) *are equivalent.*

For the proof see, for example [5, p. 72]. It is also shown there that the theorem remains true for the strong operator topology.

Two more applications of the commutant:

(27.5,h) **Problem** Show that a subalgebra $\mathcal{A} \subset B(\mathcal{H})$ is *abelian* iff $\mathcal{A} \subset \mathcal{A}'$.

(27.5,i) **Problem** Show that a von Neumann algebra $\mathcal{A}$ is *maximal abelian* iff $\mathcal{A} = \mathcal{A}'$, and that in fact a maximal commuting *-algebra of operators is necessarily a von Neumann algebra.

The notion of being maximal abelian is of great importance in quantum theory, as we have seen in Chap. 3.

(27.5,j) **Definition** A von Neumann algebra $\mathcal{A}$ is called a *factor* if and only if $\mathcal{A} \cap \mathcal{A}' = \{1_{\mathcal{A}}\}$.

It can be shown that an arbitrary von Neumann algebra can be written as a *direct integral* of factors (A direct integral is the natural generalization of a direct sum. See Supp. Sect. 21.5). Factors are classified by the structure of their system of projections. One has type $I_n$ factors, for $n = 1, \ldots, \infty$, the algebras of bounded operators in an $n$-dimensional Hilbert space, but also more exotic type II and type III factors, which play a role in quantum theories of systems with an infinite number of degrees of freedom.

*Remarks on the relation between $C^*$- and von Neumann algebras*: Operator $C^*$-algebras can be defined as $*$-subalgebras of $B(\mathcal{H})$, closed in the operator norm topology, which is stronger than the strong and weak operator topologies, so von Neumann algebras are $C^*$-algebras. For a von Neumann algebra its $C^*$-properties are secondary. A $C^*$-algebra $\mathcal{A}$ generates a unique von Neumann algebra $\hat{\mathcal{A}}$ as its weak closure.

One important difference comes from the fact that $C^*$-algebras may not have sufficiently many projections, or may not have nontrivial projections at all. It is quite possible that the spectral projections of a selfadjoint operator which belongs to a certain $C^*$-algebra lie outside it. Here is a simple example: the function algebra $C([0, 1])$. Its elements act as multiplication operators on $L^2([0, 1], dx)$; a real function gives in this manner a selfadjoint operator, with, of course, a spectral resolution. However, the algebra $C([0, 1])$ contains only the trivial projection operators 0 and 1.

Von Neumann algebras contain all the spectral projections of their selfadjoint elements. In fact a von Neumann algebra is generated by its projections. Unbounded selfadjoint operators, very important in quantum theory, can be constructed from these projections and appear as operators *associated* with the given von Neumann algebra. In the example of $C([0, 1])$ the weak closure consists of the algebra of multiplication by measurable, essentially bounded functions, i.e. $L^\infty([0, 1], dx)$; the associated not necessarily bounded operators are represented by arbitrary measurable functions.

In the preceding section we discussed the Gelfand-Naimark theorem relating commutative $C^*$-algebras with topological spaces. There is an analogous theorem for commutative von Neumann algebras, connecting commutative von Neumann algebras with measure spaces, with a formulation that is more complicated than for the $C^*$-algebra case because we have to consider equivalence classes on both sides of the relation.

Two von Neumann algebras $\mathcal{A}_1$ in $\mathcal{H}_1$ and $\mathcal{A}_2$ in $\mathcal{H}_2$ are called *equivalent* iff they are spatially isomorphic, i.e. iff there is a unitary map $U : \mathcal{H}_1 \to \mathcal{H}_2$ such that for all $A_1$ in $\mathcal{A}_1$ there exists an $A_2$ such that $A_2 = U A_1 U^{-1}$ and that for all $A_2$ in $\mathcal{A}_2$ there exists an $A_1$ such that $A_1 = U^{-1} A_2 U$.

**(27.5,k) Theorem** *There is a one-to-one correspondence between (equivalence classes of) commutative von Neumann algebras and (equivalence classes of) measure spaces.*

It can be shown that every von Neumann algebra can be generated by a single selfadjoint operator, a fact which can be used to prove the theorem in one direction.

The measure spaces in this theorem have to be *standard Borel measure spaces*, i.e. Borel measure spaces (See Supp. Sect. 19.7), with the underlying topological space separable, metrizable and complete. In practice most measure spaces are standard in this sense; in particular all measure spaces in this book are standard. Two measure spaces $(X_1, \mathcal{B}_1, \mu_1)$ and $(X_2, \mathcal{B}_2, \mu_2)$ are *equivalent* iff there is a measurable map connecting the two measure spaces.

(27.5,l) **Problem** Consider two probability measure spaces, denoted as $(R^1, \mathcal{B}, P_1)$ and $(R^1, \mathcal{B}, P_2)$, with the $P_j$ given by positive probability densities $\rho_j(x)$. Show that these two probability spaces are equivalent, in the above sense. What is the map implementing this equivalence?

Given a measure space $(X, \mathcal{B}, \mu)$, one has the Hilbert space $L^2(X, \mu)$. The measurable (essentially) bounded functions $f$ on $X$ appear as multiplication operators on $L^2(X, \mu)$ according to $(f\psi)(x) = f(x)\psi(x)$, for all $\psi$ in $L^2(X, \mu)$. These operators form a commutative von Neumann algebra. This proves the theorem in the other direction.

References for von Neumann algebras are the books by Dixmier [11] and Schwartz [12], lecture notes by Jones [13]; the references [2] and [9] are also excellent and thorough on von Neumann algebras. For a detailed exposition of noncommutative measure theory, or 'quantum measure theory', see [14].

## 27.6 Smooth *-Algebras

What we call a *smooth *-algebra* is in the commutative case the *-algebra $C^\infty(\mathcal{M})$ of smooth functions on a differentiable manifold $\mathcal{M}$. On a compact manifold this is a Fréchet *-algebra, because its underlying vector space is a Fréchet space, a locally convex topological vector space that is metrizable and complete. For a noncompact manifold we have an LF *-algebra, with an underlying LF space, a strict direct limit of a sequence of Fréchet spaces. Such spaces appear in particular as test function spaces for tempered distributions. See Supp. Sect. 25.3.

*Reminder*: An index set $\mathcal{I}$ is called *directed* iff it is a partially ordered set in which each pair of elements has a supremum (See Supp. Sect. 18.6).

Suppose one has a system $\{A_\alpha\}_{\alpha \in \mathcal{I}}$, for a certain type of mathematical objects $A_\alpha$, together with for all pairs $\{\alpha, \beta\}$ with $\alpha \prec \beta$, maps $\phi_{\beta\alpha} : A_\alpha \to A_\beta$, such that for each triple $\alpha \prec \beta \prec \gamma$ one has $\phi_{\gamma\alpha} = \phi_{\gamma\beta} \circ \phi_{\beta\alpha}$. Then there exists a unique $A$ with maps $\phi_\alpha : A_\alpha \to A$. $A$ is called the *direct limit* or *inductive limit* of the system $\{A_\alpha\}_{\alpha \in \mathcal{I}}$. It is called a *strict inductive limit* iff all the maps $\phi_{\beta\alpha}$ are injective and homeomorphisms onto their images, and countable iff the index set is just the sequence of natural numbers. This procedure can be applied to all sorts of mathematical objects. For $\mathcal{I} = N$ and the $A_n$ Fréchet spaces, the direct limit is no longer a Fréchet space; it is (by definition) an LF space.

We assume that there are suitable noncommutative versions of these algebras, describing what we heuristically may call 'noncommutative smooth manifolds'. There is an extensive literature on general locally convex algebras, both commutative and noncommutative (see [15]), but much less on Fréchet and LF algebras.

Up until now there exists no proper Gelfand-Naimark theorem for characterizing a smooth manifold. The best result so far is the following [16]:

(27.6,a) **Theorem** (G.P. Thomas) *Let $\mathcal{M}_1$ and $\mathcal{M}_2$ be two $\sigma$-compact $C^\infty$-manifolds with $C(\mathcal{M}_1)$ and $C(\mathcal{M}_2)$ their algebras of smooth functions. Then if $\phi : C^\infty(\mathcal{M}_1) \to C^\infty(\mathcal{M}_2)$ is a $*$-algebra isomorphism, then the manifolds $\mathcal{M}_1$ and $\mathcal{M}_2$ are diffeomorphic.*

Note that a $\sigma$-compact topological space is the union of countably many compact spaces. All manifolds in this book are $\sigma$-compact.

A good impression of what is involved in further work in this direction can be obtained from [17] and [18].

(27.6,b) *Remark* Alain Connes has given an algebraic characterization of ('commutative' and 'noncommutative') Riemannian manifolds in terms of what he calls *spectral triples*. An essential ingredient in this is the notion of *spin structure*. It is hard to distill from this a prescription for a smooth manifold without additional mathematical structure on it. See his book ([19, Chap. 12]).

## 27.7  Poisson and Poisson-Frèchet $*$-Algebras, etc.

A *Poisson algebra* $\mathcal{A}$, as an abstract algebra, is an associative algebra, which is at the same time a *Lie algebra* (See Supp. Chap. 24), which means that it has an additional nonassociative multiplication. The Lie bracket is in this context called a *Poisson bracket*, with notation $\{\cdot, \cdot\}$. The Poisson bracket was discussed in the context of classical mechanics in Chap. 2, where the associative multiplication was commutative, together with its mathematical basis in Supp. Chap. 20. We repeat the main properties of the Poisson bracket here. It is a bilinear map $(a, b) \mapsto \{a, b\}$ which satisfies

$\{a, b\} = -\{b, a\}$ (*antisymmetry*)

$\{a, \{b, c\}\} + \{b, \{c, a\}\} + \{c, \{a, b\}\} = 0$ (*Jacobi identity*),

together with a compatibility relation between the associative and the Poisson multiplication

$\{a, bc\} = \{a, b\}c + b\{a, c\}$ (*Leibniz relation*),

and

$\{a, b\}^* = \{a^*, b^*\}$ (*reality condition*)

for all $a$, $b$ and $c$ in $\mathcal{A}$.

In the case of Poisson-Fréchet and Poisson-LF algebras one requires the Poisson bracket to be continuous, with respect to the underlying topological vector spaces, separately in the two variables, as for the associative product. Apart from this not much can be added to what was said in the preceding subsection.

(27.7,a) *Examples Commutative*: the smooth functions on a symplectic manifold (Classical mechanics), with the standard Poisson bracket. *Noncommutative*: the bounded operators in a Hilbert space (Quantum theory), with as Poisson bracket $\{A, B\} = i[A, B]$, with $[\cdot, \cdot]$ the standard Hilbert space commutator. Both examples are central in Chap. 12.

(27.7,b) **Problem** Check that both examples satisfy the requirements for a Poisson bracket.

## 27.8 States

The term 'state' has its origin in physics, but is now widely used in mathematics, in particular in operator algebra theory.

(27.8,a) **Definition** A *state $\omega$ on a ∗-algebra $\mathcal{A}$* is a linear functional, positive, i.e. having positive values on positive elements of $\mathcal{A}$, and normalized to unity, i.e. with $\omega(1_{\mathcal{A}}) = 1$.

Note that we call an element of $\mathcal{A}$ positive iff it can be written as $A^*A$, for some $A$ in $\mathcal{A}$. Various other definitions of positivity can be given; for a $C^*$-algebra all coincide with the one given here.

It can be shown that for $C^*$-algebras a positive state is automatically real, i.e. with $\omega(A^*) = \overline{\omega(A)}$ for all $A$ in $\mathcal{A}$, and moreover continuous. See for the proof [5, p. 49].

The system $S(\mathcal{A})$ of states on a ∗-algebra $\mathcal{A}$ forms a *convex subset* of the vector space dual $\mathcal{A}^*$ of $\mathcal{A}$, i.e. for each pair of states $\omega_1$ and $\omega_2$ and for all $\lambda$ with $0 \le \lambda \le 1$, the convex linear combination $\lambda\omega_1 + (1 - \lambda)\omega_2$ is again a state. See for the notion of dual of a vector space Supp. Sect. 21.2. For topological ∗-algebras dual means here topological dual. A state which cannot be written as a convex sum of two different states for $\lambda \ne 0, 1$, and which therefore lies in a certain sense on the boundary of the set of states, is called *extremal* or, in quantum theory, *pure*. The states not on this boundary are called *mixed*. In quantum theory the pure states describe states in what we have called the 'first level' in Chap. 3. The mixed states appeared in Chap. 11 as density operators in the second level, in quantum statistical physics.

For quantum statistical physics a particular type of state is important:

(27.8,b) **Definition** A state on a von Neumann algebra is called *normal* iff it is continuous in the $\sigma$-weak topology.

A normal state on $B(\mathcal{H})$, the algebra of bounded operators on a Hilbert space $\mathcal{H}$, is a state of the form $\omega(A) = \mathrm{Tr}(DA)$, with $D$ a density operator, the quantum ensemble in statistical mechanics introduced in Sect. 11.4. More on the background for this choice in Sect. 12.4.3.

## 27.9  States and Representations: The GNS Representation

Additional information on objects like groups and algebras is obtained by studying representations, i.e. homomorphisms into groups and algebras of linear operators, usually in Hilbert spaces.

A representation of a $*$-algebra $\mathcal{A}$ is a $*$-homomorphism $\pi$ from $\mathcal{A}$ into the algebra of linear operators of an inner product space $\mathcal{H}$, a Hilbert space in the case of a $C^*$- or von Neumann algebra.

Each state $\omega$ on $\mathcal{A}$ defines in a simple but ingenious way a representation $\pi_\omega$ of $\mathcal{A}$, the *GNS (Gelfand-Naimark-Segal) representation*. It is what is usually called an adjoint representation, obtained by letting the algebra act on itself.

The construction of $\pi_\omega$ goes as follows:

(1)  Consider as representation space in first instance a second copy of $\mathcal{A}$ and call it $\underline{A}$, with elements $\underline{a}$. The inner product on $\underline{A}$ is $(\underline{a}_1, \underline{a}_2)_\omega = \omega(a_1^* a_2)$. Define, for each $a$ from $\mathcal{A}$, $\pi_\omega(a)\underline{b} = \underline{ab}$.

(27.9,a) **Problem** Check that this $\pi_\omega$ is indeed a $*$-representation of $\mathcal{A}$.

(2)  The inner product on $\underline{A}$ is positive but in general not positive definite, i.e. there may be nonzero elements with zero length.

(27.9,b) **Problem** Consider $\mathcal{I}_\omega = \{\underline{a} \in \underline{A} \mid \omega_\pi(a^*a) = 0$. Show that this $\mathcal{I}_\omega$ is a left ideal in $\underline{A}$, i.e. that it is a linear subspace of $\underline{A}$ with $\underline{a} \in \mathcal{I}_\omega \Rightarrow \underline{ba} \in \mathcal{A}$, for all $\underline{b} \in \underline{A}$. Hint: use the Schwarz inequality.

(3)  The inner product descends to the quotient space over this ideal. The result is a pre-Hilbert space, which can be completed to a Hilbert space $\mathcal{H}_\omega$, carrying a representation $a \mapsto \pi_\omega(a)$ of $\mathcal{A}$.

(4)  There is a special unit vector $\psi_\omega$, the image of $\underline{1}_\mathcal{A}$ in the quotient space, with the property $\omega(a) = (\psi_\omega, \pi_\omega(a)\psi_\omega)$, for all elements $a$ in $\mathcal{A}$.

(27.9,c) **Problem** Show that the vector $\psi_\omega$ is *cyclic*, meaning that $\{\pi_\omega(a)\psi_\omega\}_{a \in \mathcal{A}}$ is norm dense in $\mathcal{H}_\omega$.

(27.9,d) *Remark* The GNS construction still makes sense if we only require for a state $\omega$ continuity instead of positivity. The inner product on $\underline{A}$ may then be indefinite and also degenerate. To remove the degeneracy one divides out the left ideal $\mathcal{I}_\omega = \{\underline{a} \in \underline{A} \mid \omega(\underline{b}^*\underline{a}) = 0, \forall \underline{b} \in \underline{A}\}$ and obtains as representation spacepreface $\mathcal{H}_\omega$, a complete topological vector space with a possibly indefinite but nondegenerate inner product, jointly continuous in the two variables. This is necessary for the Gupta-Bleuler formalism, used for a manifest Lorentz-covariant description of the Maxwell quantum field, as is briefly discussed in Sect. 16.7.

It can be shown that a GNS-representation $\pi_\omega$ of a $C^*$-algebra is irreducible iff $\omega$ is a pure state (For the proof of this, see [5, p. 57], and for the notion of irreducible representation Supp. Chap. 24).

# References

1. Blackadar, B.: Operator Algebras: Theory of $C^*$-Algebras and von Neumann Algebras. Springer, Berlin (2006) (A more comprehensive textbook.)
2. Kadison, R.V., Ringrose, J.R.: Fundamentals of the Theory of Operator Algebras Volume I: Elementary Theory. American Mathematical Society, Providence (1997)
3. Murphy, G.J.: $C^*$-Algebras and Operator Theory. Academic Press, New York (1990)
4. Gelfand, I., Neumark, M.: On the imbedding of normed rings into the ring of operators in Hilbert space. Mathematicheskii Sbornik (Recueil Mathématique) **12**, 197–217 (1943) (In this paper the special class of Banach ∗-algebras were introduced that were later called $C^*$-algebras. The paper is very readable. http://www.mathnet.ru/links/c36a1136c25547bccfca84d9dc734f2c/sm6155.pdf.)
5. Bratteli, O., Robinson, D.W.: Equilibrium states. Models in quantum statistical mechanics. Operator Algebras and Quantum Statistical Mechanics II, 2nd edn. Springer, Berlin (2003)
6. Dixmier, J.: $C^*$-algebras. Trans.: French Edition 1969. North-Holland/Elsevier, Amsterdam (1977) (Somewhat outdated, but still an important basic book. Written in Bourbaki style; clear, precise, dry.)
7. Arveson, W.: An Invitation to $C^*$-Algebras. Springer, New York (1976). (Corrected second printing 1998). A very readable introduction
8. Conway, J.B.: A Course in Operator Theory. American Mathematical Society, Providence (2009). Another readable introduction
9. Takesaki, M.: Theory of Operator Algebras I. Springer, New York (1979). Second printing 2001 (The first volumes of these two series are solid books of reference on operator algebra theory.)
10. Murray, F.J., von Neumann, J.: On rings of operators I. Ann. Math. **37**, 116–229 (1936) (The first paper in a series which can be regarded as a landmark in twentieth century mathematics. This paper gives the basic properties of what are now rightly called 'von Neumann algebras: the division into types I, II, and III is discussed, and in particular factors not of type I are found.)
11. Dixmier, J.: Von Neumann Algebras. Trans.: Second French Edition 1969. North-Holland, Amsterdam (1981) (For this book the same qualifications can be given as for Dixmier's book on $C^*$-algebras, except that it is more outdated because of the emergence of the so-called 'modular theory' after its publication.)
12. Schwartz, J.T.: $W^*$-Algebras. Gordon and Breach, New York (1967) (A very readable introduction to the subject.)
13. Jones, V.F.R.: Von Neumann Algebras. Lecture Notes 2010. http://www.math.berkeley.edu/~vfr/MATH20909/VonNeumann2009.pdf (Another useful, readable introduction.)
14. Hamhalter, J.: Quantum Measure Theory. Kluwer, Dordrecht (2003)
15. Helemskii, A.Y.: Banach and Locally Convex Algebra. Oxford University Press, Oxford (1993) (A solid but heavy going general work of reference.)
16. Thomas, E.G.F.: Characterization of a manifold by the ∗ -algebra of its $C^\infty$ functions. Preprint Mathematical Institute, University of Groningen. Unpublished (Erik Thomas died on September 13, 2011. A remarkable theorem, which uses standard functional analysis. Copies of this note can be obtained by sending an e-mail to p.j.m.bongaarts@xs4all.nl.)
17. Azmi F.M.: Characterization of continuous functions on open connected subset of $R^n$. JP J. Geometry Topology **7**, 235–248 (2007). http://faculty.ksu.edu.sa/fazmi/Characterization%20of%20cont%20functions/cont-func-n-send.pdf
18. Azmi, F.M.: Characterization of special $C_p^*$-algebra and special differential $p$-Fréchet ∗ algebras. Far East J. Math. Sci. **31**, 99–112 (2008)
19. Connes, A.: Noncommutative Geometry. Academic Press, Boston (1994). http://www.alainconnes.org/docs/book94bigpdf.pdf (One of the great mathematics books written in the last twenty years. "The reader of the book should not expect proofs of theorems. This is much more a tapestry of beautiful mathematics and physics which contains material to intrigue readers ..." (Vaughan Jones). "... a long discourse or letter to friends" (I.E. Segal)

# List of Authors Cited

After each name there is a reference to the full annotated bibliographies appearing at the end of most of the chapters. Only the surnames of the author(s), without first names or initials, together with the year of publication, appear in this list. After this a number between brackets, like for example, '(12)' referring to the Reference section of Chapter 12, where complete bibliographic information can be found on the item in question, with additional information, e.g. initials or full first names, just as used by the authors, on information on possible translations into English of French and German items, and on internet availability. In this list (P) refer, of course, to the Reference section of the Preface. In the case of more than two authors we write the name of the first author followed by 'et al.'.

**A**

| | |
|---|---|
| Abraham, Marsden 2008 | (2) |
| Aschenbach 2014 | (17) |
| Akhiezer, Glazman 1933 | (21) |
| Anderson, C. 1993 | (16) |
| Anderson, M.H. et al. 1995 | (11) |
| Araki 1999 | (16) |
| Arnold 1997 | (2) |
| Arveson 1976, 1998 | (27) |
| Aspect et al. 1981, 1982 | (17) |
| Aspect 1999 | (17) |
| Azmi 2007, 2008 | (27) |

**B**

| | |
|---|---|
| Baggott 2013 | (17) |
| Baker 2003 | (24) |
| Balian 2007 | (10) |
| Barut, Raczka 1986 | (24) |
| Batchelor 1980, 1984 | (13) |
| Bayen et al. 1978 | (13) |
| Bell 1964, 1966, 1987, 2004 | (17) |
| Ben-Naim 2008 | (10) |

© Springer International Publishing Switzerland 2015
P. Bongaarts, *Quantum Theory*, DOI 10.1007/978-3-319-09561-5

Dugundji 1966 (18)
Dunford, Schwartz 1988 (21)
Dijksterhuis 1986 (1)

**E**

Earman, Mosterin 1999 (17)
Ehrenfest, P. and T. 1990 (10)
Einstein 1905 (11), (13), (15), (16)
Einstein et al. 1952 (15)
Einstein 1921 (15)
Einstein 1925 (11)
Einstein et al. 1935 (17)
Esposito 2014 (13)
Everett 1957 (17)

**F**

Faddeev, Merkuriev 2010 (14)
Faddeev, Yakubovskiĭ 2009 (P)
Farmelo 2003 (P)
Farmelo 2009 (12)
Fedosov 1994 (13)
Feynman 1948 (13)
Feynman, Hibbs 1965 (13)
Fock 1932, 1934 (9)
Fredenhagen 1994/1995 (9)
French 1968 (15)

**G**

Gallavotti 1999 (10)
Gallier 2011 (23)
Gårding, Wightman 1964 (16)
Gelfand, Neumark 1943 (27)
Gerstenhaber 1966 (13)
Gieres 2000 (26)
Gibbs 1902, 1960 (10)
Gilmore 1974, 2006 (24)
Glashow 1961 (16)
Gleason 1957 (12)
Glimm, Jaffe 1987 (16)
Goldstein 2001 (2)
Gokcen, Reddy 1966 (10)
Gracia-Bondía et al. 2001 (12)
Grandy 1987, 2008 (10)
Granström 2006 (12)
Greub 1978 (23)
Groenewold 1946 (13)
Gross, Wilczek. 1974 (16)
Griffith 1995 (3)
Gustafson, Sigal 2006 (21)
Guth 1981 (17)

**H**

Haag, Kastler 1964 (12), (16)
Hall 2003 (24)
Hall 2013 (P)

Halmos 1950, 1974                                            (19)
Hamhalter 2003                                               (27)
Hannabuss 1997                                               (P), (3)
Heine 2007                                                   (24)
Heisenberg 1925                                              (13)
Helemskii 1993                                               (27)
Helmberg, Gilbert 1969, 2008                                 (21)
Herzberg, Gerhard 2010                                       (8)
Humphreys 1972                                               (24)
Hunziker, Sigal 2000                                         (14)

**I**
Ising 1925                                                   (10)
Ito 1993                                                     (P)

**J**
Jancel 1969                                                  (11)
Jänich 1984                                                  (18)
Jaynes 1957                                                  (10)
Jones, H.F. 1990                                             (24)
Jones, Vaughan F.R. 2010                                     (27)

**K**
Kac 1949                                                     (13)
Kadison, Ringrose 1997                                       (27)
Kato 1951, 1966, 1995                                        (14)
Kelley 1955, 1975                                            (18)
Khinchin 1943, 1949                                          (10)
Kibble 1997                                                  (2)
Klenke 2006                                                  (22)
Kobayashi, Nomizu 1963, 1969, 2009                           (20)
Kolmogorov 1956                                              (22)
Kontsevich 2003                                              (13)
Kostant 1977                                                 (13)
Koszul 1960                                                  (20)
Kuhn 1978, 1987                                              (11)

**L**
Landsman 2006                                                (16)
Le Bellac 2006                                               (3)
Lee 2002, 2012                                               (20)
Leites 1980                                                  (13)
Lieb, Yngvason 2002                                          (10)
Lighthill 1958                                               (25)

**M**
Martin 1959                                                  (13)
McCormmach 1982                                              (11)
Mehra, Rechenberg 1982-2000                                  (3)
Merzbacher 1997                                              (3)
Messiah 1999                                                 (3), (15)
Minkowski 1909                                               (15)
Münster 1969                                                 (10)
Murphy 1990                                                  (27)
Murray, von Neumann 1936                                     (27)

**N**

Nelson 1964      (13)
Newton, Isaac 1962      (2)
Newton, Roger G. 2013      (14)

**P**

Pais 1988      (1)
Pedersen 1989      (18)
Penrose 2004      (P)
Perrin 1909, 1910      (13)
Planck 1897, 2010      (10)
Planck 1901      (11)
Polchinski 2005      (17)
Prugovecki 1981      (21)

**R**

Reed, Simon 1972      (11), (13), (21)
Reed, Simon 1979      (14)
Richards, Youn 1995      (25)
Rieffel 1990, 1994      (13)
Riemann 1854      (20)
Rindler 1991      (15)
Roberts, Roepstorff 1969      (16)
Roepstorff 1996      (13)
Rogers 2007      (13)
Rosenthal 2006      (22)
Ruelle 1999, 2004      (10)
Rutherford 1911      (3), (7)

**S**

Sagan 2001      (9)
Schmüdgen 2012      (21)
Scholz 2007      (13)
Schrödinger 1926      (7)
Schrödinger 1935, 1980      (17)
Schulman 2005      (13)
Schwartz, Jacob T. 1967      (27)
Schwartz, Laurentz 1966      (25)
Schwinger 1951      (13)
Schwinger 2012      (16)
Segal 1947      (12), (16)
Segal 1967      (13)
Sellar, Yeatman 1930      (17)
Shannon 1948      (10)
Shimony 2009      (17)
Smolin 2007      (17)
Sobolev 1936      (25)
Streater, Wightman 2000      (9), (16)
Strocchi 2005      (P)

**T**

Takesaki 2001      (27)
Takhtajan 2008      (P)
Taylor 2006      (14)
Temple 1955      (25)

Teschl 2009                     (P), (14)
Thomas no date                  (20), (27)
Treves 1967                     (25)
Trotter 1959                    (13)
Tuynman 2004                    (13)

**V**
van der Waerden 1968            (3)
van Hove 1951                   (13)
Van Vliet 2008                  (10)
Varadarajan 2009                (24)
Veltman 2003                    (16)
von Neumann 1932                (10)
von Neumann 1996                (12), (21)
Verbeure 2011                   (11)

**W**
Walter 1999                     (15)
Warner 1983                     (20), (24)
Weinberg 2005                   (16)
Weiner 2003                     (10)
Wess, Zumino 1974               (17)
Wess, Bagger 1992               (17)
Weyl 1927, 1931, 2003           (3), (13)
Weyl 1952                       (13)
Wick et al. 1952                (9)
Wiener 1923                     (13)
Woit 2007                       (17)
Wyss 1972                       (16)

**Y**
Yang 1952                       (10)
Yang, Mills 1954                (16)
Yukawa 1935                     (16)

**Z**
Zee 2010                        (16)
Zeilinger 1999                  (17)

# Index of Persons

Short biographies of all scientists in the following list can be found on Wikipedia pages. Most of these can be recommended.

Names of scientists, still alive at the moment of writing this book, have not been included. Names of scientists may appear in expressions like "Schrödinger equation". Such expressions appear in the Subject Index.

© Springer International Publishing Switzerland 2015
P. Bongaarts, *Quantum Theory*, DOI 10.1007/978-3-319-09561-5

# Subject Index

© Springer International Publishing Switzerland 2015
P. Bongaarts, *Quantum Theory*, DOI 10.1007/978-3-319-09561-5

Printed in the United States
By Bookmasters